Machine Learning, IoT and Artificial Intelligence for Sustainable Development

Machine Learning, IoT and Artificial Intelligence for Sustainable Development

Editors

Mourade Azrour
Azidine Guezzaz
Imad Zeroual
Azeem Irshad
Jamal Mabrouki
Said Benkirane
Shehzad Ashraf Chaudhry

MDPI

Basel • Beijing • Wuhan • Barcelona • Belgrade • Novi Sad • Cluj • Manchester

Editors

Mourade Azrour
Department of Computer
Moulay Ismail of Meknes
Errachidia
Morocco

Azidine Guezzaz
Computer Sciences
Cadi Ayyad University
Marakesh
Morocco

Imad Zeroual
Department of Computer
Moulay Ismail Univ. of Meknès
Errachidia
Morocco

Azeem Irshad
Higher Education Department
Govt. of the Punjab
Punjab
Pakistan

Jamal Mabrouki
CERNE2D, Faculty of Science
Mohammed V University
Rabat
Morocco

Said Benkirane
Computer Sciences
Cadi Ayyad University
Essaouira
Morocco

Shehzad Ashraf Chaudhry
College of Engineering
Abu Dhabi University
Abu Dhabi
United Arab Emirates

Editorial Office
MDPI
St. Alban-Anlage 66
4052 Basel, Switzerland

This is a reprint of articles from the Special Issue published online in the open access journal *Sustainability* (ISSN 2071-1050) (available at: www.mdpi.com/journal/sustainability/special_issues/82QC2GZJQA).

For citation purposes, cite each article independently as indicated on the article page online and as indicated below:

Lastname, A.A.; Lastname, B.B. Article Title. *Journal Name* **Year**, *Volume Number*, Page Range.

ISBN 978-3-0365-9926-7 (Hbk)
ISBN 978-3-0365-9925-0 (PDF)
doi.org/10.3390/books978-3-0365-9925-0

Contents

About the Editors

Mourade Azrour

Mourade Azrour received his PhD from the Faculty of Sciences and Techniques, Moulay Ismail University of Meknes, Morocco. He received his MS in computer and distributed systems from the Faculty of Sciences, Ibn Zouhr University, Agadir, Morocco, in 2014. Mourade currently works as a computer sciences professor at the Department of Computer Science, Faculty of Sciences and Techniques, Moulay Ismail University of Meknès. His research interests include authentication protocols, computer security, the Internet of Things, smart systems, machine learning, and so on. Mourade is a member of the scientific committee of numerous international conferences. He is also a reviewer of various scientific journals. He has published more than 60 scientific papers and book chapters. Mourade has edited two scientific books, *"IoT and Smart Devices for a Sustainable Environment"* and *"Advanced Technology for a Smart Environment and Energy"*. Finally, he has served as a guest editor in the journals *EAI-Endorsed Transactions on the Internet of Things*, *Tsinghua Science and Technology*, *Applied Sciences* MDPI, and *Sustainability* MDPI.

Azidine Guezzaz

Azidine Guezzaz received his Ph.D. from Ibn Zohr University in Agadir, Morocco, in 2018. He obtained his Master in Computer and Distributed Systems from the Faculty of Sciences, Ibn Zouhr University, Agadir, Morocco, in 2013. He is currently an associate professor of computer science and mathematics at Cadi Ayyad University in Marrakech, Morocco. His main fields of research interest are computer security, cryptography, artificial intelligence, intrusion detection, and smart cities. He is a member of the scientific and organizing committees of various international conferences. Azidine is a guest editor of Special Issues in the *Tsinghua Science and Technology Journal*, *Sustainability Journal*, and *EAI-Endorsed Transactions on the Internet of Things Journal*. He is also an editor of some books and a reviewer of various scientific journals.

Imad Zeroual

Imad Zeroual is currently an associated professor in the Department of Computer Science, Faculty of Sciences and Technology, Moulay Ismail University. He received my Ph.D. degree in computer science from Mohamed First University in 2018. His areas of research are artificial intelligence and data science. He is primarily working on natural language processing, machine learning, information retrieval and extraction, and language teaching and learning.

Azeem Irshad

Azeem Irshad received his Master's degree from Arid Agriculture University, Rawalpindi, Pakistan. Then he completed his PhD from the International Islamic University, Islamabad, Pakistan. He authored more than 60 international journal and conference publications, including 40 SCI-E journal publications. He has served as a reviewer for more than 42 reputed journals, including the *IEEE Systems Journal*, IEEE Communications Magazine, *IEEE Transactions on Industrial Informatics*, *IEEE Consumer Electronics Magazine*, *Computer Networks*, *Information Sciences*, *CAEE*, *Cluster Computing*, *AIHC*, *Journal of Supercomputing*, and *Wireless Personal Communications*, among others. His research interests include the strengthening of authenticated key agreements in SIP multimedia, IoT, WBAN, TMIS, WSN, ad hoc networks, e-health clouds, and multi-server architectures.

Jamal Mabrouki

Jamal Mabrouki is a researcher and expert in water science and technology. He is also an engineer in environment and climate. Jamal is working on the project on migration and water and the role of water governance in migration policy in Africa with the cooperation of MedYWat and the World Bank. He is currently a researcher for the environment and climate program at ECOMED in Morocco, where he started as the coordinator of the project "Adaptation of Citizens to Climate Change".

Said Benkirane

Said Benkirane obtained his engineering degree in networks and telecommunications in 2004 from INPT in Rabat, Morocco. He obtained his Master degree in Computer and Network Engineering in 2006 at the USMBA University of Fez and his PhD in Computer Science in 2013 at the UCD University of El-Jadida, Morocco. He worked as a professor in 2014 at ESTE Cadi Ayyad University. His areas of research are artificial intelligence, multi-agent systems, and system security. He also works on various wireless networks (VANET, WSN, etc.). He is president of the Robotics and Artificial Intelligence Team, and he is also an active reviewer in several high-quality journals.

Shehzad Ashraf Chaudhry

Shehzad Ashraf Chaudhry received the Master's and Ph.D. degrees (Hons.) from International Islamic University, Pakistan. Currently, he is an Associate Professor in cybersecurity engineering with the Department of Computer Science and Information Technology, College of Engineering, Abu Dhabi University, Abu Dhabi, United Arab Emirates. Before this, he was with Istanbul Gelisim University, Turkey; the University of Sialkot, Pakistan; and the International Islamic University, Islamabad, Pakistan. He has supervised more than 40 graduate students in their research. He works in the field of information and communication security.

sustainability

Article

A Novel Machine Learning Approach for Solar Radiation Estimation

Hasna Hissou [1], Said Benkirane [2], Azidine Guezzaz [2,*], Mourade Azrour [3,*] and Abderrahim Beni-Hssane [1]

1 Faculty of Science, Science and Technology Research Structure, Chouaïb Doukkali University, El Jadida 24000, Morocco; hissou.h@ucd.ac.ma (H.H.); beni-hsaane.a@ucd.ac.ma (A.B.-H.)
2 Technology Higher School Essaouira, Cadi Ayyad University, Essaouira 44000, Morocco; said.benkirane@uca.ma
3 IDMS Team, Faculty of Sciences and Techniques, Moulay Ismail University of Meknès, Errachidia 25000, Morocco
* Correspondence: a.guezzaz@uca.ma (A.G.); mo.azrour@umi.ac.ma (M.A.)

Abstract: Solar irradiation (Rs) is the electromagnetic radiation energy emitted by the Sun. It plays a crucial role in sustaining life on Earth by providing light, heat, and energy. Furthermore, it serves as a key driver of Earth's climate and weather systems, influencing the distribution of heat across the planet, shaping global air and ocean currents, and determining weather patterns. Variations in Rs levels have significant implications for climate change and long-term climate trends. Moreover, Rs represents an abundant and renewable energy resource, offering a clean and sustainable alternative to fossil fuels. By harnessing solar energy, we can actively reduce greenhouse gas emissions. However, the utilization of Rs comes with its own challenges that must be addressed. One problem is its variability, which makes it difficult to predict and plan for consistent solar energy generation. Its intermittent nature also poses difficulties in meeting continuous energy demand unless appropriate energy storage or backup systems are in place. Integrating large-scale solar energy systems into existing power grids can present technical challenges. Rs levels are influenced by various factors; understanding these factors is crucial for various applications, such as renewable energy planning, climate modeling, and environmental studies. Overcoming the associated challenges requires advancements in technology and innovative solutions. Measuring and harnessing Rs for various applications can be achieved using various devices; however, the expense and scarcity of measuring equipment pose challenges in accurately assessing and monitoring Rs levels. In order to address this, alternative methods have been developed with which to estimate Rs, including artificial intelligence and machine learning (ML) models, like neural networks, kernel algorithms, tree-based models, and ensemble methods. To demonstrate the impact of feature selection methods on Rs predictions, we propose a Multivariate Time Series (MVTS) model using Recursive Feature Elimination (RFE) with a decision tree (DT), Pearson correlation (Pr), logistic regression (LR), Gradient Boosting Models (GBM), and a random forest (RF). Our article introduces a novel framework that integrates various models and incorporates overlooked factors. This framework offers a more comprehensive understanding of Recursive Feature Elimination and its integrations with different models in multivariate solar radiation forecasting. Our research delves into unexplored aspects and challenges existing theories related to solar radiation forecasting. Our results show reliable predictions based on essential criteria. The feature ranking may vary depending on the model used, with the RF Regressor algorithm selecting features such as maximum temperature, minimum temperature, precipitation, wind speed, and relative humidity for specific months. The DT algorithm may yield a slightly different set of selected features. Despite the variations, all of the models exhibit impressive performance, with the LR model demonstrating outstanding performance with low RMSE (0.003) and the highest R2 score (0.002). The other models also show promising results, with RMSE scores ranging from 0.006 to 0.007 and a consistent R2 score of 0.999.

check for updates

Citation: Hissou, H.; Benkirane, S.; Guezzaz, A.; Azrour, M.; Beni-Hssane, A. A Novel Machine Learning Approach for Solar Radiation Estimation. *Sustainability* **2023**, *15*, 10609. https://doi.org/10.3390/su151310609

Academic Editor: Ljubomir Jankovic

Received: 22 May 2023
Revised: 20 June 2023
Accepted: 3 July 2023
Published: 5 July 2023

Keywords: sustainable energy; solar radiation; times series; machine learning; feature selection; forecasting

1. Introduction

Rs refers to the energy that is emitted by the Sun and travels through space to reach the Earth. It consists of electromagnetic waves, including visible light, ultraviolet (UV) rays, and infrared (IR) radiation. The Sun emits solar radiation in all directions, and a small fraction of it reaches the Earth's atmosphere. As the radiation passes through the atmosphere, it may be absorbed, scattered, or reflected via gases, particles, and clouds. Eventually, a portion of the solar radiation reaches the Earth's surface.

All life is powered by the Sun. It keeps the thermal energy and power equilibrium in the Earth's atmospheric conditions and ecological systems. Even a minor fluctuation in the Sun's radiation emission would have a notable effect on the climate of the Earth [1]. It influences Earth's climate system, influencing temperature variations, atmospheric circulation, and weather system formation, through heating of the atmosphere, oceans, and land. Its distribution across the globe contributes to regional climate differences and temperature gradients between the equator and poles. In the Earth's water cycle, solar radiation plays a crucial role, in providing the energy needed for processes like evaporation, condensation, and precipitation, which, in turn, affect mean sea level fluctuations. It also fuels atmospheric instability, which leads to the formation of severe weather phenomena such as storms, hurricanes, and extreme weather events, through the heating of the Earth's surface and the creation of pressure differences. Moreover, solar radiation stands as the most abundant and readily available energy source on Earth, serving as the primary energy source for natural processes and ecosystems, as well as representing the most environmentally friendly and sustainable form of energy [2]. It is also non-polluting, highly accessible, safe, and has the potential to slow the intensification of greenhouse effects [3,4]. Through harnessing solar radiation through technologies like solar panels and solar thermal systems, it can be converted and used in many ways, such as generating electricity, pumping water, and heating and purifying air and water [5]. Solar radiation at the planet's surface consists of three main components: direct radiation, which is sunlight that reaches Earth's surface without scattering or absorption; diffuse radiation, which is sunlight that has been scattered by the atmosphere and arrives from different directions; and additional radiation, resulting from the reflection and scattering of sunlight by the surrounding environment. Understanding these components is crucial for optimizing solar energy systems, designing efficient lighting in buildings, and studying microclimates. Insolation refers to the total ground radiation [6]. The world is dealing with serious issues such as industrial air pollution, global warming, and environmental destruction. Due to the detrimental environmental impact of using non-renewable resources derived from ancient organisms, such as coal, oil, and natural gas, which have harmed the environment, it has become crucial to seek alternative solutions. Thankfully, clean energy, such as solar energy, offers a pathway through which to address these issues effectively and create a more sustainable future. By transitioning to clean energy sources, we can mitigate climate change, improve air quality, enhance energy security, preserve ecosystems, alleviate energy poverty, and promote sustainable economic growth. Embracing clean energy technologies is crucial in tackling these pressing global problems and fostering a more resilient and environmentally friendly world [7]. Rs data must be available to assess the solar energy capacity of a specific area and integrate it into an electrical network [4]. Unfortunately, assessing the potential of Rs as a renewable energy source can be challenging for most weather stations worldwide. This is primarily due to the fluctuating and intermittent nature of the Rs resource and its lack of steadiness and non-controllability, which limits access to reliable data. Also, expensive equipment is required in order to measure it. These characteristics aggravate the situation. It makes the management of the grid more intricate, disturbs the balance between production and consumption, causes variations in voltage, and raises concerns about quality and stability. The initial cost of installing solar power systems, although decreasing, can still be relatively high, posing a financial barrier for some individuals or businesses. Therefore, it is essential to evaluate Rs effectively using other meteorological factors, including relative humidity, ambient temperature, wind velocity, cloud cover, and other parameters [4,8]. For the purpose of estimating Rs from these readily available weather data, quite a few methods have been proposed, such as physical models, with complex structures due to complex conditions of the environment [9]. Empirical models generate a regression-based formula, either linear or not, that is simple with limited precision [10]. Statistical models, built on statistical correlation, are more accurate, but cannot fully express the nonlinear association between Rs and other factors [11,12]. ML models are of great interest to researchers worldwide because they can solve highly nonlinear problems with high accuracy compared to the other models [13–15]. All ML methods, mainly supervised models, typically require a compromise between model accuracy and complexity [16,17]. Accordingly, determining the optimum input aggregation for prediction models is indispensable. It avoids impertinent or extra information, while exclusively keeping the most required features. This mechanism is known as Feature Selection (FS) [8]. It lowers computational costs, improves performance and over-fitting issues, and enhances multicollinearity problems and model complexity [8,18,19]. The instructions for the FS technique include generating subsets, evaluating them, setting stopping criteria, and validating results [20]. Pertaining to this article, we propose a framework for feature selection (FS) aimed at categorizing different lag values. Our objective is to investigate how FS models can enhance forecasting quality in the field of Rs. Our research comprises two significant contributions:

- We employ an FS method with which to pick out essential feature sets from the initial feature sets using various models;
- We measure each model's feature importance score, RMSE, and R2 against the others by assessing their performance on an NCEP (National Centers for Environmental Prediction) dataset.

The choice of MVTS is driven by its ability to handle multiple variables and their complex relationships, capturing their interdependencies and improving accuracy. By leveraging MVTS analysis, we aim to contribute to the existing research and demonstrate its effectiveness in accurately estimating solar radiation.

The subsequent sections of this paper are structured as follows: Section 2 provides an overview of the field's historical background and its current state. Section 3 elucidates the methodology used in this research and presents the introduced model. Section 4 outlines the working environment, presents the findings, and engages in a thorough discussion. Lastly, Section 5 accentuates the key findings and outlines potential avenues for future research.

2. Related Works

Many researchers have proposed and published research studies on Rs estimation using ML models. However, only some studies have comprehensively examined the complete procedure of developing ML models. In particular articles, the discussion on FS methods was limited to a brief overview [1,8,16,21–28].

In their study, Diagne et al. [19] explored an approach that combined statistical, satellite-based, and numerical weather prediction (NWP) techniques. They also analyzed the proposed techniques' application conditions and spatial/temporal resolution ranges. Yadav and Chandel [20] presented forecasting models for different time horizons (short-term, medium-term, and long-term) of solar irradiation (Rs). They also evaluated methods used for selecting input parameters in these forecasting models. Kumar et al. [21] summarized widely utilized empirical regression models and the ANN model. Their findings unequivocally demonstrated that ANN defeated empirical regression models. Meenal, Selvakumar, and Pang et al. analyzed diverse ML models. They discovered that, although the ANN did not exhibit high predictive accuracy, it did propose a means for enhancing algorithm quality [16,22]. Voyant et al. examined the performance of ANN, SVM, and tree-based models, noted that they produced comparable accuracy, and recommended using combined models [8]. In separate studies, both Chen et al. and Olatomiwa et al. used SVM for their research. The results indicated that its accuracy varied depending on the kernel functions used, and the optimized SVM demonstrated successful outcomes [23,29]. Mohanty et al. [26] conducted a study on the strengths and weaknesses of three models: an Adaptive Network-based Fuzzy Inference System (ANFIS), a radial basis function neural network (RBF-NN), and a multi-layer perception (MLP) model [30,31]. The ANFIS model was the focal point of discussion among the researchers, who emphasized its distinctive attributes as a hybrid intelligent model. The research team showcased its integration of conventional mathematical approaches, underscoring its uniqueness. It combines fuzzy logic and neural networks and improves learning and adaptability capabilities, yielding better results. Hedar et al. created an entirely new hybrid ML model that employs an auxiliary numerical data model with which to assess the accuracy of GHI predictions. The suggested hybrid approaches employ FS, classification, regression ML paradigms, and NWP models. Upon implementation on a dataset, the hybrid model lessened the RMSE [25]. Guermoui et al. [24] thoroughly investigated hybrid machine-learning techniques and defined five classifications: generalized, cluster-based, decomposition-based, decomposition-clustering-based, and ensemble learning methods incorporating evolutionary techniques. In their study, Ağbulut, Gürel, and Biçen [32] conducted a comparison of four ML algorithms (SVM, k-NN, DL, ANN) for predicting daily global solar radiation. The findings revealed that, in general, ANN outperformed DL, SVM, and k-NN in terms of prediction accuracy, while k-NN exhibited the least favorable performance among the algorithms assessed. Huang et al. developed a comprehensive ensemble of twelve ML methods, including a stacking model that combines the strengths of various algorithms in order to predict and compare daily and monthly Rs measurements accurately. According to the results, the XGBoost and stacking models, which combine RF, Gaussian Process Regression (GPR), GBRT, and XGBoost, exhibited superior predictive performance [1]. Guermoui, Bouchouicha, Bailek, and Boland [32] introduced a novel integrated model that combines a decomposition technique with an Extreme Learning Machine for predicting photovoltaic power generation. The performance of the proposed model was assessed using data from three distinct solar photovoltaic power plants situated in different locations with varying climatic conditions. The results indicated that the normalized error consistently remained below 10%, and the correlation coefficient exceeded 99% across the forecasting horizons. These findings demonstrate the effectiveness and accuracy of the proposed integrated model in forecasting photovoltaic power generation.

The following diagram (Figure 1) depicts the categorization of generalized models employed in forecasting Rs, while the Table 1 examines and assesses recent studies on Rs prediction utilizing machine learning techniques.

Table 1. Categorization of contemporary research pertaining to the prediction of Rs through the utilization of machine learning techniques.

Contribution	Date	Forecasting Model	Geographical Position	Optimal Model	Findings
[33]	2022	The hybrid CXGBRFR framework integrates deep learning CNN, XGB (Extreme Gradient Boosting) + RF, and bird-inspired models like HHD-BN (Harris Hawks Deep Belief Network), DNN (Deep Neural Network), ANN, ELM (Extreme Learning Machine), and MARS (Multivariate Auto-Regressive Spline models)	Australia daily	deep hybrid CXGBRFR	Correlation coefficient (r): deep hybrid CXGBRFR: 0.941–0.962 ANN/ELM: 0.934–0.956/0.954 DBN: 0.495–0.911 DNN: 0.922–0.941 MARS: 0.928–0.935 Legate's and McCabe's Index: deep hybrid CXGBRFR: 0.943–0.962, ANN: 0.933–0.958 ELM: 0.931–0.955 DBN: 0.493–0.911 DNN: 0.922–0.941 MARS: 0.928–0.942

Table 1. *Cont.*

Contribution	Date	Forecasting Model	Geographical Position	Optimal Model	Findings
[34]	2022	Seasonal Auto-Regressive Integrated Moving Average (SARIMA), K-Nearest Neighbors (KNN) Recursive Neural Network-Long Short-Term Memory (RNN-LSTM)	UAE daily	RNN-LSTM and KNN	RNN-LSTM and KNN outperform SARIMA. RNN-LSTM and KNN perform similarly RNN-LSTM slightly outperforms KNN
[27]	2022	SVM and Corrected-SVM	Ghardaia, Algeria	C-SVM	RMSE = 11.35% rRMSE = 1.713 MJ/m^2, MABE = 1.623 MJ/m^2 r = 12.61%
[28]	2022	LM, SCG, and RP	6 locations from Tamil Nadu, India	LM	LM: R = 0.9376 for training data, 0.9340 For testing data.
[35]	2022	ANN, CNN, RNN, SVR, PR RF	4 locations in Nigeria	RNN	Deep learning outperforms RNN: r = 0.9546, RMSE= 82.22 W/m^2, MAE = 36.52 W/m^2
[36]	2022	DL, SMGRT, and ANFIS	Isparta, Turkey	SMGRT	SMGRT is the best MSE = 1.878 R2 = 0.960 MBE = 0.156 RMSE = 1.371
[37]	2022	RNN, LSTM, and GRU	5 cities Bangladesh	GRU	MAPE = 19.28%
[1]	2021	GPR, RF GBRT, XGBoost {RF, GPR, XGBoost, GBRT}	12 sites in China	{GBRT, XGBoost, GPR, RF} XGBoost	Daily predictions: Stacking model outperforms Monthly predictions: Comparable performance
[5]	2021	MLP (XE MLP) SVR MLR LightGBM	Fez, Morocco	LightGBM SVR	LightGBM: Coefficient of determination R2 = 0.9377, RMSE = 0.4827 kWh/m^2 MAE = 0.3614 kWh/m^2
[38]	2021	22 empirical models RF, MLP, bagged trees, boosted trees	5 locations in Morocco.	RF	r ranges from 0.8753 to 0.9620, normalized mean absolute error (nMAE) ranges from 5.84 to 11.81%, negative root mean square error (nRMSE) ranges from 7.85 to 15.33%.
[39]	2021	SVM RF	India	RF	RF: MSE = 0.750 R2 = 0.97 SVM: MSE = 0.867—R2 = 0.9385
[4]	2021	K-Nearest Neighbors (k-NN) ANN, DL, SVM	4 Turkish stations	ANN	MBE = 0.195 MJ/m^2 RMSE = 2.157 MJ/m^2—rRMSE = 14.10% T statistic = 1.280 MJ/m^2—Mape = 15.92% MABE = 1.597—R2 = 0.9320%
[22]	2020	ANN, RNN	Tuscaloosa, Alabama in the USA	RNN with higher computational costs than ANN.	RNN: better prediction results Cloud cover impacts GSR prediction. RMSE = 7.64%, Normalized Mean Bias Error (NMBE) = 0.2%
[40]	2020	MLPd and RBF	Ghardaia in Algeria	MLP	MLP demonstrates slightly superior performance.
[41]	2020	ANN	Sapporo, Tateno, Fukuoka, Ishigakijima, and Minamitorishima in Japan	NA	Monthly diffuse, direct, and GRS forecasts are extremely accurate. All locations have a R2 of 0.988 or higher.

Table 1. *Cont.*

Contribution	Date	Forecasting Model	Geographical Position	Optimal Model	Findings
[42]	2020	M5Tree, CatBoost, and XGBoost SVM, RF	15 provinces in China	SVM	CatBoost outperforms.
[43]	2019	Naive Bayes 2 days ahead global horizontal irradiance	Austin, TX in USA	Naive Bayes	Various weather type: MBE = 2.73%, r = 86.33% Clear days: RMBE = 1.49%, r = 99.85%.
[44]	2019	RF, M5, MARS, CART	India (Gorakhpur side)	RF	RF: highest accuracy, CART: lowest accuracy.
[45]	2019	SVR, ANN, and DT	4 provinces in turkey	NA	Boosting improves prediction performance RMSE between 4.6 and 14.6%
[46]	2019	SVR, GPR, MLP, and Extreme Learning Machines (ELM)	Toledo in Spain	ELM	Satellite measurements improved predictability by increasing input parameters. ELM: RMSE = 60.60 W/m^2, r2 = 96%
[47]	2019	SP, ANN, and R	Odeillo in France	RF	nRMSE: 19.65% (GHI—first hour ahead), 27.78% (GHI—sixth hour ahead), 34.11% (Beam Normal Irradiation—first hour ahead), 49.08% (Beam Normal Irradiation—sixth hour ahead), 35.08% (Diffuse Horizontal Irradiation—first hour ahead) 49.14% (Diffuse Horizontal Irradiation—sixth hour ahead).
[48]	2019	SVR	Gurugram in India	NA	Performance SVR is influenced by the air temperature (the most significant parameter) RMSE = 14.3 MJ/m^2
[49]	2019	ANN, k-NN, empirical models	Fez -Morocco	KNN Hybrid model	k-NN: rRMSE = 0.2027 R2 = 0.9663. Hybrid model (k-NN—ANN): rRMSE = 0.1785, R2 = 0.9750.
[50]	2018	Radial basis function (RBF) MLP GPR	Ghardaïa—Algeria. Daily	GPR	MBE = 0.1861 kWh/m^2 nRMSE = 5.2%, r = 0.9842 RMSE = 0.3194 kWh/m^2,
[51]	2018	ANN Regression Analysis	4 stations in Turkey Monthly	ANN	ANN: R2 = 0.961, RMSE =0.14
[52]	2018	SVM XGBoost	China Daily	XGBoost	RMSE = 0.9238 kWh/m^2 R2 = 0.7530, XGBoost MAE = 0.6925 kWh/m^2—training phase = 3.02 s—testing phase = 0.05 s
[2]	2018	GPR	Mashha Iran Daily, Monthly	NA	Daily: RMSE = 0.16 MAPE = 1.97%, Model Efficiency (EF) = 0.99
[16]	2018	SVM ANN	India Monthly	SVM	SVM > ANN ANN: more accurate with long training time for large dataset. R2(ANN) = 0.9968 R2 (SVM) = 0.9912
[53]	2017	ANFIS, SVM, ANN	6 provinces in Mexico daily	SVM	RMSE = 2.578, R2 = 0.689 MAE = 1.97
[54]	2017	ANN	13 different stations	ANN	rMBE < 4% R2 = 0.64 r = 0.800 rRMSE = 13%

Table 1. *Cont.*

Contribution	Date	Forecasting Model	Geographical Position	Optimal Model	Findings
[55]	2017	MLP ANFIS SVM DT	Egypt Daily	MLP	MLP > ANFIS > SVM > DT
[56]	2016	ANN	Italy Monthly	ANN	MAPE = 1.67% to 4.25% based on the type and number of inputs
[57]	2016	Generalized Regression Neural Network (GRNN) Radial Basis Neural Network (RBNN) MLP	12 Sites (China) Daily	MLP	RBNN > GRNN > MLP R2 = 0.86 MAE = 0.425 kWh/m^2 RMSE = 0.5388 kWh/m^2
[58]	2016	ANN, ANFIS Gene Expression Programming (GEP)	Karmen, Iran Daily	ANN	R2 = 0.935
[7]	2015	SVR, Empirical	2 provinces (Iran)	SVR	RMSE = 0.4515 kWh/m^2 R2 = 0.9330
[21]	2015	Regression models. ANN	(1 month)	ANN	ANN > Regression models.
[59]	2015	Linear techniques SVM	Italy 1 Day	SVM	SVM > linear model
[60]	2015	k-NN	USA 30 min	K-NN	k-NN > persistence Enhancements in the forecast between 10% and 25%
[61]	2015	ANN—k-NN—SVR Autoregressive models persistence	Italia hourly	SVR	SVR > ANN > AR > k-NN > persistence

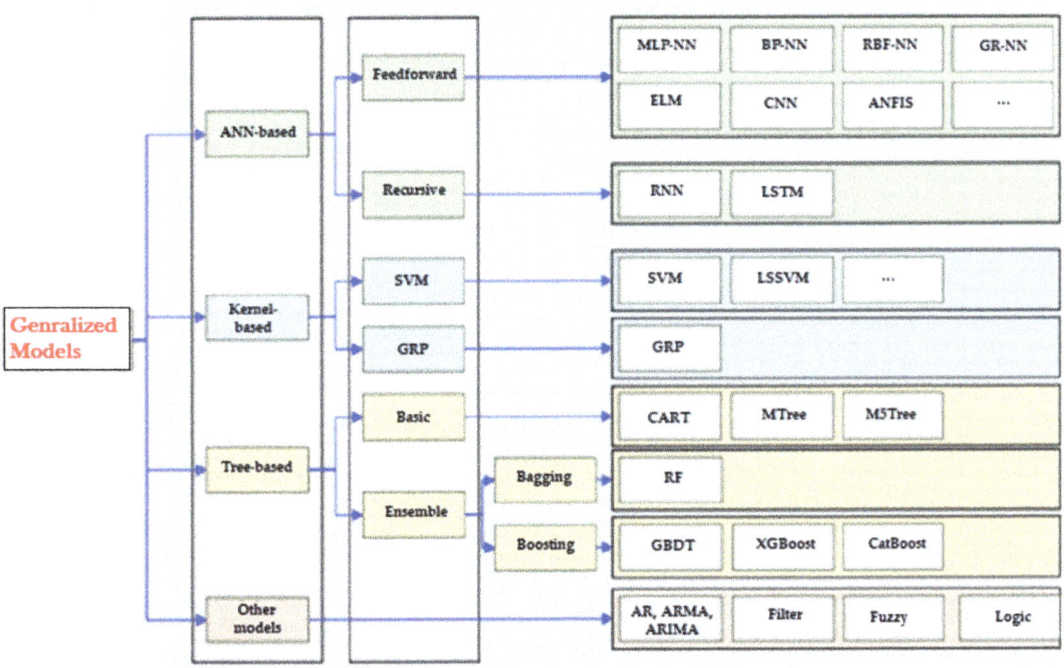

Figure 1. Categorization of the generalized models employed in predicting Rs.

Finally, the comparison of various models reveals interesting insights into their performance. The deep hybrid CXGBRFR model consistently demonstrates high correlation coefficients, ranging from 0.941 to 0.962, outperforming other models such as ANN/ELM

(0.934–0.956/0.954), DBN (0.495–0.911), DNN (0.922–0.941), and MARS (0.928–0.935). RNN-LSTM and KNN exhibit comparable performance, with RNN-LSTM slightly outperforming KNN. Corrected SVM shows accurate predictions, with an RMSE of 11.35% and rRMSE of 1.713 MJ/m^2, while LM exhibits a correlation coefficient (R) of 0.9376 for training data and 0.9340 for testing data. Deep learning models, particularly RNN, outperform others, with an RMSE of 82.22 W/m^2 and an MAE of 36.52 W/m^2. Additionally, other models, such as SMGRT, GRU, LightGBM, RF, SVM, MLP, GPR, XGBoost, GHI, ANFIS, SVR, and Naive Bayes, show varying levels of performance across different evaluation metrics. While the accuracy of the ANN and ARIMA approaches is nearly equal, ANN has the advantage of being more flexible. Merged and generalized models surpassed conventional empirical models. Additionally, combined models tended to be more accurate than generalized models, using the same input parameters. Single-stochastic algorithm methods, such as ANN and ARIMA, are progressively becoming less relevant. Because the accuracy of the predictions is contingent on the integrity of the training data, choosing the best input is critical for FS. Through removing unnecessary or redundant information and retaining only the most important features, FS reduces computing costs and solves overfitting problems. It also aids in multicollinearity problems. There are three FS methods: filter, wrapped, and embedded [15,62]. Overall, this comparative analysis of the models highlights their strengths and weaknesses, providing valuable insights for understanding their capabilities in analyzing the given data, as well as for future model selection and application.

3. Our Proposed Approach

In this section, we will introduce the proposed model, as depicted in Figure 2, and an overview of the methodology used in this study.

We begin by identifying the data requirements and collecting the data. We analyze the information for both quantity and quality. Second, we standardize the data from different formats, correct errors, and expand it, adding more dimensions if necessary. We reduce noise and ambiguities, sample from large databases, select attributes that identify the most significant attributes, and reduce dimensions by implementing various strategies (feature engineering). As we possess a model based on time series, the initial step in transforming the data involves framing it as a supervised learning problem using the sliding window technique.

The steps for processing in our study are described as follows and shown in the flowchart in Figure 2:

1. Import records from CSV files;
2. To ascertain the crucial correlation among all the features employed in the training process, we first need to determine the number of features in the training set;
3. After determining the number of features in the previous step, the Recursive Feature Elimination (RFE) technique is applied to identify the features from the CSV files that exhibit the strongest correlation;
4. In order to divide the dataset into distinct folds for training and testing, we must indicate the number of folds (in this case, ten folds are chosen);
5. Divide the dataset into numerous folds, with one allocated for testing and the remainder for training, through k-fold cross-validation;
6. Train the data using various algorithms (RF, DT, LR, Pr, GBM) to then train the model using the created training dataset. It is then used to test the rest of the dataset compared to the randomly selected feature;
7. In the next step, the trained model is applied to the test dataset, and various metrics are computed in order to assess the accuracy and efficacy of the model;
8. After calculating the scoring metrics, the final results are displayed and graphed.

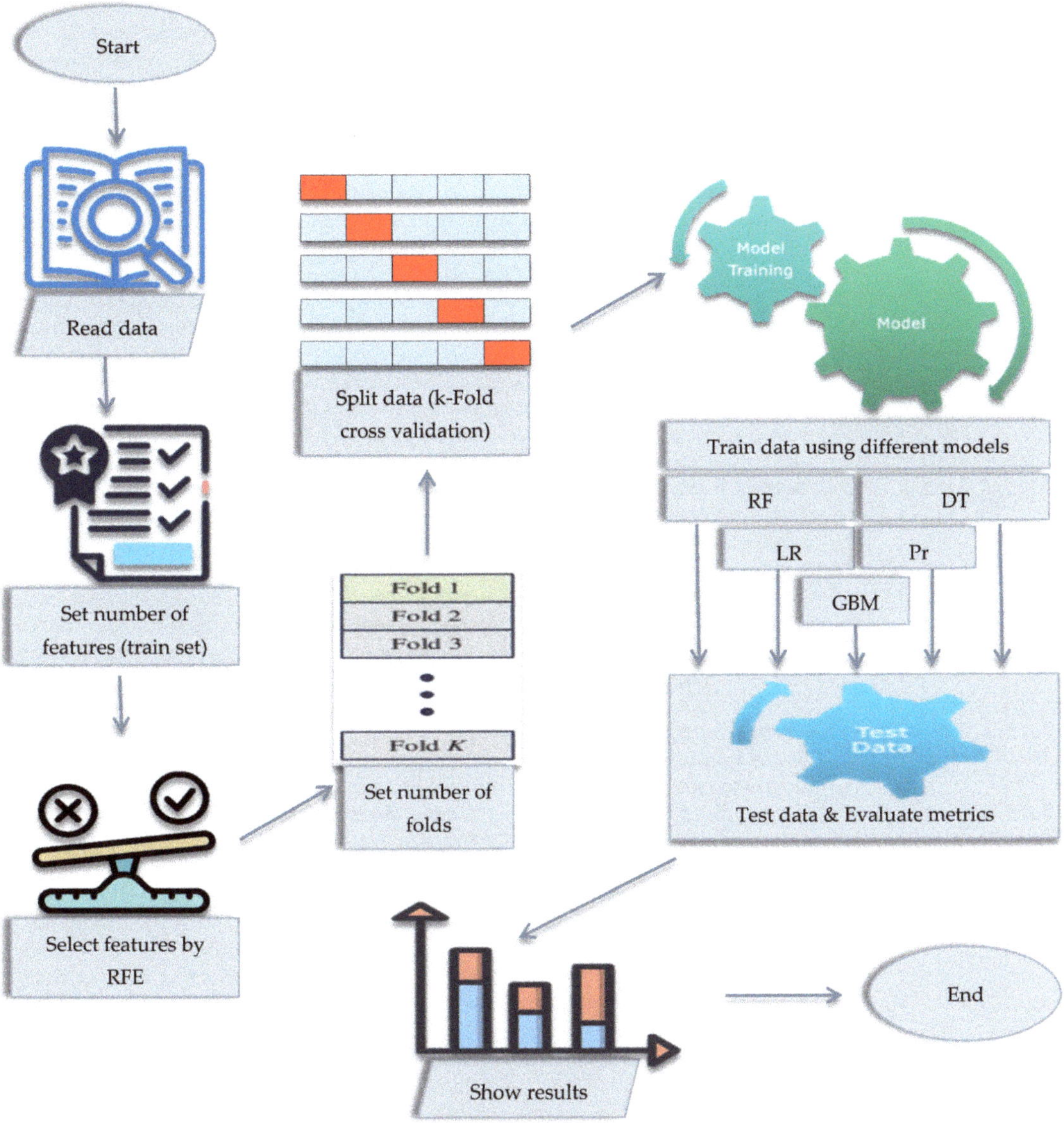

Figure 2. Our proposed model.

4. Experimental Study and Results

The proposed ML approach focuses on identifying the most relevant features that contribute significantly to solar radiation estimation. Through iteratively eliminating less informative features, the RFE algorithm helps in building a more accurate predictive model. Through utilizing only the most influential features, the model can better capture the underlying patterns and relationships in the data, leading to improved accuracy compared to traditional methods that may consider all features. It helps in mitigating the risk of overfitting, which occurs when a model performs well on the training data but fails to generalize to new, unseen data. Through eliminating less informative features, RFE prevents the model from being overly complex and overly sensitive to noise in the training data. This reduction in overfitting enhances the model's ability to provide accurate estimations on unseen data, improving its overall reliability. The reduction in the number of features not only simplifies the modeling process, but also improves computational efficiency. With a smaller feature space, the model requires fewer computational resources and less training time, making it more efficient compared to other methods that consider all available features. This provides transparency and interpretability in the Rs estimation process. In identifying the subset of features that contribute most to the prediction, it helps in understanding the key factors driving the solar radiation patterns. This interpretability allows domain experts to gain insights and make informed decisions based on the selected features. These advantages make it a promising approach when compared to existing methods, leading to more accurate and efficient predictions of Rs.

The conversion from time series to lag values, as employed in our research, offers several advantages over existing methods. It helps capture temporal dependencies and patterns in the data, incorporating past observations as predictors. This enables the model to better capture the historical context and temporal dynamics of the data, leading to improved prediction accuracy compared to methods that do not consider lag values. Additionally, the conversion to lag values facilitates feature engineering through creating additional informative variables. By including lagged features as predictors, we provide the model with a richer set of inputs, capturing the relationship between current and past observations and enabling more accurate predictions. Furthermore, the conversion to lag values can enhance computational efficiency through reducing the dimensionality of the data. Transforming the time series into a matrix of lagged features streamlines the modeling process and reduces computational complexity, leading to improved efficiency compared to methods that consider the entire time series. This approach also provides flexibility in handling various time series data. It allows for the incorporation of different lag lengths or time intervals, enabling the model to capture different time dependencies present in the data. This adaptability enables us to tailor the lagged feature representation to different temporal patterns and achieve more accurate predictions for specific time series datasets. Moreover, the conversion to lag values enhances the interpretability of the model. By analyzing the importance and influence of past observations on current predictions, we gain insights into the temporal dynamics and relationships within the time series. This interpretability enables better understanding and interpretation of the model's predictions.

The model parameters are carefully selected and tuned in order to achieve the best performance in estimating solar radiation. We employ a systematic approach where we define a list of hyperparameter values and test them one by one. The model is trained and evaluated using the chosen evaluation metric for each hyperparameter value. The hyperparameter configuration that yields the best performance metric is selected as the optimal choice. This iterative process of parameter tuning and evaluation allows us to optimize the model and ensure accurate and reliable predictions. We place great importance on this step, in order to enhance the overall effectiveness of our developed model.

4.1. Environment Description

For this ongoing investigation, we performed examinations utilizing data compiled over a period of 36 years (1979–2014), sourced from The National Centers for Environmental Prediction. This dataset encompasses 12,988 entries of daily humidity, minimum and maximum temperatures, longitude, latitude, elevation, wind, precipitation, and sunlight. Our investigation was conducted and completed on a portable computer, provided with a Core-i5 3437U CPU (2.4 GHz)—DDR3 memory capacity of 16 GB; the operating system employed during the experiment was Windows 10 Professional (64-bit), and the model was trained using Python version 3.9.7. In order to evaluate it, k-fold was repeated three times, with ten folds across all repetitions; cross-validation, a statistical technique, employs limited samples for resampling purposes. The k-fold cross-validation procedure blindly divides the dataset into k non-overlapping folds, as shown in Figure 3. The held-back test set ensures that each fold is used only once. At the same time, the remaining sections are continuously merged to compose the training set. Performance metrics are computed and preserved on the test set. At the same time, the remaining folds are repeatedly joined to form the training set. Performance metrics are calculated and saved on the test set.

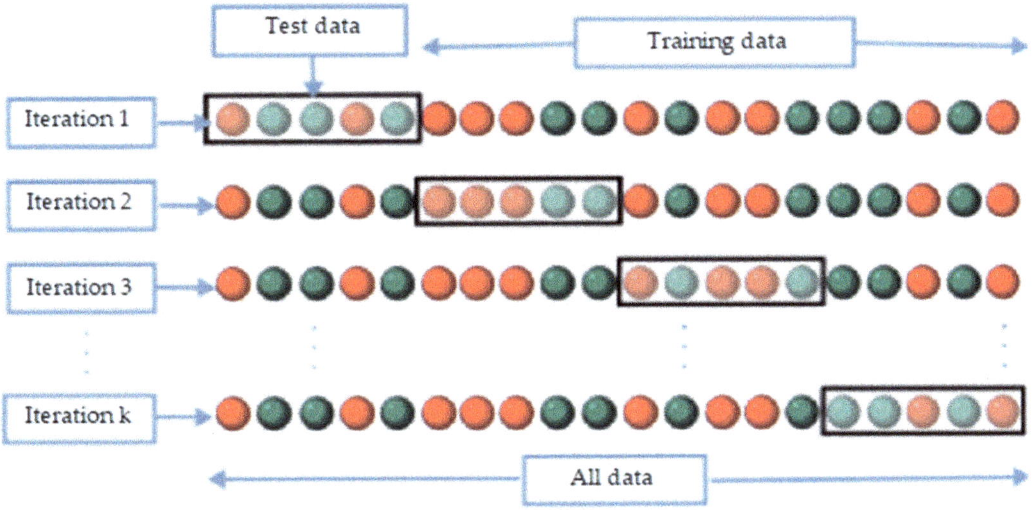

Figure 3. K-fold cross-validation method.

Performing the method several times for the designated number of folds is essential in k-fold cross-validation. The average performance metrics are provided after fitting and evaluating k models on the corresponding hold-out test sets. This technique offers the benefit of minimizing common mistakes and enhancing the anticipated model performance.

The MAE is a metric that quantifies the difference in inaccuracies between two instances of an identical event taking place. It compares the anticipated outcome with the observed outcome, denoted as X vs. Y, where the values of X and Y are identical. The computation of MAE is as follows:

$$\text{MAE} = \frac{\sum_{i=1}^{n} |y_i - x_i|}{n} = \frac{\sum_{i=1}^{n} |e_i|}{n} \tag{1}$$

It computes the average absolute deviation observed between anticipated and target values. An upward pattern implies a reduction in the measure. The feedback data substantially vary from the training data. A decreasing trend indicates an increase in the metric. This indicates that the model's training is efficient. The importance of the feature assigns a rating to each feature in the dataset according to its significance. The ratings explain the "importance" of each feature. A better score signifies that the characteristic holds greater significance and will exert a more potent influence on the model. There are multiple methods with which to calculate the importance of features. This write-up presents an outline of the Gini importance technique utilized in Scikit-learn for evaluating the impurity of nodes. The weight of a node is determined using the proportion of samples that arrive at it relative to the total number of samples. The reduction in the impurity of a node is known as feature importance, which is equivalent to the probability of the node. When a decision tree has two child nodes, the formula is as follows:

$$\text{ni}_j = w_j C_j - w_{\text{left}(j)} C_{\text{left}(j)} - w_{\text{right}(j)} C_{\text{right}(j)} \tag{2}$$

where:

ni_j represents the importance of node j, w_j is the weighted count of samples reaching node j, C_j measures the impurity measure at node j, left(j) signifies the left child node of node j, and right(j) the right child node of node j.

Formula (2) evaluates the feature importance for every DT through considering the significance of the node *j* (*nj*). An exclusive attribute can be utilized in every branch of the tree. Hence, we assess the importance of features by making use of Equation (3).

$$fi_i = \frac{\sum_{j:node\ j\ splits\ on\ feature\ i} ni_j}{\sum_{k \in all\ nodes} ni_k} \tag{3}$$

where:

fi_i: Importance of feature i

ni_j: Importance of node j

Through dividing each feature's importance by the total importance value, these values can be standardized to a numerical range that falls between 0 and 1. This can be achieved by utilizing Equation (4).

$$norm\ fi_i = \frac{fi_i}{\sum_{j \in all\ features} fi_j} \tag{4}$$

The Min–Max normalization technique is employed in order to avoid the negative impact of heavy weights. This method is a linear approach that maintains the associations among the initial data points. It recognizes the smallest and largest values of characteristic X as (min_X) and (max_X). The process involves calculating the value v'i of X within the range [new_min_X, new_max_X] through transforming the value vi using Equation (5). Equation (3) demonstrates the normalization formula utilized for range transformation.

$$v'_i = \frac{v_i - min_X}{max_X - min_X}(new\_max_X - new\_min_X) + new\_min_X \tag{5}$$

If the normalization of a following input instance goes beyond the range of the primary data for X, an "out-of-bounds" error is indicated [63]. The functioning of RFE involves developing prediction models, evaluating features, eliminating those with minor significance, and repeating this process until the desired number of features is attained. RFE is a wrapper-based FS method that uses a filter-based FS method within its internal process. Fundamentally, it utilizes unique ML algorithms. RFE encompasses and applies these algorithms in order to aid in the feature selection process. The FS method based on filters evaluates each characteristic and selects those with the most elevated (or lowest) ranking.

RFE employs a technique to choose a subset of features from the training set through eliminating irrelevant features until the optimum number of features is obtained. This involves the following steps [64]:

- Training the ML algorithm implemented in the heart of the model;
- Ranking the features based on their importance;
- Eliminating the insignificant features and providing the model with further training;
- Continuing the procedure until the intended quantity of features is chosen;
- Creating a metric of importance for variables that sorts the predictors according to their relevance once the entire model has been built;
- In every cycle, the model is reconstructed after eliminating the least significant predictors.

4.2. Discussion of Results

As illustrated in the chart below, we are dealing with an MVTS model, where we can observe periodicity in each parameter. A collection of datasets where two or more variables are observed each time refers to an MVTS. While most time series analysis techniques focus on univariate data, which is simpler to comprehend and handle, MVTS analysis, on the other hand, is often more challenging to manipulate and model. It concurrently deals with multiple time series, typically more intricate than univariate analysis.

The initial data preprocessing phase involves understanding each column's data type and identifying missing values, duplicates, and errors. Several preprocessing steps can be taken. Firstly, the missing values need to be identified and handled. Secondly, duplicates should be identified and removed. Thirdly, it is crucial to identify and manage errors, such as outliers, using statistical methods, and decide whether to replace or remove them. Fourthly, data should be stored in the appropriate data type, and data types should be converted if necessary. Fifthly, columns should be renamed with meaningful names to facilitate analysis. Sixthly, irrelevant columns that are not necessary for analysis should be removed. Finally, numeric variables are selected and standardized using the Standard Scaler method from Scikit-learn. The selected columns include MaxTemperature, MinTemperature, Precipitation, Wind, RelativeHumidity, and Solar. The appropriate transform method fits the scaler to the data and transforms the numeric variables, ensuring the data are ready for analysis.

Standardization is a data preprocessing technique that transforms data into a standard format in order to allow for a fairer comparison between them. It is helpful for many ML techniques, as it can improve model performance and reduce biases introduced by variables with different scales. This technique involves centering the data around zero and scaling them to the same range. Specifically, for each variable, Standardization includes subtracting the mean of the variable from each observation and dividing the outcome by the variable's standard deviation. This process transforms the variable into a distribution centered on 0 with a variance of 1 [65]. We used the Standard Scaler method from the Sklearn preprocessing module to perform Standardization.

Seasonal adjustment is a common technique used in time series analysis in order to remove the effects of seasonal patterns from a time series dataset. Seasonal patterns are recurring patterns within a fixed period, such as a month, a quarter, or a year. By performing a seasonal difference on the time series data, we can eliminate the seasonal pattern and focus on the data's underlying trends and irregular components. In this case, the code uses a lag of 12 months or one year to perform the seasonal difference, subtracting each value from the value 12 months prior. This will help to remove any recurring patterns that occur yearly. Thereafter, we trim off the first year of empty data (since the first 12 months of differenced data will be NaN) and save the differenced dataset. After performing the seasonal difference, the differenced data for the variables from July 2013 to July 2014 is plotted in the line graphs below (Figure 4). This allows us to inspect the data and see the seasonal adjustment visually. Overall, seasonal adjustment is an essential step in time series analysis, as it removes the effects of seasonal patterns from the time series data. This enables us to gain a deeper insight into the fundamental trends and patterns within the data, thereby enhancing the precision of our forecasts and predictions.

As we have an MVTS problem, transforming data involves converting a time series, which follows a chronological order, into a supervised learning task that includes input and output patterns (X, Y) using the sliding window method. This allows an algorithm to understand how to anticipate the output based on the input patterns. The sliding window technique involves utilizing the preceding to anticipate the succeeding time steps. It is sometimes referred to as the window method in specific texts. In statistics, it is known as a lag or lagging method. It proves to be beneficial in decreasing the time complexity of particular issues. The approach applies to solving almost any problem that satisfies the condition of being capable of adding items consecutively or simultaneously into a single variable. The sliding window strategy is adaptable for both univariate and MVTS analysis (Figure 5).

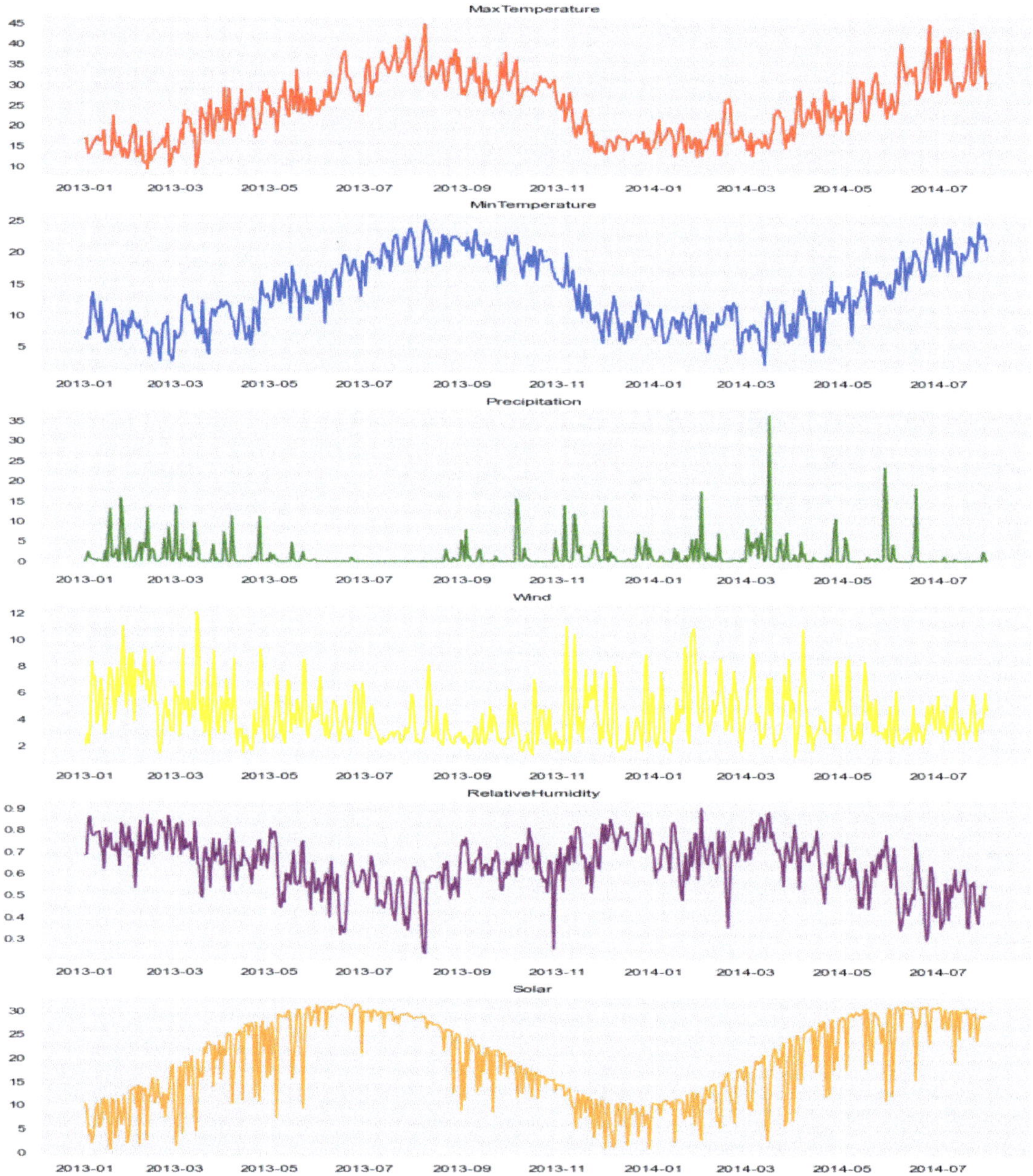

Figure 4. Graphical representation of environmental variables.

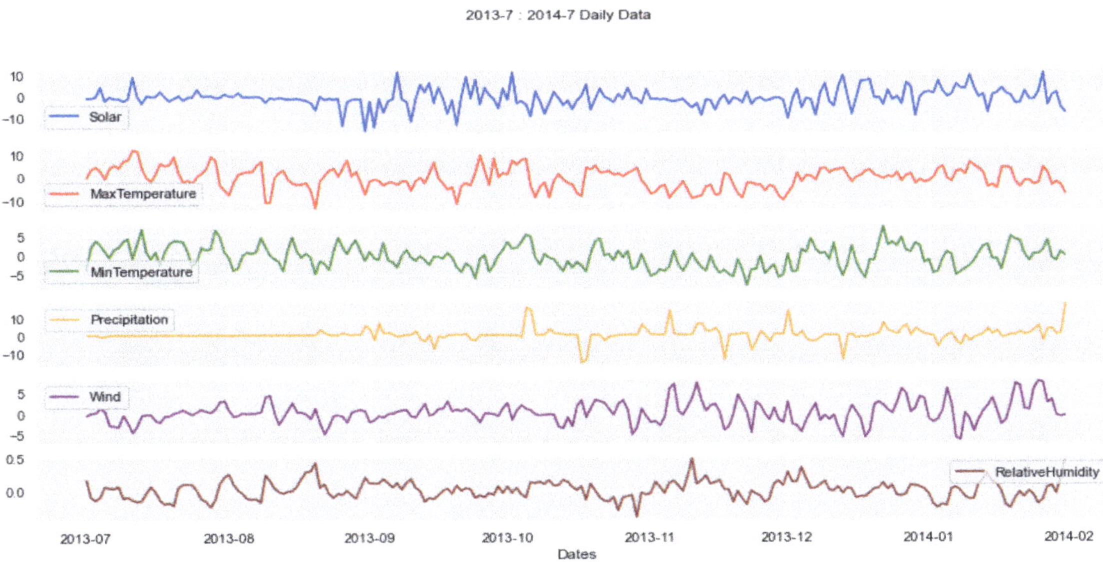

Figure 5. The differenced data for the different variables from July 2013 to July 2014.

Feature selection is an important process in preparing data. In this case, feature selection is performed using the RF Regressor algorithm. The selected features are as follows:

- MaxTemperature of months 12, 10, 4, and 1;
- MinTemperature of months 12, 11, 10, 6, 4, 3, 2, and 1;
- Precipitation of months 12, 11, 10, 7, and 1;
- Wind speed of months 12, 11, 7, 6, and 1;
- Relative humidity of months 11 and 3.

These features were selected using the RF Regressor algorithm, to build a model that uses a set of DTs to predict continuous values. The algorithm analyzes feature importance to evaluate the effect of each feature on the target variable and select the most critical features for prediction. Feature importance ranking using RF is shown in Figure 6:

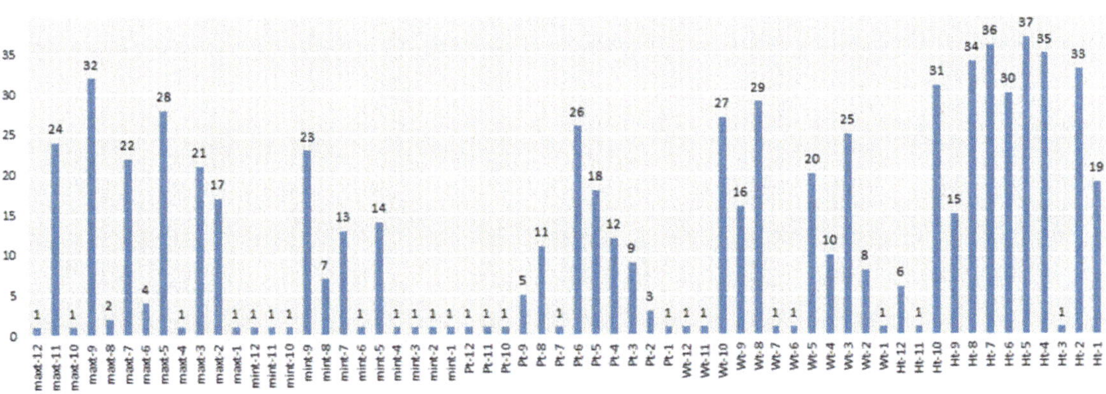

Figure 6. Feature importance ranking using RF Regressor.

The DT algorithm is an ML model that builds a tree-like structure in order to classify or forecast a target variable based on input features. The algorithm recursively splits the data based on the feature that results in the highest information gain, which measures the reduction in entropy after the split. The features with the highest information gain are considered the most important for prediction. In this case, the DT algorithm was used for FS, and the picked features are:

- MaxTemperature of months 12, 10, 4, and 1;
- MinTemperature of months 12, 10, 6, 5, 4, 3, 2, and 1;
- Precipitation of months 12, 9, 8, 6, and 1;
- Wind speed of months 12, 9, 6, and 1;
- Relative humidity of months 11, 9, 5, and 2.

It is interesting to note that some features picked with the DT algorithm differ from those selected with the RF Regressor algorithm. In conclusion, the DT algorithm selected essential features for predicting solar radiation based on their information gain. However, the selected features may differ from those selected according to the algorithm used in the core of RFE, and it is important to compare and assess the performance of different techniques and FS methods.

In order to compare the lr, RF, DT, Pr, and GBM performance models, a box and whisker plot is presented for the RMSE and R2 evaluation metrics (Figure 7).

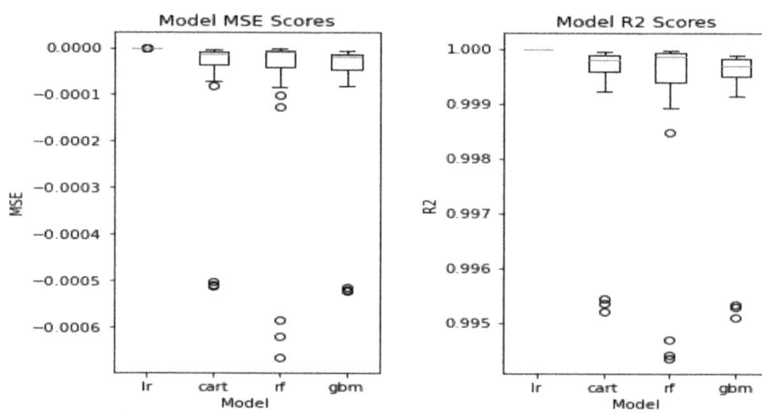

Figure 7. Lr, RF, DT, Pr and GBM performance models for MSE and R2 evaluation metrics.

Based on the RMSE and R2 scores, the LR model appears to have the most outstanding performance, of 0.003 (0.002). Nevertheless, the other models also show impressive performance, with RMSE scores ranging from 0.006 to 0.007 and consistent R2 scores of 0.999.

5. Conclusions

FS is a critical phase in preparing data for ML models; because selecting irrelevant or redundant features can lead to overfitting or poor model performance, choosing the right features is critical for building accurate and robust models.

Recursive Feature Elimination is used for FS. It recursively removes features from the dataset and constructs a model using the remaining features until the desired number of features is reached.

The approach used in our research work offers several advantages: It improves the model interpretability and enhances model performance. It focuses on the most informative features, leading to more accurate predictions, and takes into account feature interactions and dependencies. It can handle multicollinearity issues through iteratively eliminating redundant features, ensuring that the final feature set is independent and representative.

RFE offers flexibility in algorithm selection; it is a versatile technique that can be used with different ML algorithms. It is not limited to a specific algorithm and can be applied with various models, such as linear regression, support vector machines, or random forests.

It also automates the FS process. It provides a ranking of the importance of each feature, allowing researchers to gain insights into the relative significance of different variables in the model's performance. In this case, RFE is used in conjunction with the different algorithms to identify the most essential features for prediction. The advantage of using RFE is that it considers the interaction between features rather than simply evaluating each feature in isolation.

The RF Regressor algorithm analyzes feature importance in order to evaluate the effect of each feature on the target variable and select the most critical features for prediction. On the other hand, the DT algorithm is a model that builds a tree-like structure in order to classify or estimate a target variable derived from a set of input features. The features with the highest information gain are considered the most important for prediction. As we can see from the results, the two algorithms selected different sets of features, which may be due to the different methods used for evaluating feature importance. However, we have observed common patterns in the FS across different models, indicating their significance in accurately estimating solar radiation. Some of the key features consistently identified as important for accurate Rs estimation include:

- MaxTemperature of months 12, 10, 4, and 1: The maximum temperature during these months likely captures seasonal variations and their impact on solar radiation levels;
- MinTemperature of months 12, 10, 6, 4, 3, 2, and 1: The minimum temperature during these months provides insights into the daily temperature range, which can influence solar radiation patterns;
- Precipitation of months 12 and 1: The amount of precipitation during these months may affect cloud cover and atmospheric conditions, impacting solar radiation levels;
- Wind speed of months 12, 6, and 1: The wind speed during these months is an indicator of atmospheric dynamics, which can influence the dispersion of clouds and affect solar radiation availability;
- Relative humidity of month 11: The relative humidity in month 11 likely represents a critical period for moisture content in the air, which can affect solar radiation absorption and scattering.

These features highlight the importance of considering meteorological factors, such as temperature, precipitation, wind speed, and relative humidity, in accurately estimating solar radiation. By incorporating these influential features, our approach captures the relevant environmental dynamics and improves the precision of solar radiation estimation.

In order to assess and contrast the effectiveness of various models, the RMSE and R2 evaluation metrics were used, and a box and whisker plot was created to visualize the results. LR had the top RMSE and R2 scores (0.003, 0.002), followed closely by other models, such as RF, DT, Pr, and GBM (RMSE scores ranging from 0.006 to 0.007 and consistent R2 scores of 0.999). This suggests that an ensemble of regression models can help improve the accuracy of predictions for complex problems. Compared to other research, Sivanandam and Deepa discovered that an ensemble of regression models, including linear regression, random forest, and Gradient Boosting, outperformed individual models in predicting housing prices [66]. Additionally, Kumar and Singh compared machine learning models for predicting stock prices and found that the Gradient Boosting and random forest models outperformed other models [67].

The novelty of the proposed approach lies in several aspects. Firstly, it introduces a novel framework that integrates various models, offers insights into previously unexplored aspects, and challenges existing theories. It emphasizes the importance of feature selection and model evaluation within the context of RFE, shedding light on the factors influencing feature rankings and prediction performance. This contributes to a deeper understanding of the underlying mechanisms of feature selection. Secondly, the study provides insights into the suitability of LR, RF, DT, CART, and GBM models for MVTS analysis, which expands the knowledge base for solving complex problems in this domain. Lastly, by showcasing the effectiveness of RFE as a feature selection technique, the research offers a practical approach to enhancing the performance of predictive models in complex problem domains.

In conclusion, this study presents novel insights, contributes to existing knowledge in feature selection and model evaluation, and provides a practical approach to addressing challenges in MVTS analysis.

Our approach holds significant potential for real-world applications in various areas related to solar energy systems. It can assist in the planning and design of solar energy systems. Thus, the developers can optimize the placement and capacity of solar panels, maximizing energy production and ensuring optimal system efficiency. This leads to more cost-effective and sustainable solar energy installations. It can contribute to providing accurate solar radiation forecasts at different spatial and temporal resolutions. This information allows grid operators to balance the fluctuating solar energy supply with demand, optimize grid stability, and reduce reliance on conventional energy sources. Understanding solar radiation patterns and variability is essential for environmental impact studies. Reliable solar radiation estimation enables both researchers and policymakers to assess the environmental effects of solar energy systems, analyze their impact on ecosystems, and develop mitigation strategies.

Before widely adopting our approach, it is important to consider its limitations and potential drawbacks. These include the requirement for accurate and comprehensive input data, sensitivity to the chosen model, potential limitations in generalization to different geographical locations and climate conditions, the dynamic nature of solar radiation that may not be fully captured, the trade-off between interpretability and accuracy, the need for sufficient computational resources, and the necessity for validation and benchmarking against existing methods. Addressing these limitations will contribute to the robustness and suitability of our approach for real-world applications in solar radiation estimation.

To further enhance the validity and applicability of our findings, future research should consider additional evaluation metrics, explore alternative feature selection techniques, and investigate the generalizability of our approach to different datasets and problem contexts. By continuing to advance the feature selection and model evaluation field, we can improve the accuracy and robustness of ML models in solving real-world challenges.

Author Contributions: H.H. is the main author, who manages the contribution and gives the detailed description of the research work. S.B. writes the abstract and introduction, and analyzes the related works section. A.G. evaluates the results obtained from implementation and illustrates the figures. M.A. and A.B.-H. participate in the implementation of the model, prepare the final manuscript, and correct the English language. All authors have read and agreed to the published version of the manuscript.

Funding: This research received no external funding.

Institutional Review Board Statement: Not applicable.

Informed Consent Statement: Informed consent was obtained from all subjects involved in the study.

Conflicts of Interest: The authors declare no conflict of interest.

References

1. Huang, L.; Kang, J.; Wan, M.; Fang, L.; Zhang, C.; Zeng, Z. Solar Radiation Prediction Using Different Machine Learning Algorithms and Implications for Extreme Climate Events. *Front. Earth Sci.* **2021**, *9*, 596860. [CrossRef]
2. Rohani, A.; Taki, M.; Abdollahpour, M. A novel soft computing model (Gaussian process regression with K-fold cross validation) for daily and monthly solar radiation forecasting (Part: I). *Renew. Energy* **2018**, *115*, 411–422. [CrossRef]
3. Zhang, Y.; Cui, N.; Feng, Y.; Gong, D.; Hu, X. Comparison of BP, PSO-BP and statistical models for predicting daily global solar radiation in arid Northwest China. *Comput. Electron. Agric.* **2019**, *164*, 104905. [CrossRef]
4. Ağbulut, Ü.; Gürel, A.E.; Biçen, Y. Prediction of daily global solar radiation using different machine learning algorithms: Evaluation and comparison. *Renew. Sustain. Energy Rev.* **2021**, *135*, 110114. [CrossRef]
5. Chaibi, M.; Benghoulam, E.M.; Tarik, L.; Berrada, M.; Hmaidi, A.E. An Interpretable Machine Learning Model for Daily Global Solar Radiation Prediction. *Energies* **2021**, *21*, 7367. [CrossRef]
6. Boutahir, M.K.; Farhaoui, Y.; Azrour, M.; Zeroual, I.; El Allaoui, A. Effect of feature selection on the prediction of direct normal irradiance. *Big Data Min. Anal.* **2022**, *5*, 309–317. [CrossRef]

7. Piri, J.; Shamshirband, S.; Petković, D.; Tong, C.W.; ur Rehman, M.H. Prediction of the solar radiation on the Earth using support vector regression technique. *Infrared Phys. Technol.* **2015**, *68*, 179–185. [CrossRef]
8. Voyant, C.; Notton, G.; Kalogirou, S.; Nivet, M.-L.; Paoli, C.; Motte, F.; Fouilloy, A. Machine learning methods for solar radiation forecasting: A review. *Renew. Energy* **2017**, *105*, 569–582. [CrossRef]
9. Rigollier, C.; Lefèvre, M.; Wald, L. The method Heliosat-2 for deriving shortwave solar radiation from satellite images. *Sol. Energy* **2004**, *77*, 159–169. [CrossRef]
10. Huertas-Tato, J.; Aler, R.; Galván, I.M.; Rodríguez-Benítez, F.J.; Arbizu-Barrena, C.; Pozo-Vázquez, D. A short-term solar radiation forecasting system for the Iberian Peninsula. Part 2: Model blending approaches based on machine learning. *Sol. Energy* **2020**, *195*, 685–696. [CrossRef]
11. Shadab, A.; Said, S.; Ahmad, S. Box–Jenkins multiplicative ARIMA modeling for prediction of solar radiation: A case study. *Int. J. Energy Water Res.* **2019**, *3*, 305–318. [CrossRef]
12. Alsharif, M.; Younes, M.; Kim, J. Time Series ARIMA Model for Prediction of Daily and Monthly Average Global Solar Radiation: The Case Study of Seoul, South Korea. *Symmetry* **2019**, *11*, 240. [CrossRef]
13. Hocaoğlu, F.O. Novel analytical hourly solar radiation models for photovoltaic based system sizing algorithms. *Energy Convers. Manag.* **2010**, *51*, 2921–2929. [CrossRef]
14. Ghimire, S.; Deo, R.C.; Downs, N.J.; Raj, N. Global solar radiation prediction by ANN integrated with European Centre for medium range weather forecast fields in solar rich cities of Queensland Australia. *J. Clean. Prod.* **2019**, *216*, 288–310. [CrossRef]
15. Bouzgou, H.; Gueymard, C.A. Minimum redundancy—Maximum relevance with extreme learning machines for global solar radiation forecasting: Toward an optimized dimensionality reduction for solar time series. *Sol. Energy* **2017**, *158*, 595–609. [CrossRef]
16. Meenal, R.; Selvakumar, A.I. Assessment of SVM, empirical and ANN based solar radiation prediction models with most influencing input parameters. *Renew. Energy* **2018**, *121*, 324–343. [CrossRef]
17. Yadav, A.K.; Malik, H.; Chandel, S.S. Application of rapid miner in ANN based prediction of solar radiation for assessment of solar energy resource potential of 76 sites in Northwestern India. *Renew. Sustain. Energy Rev.* **2015**, *52*, 1093–1106. [CrossRef]
18. Zhou, Y.; Liu, Y.; Wang, D.; Liu, X.; Wang, Y. A review on global solar radiation prediction with machine learning models in a comprehensive perspective. *Energy Convers. Manag.* **2021**, *235*, 113960. [CrossRef]
19. Diagne, M.; David, M.; Lauret, P.; Boland, J.; Schmutz, N. Review of solar irradiance forecasting methods and a proposition for small-scale insular grids. *Renew. Sustain. Energy Rev.* **2013**, *27*, 65–76. [CrossRef]
20. Yadav, A.K.; Chandel, S.S. Solar radiation prediction using Artificial Neural Network techniques: A review. *Renew. Sustain. Energy Rev.* **2014**, *33*, 772–781. [CrossRef]
21. Kumar, R.; Aggarwal, R.K.; Sharma, J.D. Comparison of regression and artificial neural network models for estimation of global solar radiations. *Renew. Sustain. Energy Rev.* **2015**, *52*, 1294–1299. [CrossRef]
22. Pang, Z.; Niu, F.; O'Neill, Z. Solar radiation prediction using recurrent neural network and artificial neural network: A case study with comparisons. *Renew. Energy* **2020**, *156*, 279–289. [CrossRef]
23. Chen, J.-L.; Liu, H.-B.; Wu, W.; Xie, D.-T. Estimation of monthly solar radiation from measured temperatures using support vector machines—A case study. *Renew. Energy* **2011**, *36*, 413–420. [CrossRef]
24. Guermoui, M.; Melgani, F.; Gairaa, K.; Mekhalfi, M.L. A comprehensive review of hybrid models for solar radiation forecasting. *J. Clean. Prod.* **2020**, *258*, 120357. [CrossRef]
25. Hedar, A.-R.; Almaraashi, M.; Abdel-Hakim, A.E.; Abdulrahim, M. Hybrid Machine Learning for Solar Radiation Prediction in Reduced Feature Spaces. *Energies* **2021**, *14*, 7970. [CrossRef]
26. Mohanty, S.; Patra, P.K.; Sahoo, S.S. Prediction and application of solar radiation with soft computing over traditional and conventional approach—A comprehensive review. *Renew. Sustain. Energy Rev.* **2016**, *56*, 778–796. [CrossRef]
27. Guermoui, M.; Abdelaziz, R.; Gairaa, K.; Djemoui, L.; Benkaciali, S. New temperature-based predicting model for global solar radiation using support vector regression. *Int. J. Ambient. Energy* **2022**, *43*, 1397–1407. [CrossRef]
28. Geetha, A.; Santhakumar, J.; Sundaram, K.M.; Usha, S.; Thentral, T.T.; Boopathi, C.S.; Ramya, R.; Sathyamurthy, R. Prediction of hourly solar radiation in Tamil Nadu using ANN model with different learning algorithms. *Energy Rep.* **2022**, *8*, 664–671. [CrossRef]
29. Olatomiwa, L.; Mekhilef, S.; Shamshirband, S.; Mohammadi, K.; Petković, D.; Sudheer, C. A support vector machine–firefly algorithm-based model for global solar radiation prediction. *Sol. Energy* **2015**, *115*, 632–644. [CrossRef]
30. Guezzaz, A.; Benkirane, S.; Azrour, M.; Khurram, S. A Reliable Network Intrusion Detection Approach Using Decision Tree with Enhanced Data Quality. *Secur. Commun. Netw.* **2021**, *2021*, 1230593. [CrossRef]
31. Guezzaz, A.; Azrour, M.; Benkirane, S.; Mohy-Eddine, M.; Attou, H.; Douiba, M. A Lightweight Hybrid Intrusion Detection Framework using Machine Learning for Edge-Based IIoT Security. *Int. Arab. J. Inf. Technol.* **2022**, *19*, 5. [CrossRef]
32. Goliatt, L.; Yaseen, Z.M. Development of a hybrid computational intelligent model for daily global solar radiation prediction. *Expert Syst. Appl.* **2023**, *212*, 118295. [CrossRef]
33. Ghimire, S.; Deo, R.C.; Casillas-Pérez, D.; Salcedo-Sanz, S. Boosting solar radiation predictions with global climate models, observational predictors and hybrid deep-machine learning algorithms. *Appl. Energy* **2022**, *316*, 119063. [CrossRef]

34. Etxegarai, G.; López, A.; Aginako, N.; Rodríguez, F. An analysis of different deep learning neural networks for intra-hour solar irradiation forecasting to compute solar photovoltaic generators' energy production. *Energy Sustain. Dev.* **2022**, *68*, 1–17. [CrossRef]
35. Bamisile, O.; Oluwasanmi, A.; Ejiyi, C.; Yimen, N.; Obiora, S.; Huang, Q. Comparison of machine learning and deep learning algorithms for hourly global/diffuse solar radiation predictions. *Int. J. Energy Res.* **2022**, *46*, 10052–10073. [CrossRef]
36. Üstün, İ.; Üneş, F.; Mert, İ.; Karakuş, C. A comparative study of estimating solar radiation using machine learning approaches: DL, SMGRT, and ANFIS. *Energy Sources Part A Recovery Util. Environ. Eff.* **2022**, *44*, 10322–10345. [CrossRef]
37. Faisal, A.N.M.F.; Rahman, A.; Habib, M.T.M.; Siddique, A.H.; Hasan, M.; Khan, M.M. Neural networks based multivariate time series forecasting of solar radiation using meteorological data of different cities of Bangladesh. *Results Eng.* **2022**, *13*, 100365. [CrossRef]
38. Bounoua, Z.; Chahidi, L.O.; Mechaqrane, A. Estimation of daily global solar radiation using empirical and machine-learning methods: A case study of five Moroccan locations. *Sustain. Mater. Technol.* **2021**, *28*, e00261. [CrossRef]
39. Meenal, R.; Michael, P.A.; Pamela, D.; Rajasekaran, E. Weather prediction using random forest machine learning model. *Indones. J. Electr. Eng. Comput. Sci.* **2021**, *22*, 1208. [CrossRef]
40. Khelifi, R.; Guermoui, M.; Rabehi, A.; Lalmi, D. Multi-step-ahead forecasting of daily solar radiation components in the Saharan climate. *Int. J. Ambient. Energy* **2020**, *41*, 707–715. [CrossRef]
41. Kurniawan, A.; Shintaku, E. Estimation of the Monthly Global, Direct, and Diffuse Solar Radiation in Japan Using Artificial Neural Network. *Int. J. Mach. Learn. Comput.* **2020**, *10*, 253–258. [CrossRef]
42. Fan, J.; Wang, X.; Zhang, F.; Ma, X.; Wu, L. Predicting daily diffuse horizontal solar radiation in various climatic regions of China using support vector machine and tree-based soft computing models with local and extrinsic climatic data. *J. Clean. Prod.* **2020**, *248*, 119264. [CrossRef]
43. Kwon, Y.; Kwasinski, A.; Kwasinski, A. Solar Irradiance Forecast Using Naïve Bayes Classifier Based on Publicly Available Weather Forecasting Variables. *Energies* **2019**, *12*, 1529. [CrossRef]
44. Srivastava, R.; Tiwari, A.N.; Giri, V.K. Solar radiation forecasting using MARS, CART, M5, and random forest model: A case study for India. *Heliyon* **2019**, *5*, e02692. [CrossRef]
45. Basaran, K.; Özçift, A.; Kılınç, D. A New Approach for Prediction of Solar Radiation with Using Ensemble Learning Algorithm. *Arab. J. Sci. Eng.* **2019**, *44*, 7159–7171. [CrossRef]
46. Cornejo-Bueno, L.; Casanova-Mateo, C.; Sanz-Justo, J.; Salcedo-Sanz, S. Machine learning regressors for solar radiation estimation from satellite data. *Sol. Energy* **2019**, *183*, 768–775. [CrossRef]
47. Benali, L.; Notton, G.; Fouilloy, A.; Voyant, C.; Dizene, R. Solar radiation forecasting using artificial neural network and random forest methods: Application to normal beam, horizontal diffuse and global components. *Renew. Energy* **2019**, *132*, 871–884. [CrossRef]
48. Bhola, P.; Bhardwaj, S. Estimation of solar radiation using support vector regression. *J. Inf. Optim. Sci.* **2019**, *40*, 339–350. [CrossRef]
49. Marzouq, M.; Bounoua, Z.; El Fadili, H.; Mechaqrane, A.; Zenkouar, K.; Lakhliai, Z. New daily global solar irradiation estimation model based on automatic selection of input parameters using evolutionary artificial neural networks. *J. Clean. Prod.* **2019**, *209*, 1105–1118. [CrossRef]
50. Guermoui, M.; Gairaa, K.; Rabehi, A.; Djafer, D.; Benkaciali, S. Estimation of the daily global solar radiation based on the Gaussian process regression methodology in the Saharan climate. *Eur. Phys. J. Plus* **2018**, *133*, 211. [CrossRef]
51. Yıldırım, H.B.; Çelik, Ö.; Teke, A.; Barutçu, B. Estimating daily Global solar radiation with graphical user interface in Eastern Mediterranean region of Turkey. *Renew. Sustain. Energy Rev.* **2018**, *82*, 1528–1537. [CrossRef]
52. Fan, J.; Wang, X.; Wu, L.; Zhou, H.; Zhang, F.; Yu, X.; Lu, X.; Xiang, Y. Comparison of Support Vector Machine and Extreme Gradient Boosting for predicting daily global solar radiation using temperature and precipitation in humid subtropical climates: A case study in China. *Energy Convers. Manag.* **2018**, *164*, 102–111. [CrossRef]
53. Quej, V.H.; Almorox, J.; Arnaldo, J.A.; Saito, L. ANFIS, SVM and ANN soft-computing techniques to estimate daily global solar radiation in a warm sub-humid environment. *J. Atmos. Sol.-Terr. Phys.* **2017**, *155*, 62–70. [CrossRef]
54. Marzo, A.; Trigo-Gonzalez, M.; Alonso-Montesinos, J.; Martínez-Durbán, M.; López, G.; Ferrada, P.; Fuentealba, E.; Cortés, M.; Batlles, F.J. Daily global solar radiation estimation in desert areas using daily extreme temperatures and extraterrestrial radiation. *Renew. Energy* **2017**, *113*, 303–311. [CrossRef]
55. Hassan, M.A.; Khalil, A.; Kaseb, S.; Kassem, M.A. Potential of four different machine-learning algorithms in modeling daily global solar radiation. *Renew. Energy* **2017**, *111*, 52–62. [CrossRef]
56. Alsina, E.F.; Bortolini, M.; Gamberi, M.; Regattieri, A. Artificial neural network optimisation for monthly average daily global solar radiation prediction. *Energy Convers. Manag.* **2016**, *120*, 320–329. [CrossRef]
57. Wang, L.; Kisi, O.; Zounemat-Kermani, M.; Salazar, G.A.; Zhu, Z.; Gong, W. Solar radiation prediction using different techniques: Model evaluation and comparison. *Renew. Sustain. Energy Rev.* **2016**, *61*, 384–397. [CrossRef]
58. Mehdizadeh, S.; Behmanesh, J.; Khalili, K. Comparison of artificial intelligence methods and empirical equations to estimate daily solar radiation. *J. Atmos. Sol.-Terr. Phys.* **2016**, *146*, 215–227. [CrossRef]
59. De Felice, M.; Petitta, M.; Ruti, P.M. Short-term predictability of photovoltaic production over Italy. *Renew. Energy* **2015**, *80*, 197–204. [CrossRef]

60. Pedro, H.T.C.; Coimbra, C.F.M. Nearest-neighbor methodology for prediction of intra-hour global horizontal and direct normal irradiances. *Renew. Energy* **2015**, *80*, 770–782. [CrossRef]
61. Lazzaroni, M.; Ferrari, S.; Piuri, V.; Salman, A.; Cristaldi, L.; Faifer, M. Models for solar radiation prediction based on different measurement sites. *Measurement* **2015**, *63*, 346–363. [CrossRef]
62. Demirhan, H. The problem of multicollinearity in horizontal solar radiation estimation models and a new model for Turkey. *Energy Convers. Manag.* **2014**, *84*, 334–345. [CrossRef]
63. Al Shalabi, L.; Shaaban, Z. Normalization as a Preprocessing Engine for Data Mining and the Approach of Preference Matrix. In Proceedings of the 2006 International Conference on Dependability of Computer Systems, Szklarska Poreba, Poland, 25–27 May 2006; pp. 207–214. [CrossRef]
64. Kuhn, M.; Johnson, K. *Applied Predictive Modeling*; Springer: New York, NY, USA, 2013.
65. Hastie, T.; Tibshirani, R.; Friedman, J.H. *The Elements of Statistical Learning: Data Mining, Inference, and Prediction*, 2nd ed.; Springer Series in Statistics; Springer: New York, NY, USA, 2009.
66. Sivanandam, S.N.; Deepa, S.N. Hybrid models using support vector regression for stock price prediction. *J. Appl. Res. Technol.* **2014**, *12*, 205–214.
67. Singh, S.; Madan, T.K.; Kumar, J.; Singh, A.K. Stock Market Forecasting using Machine Learning: Today and Tomorrow. In Proceedings of the 2019 2nd International Conference on Intelligent Computing, Instrumentation and Control Technologies (ICICICT), Kannur, India, 5–6 July 2019; pp. 738–745. [CrossRef]

sustainability

MDPI

Article

A Framework and IoT-Based Accident Detection System to Securely Report an Accident and the Driver's Private Information

Amal Hussain Alkhaiwani [1,*] and Badr Soliman Alsamani [2,*]

1 Computer Science Department, College of Computer and Information Sciences, Imam Mohammed Ibn Saud Islamic University (IMSIU), Riyadh 11432, Saudi Arabia
2 Information Systems Department, College of Computer and Information Sciences, Imam Mohammed Ibn Saud Islamic University (IMSIU), Riyadh 11432, Saudi Arabia
* Correspondence: aalkhaiwani@sm.imamu.edu.sa (A.H.A.); bsalsamani@imamu.edu.sa (B.S.A.)

Abstract: Road traffic accidents in Saudi Arabia have become a serious issue because many of these accidents lead to deaths, injuries, and financial losses. Human lives are often lost in road accidents due to the delay in accident detection by medical assistance. In fact, the accident's location and the driver's personal information are considered critical information that plays a vital role in preserving human life. Additionally, previous studies have found a limitation in the encryption of sensitive data; in fact, a leak of private information is thought to be one of the challenges that restrict the use of IoT devices. To resolve this problem, this research presents an intelligent security framework, and an Internet-of-Things-based system is proposed for immediate accident detection. Thus, this system requires the highest level of security and privacy to maintain the driver's privacy. Moreover, the design science research methodology was followed to design and evaluate the artifacts. Thus, the study's research resulted in the ability to design a secure and effective IoT-based system to detect and report a car accident instantly. In addition, the message is encrypted using Elliptic Curve Integrated Encryption and sent through Message Queuing Telemetry Transport over GSM. The study's overall results show the flexibility with which the proposed artifact can be used for other purposes related to the IoT security framework to send and encrypt critical information.

Keywords: Internet of Things; security; design science research; GSM; GPS; elliptic curve integrated encryption scheme

check for updates

Citation: Alkhaiwani, A.H.; Alsamani, B.S. A Framework and IoT-Based Accident Detection System to Securely Report an Accident and the Driver's Private Information. *Sustainability* **2023**, *15*, 8314. https://doi.org/10.3390/su15108314

Academic Editors: Mourade Azrour, Azidine Guezzaz, Imad Zeroual, Azeem Irshad, Jamal Mabrouki, Said Benkirane and Shehzad Ashraf Chaudhry

Received: 13 April 2023
Revised: 13 May 2023
Accepted: 16 May 2023
Published: 19 May 2023

1. Introduction

Nowadays, the number of road accidents has been increasing around the world, especially in Saudi Arabia. The number of cars on the road is increasing with the growing population, which contributes to the serious number of accidents that occur daily. Saudi Arabia has been listed as a country with a high number of road accidents [1]. According to statistics, 330,454 automobile accidents occurred in Saudi Arabia in 2022. The World Health Organization predicts that by 2030, traffic accidents may cause 500 million injuries and 13 million fatalities worldwide. If quick action is not taken, this prediction will come true [2]. In fact, the lack of immediate accident detection, which might save a person's life by a few seconds, is the most common reason for a driver's death when they are involved in an accident. Once an accident occurs, the passengers' lives are put at risk. An accident detection system would detect the accident in a faster way and with a shorter response time, which can provide the variance between life and death in a matter of minutes or seconds. According to statistics, even one minute of rapid response to accidents can save 6% of lives [3]. Therefore, every car should have an intelligent device that not only detects road accidents but also immediately alerts the first responders with the required information. In today's computing world, the IoT has been reaching unexpected bounds. It is a concept that has the potential to influence not only human lives but also how they function [3], where the selected IoT devices are the optimal solutions, which have the potential to play a critical role in promoting sustainability by optimizing resource usage and promoting sustainable practices in various industries. As such, it is essential to continue to develop and integrate IoT technology into sustainability initiatives to help build a more sustainable future. In addition, the IoT is an emerging technology that intersects with many fields, including science, industry, engineering, and policy. The IoT refers to a variety of products, including sensors [4]. Smart sensors are at the heart of the IoT, without which it would not exist. For communication, these sensors form a huge network. They record minute details of their surroundings and communicate this vital information to one another. Relevant actions are then taken in response to the information obtained [3].

Today, the IoT has improved systems to pave the way for intelligent technology to detect or monitor if human lives are in critical situations [5]. An example of these improved systems is an intelligent transportation system (ITS) that uses IoT sensors to detect accidents and traffic jams. Furthermore, there are a variety of IoT applications, such as home automation systems, that can control a home's electronic devices from a mobile phone. The advent of IoT technology has simplified the prediction of natural disasters and the reporting of temperature fluctuations by monitoring the environment using sensors. Additionally, the IoT is used in healthcare facilities for monitoring patients' health parameters and activities [6].

In Ref. [7], it is indicated that the number of IoT devices will reach 75 billion by 2025, up from approximately 7 billion in 2021. As a result, the internet network will soon become even more complex. In addition, the IoT has three common layers: perception (i.e., sensing interface domain), network (i.e., networking domain), and application layers (i.e., cloud domain) from bottom to top [8]. Each IoT layer is designed to perform a specific function. The perception layer works to gather information using IoT objects. This layer includes RFID tags, sensors, and cameras, which are responsible for collecting information. The network layer transmits the data gathered by the physical layer, also known as the IoT's heart. The application layer is the third layer. This layer's goal is to work as a link between industrial technology and the IoT's social demands [6].

The IoT has become an essential component of potential solutions that span from industrial to daily human life. This new technology is appealing for the facilitation of human life, as it adds a new dimension of knowledge to artifacts and automates decisions [7]. However, the authors in Ref. [9] present some of the IoT's vulnerabilities, which are unprotected network services, not enough authentication/authorization attempts, privacy-violating concerns, unsatisfactory security configurability, and insufficient transport encryption/integrity verification. The latter, due to insufficient encryption/integrity operations, may transport unencrypted data and credentials. A leak of confidential data is considered one of the challenges that limit the utilization of IoT devices. Thus, the contribution of the proposed research is as follows:

- To develop an intelligent IoT framework for instantly and securely detecting and reporting car accidents;
- To develop an IoT product that considers security and privacy requirements for protecting critical information;
- To respond as soon as possible to injured people before the situation becomes critical;
- To evaluate and apply lightweight cryptography (LWC) for preserving and encrypting sensitive information.

Hence, our proposed framework is also designed to contribute to sustainability by improving road safety, reducing traffic congestion, protecting privacy, and promoting more efficient use of resources. This paper is organized into seven main sections. First, Section 1 is the introduction, which includes the research motivation, study issues, and study objectives. Section 2 provides the related work, and this section reviews earlier research pertinent to IoT-based accident detection systems, IoT security and privacy, and LWC to determine which problems have already been addressed and which still need to be studied further. In Section 3, the methodology is provided by outlining the study's phases, case studies, and their analysis, as well as the software and the design of the components employed; this section illustrates the research approach. Section 4 provides an analysis of the requirements and the design; this section discusses the requirements, explains how to assess the environment and difficulties associated with IoT security artifacts, and provides the architecture and aims of the system's design. Then, Section 5 covers the phases of the study, detailing how to run the stages of the IoT artifact, implement the elliptic curve integrated encryption scheme (ECIES) of the LWC algorithms, and provide the overall design of the proposed artifact. Next, the evaluation and results are provided in Section 6, covering the study's findings and detailing how to evaluate the IoT security artifact, providing an analysis of the test results and the knowledge base that will be contributed. Finally, Section 7 is the conclusion and provides future work. The main conclusions of these case studies are outlined in this chapter, along with recommendations and ideas for further study.

2. Related Works

2.1. IoT Accident Detection System

Vehicle tracking systems are utilized in a variety of industries around the world, including vehicle location tracking, stolen vehicle tracking, and fleet management [10]. The author of [11] proposed a simple vehicle tracking system that can be used in different situations and cases, such as if the car is stolen (i.e., theft detection) or when parents need to track their children's school bus. The vehicle tracking system used GPS and GSM technology to track and send SMS messages. In addition, the author concluded the research with future work by monitoring some parameters of the vehicle, such as an LPG gas sensor and gas leak alarm.

The IoT enables the tracking and monitoring of a variety of operations (for example, an IoT-based framework for vehicle overspeed detection). The concept of the proposed system is to estimate the time required for a certain vehicle to travel from its starting point to its destination. The smart vehicle overspeed detector assesses a vehicle's speed using radar, according to the data. This information is gathered and wirelessly transmitted to the appropriate authorities at a remote location using IoT technology. The device includes a GPS-sensing module with a transmitter and receiver that works in tandem with an electronic tracking device to determine a vehicle's speed. If a speeding car is detected, the proposed device emits a buzzer signal to alert the authorities. The accuracy of the speed tracking is predicted by the Connection App, which uses radar, to be between 40 and 80 percent, depending on the internet speed and connectivity [12].

Moreover, the authors of [13] proposed a fully functional system that works to detect accidents using a shock sensor and sends SMS messages through a GSM module. The reliability test of the proposed system showed that the system is robust, available, valuable, operative, and workable, especially when the IoT device continues to send continual crash notifications until it confirms receipt by the authorities.

The authors of [14] proposed a system that can identify accidents through an accelerometer and GPS in conjunction with a smartphone. On the server, there is data pertaining to the accident. Further, the system is more dependable than others, but failure may still occur in the event of a server failure. Additionally, Khot et al. [15] developed a smartphone-based system that employs an accelerometer to recognize accidents and report the location of such accidents to emergency personnel. Once more, the single point of failure in this system increases the likelihood of an erroneous accident alert. Furthermore, in another paper, Chaturvedi et al. [16] proposed an accident detection and reporting system that recognizes incidents with the aid of one sensor. A location is delivered to the police station when an accident occurs. Moreover, in Ref. [17], a prototype for automatic accident detection using the Vehicular Adhoc Network (VANET) and the Internet of Things (IoT) is shown. It uses mechanical and medical sensors installed in the car to detect accidents and the severity of emergency situations.

2.2. IoT Privacy and Security

The IoT facilitates smart devices becoming more traceable, and, in turn, privacy and security issues have multiplied. In Ref. [18], a single IoT environment that was extensively researched is presented, and the findings relating to privacy, security, and defect detection are provided. Thus, the study of the numerous security challenges related to the IoT is extremely important. One target objective of IoT security is to ensure the privacy and confidentiality of all users, as well as enhanced protection, infrastructure, and a guarantee of the availability of various services provided by the IoT ecosystem. Regardless, issues concerning security and data privacy must be addressed. Appropriate measures must be taken to make the user feel at ease about being a part of the IoT system and sharing private information. It must be clearly defined who owns the data and where it will not be utilized without their permission, especially if the data are shared via the Internet [6].

Thus, IoT privacy and security are the main aspects that must be emphasized. For achieving the security and privacy requirements, there should be data confidentiality, access control, an IoT network, privacy, and trust among users and smart devices, including the enforcement of security and privacy policies [8]. Moreover, the Open Web Application Security Project's (OWASP) IoT Top 10 list named the most popular issues related to the IoT, including privacy concerns surrounding what would happen if critical data were exposed or viewed by unauthorized users. Another IoT issue concerns a lack of data transport encryption, which may cause data leakage or lead to a device or user account being completely compromised [19].

In addition, the authors of [20] presented the most critical threats to IoT devices, which are identification, such as the name and address of the individual entity, and sensitive information, including localization and tracking. The latter, localization and tracking, are considered to be the threat of identifying an individual's position using various methods, such as GPS, internet traffic, or smartphone location [20]. The authors reviewed a total of 122 original research papers on IoT privacy that summarized the top concerns about IoT privacy and security. In fact, location tracking and sharing private information are considered the IoT's top vulnerabilities and concerns that must be solved. Furthermore, they presented an optimal solution for preserving privacy by performing cryptographic techniques, privacy awareness, access control, and data minimization. First, for cryptographic techniques, the researchers may have spent many years suggesting new privacy-preserving strategies. The encryption procedure remains the lead technology in most currently proposed solutions [19].

2.3. Lightweight IoT Cryptography

IoT devices are prone to malicious attacks and data theft due to the fact that critical information is transmitted across public networks. Advanced technology is required to safeguard the system. Thus, cryptographic algorithms are a useful way to ensure data security in the IoT. On the other hand, many IoT devices are still insufficiently able to provide such strong solutions. As a result, algorithms must consume less energy while preserving their efficiency to be used in the IoT [6].

In Ref. [7], the author found that most IoT devices do not currently use robust encryption or authentication techniques in their connections, which can corrupt the transmission of data and lead to eavesdropping attacks. For example, many implantable medical devices, such as ECG pacemakers, are affected in terms of memory, processing capabilities, and power consumption because they rely on embedded microprocessors and integrated circuits (ICs). The failure to use effective encryption on these devices can have major consequences, such as unauthorized access to sensitive information and device failure [6].

In fact, cryptographic algorithms are used to ensure information is kept secret and secure through transmission without exposing it to alteration. These algorithms are divided into two categories: (i) symmetric-key algorithms and (ii) asymmetric-key algorithms. Cryptographic algorithms that use the same cryptographic keys are known as symmetric-key algorithms. Pairs of keys are used in asymmetric cryptography, including public keys, which are available publicly, and private keys, which are only known by the owner [21].

However, IoT devices are usually small in size with limited computing capacity, memory, and power resources. Applying conventional cryptographic algorithms to IoT devices that require intensive resources is difficult. Therefore, designing lightweight protection schemes for the IoT becomes the optimal and recommended solution [22].

This research attempts to provide an in-depth and up-to-date analysis of lightweight cryptographic primitives and which ones are best to use (asymmetric, symmetric, or a combination of the two) to provide high-level data protection.

In Ref. [22], the authors provided a comprehensive evaluation and comparison among the types of lightweight cryptographic primitives and their performance. Four different types of lightweight cryptographic primitives can be used: (i) lightweight block ciphers (LWBCs), (ii) lightweight stream ciphers (LWSCs), (iii) lightweight hash functions (LWHFs), and (iv) elliptic curve encryption (ECC). A block cipher is a type of symmetrical cipher that processes an entire block at once. Substitution–permutation networks (SPNs) and Feistel block ciphers are two types of LWBCs. A stream cipher encrypts and decrypts "r" bits at a time. Another technique to provide security is to use LWHFs. They take an arbitrary-length message and turn it into a fixed-length "message digest". IoT security using ECC also utilizes an asymmetric cipher's advanced encryption standard (AES). These ciphers offer both authentication and confidentiality. AES requires a larger key size and higher memory consumption. Rivest, Shamir, and Adleman (RSA) and ECC are two primitives that can be utilized in a public-key cryptosystem for IoT encryption transmission messages compared to higher memory consumption.

In comparison to RSA, ECC provides the same level of security with a reduced key size. On 8-bit microcontrollers, AES is 100–1000 times faster than ECC. The computational complexity of the algorithm can be reduced to enhance execution time.

Table 1 presents a comparison of lightweight asymmetric algorithms, which are RSA and ECC, for the IoT environment based on their key size, code length, possible attacks, key generation(s), signature generation(s), and signature verification(s).

Table 1. Comparison of lightweight asymmetric algorithms.

Asymmetric Algorithm	Key Size	Code Length		Possible Attacks	Key Generation(s)	Signature Generation(s)	Signature Verification(s)	
RSA	1024 15,360	900	[20]	Module attack, man-in-the-middle, timing	0.16 679.06	0.01 9.20	0.01 0.03	[5]
Elliptic curve cryptography (ECC)	160 571	8838		Timing attacks, side channel attacks	0.08 1.44	0.15 3.07	0.23 4.53	

In addition, the authors conclude that AES is the preferred method for a symmetrical cryptography network, which provides provisioning and protection. ECC is the optimal solution in asymmetrical cryptography because it can offer authentication and nonrepudiation [22].

The authors of [23] presented secured text encryption cryptography using ECC. Moreover, the authors of [4] described that privacy and security in the IoT have proven to be one of the most complex issues in recent years. In addition, they demonstrated the current cryptographic models and security techniques used in encryption algorithms and privacy standards. The AES ensures confidentiality in most circumstances as symmetric encryption. Asymmetric encryption, digital signatures, and key management are all provided with the asymmetric algorithm RSA. For secure hash functions, the standards are utilized. In asymmetric cryptography, Diffie–Hellman (DH) and ECC are used to provide privacy techniques [3].

Furthermore, the authors of [21] presented a framework design for secure communication among IoT devices and gateway (i.e., intra-network). The authors recommended that, for the enhancement of changing asymmetric cryptographic techniques, RSA should be substituted with ECC for protecting data transmission [19].

2.4. Comparison of Related Works

To summarize, numerous systems place a high priority on accident detection accuracy, as shown in Table 2, which provides a summary of previous studies. In addition, several systems are designed to respond quickly in order to reach the location of the accident. The use of smartphone sensors can also lower a system's overall cost and increase user accessibility. Researchers have offered a variety of smartphone-based alternatives. The accident detection and response systems' sensory inputs are described in detail in Table 3. As can be observed, there are currently only two different types of sensors that can be employed in systems, and no previous studies have used lightweight cryptography with IoT.

Table 2. Summary of related works' findings.

Reference	Methodology/Techniques	Measure	Evaluation
[11]	Uses speed and GPS sensors	Accuracy	Prototype
[12]	Uses GPS and GSM to detect and send messages	Accuracy	Testing and simulation
[13]	Detects the accident using one sensor	Reliability	Testing and simulation
[14]	Uses two sensors: accelerometer and GPS	Accuracy	Actual implementation
[15]	Uses an accelerometer and GPS for detecting and reporting accidents	Response time	Actual implementation
[16]	Based on one sensor: accelerometer for detecting and reporting accidents	Accuracy	Microcontroller Arduino
[17]	Uses an accelerometer for detecting and reporting accidents	Response time	Testing and simulation

Table 3. Summary of sensor types and LWC used in the related works and the proposed work.

Reference	Vibration	Airbag	GPS	Other	Total	LWC Encryption
[11]	☒	☒	☑	☒	1	☒
[12]	☒	☒	☑	☑	2	☒
[13]	☒	☒	☒	☑	1	☒
[14]	☒	☒	☑	☑	2	☒
[15]	☒	☒	☑	☑	2	☒
[16]	☒	☒	☒	☑	1	☒
[17]	☒	☒	☒	☑	2	☒
Proposed work	☑	☑	☑	☒	3	☑
	☑	Include	☒	Is not include		

The following table summarizes several researchers who have proposed works or platforms for automatically locating accidents and informing authorities about them. Despite a few developed proprietary solutions, most of the systems rely on smartphones. Typically, these latter systems usually require manual activation and restrict calls to call centers. The proposed system in this research is composed of the following elements: navigation, communication, accident detection encryption, and rescue. It makes use of three sensors, including accuracy, response time, and encryption of tracking information, for accident detection, which are key components of our technology.

3. Research Design

The appropriate research methodology for this study is to use design science research (DSR). DSR is a type of research paradigm that allows researchers to provide answers addressing human issues and to create valuable artifacts. Furthermore, DSR is used to grow the existing knowledge base [21]. In fact, the developed artifacts are both helpful and challenging to understand, but they solve real-life problems [24].

3.1. Research Methodology

The DSR framework comprises three research cycles: relevance, rigor, and design. First, the relevance cycle works to take the requirement inputs from the contextual environment into the research and initiate the research artifacts to be presented in the testing field. Second, the rigor cycle works to provide the background theories and methods according to the domain experience and expertise in the foundation's knowledge base. Thus, the rigor cycle acquires the new knowledge generated by research and uses it to grow the knowledge base. Third, the central design cycle works as a tighter loop of research activity for constructing and evaluating design artifacts and processes. In addition, the work with these three cycles in a research project clearly defines the positions and differentiates design science from other research paradigms [25]. Figure 1 defines the general combined DSR framework published in Refs. [25,26].

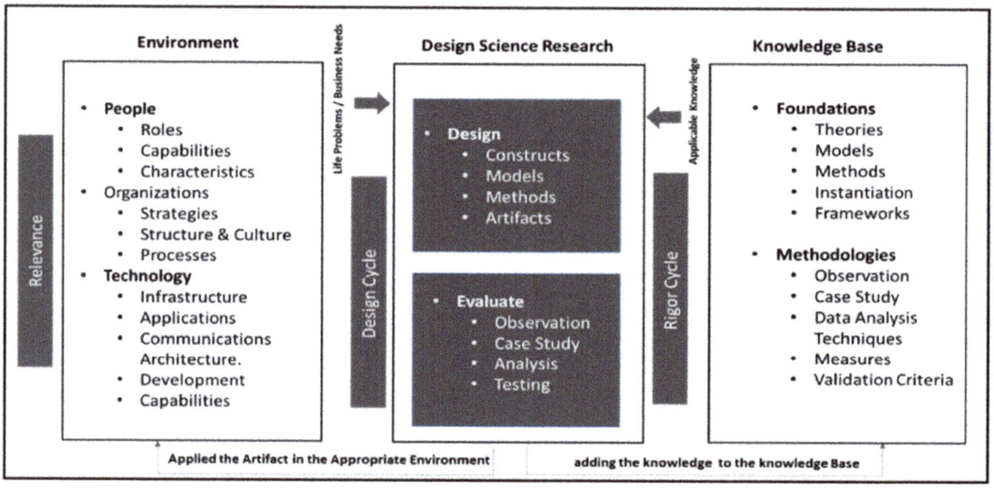

Figure 1. Design science research framework (based on on Hevner, 2007; Hevner., 2004 [25,26]).

3.2. Cycle Design

To explain how the cycles work in this study, Figure 2 presents the roadmap stages and cycles of the research methodology to initiate the process and understand the proposed research, which can be observed in the following:

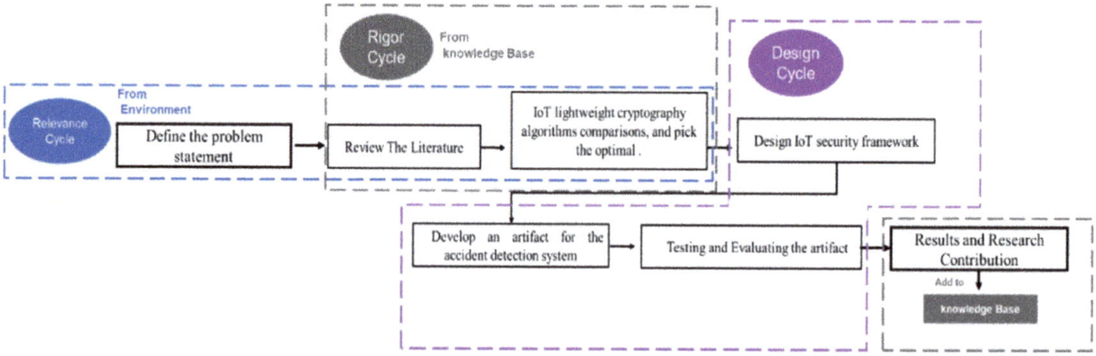

Figure 2. The proposed research roadmap reflects the research methodology.

1. Relevance cycle: The first stage is to define the problem in context and the requirements: a secure and effective IoT-based system to detect and report car accidents instantly while ensuring privacy. In this cycle, as explained in the figure, the IoT security framework literature is searched, and the requirements needed to implement the proposed artifact are determined. Thus, this cycle specifies the scope of the research and the literature review needed for this work.
2. Rigor cycle: This cycle uses a literature review of previous studies that have focused on similar issues. Moreover, the information gathered from the literature will then be used to provide a comprehensive comparison of IoT LWC algorithms. Equally important, this cycle also provides a summary of all studies on IoT security frameworks related to detecting and reporting car accidents instantly.
3. Design cycle: A comparison between the IoT and LWC algorithms will assist in delivering an IoT security framework considering the delivery of sensitive information by message and the authentication processes between the sender and recipient to prevent spoofing by an attacker. Thus, the proposed research follows the DSR process for developing and evaluating a prototype system as an IoT-based security framework artifact for an accident detection system. Additionally, the following section covers how the research methodology works, aligns with the research objectives, and develops the proposed artifact.

3.3. Development of the Study Using the Research Approach

The development of the study using the research methodology is covered in this section. The section above provides a brief explanation of the research methodology's cycles. In addition, this section describes the cycle as it appears in the research study. In addition, Figure 3 demonstrates how the three cycles of the DRS technique map to the proposed study's components.

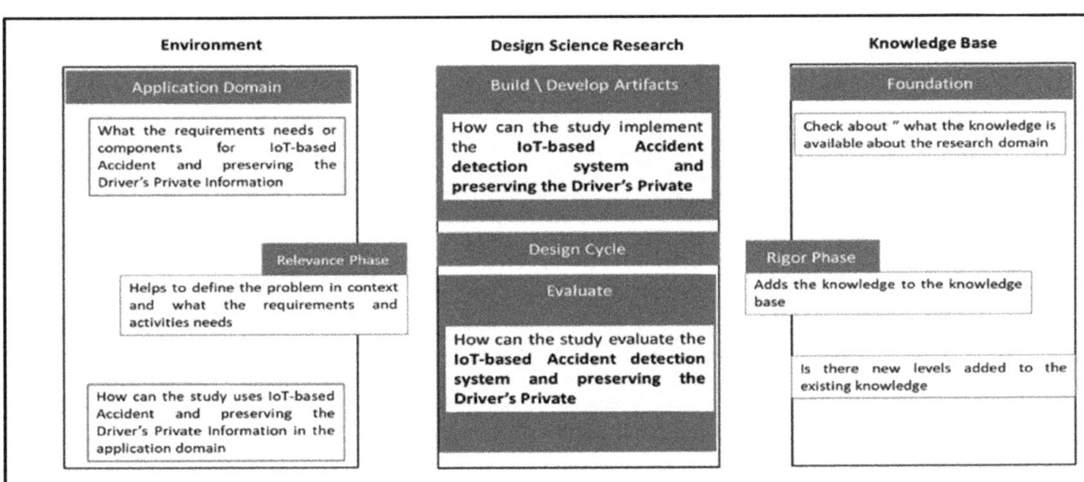

Figure 3. Mapping of the elements of the proposed research to the DSR.

The main elements of the design science research involved in the cycles.

1. Environment (relevance cycle): With this element, the research problem and motivation are clearly defined. In addition, the value of the solution is justified, as mentioned in Section 1, as well as in all related publications' research outputs sections that identify the problem and clearly motivate the research direction toward a solution. The related work is divided into three major sections: IoT accident detection, IoT security and privacy, and the IoT–LWC method. The sections were divided by the research's objectives to

define the problem and present the previous studies mentioned [3,5–8,10–24]. Equally important, the researcher conducted face-to-face interviews with experts in the fields of cryptography and the IoT as well as observational studies to gather data for this study.

2. Designs Science Research: This element is divided into two sections.
    - Build/develop the artifact: This section shows the design and development of the research artifact. Here, the theoretical knowledge obtained from the environment and knowledge base is applied to the creation of an artifact. To address the identified problems, this includes both functionality and architecture. In addition, the artifact is designed and tested, utilized in simulations, or used in other ways as a solution to the perceived problems [26,27]. The proposed solution of this research study was to provide a secure and effective IoT-based system to detect and report car accidents instantly. This used RPM, which relates to other technologies such as IoT sensors (airbag and vibration) and GSM and GPS modules, to detect accidents faster.
    - Evaluate: In this section, an analysis is performed as to how well the research artifact meets the objectives highlighted once the problem has been identified [27]. The evaluating and testing section includes the sections that design and implement the overall proposed artifact. Additionally, improvements can be made to the artifact to ensure the IoT-based security framework for the accident detection system meets the objectives of detecting and reporting a car accident instantly, preserving the driver's private information securely.
3. Knowledge Base:

The last activity is disseminating the obtained results and their effectiveness in solving the problems identified in relevant forums and publishing outlets. The research results are added to the knowledge base [27]. Thus, the proposed solution applies a secure and efficient IoT–LWC artifact for detecting accidents and encrypting the driver's private information in an effort to present and achieve a valuable framework that can prevent location tracking and preserve the privacy of critical information during an accident.

4. Requirement Analysis

For the requirement analysis, information was acquired from a variety of publications and proposed prototypes. To gather the required information on measurements, technologies in use, and challenges in the current approach, as well as to determine more concerning the level of awareness and the challenges that will be encountered when designing an IoT security framework for detecting accidents and encrypting driver's data.

4.1. The Overall Conditions and Challenges for IoT–LWC (Research Study Challenges)

This study's first objective was to determine the optimal solution for preserving privacy by performing cryptographic techniques. The data were used to provide insight into this research using document analysis and a literature review for IoT–LWC. According to the research published in Ref. [27], an in-depth and up-to-date, comprehensive comparison of simple cryptography techniques was presented. The publication presented 54 LWC algorithms in their respective classes, including five different ECC cryptography algorithms, 21 lightweight block ciphers, 19 stream ciphers, and nine lightweight hash functions. The efficiency of the hardware and software, chip size, energy and power consumption, throughput, latency, and figure of merit of the ciphers were compared (FoM). The research concluded with a comparison of AES and ECC as the best lightweight cryptographic primitives to use based on published research in the area of portable cryptography [28]. To protect the privacy of the IoT through messaging transmission, the study used ECC combined with AES in the ECIES algorithm. The length and transmission duration of the message could be a challenge.

The challenge and objective addressed in this research were also how to apply the ECIES–LWC algorithm for preserving and encrypting sensitive information. The second objective of this research was to determine how to apply the IoT-based LWC in the artifact to securely design an IoT framework for detecting accidents and encrypting the driver's private information.

Challenges Faced in an IoT-Based ECC LWC

The following challenges are divided into two sections.

- Application of the ECIES algorithm:
    1. Time consumption when sending a message: This research explored the use of ECC to improve data privacy and security in the IoT. Due to its effectiveness and performance in terms of time and energy, it is intended to demonstrate that ECC is relevant for the IoT. To do so, this study also emphasizes earlier studies on lightweight ECC to demonstrate the significance of ECC and to assess the effectiveness of techniques, as described in Section 2.3 (Lightweight IoT Cryptography), to quickly report an accident to first responders and encrypt messages. Thus, this study focused on the ECIES algorithm key size and compared it with the RSA algorithm in terms of time consumption in the generation of the RSA key size. Furthermore, a comparison between the ECIES algorithm and the RSA method is covered in the evaluation.
    2. SMS message length: The message content must be kept to a minimum via a maximum SMS message length for mobile devices to achieve the objective of sending a message instantly while also being encrypted by ECC. In fact, the message can only be 160 characters long at most. For instance, if an accident occurs, a location is sent through an SMS message to first responders. For this reason, only the longitude and latitude are encrypted to keep the message short, and only one SMS is sent. However, the encrypted message is still more than 160 characters [28], and to solve this challenge, the message is sent from the device controlling the signals, encrypted, and then sent to the CServer as MQTT. The message is then sent from the CServer to the first responders.
- IoT-based Accident Detection System

The challenges of accident detection techniques and promptly alerting first responders to incidents are provided in this section. Thus, a few sensors are fitted to the vehicles to collect information on the location and direction of the car in the event of an accident. To illustrate, the airbag sensor (AS) and vibration sensor (VS) can be used as examples. The first responders receive information about the accident's location using GPS/GPRS modems. Furthermore, the sensors are examined and evaluated in this study to ensure their functionality. If an incident occurs, the vibration in the VS increases above its maximum level, or the airbag explodes out of its site (i.e., bursts). The GSM module then receives this information. Later, the GSM module transmits a message to the first responders. In brief, the integration of all of the proposed artifact's functionalities with the encryption process was the greatest challenge of this research.

4.2. Requirement Specification

In this section, different hardware, software, and platforms are required for deploying the proposed system.

4.2.1. Software Requirements

- Raspberry Pi OS Raspbian: The Raspbian operating system supports Raspberry Pi. Based on Debian, Raspbian is a free operating system designed specifically for the Raspberry Pi device [29].
- PyCharm IDE Software2022.3.2: PyCharm is a Python IDE developed by JetBrains, the organization that also produces the well-known IntelliJ IDEA IDE for Java. Any developer who wants to construct Python applications, including web applications, data science applications, or even simply basic Python scripts, is encouraged to use

PyCharm [30]. It has a code editor with tools such as automatic indentation, brace matching, and syntax highlighting. A "Python file" is written code or a program.
- The Central Server (CServer)—"Website-Database Server": All information on an accident is kept in a Microsoft SQL database, and MySQL was adopted as the database management system (DBMS). In addition, the website develops client-side interfaces using HTML, CSS, and Bootstrap. For server-side programming, it uses PHP [31].
- MQTT (message queuing telemetry transport): An IoT ecosystem uses the MQTT protocol for communication, which runs on top of the transport control protocol. Moreover, MQTT is considered a lightweight machine-to-machine communication protocol, and it was developed by IBM. It is used to send data from sensors to the server [32].
- Fritzing Software 0.9.10: The Fritzing software package can be helpful during development phases, such as the assembly of the prototype based on the scheme in the mock-up sketch and boards, as well as the automatic generation of schematic diagrams and the PCB [33].
- Twilio SMS Massage Service: The use of virtual phone numbers to deliver SMS messages is made possible by Twilio's cloud-based messaging technology [34].

4.2.2. Hardware Requirements

- Raspberry Pi Model B: The newest and fastest Raspberry Pi model, the Raspberry Pi 4 Model B, was used to build an IoT ecosystem. It features different RAM capacities (2, 4, or 8 GB), a USB-C port for power, a MicroSD card slot for the operating system and file storage, two micro-HDMI ports, and the option to connect to the Internet wirelessly or with an ethernet cable. The features of the Raspberry Pi 4 Model B used in this experiment are listed in Table 4 [35].

Table 4. Raspberry Pi 4 Model B specifications.

Raspberry Pi 4 Model B Specifications	
Operating system	Raspbian
Internet connectivity	Wi-Fi
Card size	4 GB

- Airbag sensor: An airbag is a type of occupant restraint system and a vehicle safety component. It is composed of an inflatable fabric bag, an impact sensor, an inflation module, and an airbag cushion. If an accident causes the airbag to blow, the sensor detects it. If the airbag bursts, the airbag sensor detects that an accident has occurred [36]. The digital airbag sensor in this research requires three wires for connection: a ground pin wire (GND), a voltage pin (+5 Vcc) to supply the circuit with the required electricity, and an input pin to communicate with the Raspberry Pi.
- Vibration sensor (VS): This is a transducer that transforms vibrations into an electrical equivalent, such as a voltage, such as one that uses a laser or piezoelectric crystal. It is also known as a vibration transducer. It detects the vibration of a car, which is a specific and substantial vibration. However, vibration sensors are used to measure and analyze linear velocity, displacement, proximity, and various shock triggers [37].
- GPS and GSM modules: A radio navigation system called GPS, or "global positioning system", enables land, sea, and aerial users to pinpoint their precise location, speed, and time at any time, day or night, in any weather, anywhere in the world. An accident's location will be known. In addition, it is a digital mobile telephony system that is widely utilized in Europe and other parts of the world. It stands for global system for mobile communication [38]. Furthermore, the GSM and GPS modules are utilized as a security system to improve the project and make it more secure by calling or texting the user's phone number if there is a car theft attempt [39].
- Power supply: This is an electronic device that supplies electric energy to an electrical load [38].

4.2.3. Proposed System

This research addresses the challenges by developing an intelligent security framework and an IoT-based system for immediate accident detection. It offers the following primary actions: instant detection of an accident and sharing of the driver's personal health record and other crucial information with the first responders (for example, Najm and ambulance) while ensuring privacy. Figure 4 explores the IoT security framework accident detection system artifact proposed in this research.

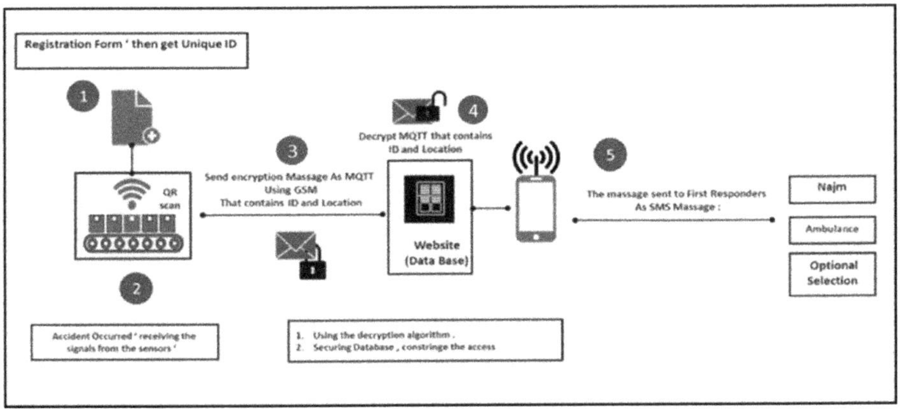

Figure 4. Accident detection system conceptual block diagram design.

First, the device's unique ID is a code generated when the driver fills out their vehicle owner information record and personal health record. Once an accident occurs, the AS and VS that work together to report the accident instantly send the signal to the RPM to retrieve the location from the GPS. Next, the GSM module enables the system to send sensitive data about the accident using the message queue telemetry transport (MQTT) protocol to the CServer. The message also includes coordinates from the GPS module and a unique ID that connects to the system. The message must apply the proposed ECIES algorithm to protect and preserve its security and privacy. Then, to prevent a potential attack, there is an authentication phase between the sender (i.e., the IoT device) and the recipient to check that the alerted accident is real. After that, the Cserver sends a message as an SMS to the first responders, who are then able to view the data.

Eventually, the accident detection system's artifact prototype is tested and evaluated under real-life conditions, and its performance is evaluated using the system's experimental setup.

4.3. System Design

4.3.1. Design Goal

An intelligent security framework and an IoT-based system are proposed as solutions to this issue for instant accident detection. Its main features are instant accident detection, privacy protection, and the ability to share a driver's personal health information with first responders, such as an ambulance and Najm.

4.3.2. Security Issue

This study focuses on two aspects: MQTT privacy transmitted by the artifact and critical information sent as an SMS provided by the website. In any scenario, the system and method of sharing messages should be secured. In fact, it is intended that only authenticated and authorized first responders on the website side will have the right to receive critical information. On the other hand, by creating an artifact that offers a secure and efficient IoT-based system to instantaneously detect and report accidents, this research protects the privacy of drivers' information during SMS transmission.

4.4. *System Architecture*

The proposed architecture securely designs an IoT framework for detecting accidents and encrypting drivers' private information. Accordingly, the process flow is shown in the following diagram (Figure 5).

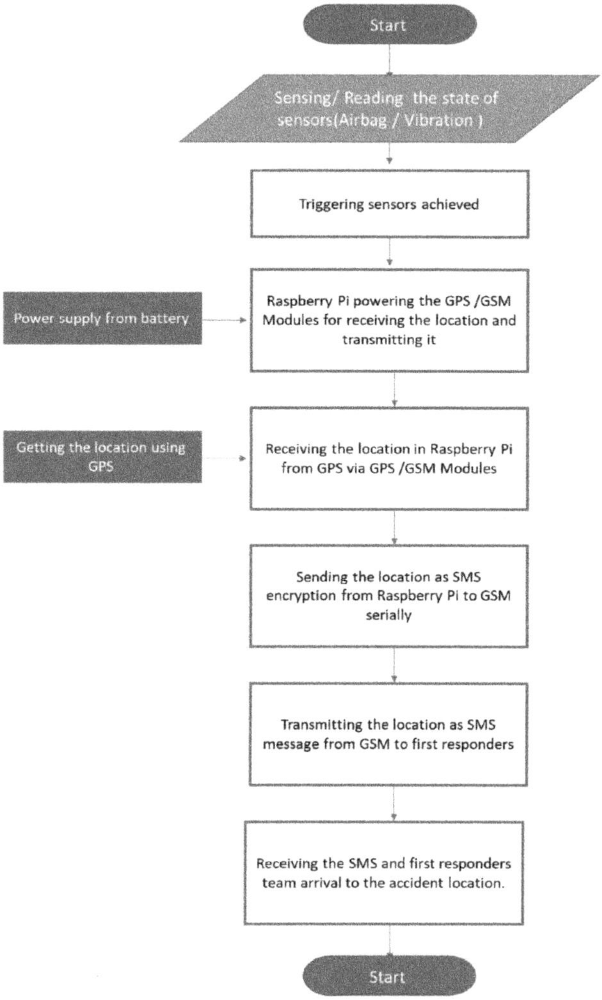

Figure 5. Flowchart for intelligent IoT-based accident detection.

System flowcharts are used to display the proposed project's processes, from the start of accident detection to the end, when the CServer sends an accident report to the first responders.

4.4.1. Block Diagram

A block diagram displays a complex system or process, such as an electronic circuit, in schematic form, along with a general layout of its pieces or components. In addition, adopting this type of diagram will simplify the IoT-based security framework to detect and report a car accident instantly as a block diagram.

The proposed architecture, as shown in Figure 6, is composed of the following elements: first responders (e.g., Najm and ambulance), relationships among them, A9G

GSM/GPS modules, tower, AS, VS, power supply, Raspberry Pi 4 Model B, and the CServer.

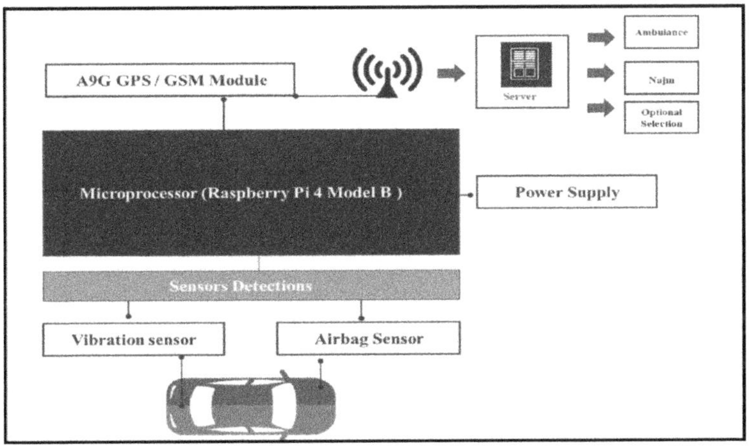

Figure 6. Block diagram of intelligent IoT-based accident detection.

4.4.2. Sequence Diagram

A graphic that illustrates interactions among items and captures how activities are carried out is called a sequence diagram (Figure 7). Therefore, it demonstrates how and in what order various items interact. Additionally, a time-sequenced display of item interactions is provided.

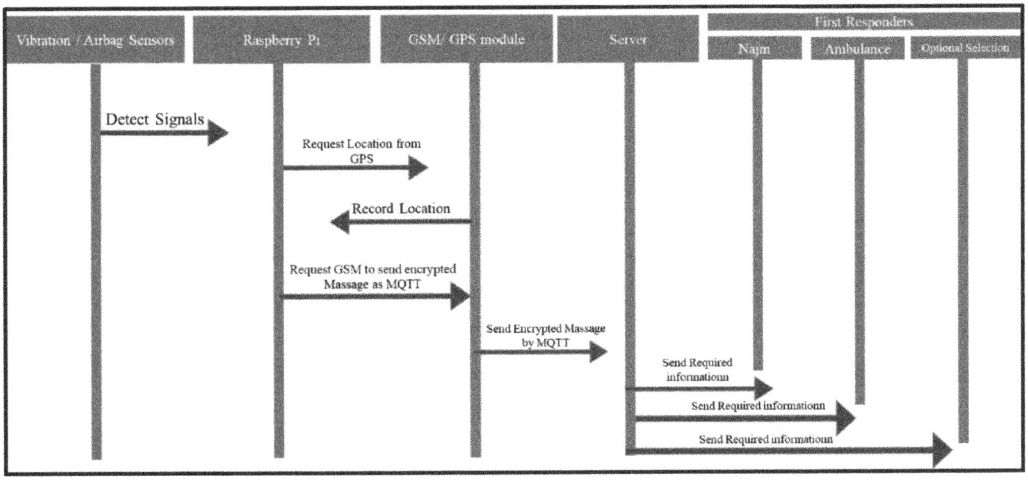

Figure 7. Sequence diagram of automated requests.

5. Implementation

This section discusses the implementation of the prototype using the RPM, sensors, GSM/GPRS, and other modules to communicate with the CServer (website and database server), and the results are presented. The following sections show the implementation of an IoT-based intelligent security framework for immediate accident detection artifacts.

Development Tools

A discussion of the development tools used for the prototypes is provided in this section in phases. Each phase contains the tools used to implement the prototype of the proposed system.

1. Phase 1: QR Code Scanning (Website—Database Server):

This section covers how the required information from the driver will be entered and saved by scanning a QR code generated by the system. After that, the data is stored in the database. Thus, a new record and unique ID are established for the device that was used throughout the accident. Moreover, the system provides an option if the driver needs to share the accident data with any of their family members or their family doctor during the accident.

Additionally, MySQL is the database management system (DBMS) that manages the entire system. Detailed information on the different tables used in the proposed system is as follows:

- Driver details table: This table includes all driver-related data, including Driver_Name, Driver_Address, Driver_Email, and Driver_Phone number;
- Vehicle table: Vehicle information, including Vehicle_ID, Vehicle_Name, and Vehicle_Insuracne, is available and provided in this table;
- Diseases (medical records) table: This table contains data on all diseases and the health insurance of the driver;
- Device table: This table contains all data related to the device, such as Device_ID, Driver_ID, Vehicle_ID, and the private keys utilized in the decryption process. The system generates the QR code for each vehicle to allow the driver to input the required information;
- Accident table: This table is used to keep track of all details of accidents, such as Driver_ID, Device_ID, Location_ latitude, and Location_ longitude, including those that were used during the notification phase.

2. Phase 2: IoT Accident Detection Sensors
    - Airbag sensor (AS): While developing the circuit, an AS is linked to the RPM. Thus, the AS works to detect the signals when an airbag is bursting. Furthermore, the AS can be added to the breadboard and connected to the microprocessor to complete the circuit;
    - Vibration sensor (VS): The second sensor used in this research is the VS, which works to detect signals when it is activated. It also employs a digital VS. The sensor's vibration frequency ranges from 40 Hz to 65,535 Hz when it is attached to the RPM. In addition, the shock sensitivity of the VS was modified by selecting a time alert to detect the shock. Periodically shaking the sensor causes it to automatically produce "1" when working as a digital signal (as an indicator of shaking or movement).
3. Phase 3: Accident Transmission Encryption Message
    - Raspberry Pi Microprocessor (RPM): In this research, the system uses a Raspberry Pi 4 Model B connected to a CServer, an AS, a vibration sensor, GSM/GPRS, and other devices to detect and report accidents instantly. The Raspberry Pi receives the signals, requests the location of the accident, encrypts the message, and sends the encrypted message to the CServer; thus, the IoT devices connect to the Raspberry Pi through input pins (Figure 8), and the CServer retrieves the data through MQTT.
    - GPS and GSM module: The A9 GSM/GPRS module is a tiny GSM modem that can be used in numerous IoT projects. In this research, the GSM was used to send the encrypted message to the CServer via MQTT. Additionally, the exact location of the vehicle is also determined through the GSM/GPRS and GPS tracking systems, which also use GPS technology. In this research, we proposed a GPS

technology system to capture the location. Additionally, the connection pins of the A9G GSM/GPRS+GPS were as follows (Table 5).

Table 5. A9G-TTL pin connection.

TTL	A9G
5V	USB
RX	TX1
TX	RX1

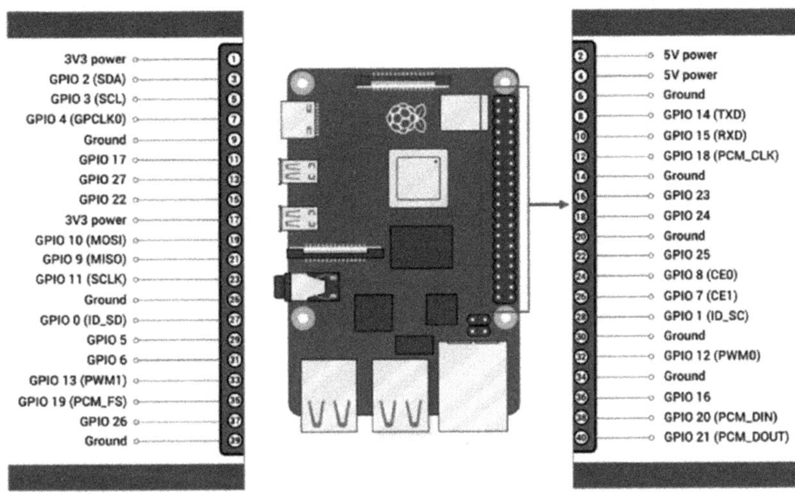

Figure 8. RPM pins.

4. Phase 4: Notifications

The Central Server (CServer) "Website-Database Server":

The Raspberry Pi CPU sends an encrypted message to the main server via MQTT when an accident is detected. Next, the CServer instantly decrypts the message and sends an SMS message containing the website URL to the first responders, informing them of the accident as a report so they may respond as soon as possible. In addition, the CServer is responsible for notifying any family members or the driver's family doctor that have been stored in the database. Additionally, the accident table in this phase will also be updated at the same time based on the accident information.

6. Evaluating the Phases (As Experimental Implementation)

1. Phase 1: QR Code Scanning (Data Entry) Implementation

The registration of vehicles is the goal of this phase. The owner of the vehicle must install the IoT device to get their vehicle ready for this system. The first step after setting up the device is for the owner to scan the QR code to complete the registration process and create the Vehicle ID, which will be kept in the database as shown in the following (Figure 9).

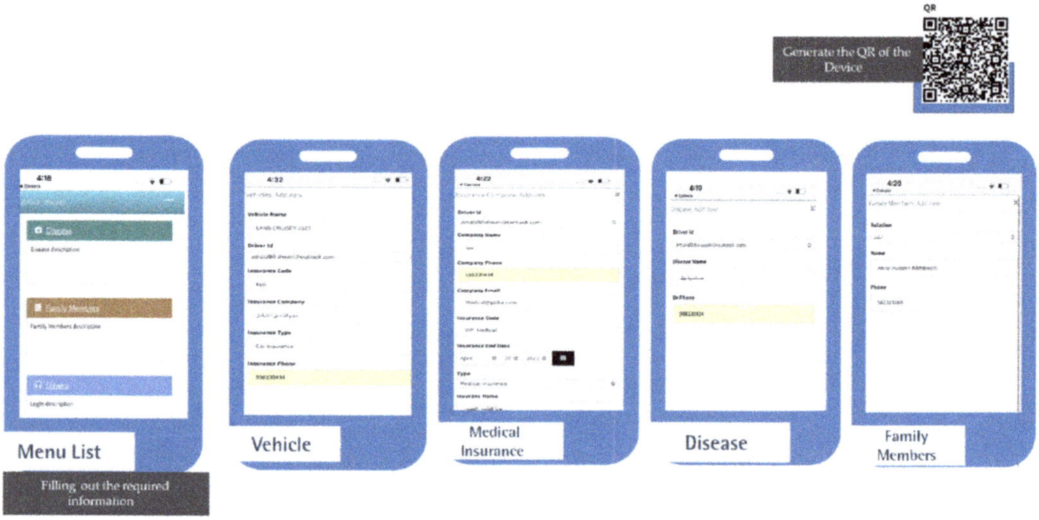

Figure 9. QR scanning and filling in the required information through the website registration.

2. Phase 2: IoT Accident Detection Sensor Implementation

Once an accident occurs, the AS and VS work together to promptly and instantly report the accident and send the signal to the RPM. The circuit is built by the RPM using a breadboard and wiring system. For optimal operation, each component needs to be connected to both GND (ground) and VCC (5 volts). The VS will send analogy signals after identifying the accident to the Raspberry Pi through a wire connected to pin #23, and the AS through a cable linked to pin #24 (Figure 10).

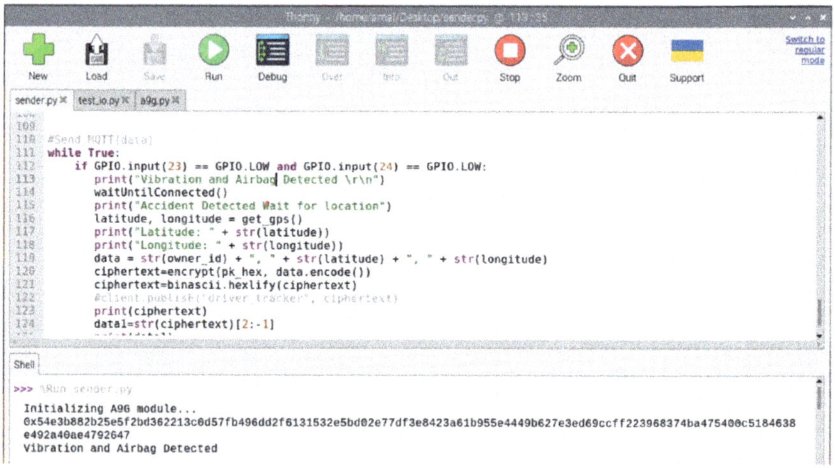

Figure 10. The AS and VS outputs are activated.

3. Phase 3: Accident Transmission Encryption Message

Next, the RPM sends a wire (pin #4) request to GPS to acquire the accident location. The GSM/GPRS USB model is connected to the GPS chip's RX (receiver) pin. The GPS will then save the location, providing its longitude and latitude, and send it back to the Raspberry Pi via the GSM/GPRS USB model that is linked to the TX (transmit) pin. In addition, upon GSM module activation through the RPM, the encryption feature will be performed. The ECC model generates random values, creates private and public keys, and encrypts the message using the ecies.utils key-generating and encryption libraries. In brief, the ECIES algorithm is being used to implement the data encryption module (Figure 11). Thus, it will use the Message Queue Telemetry Transport (MQTT) Protocol to send the encrypted message to the CServer. Moreover, the encrypted message contains a unique Driver_ID and GPS coordinates.

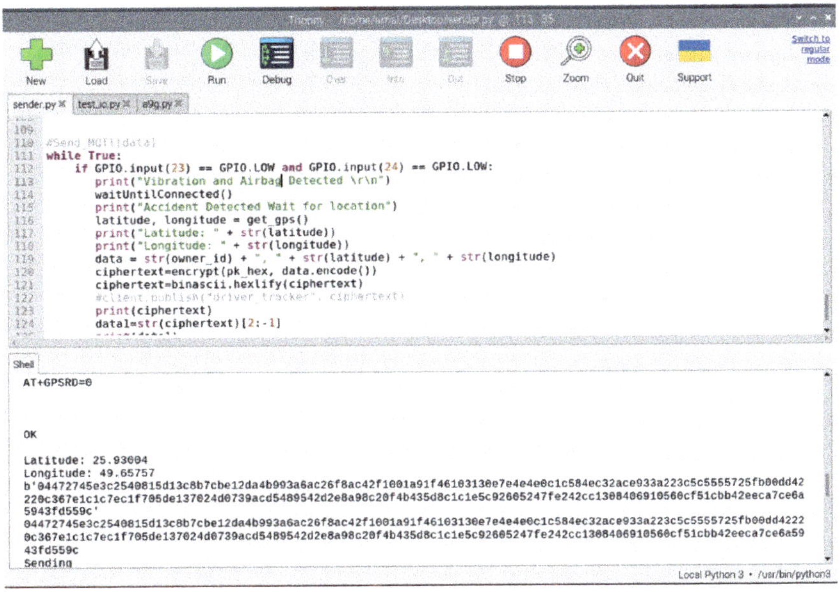

Figure 11. The GPS records the location (longitude and latitude) and encrypts the message.

4. Phase 4: Notifications to First Responders and Accident Details

The CServer receives an encrypted notification about the accident each time it occurs. Following this, CServer utilizes the private key to decrypt the message and displays the message's contents, including the accident's location on a map and the required information on the driver (Figure 12). Thus, the required information is then sent to the first responders via SMS messages, as shown in the figure below, via the CServer (Figure 13).

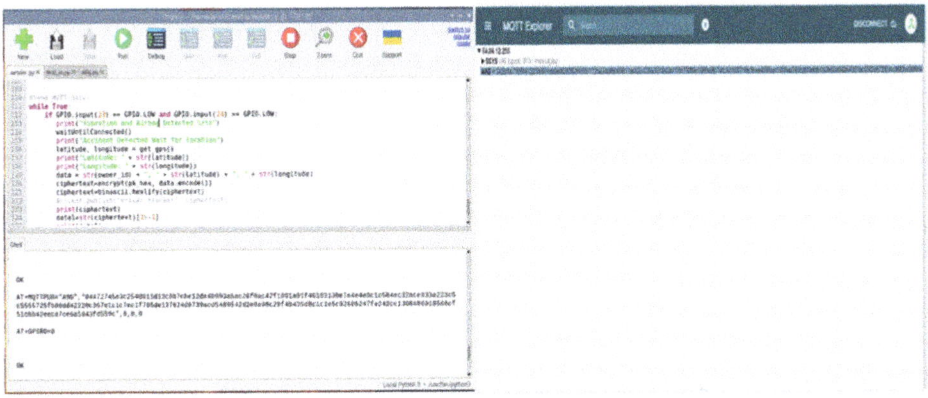

Figure 12. The encrypted message is sent through MQTT.

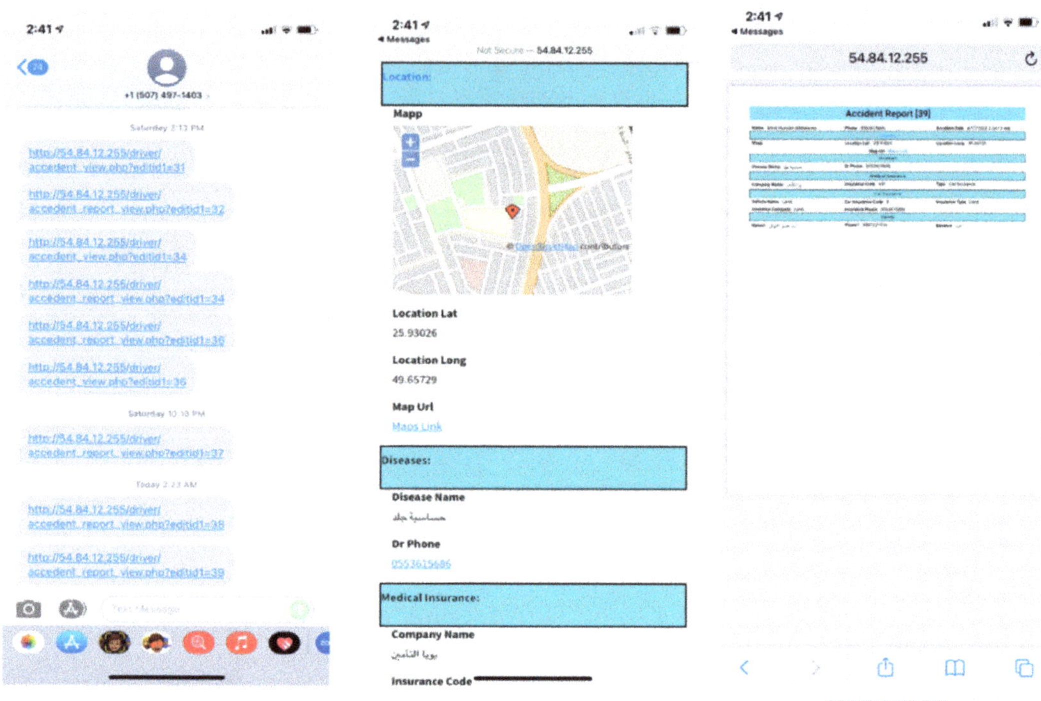

Figure 13. The CServer decrypts the message and sends the SMS message to the first responders.

The following (Figures 14 and 15) presents the complete artifact components connected to the IoT-based system for instant accident detection.

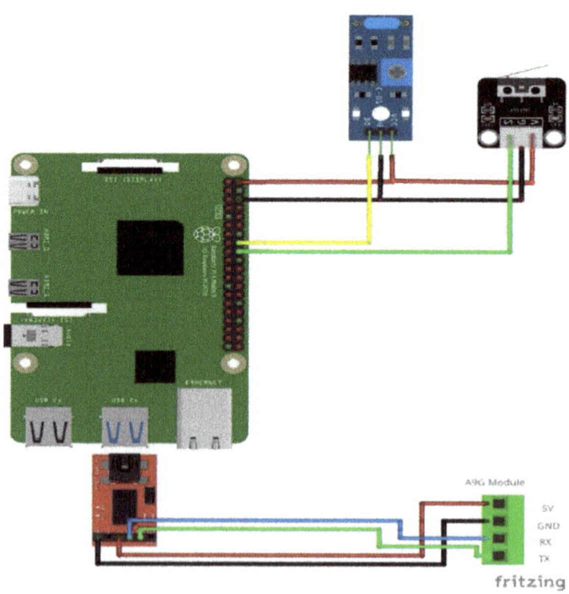

Figure 14. The complete circuit implementation with the CServer.

Figure 15. The complete artifact.

Finally, from the time the driver installs the IoT device on the vehicle until an accident occurs, all information related to the accident and the driver is stored in the database.

6.1. Overall Outcomes/Results

The result was instant car accident detection and reporting technology based on the IoT. In addition, first responders (including Najm and ambulances) were able to instantly locate the vehicle involved in the accident and provide medical help, saving their lives. Thus, the IoT-based accident detection system assists in locating the precise location of the accident and relaying this information to the appropriate first responders so that the injured person may obtain assistance as soon as possible.

The system's functionality was confirmed in the preliminary findings, as described in this section. Table 6 shows the expected and actual values of the functions performed by the proposed artifact.

Table 6. Expected and actual functions of the artifact.

Functional Requirement	Test Conducted	Expected	Actual
QR scanning	The QR barcode is read, and the driver's information is entered.	The driver can complete all fields with the required information.	The QR is generated by the system. Then, the driver can complete the required information. When the recorded information is complete, the device's unique ID is created, which can be used during an accident.
AS (detection)	The closing of the airbag's switch is detected.	The airbag communicates with the microprocessor once the airbag switch has been closed or depressed (i.e., blown).	When the airbag button is pressed, the airbag light turns on and sends a signal to the microprocessor to let it know that it has done so.
VS (detection)	Once any pressure or shaking is detected.	The VS should send the signal to the microprocessor by delivering a signal as soon as it moves.	The microprocessor is quickly informed that a vibration just occurred when shaking the VS.

Table 6. *Cont.*

Functional Requirement	Test Conducted	Expected	Actual
Locating the position of the vehicle or accident (location recording)	The GPS responds to the signals it receives from the microprocessor.	Once the GPS receives the signals from the microprocessor, it should locate the current position of the accident.	The GPS locates the current position when it receives a request from the microprocessor.
Encrypting the message using the ECIES algorithm	Sending the location through the GPS to the microprocessor for application of the ECIES algorithm.	The ECIES algorithm should be utilized by the microprocessor to apply the encryption message through the ECIES utilities library, and the microprocessor should be able to identify the device's unique ID and GPS location through the encryption message.	GPS provides the device's unique ID, which the microprocessor recognizes, together with the current location coordinates (longitude and latitude), which are used to encrypt the message.
GSM is operational and transmitting an encrypted message	Transmitting the encryption message using GSM and MQTT.	With MQTT, GSM transmits an encrypted message to the CServer.	The GSM sends the MQTT to the CServer after receiving the encrypted message.
Receiving the message and decrypting it into the Cserver	Decrypting the message and examining the accident details.	Using the database's private keys, which are created using the encryption procedure, the message is decrypted, and the accident details are displayed in full on the website's interface.	The accident's details are displayed on the website's interface, and the messages are decrypted.
CServer and website display the accident details on the web page	To see the driver's detailed information.	After decrypting, the required information about the accident is loaded on the site.	The required information about the accident is displayed on the website's interface.

As briefly stated in the table, the major features were verified, and it was ensured that the flow of all microprocessor system units (such as sensors, GSM, GPS, database server, and ECIES IoT encryption algorithm library) is successful and efficient.

6.2. Case Study and System Limitations

The following (Table 7) provides case studies that were tested during the proposed system's implementation. In general, human-designed systems can provide accurate readings up to a particular threshold, but beyond that point, they will have certain limitations. The table of work in this study describes the sensors' limitations and their intended use in combination to obtain an accurate value. In addition, the following two situations contrast the size of the keys and the time consumed with the RSA and ECC during the encryption process.

Table 7. Case studies tested during the proposed system's implementation.

Case Number	Case	Sensor Test	Output/Result
Case 1	In this case, consider the extracted real vibration value with a system that only has one sensor.	The VS rate is 2200 Hz	Accidents that happen at vibrations lower than this one are not detectable by this system because the detectable vibration is up to 4000 Hz.
Case 2	In this case, consider the extracted real vibration value with a system that only has one sensor.	The VS rate is 4000 Hz	The vibration is activated. The mechanism is activated. However, the accuracy of one sensor's operation is not enough, and a notification is sent even if it turns out to not have been an accident.

Table 7. *Cont.*

Case Number	Case	Sensor Test	Output/Result
Case 3	In this case, consider the possibility of the AS being activated either as a result of an accident or as a result of manufacturing defects.	The AS bursts	The airbag bursts. The system is triggered in any case, and a message is sent.
Case 4	In this case, consider the possibility of two sensors being triggered simultaneously.	The VS rate is 4000 Hz, and the airbag bursts	When the first and second sensors (VS and AS) are activated (hertz value), an accident is recognized (triggered).
Case 5	In this case, the accident is detected by the sensors, and a message is encrypted using the RSA public/private key size (30) via the ECIES algorithm.	The size of the RSA key is 3072	The private and public keys are generated with RSA, and the key size is determined using a time-consuming process.
Case 6	In this case, the accident is detected by the sensors, and a message is encrypted using the ECC public/private key size (256) via the ECIES algorithm.	ECC key size is 256	The private and public keys are generated faster than the RSA key size.

Cases 5 and 6 used an experimental test that compared ECIES and RSA, utilizing the Python, ECIES, and RSA libraries. The National Institute of Standards and Technology (NIST) provided the security strengths of the symmetrical block cipher and asymmetric-key algorithms, shown in (Table 8).

Table 8. Comparable key lengths for ECIES and RSA in bits [40].

Security Level	ECIES Key Length	RSA Key Length
80	160	1024
112	224	2048
128	256	3072
192	384	7680
256	512	15,360

As a result, the test functions were analyzed using the ECIES Python Library, which utilizes the secp256k1 curve, which implies that the ECC key size is 256 at the security level of 128. In light of this, 3072 is the RSA key size presented in the ECC domain parameters with recommended properties [41]. In addition, Table 9 explains the computational time required to generate the ECIES and RSA's public and private keys and to perform the encryption and decryption operations. Furthermore, it was presumed that messages in plaintext, including the location of the Device ID, including longitude and latitude, were sent. The researcher's location was utilized in this case in the following format: b'ID:7865409, LAT:25.933333, LOG:49.666667.

Table 9. Comparison between ECIES and RSA.

Key Size Length		Key Generation Running Time (Sec)		Encryption Process Running Time (Sec)		Decryption Process Running Time (Sec)	
ECC	RSA	ECC	RSA	ECC	RSA	ECC	RSA
256	3072	0.0021	1.887	0.006	0.002	0.003	0.009

According to the comparison between ECIES and RSA, Table 9 shows that ECIES generates keys faster than the RSA algorithm. As the private key is created at random and the public key is a point on a curve, ECIES has the benefit of being able to generate both keys in a short time. Moreover, the RSA algorithm encrypts data faster than an algorithm based on ECIES. However, it is noted that the difference between the two computational times is comparatively shorter than one second, and this is not a large difference. In addition, the ECC algorithm's decryption process is faster than the RSA process. As a result, the outcomes show that for systems based on ECC, thanks to the mathematical strategies and benefits provided by the curves, ECIES represents an effective substitute for RSA-based cryptosystems. Particularly since embedded systems lack the memory and processing power necessary to complete the calculations required by RSA-based cryptosystems with large numbers, ECIES cryptosystems are ideal for IoT-embedded systems.

Thus, to compare this suggested work to the prior studies, it must be noted that this system was created with privacy in mind and that the proper security mechanisms are in place to protect the data transmitted by the system, which has been overlooked in previous studies. Additionally, there are benefits to having two sensors to provide an accurate alert and a faster response time.

6.3. Comprehensive Remarks on the Overall Proposed System (Adding Knowledge Base)

Utilizing an IoT security framework to encrypt, identify accidents, and detect accidents is our proposed artifact development in this research. Furthermore, it is advantageous for developing a framework for IoT security that can be applied to other purposes. In addition, we have successfully finished the coding of the sensors, the CServer (database server), GPS, and GSM to complete the process of sensing the accident and informing about it. In addition, the procedure of collecting signals from the sensor, using GPS to determine the current location by providing the location's longitude and latitude, and encrypting them using the ECIES algorithm on the CPU. Following this, the MQTT-encrypted message is sent through GSM to the CServer. In addition, the CServer will decrypt the message after receiving it to deliver the accident information, which will then be shown on the website. The accident information will then be sent to the first responders via SMS messages as URL links through the CServer.

Last but not least, the proposed system device may be easily modified and will benefit society by enabling improved outcomes, such as accident detection, if it is operating as planned to deliver an intelligent IoT security framework. Preventing location monitoring and the numerous deaths that can occur as a result of a delayed notification process, preventing the people's injuries from growing worse as soon as possible when the first responders are informed about the accident in detail instantly.

7. Conclusions and Future Work

The number of vehicles has dramatically increased recently in countries such as Saudi Arabia. In addition, accident rates have increased as a result of the increased traffic. Many accident detection devices exist, but a significant number of fatalities still occur. This problem is caused, at least in part, by insufficient automatic accident detection, insufficient warning, and inefficient emergency response routing, which obstruct the proper response to catastrophic accidents. The lack of applicable technologies because of financial and capacity restrictions on retrofitting just makes the situation worse. This research provides an intelligent security framework to handle these problems, and an IoT-based security framework solution is suggested for accident detection instantly.

This research demonstrated that making use of a range of various sensors can improve the accuracy with which a traffic accident is detected while preserving the privacy of the driver's critical information. Thus, the proposed system detects the location of an accident instantly, encrypts the critical information about the accident, and then transmits that information to first responders so that they can save lives.

Indeed, the results showed that the automatic IoT-based security framework accident detection system, which also protects the privacy of the driver's personal information, performed as expected. The main advantage of this research is that it reduces the number of false alarms about accident reports. Moreover, RSA-based cryptosystems can be effectively replaced by ECIES. ECIES cryptosystems are particularly well suited for IoT-embedded systems since such systems lack the memory and processing power needed to finish the calculations needed by RSA-based cryptosystems with different key sizes. Another advantage is that the artifact might be used as a model for the implementation of similar artifacts in other contexts related to applying an intelligent security framework and the Internet of Things (IoT). A similar artifact, such as one created for automatically predicting plantation location and watering, where data protection is required, might be produced using the prototype as a guide to meet various public or commercial objectives.

Furthermore, there were limitations to the research due to conducting the initial evaluation of the system in a simulated environment. One of the study's limitations is that the system currently cannot calculate the closest responders (e.g., Najm and ambulances) to the accident through the nearest first responders with the shortest response time, computed using the distance matrix API, allocated to the user. In addition, the driver of the first responders receives an SMS specifying the user's location. Further, the system currently does not allow the user to receive information about the probable arrival time.

In the future, there are a number of important recommendations, citing the data that were gathered, including the need for the system to calculate the distance to identify the first responders (for example, Najm and ambulances) who are nearest to the accident and to send an urgent SMS that contains the location to provide assistance instantly. Future advancements in this technology may involve attaching a camera module to take pictures of the accident and sending the pictures to a server. In the future, big data processing can be used to examine some road accident results using the data gathered on the server. Additionally, in the not-too-distant future, system security and privacy will be improved, and these challenges will be fully addressed in upcoming efforts, especially on the website and the General Data Protection Regulation (GDPR).

Author Contributions: Conceptualization, A.H.A.; Methodology, A.H.A. and B.S.A.; Validation, A.H.A. and B.S.A.; Formal analysis, A.H.A.; Investigation, A.H.A. and B.S.A.; Resources, A.H.A.; Writing—original draft, A.H.A.; Writing—review & editing, A.H.A. and B.S.A.; Supervision, A.H.A. and B.S.A.; Project administration, A.H.A. and B.S.A. All authors have read and agreed to the published version of the manuscript.

Funding: This research received no external funding.

Institutional Review Board Statement: Not applicable.

Informed Consent Statement: Not applicable.

Data Availability Statement: The data presented in this study are available on request from the corresponding author.

Conflicts of Interest: The authors declare no conflict of interest.

References

1. Almoshaogeh, M.; Abdulrehman, R.; Haider, H.; Alharbi, F.; Jamal, A.; Alarifi, S. Shafiquzzaman Traffic Accident Risk Assessment Framework for Qassim, Saudi Arabia: Evaluating the Impact of Speed Cameras. *Appl. Sci.* **2021**, *11*, 6682. [CrossRef]
2. Raghad, A.; Areej, S. IoT Based Accident Prevention System Using Machine Learning Techniques. Available online: https://www.researchgate.net/profile/Hala-Qudaih/publication/369595854_IoT_Based_Accident_Prevention_System_using_Machine_Learning_techniques/links/6424132192cfd54f8439c7bc/IoT-Based-Accident-Prevention-System-using-Machine-Learning-techniques.pdf (accessed on 1 April 2023).
3. Sharma, S.; Sebastian, S. IoT based car accident detection and notification algorithm for general road accidents. *Int. J. Electr. Comput. Eng.* **2019**, *9*, 4020. [CrossRef]
4. Sklavos, N.; Zaharakis, I.D. Cryptography and Security in Internet of Things (IoTs): Models, Schemes, and Implementations. In Proceedings of the 2016 8th IFIP International Conference on New Technologies, Mobility and Security (NTMS), Larnaca, Cyprus, 21–23 November 2016; pp. 1–2. [CrossRef]
5. Borgia, E. The Internet of Things vision: Key features, applications and open issues. *Comput. Commun.* **2014**, *54*, 1–31. [CrossRef]

6. Bhardwaj, I.; Kumar, A.; Bansal, M. A review on lightweight cryptography algorithms for data security and authentication in IoTs. In Proceedings of the 2017 4th International Conference on Signal Processing, Computing and Control (ISPCC), Solan, India, 21–23 September 2017; pp. 504–509. [CrossRef]
7. Fotovvat, A.; Rahman, G.M.E.; Vedaei, S.S.; Wahid, K.A. Comparative Performance Analysis of Lightweight Cryptography Algorithms for IoT Sensor Nodes. *IEEE Internet Things J.* **2020**, *8*, 8279–8290. [CrossRef]
8. Sridhar, S.; Smys, S. Intelligent security framework for iot devices cryptography based end-to-end security architecture. In Proceedings of the 2017 International Conference on Inventive Systems and Control (ICISC), Coimbatore, India, 19–20 January 2017; pp. 1–5. [CrossRef]
9. Bertino, E.; Islam, N. Botnets and Internet of Things Security. *Computer* **2017**, *50*, 76–79. [CrossRef]
10. Lee, S.; Tewolde, G.; Kwon, J. Design and implementation of vehicle tracking system using GPS/GSM/GPRS technology and smartphone application. In Proceedings of the 2014 IEEE World Forum on Internet of Things (WF-IoT), Seoul, Republic of Korea, 6–8 March 2014; pp. 353–358. [CrossRef]
11. Jethwa, A. Vehicle Tracking System Using Gps and Gsm Modem—A Review. *Int. J. Recent Sci. Res.* **2015**, *6*, 4805–4808.
12. Khan, M.A.; Khan, S.F. IoT based framework for Vehicle Over-speed detection. In Proceedings of the 2018 1st International Conference on Computer Applications & Information Security (ICCAIS), Riyadh, Saudi Arabia, 4–6 April 2018; pp. 1–4. [CrossRef]
13. Yee, T.H.; Lau, P.Y. Mobile vehicle crash detection system. In Proceedings of the 2018 International Workshop on Advanced Image Technology (IWAIT), Chiang Mai, Thailand, 7–9 January 2018; pp. 1–4. [CrossRef]
14. Nasr, E.; Kfoury, E.; Khoury, D. An IoT approach to vehicle accident detection, reporting, and navigation. In Proceedings of the 2016 IEEE International Multidisciplinary Conference on Engineering Technology (IMCET), Beirut, Lebanon, 14–16 November 2016; pp. 231–236. [CrossRef]
15. Khot, I.; Jadhav, M.; Desai, A.; Bangar, V. Go Safe: Android application for accident detection and notification. *Int. Res. J. Eng. Technol.* **2018**, *5*, 4118–4122.
16. Chaturvedi, N.; Srivastava, P. Automatic Vehicle Accident Detection and Messaging System Using GSM and GPS Modem. *Int. Res. J. Eng. Technol.* **2018**, *5*, 252–254.
17. Khaliq, K.A.; Raza, S.M.; Chughtai, O.; Qayyum, A.; Pannek, J. Experimental validation of an accident detection and management application in vehicular environment. *Comput. Electr. Eng.* **2018**, *71*, 137–150. [CrossRef]
18. Cauteruccio, F.; Cinelli, L.; Corradini, E.; Terracina, G.; Ursino, D.; Virgili, L.; Savaglio, C.; Liotta, A.; Fortino, G. A framework for anomaly detection and classification in Multiple IoT scenarios. *Futur. Gener. Comput. Syst.* **2021**, *114*, 322–335. [CrossRef]
19. OWASP Internet of Things Top Ten Project. The Open Web Application Security Project®. 2014. Available online: https://owasp.org/www-project-internet-of-things-top-10/#tab=OWASP_Internet_of_Things_Top_10_for_2014 (accessed on 5 December 2022).
20. Aleisa, N.; Renaud, K. Privacy of the Internet of Things: A Systematic Literature Review. In Proceedings of the 50th Hawaii International Conference on System Sciences, Waikoloa Village, HI, USA, 4–7 January 2017. [CrossRef]
21. Henriques, T.A.; O'Neill, H. Design science research with focus groups—A pragmatic meta-model. *Int. J. Manag. Proj. Bus.* **2023**, *16*, 119–140. [CrossRef]
22. Dhanda, S.S.; Singh, B.; Jindal, P. Lightweight Cryptography: A Solution to Secure IoT. *Wirel. Pers. Commun.* **2020**, *112*, 1947–1980. [CrossRef]
23. Keerthi, K.; Surendiran, B. Elliptic curve cryptography for secured text encryption. In Proceedings of the 2017 International Conference on Circuit, Power and Computing Technologies (ICCPCT), Kollam, India, 20–21 April 2017; pp. 1–5. [CrossRef]
24. Hevner, A.; Chatterjee, S. Design Science Research in Information Systems. In *Design Research in Information Systems*; Integrated Series in Information Systems; Springer: Boston, MA, USA, 2010; Volume 22, pp. 9–22. [CrossRef]
25. Hevner, A.R. A Three Cycle View of Design Science Research. *Scand. J. Inf. Syst.* **2007**, *19*, 4.
26. Hevner, A.R.; March, S.T.; Park, J.; Ram, S. Design Science in Information Systems Research. *Manag. Inf. Syst. Q.* **2004**, *28*, 75–105. [CrossRef]
27. Imam, R.; Areeb, Q.M.; Alturki, A.; Anwer, F. Systematic and Critical Review of RSA Based Public Key Cryptographic Schemes: Past and Present Status. *IEEE Access* **2021**, *9*, 155949–155976. [CrossRef]
28. Alphonse, R.M.; Malalatiana, R.H.; Choube, M.P. Segment optimization of short message service in telecommunication. *Innov. Technol. Methodical Res. J.* **2021**, *2*, 81–88. [CrossRef]
29. Nayak, M.; Dash, P. Smart surveillance monitoring system using raspberry PI and PIR sensor. *Indian J. Text. Res.* **2018**, *7*, 493–495.
30. Saabith, S.; Thangarajah, V.; Fareez, M. A Review on Python Libraries and IDEs for Data Science. *Int. J. Res. Eng. Sci.* **2021**, *9*, 36–53.
31. Kurniawan, D.E.; Iqbal, M.; Friadi, J.; Borman, R.I.; Rinaldi, R. Smart Monitoring Temperature and Humidity of the Room Server Using Raspberry Pi and Whatsapp Notifications. *J. Phys. Conf. Ser.* **2019**, *1351*, 012006. [CrossRef]
32. Dinculeană, D.; Cheng, X. Vulnerabilities and Limitations of MQTT Protocol Used between IoT Devices. *Appl. Sci.* **2019**, *9*, 848. [CrossRef]
33. Kryvonos, O.; Strutynska, O.; Kryvonos, M. The use of visual electronic circuits modelling and designing software fritzing in the educational process. *Zhytomyr Ivan Franko State Univ. Jo. Pedagogical Sci.* **2022**, *1*, 198–208. [CrossRef]
34. Jacobsen, R.H.; Aliu, D.; Ebeid, E. A Low-cost Vehicle Tracking Platform using Secure SMS. In Proceedings of the 2nd International Conference on Internet of Things, Big Data and Security, Porto, Portugal, 24–26 April 2017; SCITEPRESS—Science and Technology Publications: Porto, Portugal, 2017; pp. 157–166. [CrossRef]

35. Alkhudhayr, F.; Moulahi, T.; Alabdulatif, A. Evaluation Study of Elliptic Curve Cryptography Scalar Multiplication on Raspberry Pi4. *Int. J. Adv. Comput. Sci. Appl.* **2021**, *12*, 472–479. [CrossRef]
36. Narayanan, K.L.; Ram, C.R.S.; Subramanian, M.; Krishnan, R.S.; Robinson, Y.H. IoT based Smart Accident Detection & Insurance Claiming System. In Proceedings of the 2021 Third International Conference on Intelligent Communication Technologies and Virtual Mobile Networks (ICICV), Tirunelveli, India, 4–6 February 2021; pp. 306–311. [CrossRef]
37. Gautam, R.; Choudhary, S.; Surbhi Kaur, I.; Bhusry, M. Cloud based automatic accident detection and vehicle management. In Proceedings of the 2nd International Conference on Science Technology and Management, New Delhi, India, 27 September 2015; pp. 341–352.
38. Jebril, N.A.; Al-Haija, Q.A.; AlBarrak, N.; Almutlaq, G.; Alkhaiwani, A.H. Complete Microcontroller Based Vehicle Accident Detection System with Case Study for Saudi Arabia. *Wseas Trans. Commun. Arch.* **2017**, *16*, 118–130.
39. Ghanem, S.; Ghanim, S. Design and Implementation of an Integrated Vehicle Security System (IVSS). *Int. J. Ind. Sustain. Dev.* **2022**, *3*, 87–97. [CrossRef]
40. Barker, E. *Recommendation for Key Management: Part 1—General*; National Institute of Standards and Technology: Gaithersburg, MD, USA, 2020; NIST SP 800-57pt1r5. [CrossRef]
41. Brown, D. Sec 2: Recommended Elliptic Curve Domain Parameters. secg.org. 2010. Available online: https://www.secg.org/sec2-v2.pdf (accessed on 2 November 2022).

sustainability

MDPI

Article

Predictive Analytics for Sustainable E-Learning: Tracking Student Behaviors

Naif Al Mudawi [1], Mahwish Pervaiz [2], Bayan Ibrahimm Alabduallah [3,*], Abdulwahab Alazeb [1], Abdullah Alshahrani [4], Saud S. Alotaibi [5] and Ahmad Jalal [6,*]

1 Department of Computer Science, College of Computer Science and Information Systems, Najran University, Najran 55461, Saudi Arabia; naalmudawi@nu.edu.sa (N.A.M.); afalazeb@nu.edu.sa (A.A.)
2 Department of Computer Science, Bahria University, Islamabad 44000, Pakistan; mahwish.buic@bahria.edu.pk
3 Department of Information Systems, College of Computer and Information Sciences, Princess Nourah bint Abdulrahman University, Riyadh 11671, Saudi Arabia
4 Department of Computer Science and Artificial Intelligence, College of Computer Science and Engineering, University of Jeddah, Jeddah 21959, Saudi Arabia; asalshahrani2@uj.edu.sa
5 Information Systems Department, Umm Al-Qura University, Makkah 24382, Saudi Arabia; ssotaibi@uqu.edu.sa
6 Department of Computer Science, Air University, E-9, Islamabad 44000, Pakistan
* Correspondence: bialabdullah@pnu.edu.sa (B.I.A.); ahmadjalal@mail.au.edu.pk (A.J.)

Abstract: The COVID-19 pandemic has sped up the acceptance of online education as a substitute for conventional classroom instruction. E-Learning emerged as an instant solution to avoid academic loss for students. As a result, educators and academics are becoming more and more interested in comprehending how students behave in e-learning settings. Behavior analysis of students in an e-learning environment can provide vision and influential factors that can improve learning outcomes and guide the creation of efficient interventions. The main objective of this work is to provide a system that analyzes the behavior and actions of students during e-learning which can help instructors to identify and track student attention levels so that they can design their content accordingly. This study has presented a fresh method for examining student behavior. Viola–Jones was used to recognize the student using the object's movement factor, and a region-shrinking technique was used to isolate occluded items. Each object has been checked by a human using a template-matching approach, and for each object that has been confirmed, features are computed at the skeleton and silhouette levels. A genetic algorithm was used to categorize the behavior. Using this system, instructors can spot kids who might be failing or uninterested in learning and offer them specific interventions to enhance their learning environment. The average attained accuracy for the MED and Edu-Net datasets are 90.5% and 85.7%, respectively. These results are more accurate when compared to other methods currently in use.

Keywords: crowd management; human verification; machine learning; big data analytics; GA classifier; Viola–Jones

Citation: Mudawi, N.A.; Pervaiz, M.; Alabduallah, B.I.; Alazeb, A.; Alshahrani, A.; Alotaibi, S.S.; Jalal, A. Predictive Analytics for Sustainable E-Learning: Tracking Student Behaviors. *Sustainability* **2023**, *15*, 14780. https://doi.org/10.3390/su152014780

Academic Editor: Hao-Chiang Koong Lin

Received: 23 June 2023
Revised: 3 October 2023
Accepted: 6 October 2023
Published: 12 October 2023

1. Introduction

Internet usage has rapidly increased during the last ten years. People are constantly using the Internet to carry out a variety of tasks, including studying, commerce, and research. The old classroom setting has given way to the new digital phenomena where computers aid in teaching [1–4]. Today, the Internet is a great resource for finding courses, seminars, credentials, and other educational activities. The efficiency of the traditional educational strategy still used at universities and other educational institutions has been called into question by this wave of instructional materials and e-learning [5,6]. As a result, these institutions are finding it difficult to redefine and restructure their approaches to offering information and education (Association of European Universities, 1996) [7]. Given the current student population, educational institutions are scrambling to develop online learning resources that will enable computer-assisted instruction in the classroom. There appear to be two main research areas in e-learning [8], one of which focuses on the creation of effective designs and the other on the evaluation of student satisfaction and behavior in relation [9] to the course as compared to a conventional face-to-face course.

E-learning has become a necessary and timely solution, as the global COVID-19 pandemic has particularly shown [10]. The COVID-19 pandemic caused global traditional education systems to experience formerly unprecedented disturbances. At the height of the pandemic, 195 nations and more than 1.5 billion students were affected by school closings, according to UNESCO [11]. Millions of students were affected by prolonged closures of schools and colleges to stop the spread of viruses [12].

E-learning quickly emerged as a vital resource to guarantee educational continuity during this crisis [13]. Technology-driven learning platforms helped educational institutions to adapt and offer remote learning possibilities as physical classrooms became inaccessible. The availability of a variety of courses and materials on e-learning platforms ensured that learning could continue despite the restrictions put in place by the pandemic [14].

We suggested a useful approach to evaluate human behavior during e-Learning, whether in a classroom or any public area setting, which was motivated by the importance of the engagement of learners in an e-learning environment. This system's objective is to find anomalous behaviors [15] that are prohibited in educational settings. For instance, in any educational setting, sitting or standing and writing on a notebook or board are all permissible activities, but throwing objects, slapping, kicking, and taking naps are not. Any sort of e-learning environment should be able to use the suggested system to monitor and evaluate student behavior [16].

On the foundation of this idea, we offered a sophisticated predictive analytic system that can monitor and forecast student behavior in an online learning environment. The main contribution of this work is a fresh approach to analyzing and tracking the behavior of students during e-learning using a multimodal approach of feature extraction. To boost the accuracy of our system as compared to other state-of-the-art methods, we extracted features of objects with two different approaches, one at object level and one using a stick model [17] extracted from objects. Moreover, we tested our system using two different settings, one with emotion-based data and the other with action-based data.

With a focus on sustainable practices [18], this study aims to investigate and demonstrate the potential of these predictive analytics systems in e-learning environments. We intend to contribute to the long-term success of online education initiatives by focusing on the sustainability of e-learning and therefore providing a rich learning environment that may efficiently augment, and in some circumstances, replace traditional face-to-face education.

The identification of motion-based [19] elements that can be used to accurately detect pedestrian behavior is the key contribution to this research. The occlusion removal procedure is the other crucial component of this effort. If an occlusion [20] was discovered, we used the Hough transform [21] with a semi-circle to determine the pedestrian's head parts, and then we used body parts estimation [22] to roughly determine how the silhouettes were laid out. Then, approximated zones were separated and occlusion was removed.

The article's remaining sections are organized as follows: Section 2 presents related work; Section 3 covers the detailed methodology of the system followed by Section 4, where the experimental results are reported together with a comparison to comparable state-of-the-art HAR systems. Discussion on the pros and cons of the system has been presented in Section 5 and the paper is concluded in Section 6.

2. Related Work

The COVID-19 pandemic has accelerated the adoption of e-learning as an alternative to traditional classroom education [23]. As a result, educators and researchers are increasingly interested in understanding the behavior of students in e-learning environments [24].

Behavior analysis is a useful approach for studying student behavior in e-learning, as it can provide insights into factors that influence learning outcomes and inform the design of effective interventions [25]. Behavior analysis has been utilized in several types of research to look into how students behave in e-learning settings. The behavior of pupils during a computer-based training program, for instance, was examined by Kun et al. [26] using a microanalytic technique. They discovered that students who exhibited more active learning behaviors, such as taking notes and asking questions, outperformed passive learners in terms of their learning results. Similarly, Liu et al. [27] examined student behavior in a massive open online course (MOOC) using data mining tools. They discovered that students were more likely to finish the course and receive higher scores if they participated in more discussion forums and course activities. Some more research contributions are summarized in Table 1.

Table 1. Summary of state-of-the-art methods.

Reference	Objectives	Advantages	Disadvantages
[28]	Early exploration of e-learning behavior tracking.	Established the importance of engagement metrics. Paved the way for future research.	Lack of real-time monitoring, Limited emphasis on individualized feedback.
[29]	Focus on learner-centered tracking using a mixed methods approach, including surveys and behavioral analysis.	Enhanced understanding of self-regulation.	Time-consuming data analysis and difficulties in measuring subjective behaviors.
[30]	Use of predictive modeling for retention. Used method of longitudinal data collection and machine learning techniques.	Informed personalized interventions. Highlighted the role of social interaction.	Challenges in predicting non-academic behaviors and privacy concerns.
[31]	Application of data mining techniques. Performed analysis of large-scale learning data using data mining algorithms.	Insights into collaborative learning behaviors and identification of factors affecting performance.	Ethical considerations, resource-intensive data collection, and limited explanation of causality.
[32]	Integration of multimodal data sources. Data fusion of various sources, including clickstream and biometric data.	Comprehensive learner profile generation and personalized learning pathway recommendations.	Technical challenges in data fusion and limited generalizability. Privacy and security concerns.
[33]	Explore tracking student engagement. Performed surveys, interviews, and behavioral data analysis and identified factors influencing online behaviors.	Established the importance of engagement metrics. Informative for instructional design.	Limited sample size, lack of long-term data, and dependency on self-reported data.
[34]	Investigate social network influence using social network analysis and content analysis.	Provide insights into collaborative learning dynamics. Integration of social network analysis.	Limited focus on non-cognitive behaviors, ethical considerations, and incomplete data from private groups.
[35]	Examine online procrastination using surveys and analysis of online procrastination behaviors. Proposed strategies for reducing procrastination.	Implications for time management in e-learning.	Limited generalizability, self-reporting bias, and limited consideration of other behaviors.
[36]	Developed a system for giving students immediate feedback throughout an online course using a behavior analytic approach.	Identified effect of individual differences on students behavior in e-learning settings.	Results can be biased based on false feedback.

Table 1. *Cont.*

Reference	Objectives	Advantages	Disadvantages
[37]	Discovered that students who had high levels of motivation and self-efficacy were more likely to participate in active learning strategies.	Suggest parameters that will increase student performance and engagement in e-learning.	Motivation level is a derived parameter that cannot be measured accurately; it can affect performance of the system.

Most of the studies investigated in the literature focused on derived parameters like eye movement ratio, screen activity monitoring through their screen recording, and their interaction with the system but not on the real actions and emotions of students. This can predict their engagement level but not their real feelings and involvement in the subject. In this study, we take into account the conclusions of other scholars and suggest an efficient approach to examine student behavior during online learning. The basic motivation behind this work is to use the emotions and actions of students to analyze their behavior and identify prohibited actions during the class. Also, we have taken into account the variability in their behavior and have used several datasets containing a variety of activities taken in different settings to train our system to handle a variety of behaviors.

3. Proposed System Methodology

In this part, the suggested system methodology is explained. The entire workflow of the system is shown in Figure 1. The Motion Emotion Dataset (MED) [38] and Edu-Net datasets [39] have been chosen to assess the efficacy of the proposed technique in both indoor and outdoor settings, respectively. Six elements make up the system used to assess how well students behaved in an online learning environment. The complete algorithm has been presented in Algorithm 1.

Algorithm 1 Multistage processing to detect students' behavior in e-Learning.

```
Input: Image (1)
Step 1: Preprocessing Phase
        1.1 Apply Noise Removal Techniques
              1_{denoised} = Denoise(l)
        1.2 Perform Image Enhancement
              I_{enhanced) = Enhance(l_{denoised})
        1.3 Apply Object Filtering
              I_{filtered) = Filter(I_{enhanced})
Step 2: Object Extraction
        2.1 Use Viola Jones Object Detection
              Detected_Objects = ViolaJones(l_{filtered})
Step 3: Feature Extraction
        3.1 Extract Full Object Features
              Full_Object_Features = CalculateStatistics(Detected_Objects)
        3.2 Generate Stick Model Skeleton
              Skeletons = ExtractSkeletons(Detected_Objects)
        3.3 Extract Skeleton Features
              Skeleton_Features = Measure Skeleton Attributes(Skeletons)
Step 4: Training and Classification using Genetic Algorithm
        4.1 Define Genetic Algorithm Parameters
              Parameters = DefineParameters()
        4.2 Initialize Population
              Population = InitializePopulation(Parameters)
        4.3 Evaluate Fitness
              Fitness_Values = EvaluateFitness(Population)
        4.4 Genetic Operations
              New:_Population = ApplyGeneticOperators(Population, Fitness_Values)
```

Algorithm 1 Cont.

```
        4.5 Replace Population
                Population = New_Population
        4.6 Termination Criterion
                Termination = CheckTerminationCriterion()
Step 5: Classification and Result Analysis
        5.1 Select Best Features
                Best_Features = SelectBestFeatures(Population)
        5.2 Train Classifier
                Classifier = TrainClassifier(Best_Features)
        5.3 Perform Classification
                Classification_Results = ClassifyImage(I, Classifier)
        5.4 Analyze Results
                Analysis_Metrics = Analyze ClassificationResults(Classification_Results)
Output: Behavior Type[Classification_Results]
End Algorithm
```

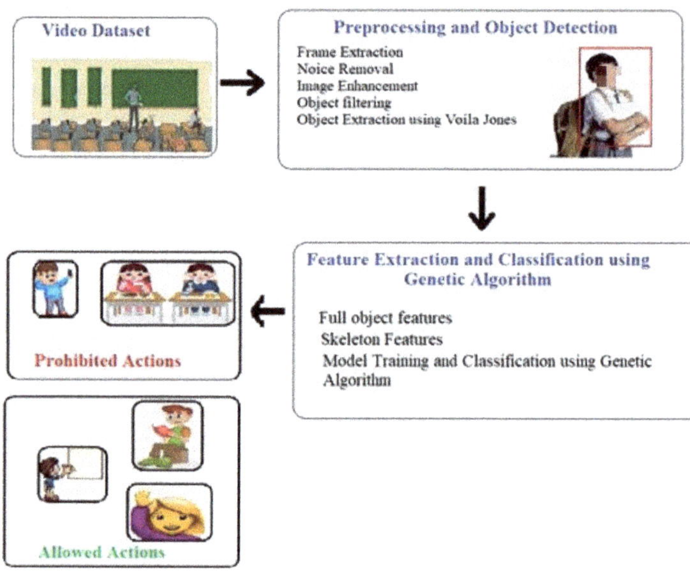

Figure 1. Proposed methodology to detect learner behavior.

To achieve accurate and successful outcomes, a strong and multifaceted technique was used in the development of our system for monitoring student behaviors in the classroom. Preprocessing was the first phase of this procedure, which was performed to isolate important classroom items and remove background noise. Objects outside of the established threshold range were eliminated, leaving just the characteristic human layout. Our dataset contains a variety of outdoor locations and objects that may have difficult shadows; thus, an additional step was included to improve the quality of the human silhouettes. In order to show human forms more accurately and clearly, shadows had to be found and then eliminated [40].

We employed the template matching technique [41] to extract exclusively human data from the image data in order to improve the accuracy of our system. These steps came together to isolate and extract human silhouettes, which served as the basis for further analysis.

Continuing with our methodology, the next critical phase involved the extraction of features from the human silhouettes utilizing conditional random fields. This step allowed for a more comprehensive understanding of the various aspects of human behavior and posture within the classroom setting. To classify the activities performed by students as either allowed or prohibited, we employed a genetic algorithm [42]. This sophisticated algorithm played a pivotal role in categorizing and analyzing student behaviors, offering a dynamic and adaptive approach to the assessment of classroom activities. By integrating these various techniques and algorithms, our system was well equipped to accurately and efficiently track and categorize student behaviors, providing educators with valuable insights and tools for maintaining a conducive and productive learning environment.

3.1. Image Preprocessing

Our dataset was in the form of videos. The next step we performed was frame extraction from the video, and then we utilized each frame to preprocess the image. As seen in Equation (1), a special median filter has been used to remove noise and smooth the video frame images that were retrieved. Then, the foreground objects' appearance was improved using image enhancement as shown in Figure 2.

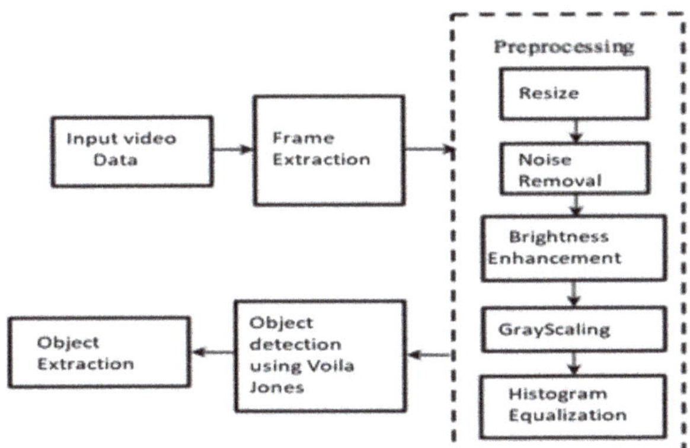

Figure 2. The process used to extract objects of interest in the image.

The median filter opted to remove the noise from the frames extracted from video data according to Equation (1).

$$(u, v) = m\{i(u + i, v + j)|(u, v) \in R\} \quad (1)$$

In the following stage, we used gamma correction as provided in Equation (2) to enhance the brightness of the image. Results from the preprocessing step are shown in Figure 3.

$$(c) = I\_out = I\_in\hat{}\gamma \quad (2)$$

Figure 3. Preprocessing. (**a**) Original image, (**b**) noise removed, and (**c**) image enhancement.

3.2. Object Extraction

Object extraction was performed using the Viola–Jones algorithm which involves Haar-like features extraction, Adaboost [43] training, and a cascade of classifier. The decision to use the Viola–Jones algorithm for human detection in the context of our research on monitoring student behaviors in the classroom was carefully examined and was influenced by a number of variables. We chose the Viola–Jones algorithm due to the specific benefits it offers within the scope of our project, despite the fact that deep learning-based algorithms have acquired significant popularity in recent years for their outstanding capabilities in object detection and categorization. The Viola–Jones technique is computationally effective and substantially faster than deep learning-based approaches. We were able to monitor and analyze student behaviors in a timely manner without adding a lot of latency due to its capacity to achieve real-time performance, even on hardware with constrained computational capabilities.

In comparison to deep learning-based techniques, the Viola–Jones algorithm also needs less training data. It can be difficult and time-consuming to gather a sizable dataset for deep learning models in a real-world classroom setting. The Viola–Jones method was a practical choice for our research because of its propensity to perform well with small datasets. Nextstep includes a set of rectangular Haar-like features defined to capture the difference between the object and background regions. Each feature was represented as the difference between the sum of pixel intensities in two rectangular regions. Using the Adaboost approach, a set of weak classifiers were trained on a set of positive and negative examples. Each weak classifier was trained to classify an image patch as containing the object or not based on a selected Haar-like feature [44], and then classifiers were combined into a cascade of strong classifiers. Each weak classifier in a strong classifier was trained to pass the positive data to the next stage while highly likely rejecting the background samples. The cascade of classifiers was applied to the input video frames by sliding a window over each frame and evaluating the objectness score for each window. The results are shown in Figure 4.

$$O(x, y) = \sum i\ \alpha iT\_i(f\_i(x, y)) \tag{3}$$

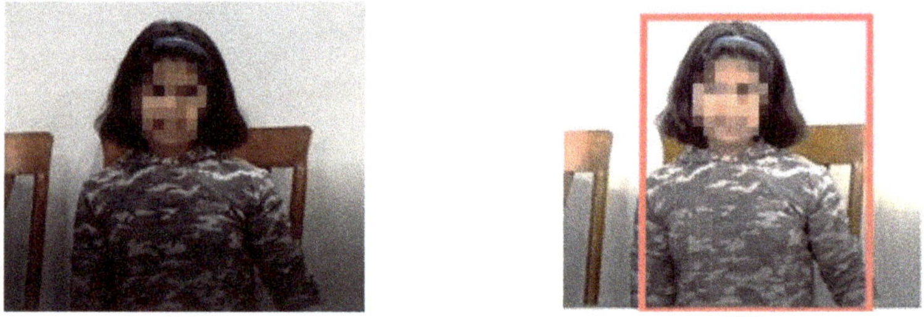

(a) **(b)**

Figure 4. Object detection using Viola–Jones. (**a**) Original image and (**b**) object detection using Viola–Jones.

3.3. Feature Extraction

The feature extraction procedure for human silhouettes discovered by the layout verification module is described in this section. Full human silhouettes' features were extracted, as well as the skeleton against each silhouette [45]. The features extraction outline is presented in Figure 5 and is divided into two directions.

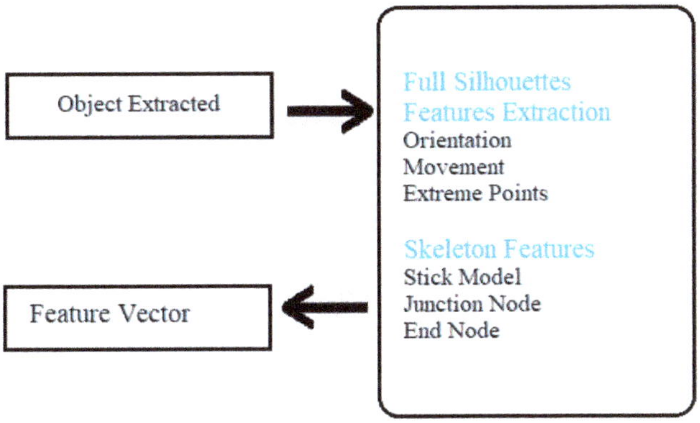

Figure 5. Features extraction for full body and skeleton levels.

3.3.1. Full Silhouette Features

The positions of each human silhouette in the current frame and the frame before it were obtained, as illustrated in Equation (4), and they were regarded as separate objects.

$$P\left(I_{o,f}\right) = I_{x,y} \in O \tag{4}$$

where the current frame is represented by f and the current silhouette by o. Movement of centroid among successive frames concerning time was used to determine the distance for each silhouette using Equation (5).

$$d = \sqrt{(x_2 - x_1)^2 + (y_2 - y_1)^2} \tag{5}$$

$$velocity = {}^{du}/_{dt} \tag{6}$$

Then, the velocity [46] of each object was computed using Equation (6) and the distance was calculated using Equation (7); these factors were then used to distinguish between the allowed actions and prohibited actions.

$$\theta = tan^{-1}({}^{y}/_{x}) \tag{7}$$

We first selected random points of the complete silhouette to describe the structure of the object, using Principal Component Analysis (PCA) [47] to determine the orientation of the object. The coordinates or position of the object should be disclosed for each data point. The dataset's covariance matrix, which illustrates the connections between various dimensions of the data points, was then computed. The principle components were then obtained by applying PCA to the dataset. These elements were eigenvectors that showed where the data's greatest variance occurs. We may determine the object's fundamental orientation by examining the first principal component, which captures the most significant change. The orientation of the object can be inferred from the first principal component's direction (See Figure 6).

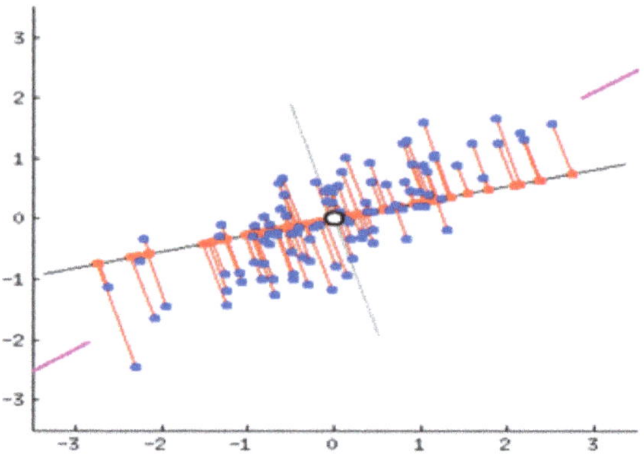

Figure 6. PCA to compute the orientation of the object.

3.3.2. Stick Model Features

Stick models were used to extract the features at the micro level. The human silhouette's skeleton was first removed, and endpoints and junction nodes were located. Endpoints and junction points were utilized to draw the stick model in Figure 7 to demonstrate it. To connect the nodes, we employed the optical flow [33,34] of each node presenting the model, as well as the distance and angle between the slants.

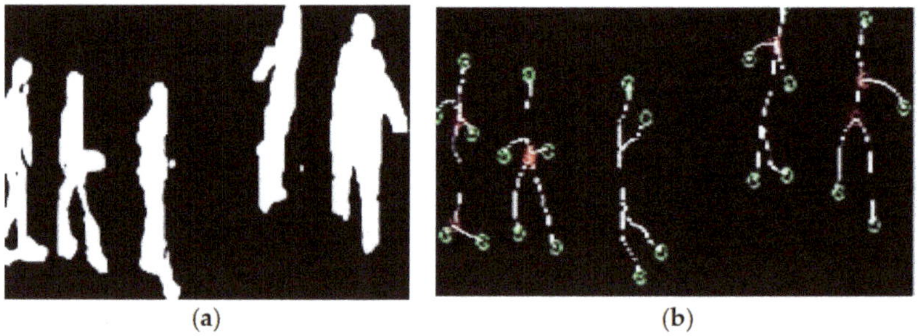

Figure 7. Stick model presentation of human silhouettes detected. (**a**) Humans detected, and (**b**) Stick Model for each detected human.

3.4. Feature Optimization and Classification

The genetic algorithm [48] was utilized as a classifier, with each data point assigned to one of several predefined categories based on its features or attributes. To achieve this, the GA creates a set of candidate classifiers, each represented by a set of parameters that define its decision boundary [49]. The fitness of each classifier is evaluated by its ability to correctly classify a set of training data, and the GA evolves the population of classifiers by selecting the fittest ones and generating new ones through crossover and mutation operations. The process continues until satisfactory classification accuracy is achieved on the training data, and the final classifier can then be used to classify new, unseen data.

The main reason for using the GA for classification is that it can search a large solution space and discover complex decision boundaries that may be difficult to find using other methods. However, the effectiveness of the GA depends on various factors such as the quality of the training data, the choice of genetic operators, and the number of parameters in the classifier. Nonetheless, the GA remains a popular and powerful technique for data classification in various domains such as image recognition and bioinformatics. The general architecture of the genetic algorithm is displayed in Figure 8. Initially, a population of the potential solution was created, where each individual represents a solution and is evaluated by the fitness function. The solution with a higher fitness value was chosen to become a parent for the next generation and parents were combined to generate a new population; mutation was performed to avoid premature convergence. The cycle repeated until a satisfactory fitness level was achieved. Once it was terminated, the individual with the highest fitness value was considered as the best solution.

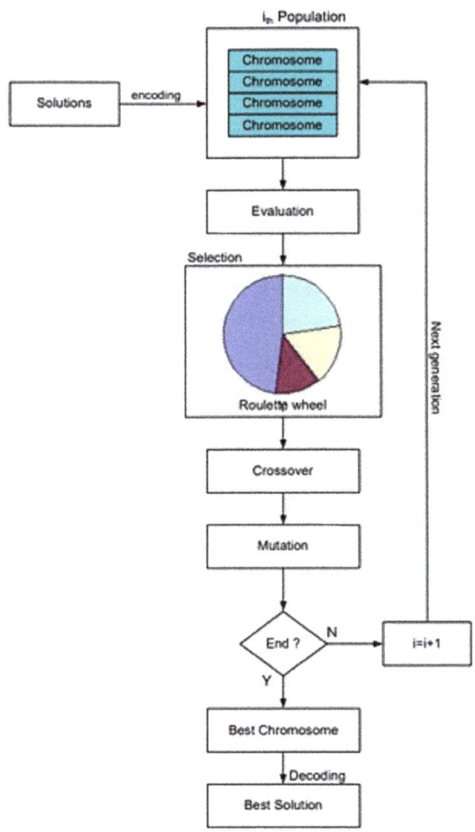

Figure 8. The architecture of the genetic algorithm with population distribution and selection.

4. Experiments and Results

This section discusses the dataset and the specifics of the research, such as the experimental setup, the performance of the suggested system, and a comparison analysis with cutting-edge techniques.

4.1. Dataset

Two different datasets were used to evaluate the performance of the system in different environments and had different actions performed by multiple persons. The first dataset is made up of around 44,000 normal and abnormal video clips divided across 31 video sequences. The videos are 554 × 235 in resolution and were recorded at 30 frames per second using a fixed video camera that was raised above and looked down on specific paths. We only selected sample data having videos annotated with human emotions. Other datasets were collected from YouTube videos and recordings from actual classrooms in a range of settings, including those with different age ranges, class standards, and rural and urban settings. There are 200 video clips in each activity category within the 7851 total video clips that make up the produced dataset. Each video clip lasts between 3 and

12 s. The video lasts for about 12 h in total. The collection comprises recordings from actual classrooms as well as real videos from YouTube that have been uploaded by users from around the world. We chose videos recorded in the actual classroom environment as our sample data. We separated the samples of each class randomly into a training set, validation set, and test set, with a ratio of 80%, 10%, and 10%, according to the established dataset division rules [50]. Complete details of both datasets are given as follows.

4.1.1. MED (Motion Emotion Dataset)

Two significant segments in various indoor-outdoor situations can be found in the MED dataset [51]. One section includes video clips that demonstrate five distinct behaviors: panic, fighting, congested area, obstacle or strange object, and neutral. The other section, on the other hand, is made up of various video sequences that provide information on six distinct emotions: anger, happiness, excitement, fear, sadness, and neutrality.

31 actor-filled video clips make up the videos, and the dataset also includes numerous motorbikes and bicycles that behave as obstacles. The remaining 40% of the dataset is utilized for testing, with the remaining 60% used for training. Figure 9 displays a few instances of MED sceneries. We have combined these emotions in two classes and categorized all emotions and behavioral videos into allowed and prohibited behavior categories.

Figure 9. Examples of different scenes of the MED dataset.

4.1.2. Edu Net Dataset

There are several videos of various e-learning-related acts on EduNet [52]. The dataset, which includes several teachers and pupils, was obtained from a classroom setting. Videos show a variety of permitted classroom behaviors, such as standing, writing on the board, raising hands, and maintaining a book in hand. Prohibited behaviors include eating, using a phone, and bouncing around during class. Figure 10 shows some examples of the EduNet dataset having multiple allowed and not allowed actions.

Figure 10. Allowed and prohibited behavior of the Edu-net dataset.

4.2. *Performance Metric and Experimental Outcome*

Precision [53] was chosen as the performance metric for our system evaluation to assess its effectiveness. Equation (8) was used to calculate precision [54],

$$\text{Precision} = t_c/(t_c + f_c), \quad (8)$$

where t_c represents the total number of prohibited actions classified correctly and fc represents the total number of false detected actions. The results of the MED and Edu-Net datasets are shown in Table 2. Classes are categorized into allowed and prohibited behavior and each subcategory has been evaluated.

Table 2. The experimental outcome of MED and Edu-Net dataset.

DataSet	**MED Dataset**							
Activity	**Allowed Behaviors**				**Prohibited Behavior**			
Actions	Happy	Sad	Excited	Neutral	Panic	Fight	Scared	Angry
Accuracy	89%	86%	89%	82%	92%	94%	89%	87%
Average Accuracy	85.75%				90.5%			
Dataset	**Edu-Net Dataset**							
Activity	**Allowed Behaviors**				**Prohibited Behaviors**			
Action	Writing on Board	Writing on Book	Reading Book	Hand Raise	Sleeping on Chair	Eating Food	Holding Mobile Phone	Fighting
Accuracy	83%	85%	89%	82%	83%	84%	89%	87%
Average	85%				86%			

The statistical analysis discussed in the preceding sections offers important insights into how well the suggested system performs in identifying and categorizing both permitted and forbidden behaviors in the MED and Edu-Net datasets. The performance metric of precision shows how well the system performs in correctly identifying actions while reducing false detections. A great overall performance is indicated by the average accuracy values for both datasets, which vary from 85.75% to 90.5%.

To visualize the results in more detail, those for each dataset are displayed separately in Figures 11 and 12. The name of each subcategory that is used to evaluate the behavior is displayed on the X-axis, while the accuracy of each behavior is displayed on the Y-axis.

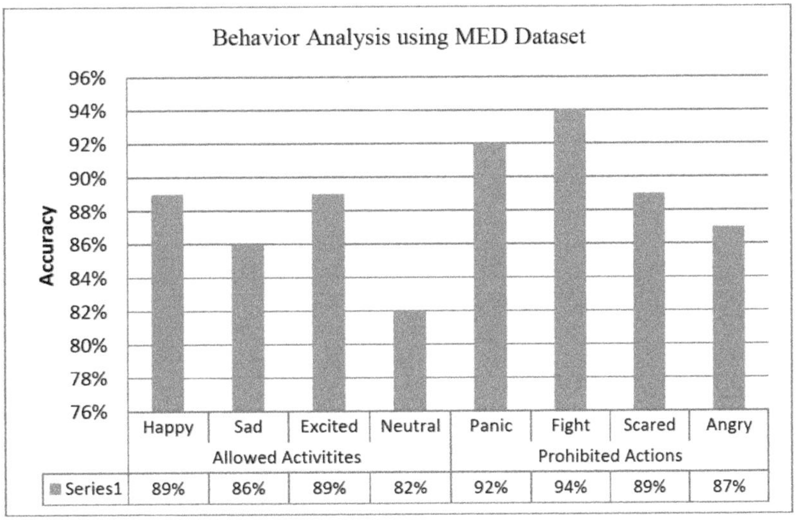

Figure 11. Results of behavior detection with MED dataset.

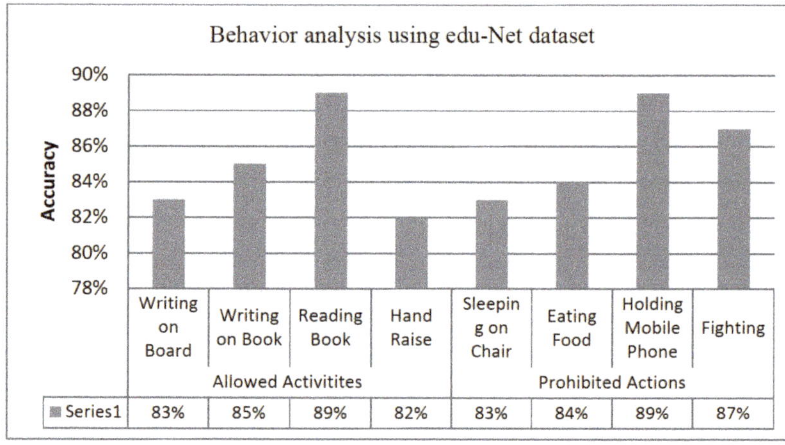

Figure 12. Results of behavior detection with Edu-Net dataset.

The area under the curve (AUC) [55], error equivalence rate (EER) [56], and decidability [57] are used to assess the performance across both datasets in greater detail. Figure 13 shows the combined performance of the two datasets. Additionally, the study is further enhanced by the use of additional performance indicators including AUC [55], EER [56], and decidability. AUC provides total performance measurements over all possible classification criteria [58]. The value of AUC might be between 0 and 1 [59]. EER is employed as a threshold parameter to indicate false acceptance and false rejection rates [60]. These metrics offer a thorough evaluation of the system's capability to distinguish between permitted and unacceptable behavior while taking into account the trade-off between erroneous acceptances and false denials. A comprehensive assessment of the system's capabilities across both datasets is provided by the combined performance measures shown in Figure 13.

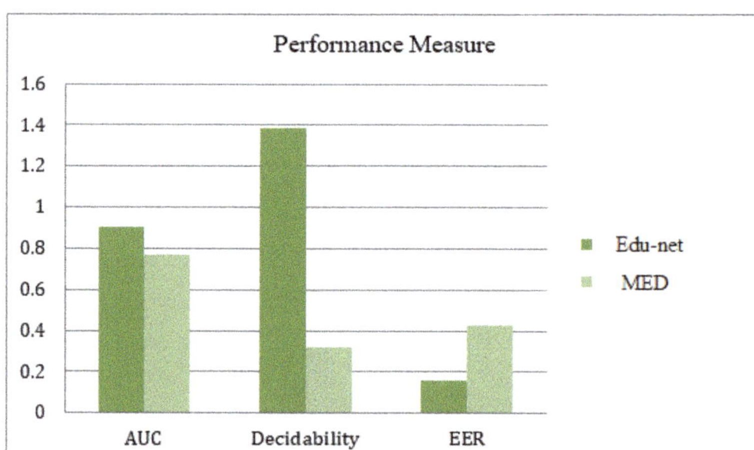

Figure 13. Performance measures of both datasets were used to evaluate the system.

The comparison with current cutting-edge approaches also reveals the advantages of the proposed technology. It performs better than conventional techniques in terms of precision and accuracy, demonstrating its promise as a cutting-edge approach to behavior identification in academic and practical settings. The significance of the study and its possible effects on enhancing security and monitoring in educational settings are highlighted by this comparison. Comparative analysis of the suggested approach and the existing method, which both used the same assessment datasets, was performed (See Table 3). Compared to other state-of-the-art methods, the proposed system performs admirably.

Table 3. Comparison of the stated system with other contemporary methods.

Dataset	Methods	Accuracy (%)	Methods	Accuracy	Methods	Accuracy
MED	Khalid [52]	84.9	Alberto et al. [59]	87.4	Proposed Method	89.2
Edu-Net	Rawashdeh et al. [60]	80.3	Fuady et al. [61]	82.2	Proposed Method	85.5

5. Discussion

Access to educational opportunities has been made simpler due to the growth of e-learning. However, concerns about student misconduct and reduced engagement have also arisen as a result of the increased use of e-learning platforms. A mechanism has been developed in place to solve this problem that looks at visual data to find students practicing unauthorized behaviors during online learning. This article offers a comprehensive overview of the system, its elements, and its functionality.

The preprocessing stage of the system, which aims to lower noise and improve image quality, is the first stage. The Viola–Jones technique [62] is then used for object detection to determine whether a person is present in the frame. By using template matching, it is possible to confirm that the identified object is a human. Each silhouette is subjected to skeleton extraction, and feature extraction is conducted for both skeleton points and human silhouettes. A genetic algorithm is then used for classification.

The system was assessed using a collection of videos of students engaging in online learning activities. The algorithm accurately identified 90.5% of the prohibited actions, including talking, using a phone, standing on a chair, and sleeping. The system's performance was also assessed in terms of detection time, and it was found that it ran in real time with a frame rate of 30 frames per second.

An important area of interest in the realm of education is the assessment of student behaviors in e-learning. Understanding and observing student behavior has become essential for teachers and educational institutions to effectively help students and improve learning outcomes as a result of the rising popularity of online learning platforms. The objective of this discussion is to critically examine student behavior assessment in online learning and its implications for educational practices.

There is a potential connection between HAR outcomes and e-learning. HAR technology could be used to track student involvement and engagement in online learning. Educational systems could assess students' levels of engagement, interaction, and participation by examining video feeds from online classes. With the help of this data, instructional strategies might be customized, and online learning could be made better overall. Gaining insights into students' involvement, participation, and learning progress is one of the main benefits of evaluating student actions in e-learning. Using this data, instructors can identify children who might be failing or uninterested in learning and offer them specific interventions to enhance their learning environment.

Additionally, e-learning assessments of student behavior enable personalized and adaptive learning processes. Educational platforms can produce data-driven recommendations and personalized feedback based on the behaviors and preferences of individual students by utilizing predictive analytics and machine learning algorithms. This tailored strategy improves learning outcomes by making sure that resources and activities are tailored to the individual needs of the students while also increasing their enthusiasm to learn.

The evaluation of student behavior in e-learning does not, however, come without difficulties and restrictions. When gathering and examining student data, privacy issues and ethical problems must be carefully considered. Educational organizations must make sure that data collecting procedures are open and that they seek student consent while protecting the security and confidentiality of the data collected. Additionally, there is a danger of placing an undue reliance on quantitative behavioral measurements, which could mean ignoring the qualitative components of student engagement and learning. To fully understand student behaviors in e-learning, a balanced strategy combining quantitative and qualitative assessment methodologies is required.

With an emphasis on sustainability, the blending technologies of AI, digital transformation, IoT, and edge computing show enormous potential in transforming the landscape of e-learning. Educational institutions can use the potential of data-driven insights to customize learning experiences for specific students by integrating AI into e-learning. This includes fast feedback, automated grading, and recommendations for tailored material. In addition, interactive components like virtual labs, simulations, and AR/VR applications can be incorporated into the digital transformation process to go beyond the limitations of traditional online lectures. These developments encourage increased comprehension and involvement among students, converting inactive learning into active participation.

Alongside these developments, the Internet of Things (IoT) can make real-time data collecting easier and provide educators with insightful data on student behavior, preferences, and performance. This data can be handled locally when used in conjunction with edge computing, allowing for quick answers and fluid communication in the e-learning environment. By minimizing the environmental impact connected with physical infrastructure and travel-related emissions, e-Learning links education with sustainability aims. In addition to revolutionizing e-learning, this comprehensive integration of technology demonstrates a dedication to open, individualized, and environmentally responsible education.

In conclusion, there are many important benefits of using predictive analytics to monitor student behavior in sustainable e-learning. First of all, it enables educators to monitor student engagement, involvement, and performance in real-time, allowing them to quickly identify at-risk pupils and step in to intervene and give the required support. Second, the approach provides students with a personalized learning experience by adapting interventions and content to their unique behaviors and requirements. This not only improves the quality of learning overall but also increases the retention of students. Additionally, predictive analytics can offer insightful data on the efficiency of different teaching methods and materials, assisting in curriculum optimization and ongoing curriculum improvement. Additionally, it lessens resource waste and dropout rates, which contribute to the sustainability of e-learning and make it an important tool in the changing environment.

Although predictive analytics in sustainable e-learning has a lot of potential, it is not without difficulties and drawbacks. The ethical use of data is one important worry, as tracking and analyzing student behavior can lead to privacy concerns, if not handled appropriately. It is essential to make sure that data are secure and anonymous. Additionally, the quantity and quality of data obtained, which can occasionally be constrained or biased, have a significant impact on how accurate predictive models are. Additionally, there is a chance that an excessive dependence on algorithms and statistics will lead to a potential disregard for the value of qualitative insights and human contact in education. The use of data-driven decision making and instructional knowledge should be balanced by educators.

Despite these obstacles, the proposed approach's importance to the area of study cannot be emphasized enough because it provides a potent instrument for improving student performance and institutional advancement while enhancing the sustainability and effectiveness of e-learning. Future research should concentrate on improving the system's flaws to boost performance. If you want to increase system coverage and decrease the effects of occlusion, think about using multiple cameras to increase the system's accuracy in detecting illicit activities. Researchers are also looking into the use of additional sensors, such as microphones. Additional research can examine the incorporation of machine learning methods like deep learning to boost the system's accuracy and detection speed.

Additionally, it is critical to recognize that the application of predictive analytics and monitoring systems for sustainable e-learning is an area that is always changing. New opportunities and difficulties will arise as technology develops and our comprehension of student behavior grows. To expand and improve these systems and guarantee their continued efficacy and morality, ongoing research and development are required. Additionally, the research of multisensor techniques and the incorporation of cutting-edge technologies like deep learning will probably improve the precision and thoroughness of these systems. The sustained success of these projects will also depend on creating a collaborative atmosphere where educators, data scientists, and policy makers can cooperate to find the ideal balance between data-driven decision making and the human aspect in education. In the end, predictive analytics has the potential to provide sustainable e-learning, but to fully realize its potential while resolving its drawbacks, constant diligence and innovation are needed.

6. Conclusions

E-learning is the top trending source of education in this era especially after the COVID-19 pandemic. Educators and researchers are paying more attention to improving e-learning systems. The behavior of students and their engagement level is the most important factor of the e-learning system. This system was implemented to identify the behavior of students in an e-learning environment. Multiple datasets were used to evaluate the performance of this system. Videos were converted into frames and then objects were segmented to narrow down the region of interest. Features for each object and its skeleton models were used to characterize the behavior of students. Datasets were divided into allowed and prohibited behaviors. Experiments were performed and an average accuracy of 89% and 85.5% was achieved on both datasets.

Author Contributions: Validation, M.P. and N.A.M.; Formal analysis, N.A.M., M.P. and A.A. (Abdulwahab Alazeb); Investigation, B.I.A.; Resources, B.I.A. and A.A. (Abdullah Alshahrani); Data curation, N.A.M.; System Implementation ,Writing—original draft, M.P .(Mahwish Pervaiz); Writing—review & editing, A.A. (Abdulwahab Alazeb), A.A. (Abdullah Alshahrani) and S.S.A.; Supervision, A.J. All authors have read and agreed to the published version of the manuscript.

Funding: Princess Nourah bint Abdulrahman University Researchers Supporting Project number (PNURSP2023R440), Princess Nourah bint Abdulrahman University, Riyadh, Saudi Arabia. The authors are thankful to the Deanship of Scientific Research at Najran University for funding this work under the Research Group Funding program grant code (NU/RG/SERC/12/6).

Acknowledgments: Princess Nourah bint Abdulrahman University Researchers Supporting Project Number (PNURSP2023R440), Princess Nourah bint Abdulrahman University, Riyadh, Saudi Arabia.

Conflicts of Interest: The authors declare that they have no conflict of interest to report regarding the present study.

References

1. James, R.J.E.; Tunney, R.J. The need for a behavioral analysis of behavioral addiction. *Clin. Psychol. Rev.* **2017**, *52*, 69–76. [CrossRef] [PubMed]
2. Miah, S.J.; Vu, H.Q.; Gammack, J.; McGrath, M. A big data analytics method for tourist behaviour analysis. *Inf. Manag.* **2017**, *54*, 771–785. [CrossRef]
3. Zhang, X.; Wen, S.; Yan, L.; Feng, J.; Xia, Y. A Hybrid-Convolution Spatial–Temporal Recurrent Network for Traffic Flow Prediction. *Comput. J.* **2022**, bxac171. [CrossRef]
4. Li, B.; Tan, Y.; Wu, A.; Duan, G. A distributionally robust optimization based method for stochastic model predictive control. *IEEE Trans. Autom. Control.* **2021**, *67*, 5762–5776. [CrossRef]
5. Matthew, T.; Banhazi, T.M. A brief review of the application of machine vision in livestock behavior analysis. *Agrárinform./J. Agric. Inform.* **2016**, *7*, 23–42.
6. Jaganeshwari, K.; Djodilatchoumy, S. An Automated Testing Tool Based on Graphical User Interface with Exploratory Behavioural Analysis. *J. Theor. Appl. Inf. Technol.* **2022**, *22*, 6657–6666.
7. Michalis, V.; Nikou, C.; Kakadiaris, I.A. A review of human activity recognition methods. *Front. Robot. AI* **2015**, *2*, 28.
8. Qian, L.; Zheng, Y.; Li, L.; Ma, Y.; Zhou, C.; Zhang, D. A New Method of Inland Water Ship Trajectory Prediction Based on Long Short-Term Memory Network Optimized by Genetic Algorithm. *Appl. Sci.* **2022**, *12*, 4073. [CrossRef]
9. Guo, F.; Zhou, W.; Lu, Q.; Zhang, C. Path extension similarity link prediction method based on matrix algebra in directed networks. *Comput. Commun.* **2022**, *187*, 83–92. [CrossRef]

10. Ferhat, A.; Mohammed, S.; Dedabrishvili, M.; Chamroukhi, F.; Oukhellou, L.; Amirat, Y. Physical human activity recognition using wearable sensors. *Sensors* **2015**, *15*, 31314–31338.
11. Xie, X.; Xie, B.; Cheng, J.; Chu, Q.; Dooling, T. A simple Monte Carlo method for estimating the chance of a cyclone impact. *Nat. Hazards* **2021**, *107*, 2573–2582. [CrossRef]
12. Gupta, N.; Gupta, S.K.; Pathak, R.K.; Jain, V.; Rashidi, P.; Suri, J.S. Human activity recognition in artificial intelligence framework: A narrative review. *Artif. Intell. Rev.* **2022**, *55*, 4755–4808. [PubMed]
13. Jiang, H.; Wang, M.; Zhao, P.; Xiao, Z.; Dustdar, S. A Utility-Aware General Framework With Quantifiable Privacy Preservation for Destination Prediction in LBSs. *IEEE/ACM Trans. Netw.* **2021**, *29*, 2228–2241. [CrossRef]
14. Long, W.; Xiao, Z.; Wang, D.; Jiang, H.; Chen, J.; Li, Y.; Alazab, M. Unified Spatial-Temporal Neighbor Attention Network for Dynamic Traffic Prediction. *IEEE Trans. Veh. Technol.* **2023**, *72*, 1515–1529. [CrossRef]
15. Xiao, Z.; Li, H.; Jiang, H.; Li, Y.; Alazab, M.; Zhu, Y.; Dustdar, S. Predicting Urban Region Heat via Learning Arrive-Stay-Leave Behaviors of Private Cars. *IEEE Trans. Intell. Transp. Syst.* **2023**, *24*, 10843–10856. [CrossRef]
16. Wang, W.; Liu, A.X.; Shahzad, M.; Ling, K.; Lu, S. Understanding and modeling of wifi signal based human activity recognition. In Proceedings of the 21st Annual International Conference on Mobile Computing and Networking, Paris, France, 7–11 September 2015; pp. 65–76.
17. Abdulmajid, M.; Pyun, J.-Y. Deep recurrent neural networks for human activity recognition. *Sensors* **2017**, *17*, 2556.
18. Ortiz, R.; Jorge, L.; Oneto, L.; Samà, A.; Parra, X.; Anguita, D. Transition-aware human activity recognition using smartphones. *Neurocomputing* **2016**, *171*, 754–767. [CrossRef]
19. Chen, G.; Chen, P.; Huang, W.; Zhai, J. Continuance Intention Mechanism of Middle School Student Users on Online Learning Platform Based on Qualitative Comparative Analysis Method. *Math. Probl. Eng.* **2022**, *2022*, 3215337. [CrossRef]
20. Xiong, Z.; Liu, Q.; Huang, X. The influence of digital educational games on preschool Children's creative thinking. *Comput. Educ.* **2022**, *189*, 104578. [CrossRef]
21. Lu, S.; Liu, M.; Yin, L.; Yin, Z.; Liu, X.; Zheng, W.; Kong, X. The multi-modal fusion in visual question answering: A review of attention mechanisms. *PeerJ Comput. Sci.* **2023**, *9*, e1400. [CrossRef] [PubMed]
22. Ann, R.C.; Cho, S.B. Human activity recognition with smartphone sensors using deep learning neural networks. *Expert Syst. Appl.* **2016**, *59*, 235–244.
23. Chen, K.; Zhang, D.; Yao, L.; Guo, B.; Yu, Z.; Liu, Y. Deep learning for sensor-based human activity recognition: Overview, challenges, and opportunities. *ACM Comput. Surv. CSUR* **2021**, *54*, 1–40. [CrossRef]
24. Qiu, S.; Zhao, H.; Jiang, N.; Wang, Z.; Liu, L.; An, Y.; Fortino, G. Multi-sensor information fusion based on machine learning for real applications in human activity recognition: State-of-the-art and research challenges. *Inf. Fusion* **2022**, *80*, 241–265. [CrossRef]
25. Li, Y.; Yang, G.; Su, Z.; Li, S.; Wang, Y. Human activity recognition based on multi-environment sensor data. *Inf. Fusion* **2023**, *91*, 47–63. [CrossRef]
26. Kun, X.; Huang, J.; Wang, H. LSTM-CNN architecture for human activity recognition. *IEEE Access* **2020**, *8*, 56855–56866.
27. Liu, X.; Shi, T.; Zhou, G.; Liu, M.; Yin, Z.; Yin, L.; Zheng, W. Emotion classification for short texts: An improved multi-label method. *Humanit. Soc. Sci. Commun.* **2023**, *10*, 306. [CrossRef]
28. Feng, W.; Hannafin, J. Design-based research and technology-enhanced learning environments. *Educ. Technol. Res. Dev.* **2005**, *53*, 5–23.
29. Liu, X.; Zhou, G.; Kong, M.; Yin, Z.; Li, X.; Yin, L.; Zheng, W. Developing Multi-Labelled Corpus of Twitter Short Texts: A Semi-Automatic Method. *Systems* **2023**, *11*, 390. [CrossRef]
30. Lu, C.; Shi, J.; Jia, J. Abnormal event detection at 150 fps in Matlab. In Proceedings of the IEEE International Conference on Computer Vision, Sydney, NSW, Australia, 1–8 December 2013.
31. Degardin, B.; Proença, H. Human Activity Analysis: Iterative Weak/Self-Supervised Learning Frameworks for Detecting Abnormal Events. In Proceedings of the IEEE International Joint Conference on Biometrics (IJCB), Houston, TX, USA, 28 September–1 October 2020.
32. Merad, D.; Drap, P. Tracking multiple persons under partial and global occlusions: Application to customers' behavior analysis. *Pattern Recognit. Lett.* **2016**, *81*, 11–20. [CrossRef]
33. Chen, T.; Chen, H. Anomaly detection in crowded scenes using motion energy model. *Multimed. Tools Appl.* **2018**, *77*, 14137–14152. [CrossRef]
34. Klingner, J. The pupillometric precision of a remote video eye tracker. In Proceedings of the ETRA 2010 (Eye Tracking Research and Applications Symposium), Austin, TX, USA, 22–24 March 2010; pp. 259–262.
35. Srichanyachon, N. EFL Learners' Perceptions of Using LMS. *TOJET Turk. Online J. Educ. Technol.* **2014**, *13*, 30–35.
36. Liang, M.; Hu, X. Recurrent Convolutional Neural Network for Object Recognition. In Proceedings of the IEEE Conference on Computer Vision and Pattern Recognition, Boston, MA, USA, 7–12 June 2015; pp. 3367–3375.
37. Jalal, A.; Mahmood, M.; Siddiqi, M.A. Robust spatiotemporal features for human interaction recognition via an artificial neural network. In Proceedings of the IEEE Conference on International Conference on Frontiers of Information Technology, Islamabad, Pakistan, 17–19 December 2018.
38. Jalal, A.; Quaid, M.A.K.; Sidduqi, M.A. A Triaxial acceleration-based human motion detection for an ambient smart home system. In Proceedings of the IEEE International Conference on Applied Sciences and Technology, Islamabad, Pakistan, 8–12 January 2019.

39. Dahlstrom, E.; Brooks, D.C.; Bichsel, J. *The Current Ecosystem of Learning Management Systems in Higher Education: Student, Faculty, and IT Perspectives*; Educause: Boulder, CO, USA, 2014.
40. Nawaratne, R.; Yu, X. Spatiotemporal anomaly detection using deep learning for real-time video surveillance. *IEEE Trans. Ind. Inform.* **2019**, *16*, 393–402. [CrossRef]
41. Oliveira, P.C.D.; Cunha, C.; Nakayama, M.K. Learning Management Systems (LMS) and e-learning management: An integrative review and research agenda. *JISTEM-J. Inf. Syst. Technol. Manag.* **2016**, *13*, 157–180. [CrossRef]
42. Ahmad, F. Deep image retrieval using artificial neural network interpolation and indexing based on similarity measurement. *CAAI Trans. Intell. Technol.* **2022**, *7*, 200–218. [CrossRef]
43. Hassan, F.S.; Gutub, A. Improving data hiding within color images using hue component of HSV colour space. *CAAI Trans. Intell. Technol.* **2022**, *7*, 56–68. [CrossRef]
44. Quaid, M.A.K.; Jalal, A. Wearable sensors based human behavioral pattern recognition using statistical features and reweighted genetic algorithm. *Multimed. Tools Appl.* **2019**, *79*, 6061–6083. [CrossRef]
45. Nadeem, A.; Jalal, A.; Kim, K. Human actions tracking and recognition based on body parts detection via an artificial neural network. In Proceedings of the IEEE International Conference on Advancements in Computational Sciences, Lahore, Pakistan, 17–19 February 2020.
46. Golestani, N.; Moghaddam, M. Human activity recognition using magnetic induction-based motion signals and deep recurrent neural networks. *Nat. Commun.* **2020**, *11*, 1551. [CrossRef] [PubMed]
47. Liu, X.; Song, M.; Tao, D.; Bu, J.; Chen, C. Random Geometric Prior Forest for Multiclass Object Segmentation. *IEEE Trans. Image Process.* **2015**, *24*, 3060–3070. [PubMed]
48. Jalal, A.; Khalid, N.; Kim, K. Automatic recognition of human interaction via hybrid descriptors and maximum entropy Markov model using depth sensors. *Entropy* **2020**, *22*, 817. [CrossRef] [PubMed]
49. Rafique, A.; Ahmad, J.; Kim, K. Automated sustainable multi-object segmentation and recognition via modified sampling consensus and kernel sliding perceptron. *Symmetry* **2020**, *13*, 1928. [CrossRef]
50. Zhang, J.; Ye, G.; Tu, Z.; Qin, Y.; Qin, Q.; Zhang, J.; Liu, J. A spatial attentive and temporal dilated (SATD) GCN for skeleton-based action recognition. *CAAI Trans. Intell. Technol.* **2022**, *7*, 46–55. [CrossRef]
51. Pervaiz, M.; Jalal, A.; Kim, K. Hybrid algorithm for multi-people counting and tracking for smart surveillance. In Proceedings of the IEEE 2021 International Bhurban Conference on Applied Sciences and Technologies (IBCAST), Islamabad, Pakistan, 12–16 January 2021.
52. Khalid, N.; Gochoo, M.; Jalal, A.; Kim, K. Modeling two-person segmentation and locomotion for stereoscopic action identification: A sustainable video surveillance system. *Sustainability* **2021**, *12*, 970. [CrossRef]
53. Cong, R.; Lei, J.; Fu, H.; Cheng, M.-M.; Lin, W.; Huang, Q. Review of Visual Saliency Detection with Comprehensive Information. *IEEE Trans. Circuits Syst. Video Technol.* **2018**, *29*, 2941–2959. [CrossRef]
54. Nadeem, A.; Jalal, A.; Kim, K. Automatic human posture estimation for sports activity recognition with robust body parts detection and entropy Markov model. *Multimed. Tools Appl.* **2021**, *80*, 21465–21498. [CrossRef]
55. Meng, J.; Li, Y.; Liang, H.; Ma, Y. Single-image Dehazing based on two-stream convolutional neural network. *J. Artif. Intell. Technol.* **2022**, *2*, 100–110. [CrossRef]
56. Liu, Y.; Wang, K.; Liu, L.; Lan, H.; Lin, L. Tcgl: Temporal contrastive graph for self-supervised video representation learning. *IEEE Trans. Image Process.* **2022**, *31*, 1978–1993. [CrossRef] [PubMed]
57. Zheng, M.; Zhi, K.; Zeng, J.; Tian, C.; You, L. A hybrid CNN for image denoising. *J. Artif. Intell. Technol.* **2022**, *2*, 93–99. [CrossRef]
58. Hu, X.; Kuang, Q.; Cai, Q.; Xue, Y.; Zhou, W.; Li, Y. A Coherent Pattern Mining Algorithm Based on All Contiguous Column Bicluster. *J. Artif. Intell. Technol.* **2022**, *2*, 80–92. [CrossRef]
59. Alberto, R.; Briones, A.; Hernandez, G.; Prieto, J.; Chamoso, P. Artificial neural network analysis of the academic performance of students in virtual learning environments. *Neurocomputing* **2021**, *423*, 713–720.
60. Rawashdeh, A.; Zuhir, A.; Mohammed, E.Y.; Al Arab, A.R.; Alara, M.; Al-Rawashdeh, B. Advantages and disadvantages of using e-learning in university education: Analyzing students' perspectives. *Electron. J. E-Learn.* **2021**, *19*, 107–117. [CrossRef]
61. Fuady, I.; Sutarjo, M.A.S.; Ernawati, E. Analysis of students' perceptions of online learning media during the COVID-19 pandemic Study of e-learning media: Zoom, Google Meet, Google Classroom, and LMS. *Randwick Int. Soc. Sci. J.* **2021**, *2*, 51–56. [CrossRef]
62. Li, T.; Fan, Y.; Li, Y.; Tarkoma, S.; Hui, P. Understanding the Long-Term Evolution of Mobile App Usage. *IEEE Trans. Mob. Comput.* **2023**, *22*, 1213–1230. [CrossRef]

sustainability

MDPI

Article

Transformer Architecture-Based Transfer Learning for Politeness Prediction in Conversation

Shakir Khan [1,2,*], Mohd Fazil [3], Agbotiname Lucky Imoize [4], Bayan Ibrahimm Alabduallah [5,*], Bader M. Albahlal [1], Saad Abdullah Alajlan [1], Abrar Almjally [1] and Tamanna Siddiqui [6]

1 College of Computer and Information Sciences, Imam Mohammad Ibn Saud Islamic University, Riyadh 11432, Saudi Arabia
2 University Center for Research and Development, Department of Computer Science and Engineering, Chandigarh University, Mohali 140413, India
3 Center for Transformative Learning, University of Limerick, V94 T9PX Limerick, Ireland
4 Department of Electrical and Electronics Engineering, Faculty of Engineering, University of Lagos, Akoka, Lagos 100213, Nigeria
5 Department of Information System, College of Computer and Information Sciences, Princess Nourah Bint Abdulrahman University, Riyadh 11564, Saudi Arabia
6 Department of Computer Science, Aligarh Muslim University, Aligarh 202002, India
* Correspondence: sgkhan@imamu.edu.sa (S.K.); bialabdullah@pnu.edu.sa (B.I.A.)

Abstract: Politeness is an essential part of a conversation. Like verbal communication, politeness in textual conversation and social media posts is also stimulating. Therefore, the automatic detection of politeness is a significant and relevant problem. The existing literature generally employs classical machine learning-based models like naive Bayes and Support Vector-based trained models for politeness prediction. This paper exploits the state-of-the-art (SOTA) transformer architecture and transfer learning for respectability prediction. The proposed model employs the strengths of context-incorporating large language models, a feed-forward neural network, and an attention mechanism for representation learning of natural language requests. The trained representation is further classified using a softmax function into polite, impolite, and neutral classes. We evaluate the presented model employing two SOTA pre-trained large language models on two benchmark datasets. Our model outperformed the two SOTA and six baseline models, including two domain-specific transformer-based models using both the BERT and RoBERTa language models. The ablation investigation shows that the exclusion of the feed-forward layer displays the highest impact on the presented model. The analysis reveals the batch size and optimization algorithms as effective parameters affecting the model performance.

Keywords: politeness prediction; conversation AI; machine learning; transfer learning

Citation: Khan, S.; Fazil, M.; Imoize, A.L.; Alabduallah, B.I.; Albahlal, B.M.; Alajlan, S.A.; Almjally, A.; Siddiqui, T. Transformer Architecture-Based Transfer Learning for Politeness Prediction in Conversation. *Sustainability* **2023**, *15*, 10828. https://doi.org/10.3390/su151410828

Academic Editors: Mourade Azrour, Azidine Guezzaz, Imad Zeroual, Azeem Irshad, Jamal Mabrouki, Said Benkirane, Shehzad Ashraf Chaudhry and Rui Xiong

Received: 22 May 2023
Revised: 12 June 2023
Accepted: 14 June 2023
Published: 10 July 2023

1. Introduction

In literature, politeness refers to good manners or etiquette in human behavior. It is also reflected in online conversations while using social networks. Politeness is the fundamental etiquette of various communication methods, either verbal, textual, or something else [1]. In Online social media, politeness in conversation is more crucial due to anonymity, the abundance of fake profiles, and usability issues. Advancement in natural language understanding and machine learning-based large language models are employed in various domains [2–5]. These language models also provide the opportunity to encode and detect the presence of politeness in a conversation or social media post. Online messaging platforms are popular and used by millions of users. Large organizations and business houses also use intelligent chatbots to communicate and resolve the queries of their customers. Therefore, if a chatbot is not polite in its language, it may lead to customer dissatisfaction which will hamper the organization's business. Impolite language in a conversation may lead to conflict and hate speeches among the users. Online social media platforms can avoid these conflicts through timely moderation of instigating language. Therefore, tracking, profiling & predicting politeness in a conversation is vital. This study presents a transformer architecture-based deep neural network model for classifying polite conversations from impolite ones.

Our Contributions

Recently, offensive speech problem has attracted researchers from across disciplines to analyze different aspects of hate speech problem. They also introduced various machine learning models to classify hate content into different categories [6,7]. Impolite language is a type of offensive content. In the existing literature, a closely related but contrasting research problem—politeness prediction, is understudied. Recently, this vital problem in textual content has received attention from researchers. They analyzed and addressed different aspects of an offensive content problem in conversational systems and chatbots [8–10]. However, existing approaches understudied the detection of politeness in social media posts. However, the politeness prediction methods don't employ the state-of-the-art transformer-based representation models like Bidirectional Encoder Representations from Transformers (BERT) [11], Robustly Optimized BERT (RoBERTa) [12]. To this end, this study presents a transformer employing deep model for politeness prediction in textual content. The preprocessed input text is passed through a BERT layer to encode the textual input into a numeric representation. It converts each word into a vector of the same length. Further, the model forwards the BERT representation to a feed-forward neural network (FFN) consisting of 2 dense layers to learn abstract feature extraction and representation. Finally, the encoded vector from FNN is passed to a sigmoid layer for classification to predict the final class of the text label.

In short, the proposed model can be summarized as employing the strength of transformer-based language architecture, feed-forward dense layer, and variable weight assignment to words based on their importance through an attention mechanism for efficient representation learning. Further, the trained representation is passed through a sigmoid layer having two neurons for politeness prediction in the content. We outline the contributions of the presented study as follows:

- Introduce a deep model for the understudied problem of politeness prediction by integrating the strength of transformer architecture-based large language models, feed-forward dense layer, and attention mechanism. Presented model learns an efficient text representation, passed to a sigmoid layer to classify into polite and impolite categories.
- Perform an in-depth investigation of the presented model by applying the initial weight assignment from transformer-based language models in politeness prediction over two benchmark datasets, including a blog dataset.
- Conduct an ablation study to investigate the impact of various neural network components like the attention mechanism towards the politeness prediction.
- Also, investigate the impact of different values of hyperparameters like batch size and optimization algorithm on the efficacy of the presented model to discover their optimal number.

The remaining manuscript is divided into the following sections. Section 2 explores the existing literature studying the different aspects of the politeness prediction. It also discusses the evolution of existing classification models presented over the year. Section 3 discusses the proposed model, including all its components. Further, Section 4 describes datasets, empirical evaluation, and analytical observations. Conclusion Section 6 presents the main findings of this study and a discussion of future research directions.

2. Literature Survey

The existing literature has analyzed different aspects of politeness in textual content generated from conversation systems and simple online posts. Thus, we group existing approaches into two categories: (i) politeness prediction methods developed for conversational systems and (ii) politeness prediction in textual content.

Researchers have presented politeness detection methods for conversational systems in the first group of approaches. These existing approaches generally employ different neural network components to propose architectures for politeness prediction [13–15]. In [14], the authors used emotions extracted from the multi-modal content to develop an end-to-end dialogue generation framework. The inclusion of sentiment made the dialogue system efficient and user-adaptive. In another interesting study, authors in [16] in-depth investigated the correlation between politeness and social interaction. In [17], the authors presented three weakly supervised machine learning models to generate diverse and polite-incorporating text. In [18], authors presented a dictionary matching and machine learning-based generative model. They used word- and sentence-level emotion captured from embeddings to generate a quality emotion dataset. Cristiana et al. [19] presented a sliding window-based strategy to compute the politeness score of each sentence in a conversation. It shows how politeness in a conversation evolves and tracks the chances of a conflict. Similarly, Zhang et al. [20] presented a politeness- and rhetoric-based strategy to early detection of derailing conversations. In another similar approach, Jonathan et al. [21] devised an unsupervised forecasting model to learn the representation of conversation dynamics and predict its probability of turning into a toxic conversation.

Authors in [8] studied the linguistic aspects of politeness. They devised lexical and syntactic features incorporating politeness-based theories like *difference* and *modality*. Authors found that polite Wikipedia authors have higher social power. In a vital work, Madaan et al. [9] presented a pipeline to convert impolite sentences into polite sentences. The proposed approach used the concept of *tag* and *generate*. In [22], the authors introduced a simple convolutional neural network-based model to predict politeness in natural language requests. In [23], researchers studied the linguistic features of politeness and presented a lexicon of politeness features called *PoliteLex*. They performed various empirical analyses to observe the politeness perception across cultures. Authors in [24] analyzed the impact of politeness in a vehicle speech interface on drivers' trustworthiness and found that polite speech interfaces have more trust among drivers. They also used activation clusters to identify the linguistic factors behind the polite text. In another method, authors in [25] presented a hierarchical transformer-based architecture to detect politeness in a goal-oriented dialog system. The authors also investigated the system efficacy against multiple baseline systems. As per our knowledge, no study in the existing literature employs the transformer-based context-incorporating language model for politeness prediction. It is the first study that integrates the strength of the context-aware language model, feed-forward neural network & attention mechanism for politeness prediction.

3. Models and Methods

The architecture of the presented model includes four layers: data-preprocessing, contextual embedding layer, feed-forward neural network layer, and output layer, as shown in Figure 1. We present a detailed introduction of these four layers in the subsequent subsections.

3.1. Data Pre-Processing

We investigate the efficacy of our model on two benchmark datasets from Wikipedia and Stack-Exchange question-answering forums provided by Mizil et al. [8]. We applied pre-processing steps to these datasets to filter noisy content and useless symbols. We performed the following pre-processing steps in sequence:

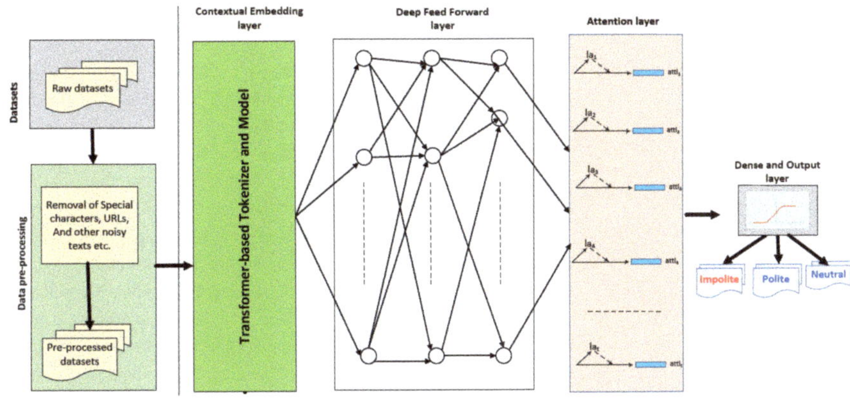

Figure 1. Workflow the model for politeness prediction.

3.1.1. Contraction Expansion

In English, many users use the contracted form of verbs, like I'm for I am, to shorten the sentence length. However, it is difficult for a language model to understand these words. Interestingly, negative contraction changes sentiment. Therefore expanding the contracted verb is essential. To this end, we identified 128 contracted verb forms like can't, could've with expansion to replace the contracted verb with their expanded form.

3.1.2. URL Filtration

In a text, URLs or hyperlinks don't contain much information. Therefore, we filtered all types of URLs.

3.1.3. Special Character Removal

The representation learning does not has any presentation for special characters. Therefore, we filter all the special characters and symbols like '#', '?' and ' ' to clean the text. We also remove the usernames (words starting with '@' symbol), replace two or more repetitions of the same character with a maximum of two, filter non-ASCII characters & numbers, and remove the extra white spaces. Finally, we convert the text into lower case to avoid the case-related issue.

3.2. Contextual Embedding Layer

Transformer-based large-language models are state-of-the-art models for text representation learning. These pre-trained models, trained over large corpora, are massive and demonstrate effective performance in different downstream tasks. For example, BERT (Bidirectional Encoder Representations from Transformers) [11,26] is trained on a large unlabelled corpus having the complete Wikipedia, approximately 2500 million words, and Book corpus, 800 million words. The two versions of the BERT: $BERT_{BASE}$ and $BERT_{LARGE}$, are used based on the complexity of the underlying problem. The first one, $BERT_{BASE}$, has 12 hidden layers along with 12 attention heads and 110 million parameters. On the other hand, the second one, $BERT_{LARGE}$, has 24 encode-decoder layers with 16 attention heads and 340 million parameters. Authors of the BERT model evaluated it on two tasks: (1) masked language modeling and (2) Next Sequence Prediction. Unlike traditional recurrent neural networks, which process information sequentially, BERT's success lies in training on a large corpus and parallel processing. Similarly, researchers have introduced many improved and domain-specific BERT, such as RoBERTa [12] and MentalBERT [27]. Training a BERT model from scratch requires massive computing resources and a large dataset. We generally fine-tune a model to avoid the requirement of computing resources and incorporate the domain-specific content in representation learning. In this study also, we fine-tune two transformer-based large language models—BERT and RoBERTa for efficient representation learning to predict the politeness label of a text. Fine-tuning refers to using the weights of an already trained network as the starting values for training a new model, like weights from BERT and RoBERTa models as the starting weight in the presented model. Further, this weight is updated for politeness prediction during the training process. This whole process of using already trained weight on one problem and updating it for another problem is called fine-tuning.

3.3. Attention-Aware Deep Feed Forward Network Layer

The encoded output from the transformer-based language model passes through a feed-forward network consisting of two dense layers. The presented model includes two layers for efficient and effective representation learning. The encoded output passes through an attention layer, which assigns weights to encoded features based on their segregating power in classifying three target classes. If the input encoded representation of a feature f is f_n, then the attention layer learns f_n' representation using Equation (1). Further, it computes the similarity using dot product between f_n' and a high-level context tensor v_h. Finally, based on the computed similarity, Equation (2) calculates the attention score α_f of each feature f. The context tensor v_h is randomly initialized and updated during the training process [28]. The attention score is multiplied with each feature to assign the relative weight. Finally, the resultant representation vector, F_n, is the weighted sum of the hidden features, as shown in Equation (3).

$$f_n' = \tanh(wf_n + b) \tag{1}$$

$$\alpha_f = \frac{exp(f_n' v_h)}{\sum_f exp(f_n' v_h)} \tag{2}$$

$$F_n = \sum_f (\alpha_f f_n) \tag{3}$$

3.4. Output Layer

Finally, the learned feature vector from the attention-aware deep FFN layer is passed to a softmax layer to classify each text into one of three categories: *polite*, *impolite*, and *neutral* for final labeling.

4. Experiment

We investigate the efficacy of the presented model over two standard datasets related to Wikipedia and Stack-exchange. This section describes evaluation phases like dataset, hyper-parameter setting, and evaluation metrics. Finally, this section ends with a discussion of experimental results and performs the comparison with the SOTA and baseline methods.

4.1. Evaluation Datasets

We use the two publicly released datasets by Mizil et al. [8] to evaluate the presented model. The first one, D_w, is a Wikipedia request dataset. It initially has 35,991 text annotated by 219 annotators from Amazon Mechanical Turk (AMT), a crowdsourcing marketplace. Finally, the constructed dataset contains 4353 requests, each having exactly two-sentence, where the second sentence is the request. The annotated dataset contains both politeness score and class. Every Wikipedia request is given one of the three labels: *polite*, *impolite*, and *neutral* depending upon the politeness score. The first row of Table 1 gives brief statistics of D_w. The authors also constructed a dataset D_s from the Stack-Exchange forum, a question-answer community. These two data source platforms are standard for conversational datasets having user-to-user request information. Table 1 presents a brief description of datasets. We can see from the table that both datasets are almost balanced.

Table 1. A brief dataset statistics.

Dataset	Dataset Size	#Polite	#Impolite	Neutral
D_w	4353	1089	1089	2175
D_s	8254	3302	1651	3301

4.2. Experimental Setting

The experiments conducted in this study are coded in Python version 3.7.12 using Keras 2.7.2 framework over the freely available Google colab notebook. All the training is performed using a five-fold cross-validation strategy, which splits the evaluation dataset into five equal parts. Under this strategy, four parts train the underlying model, whereas the left part evaluates the trained model. This procedure is conducted five times to ensure the usability of each sample in training and testing. We train the model in the batches of 16 instances for 20 epochs. *Adam* and categorical cross-entropy are used as the optimization method and loss function, respectively. The learning rate is 0.001 during the training process. Table 2 provides various hyper-parameters used in the models and underlying adjusted values.

Table 2. Parameters in the proposed model.

Hyperparameter	Value
Model learning rate	0.001
Batch size	16
Loss method	Categorical Cross-Entropy
Optimization algorithm	Adam
Dropout	0.5
Epoch	20

4.3. Experimental Results

This section presents the experimental results over the two benchmark datasets. We compare the model to SOTA and six baselines considering accuracy to establish its efficacy. We use the recent transformer-based pre-trained language models—BERT and RoBERTa for encoding the textual content to classify the text into polite, impolite, and neutral categories. The first two rows of Table 3 present the results using the BERT and RoBERTa language models. We can see from the table that our model reports an accuracy of approximately 90%.

4.3.1. Comparative Evaluation

To establish the efficacy of the presented approach, we evaluate its comparison against two SOTA and six baseline methods for politeness prediction. We constructed six baselines: two domain-specific large language models: fBERT and HateBERT and four using neural network components like LSTM and BiLSTM to analyze their impact on politeness prediction. We use GloVe embedding of 200-d as input to the embedding layer in the last four baselines. The following paragraphs discuss the SOTA and baseline models briefly.

- Aubakirova and Bansal [22]: In this paper, the authors introduced a simple neural network employing the convolutional neural network to predict the politeness in requesting sentences. Further, the authors performed network visualization using activation clusters, first derivative saliency, and embedding space transformation to analyze the linguistic signals of politeness.
- Mizil et al. [8]: In this early study, the authors presented a computational framework employing the domain-independent lexicon and syntactic features to analyze the linguistic aspect of politeness. They further trained an SVM classifier to predict politeness. They also investigated the relationship between politeness and social power.

- BiLSTM: The first baseline, BiLSTM, is a simple RNN network. It incorporates both left-to-right and right-to-left contexts during representation learning. This model has an input layer to receive the embedding-based text representation followed by a BiLSTM layer having 128 neurons. Finally, a final softmax layer classifies the text into polite and non-polite categories.
- LSTM: The second baseline is an LSTM network to compare its performance against the presented model. The baseline network has an LSTM layer with 128 neurons. It also has an input layer and a softmax layer for classification. It also uses GloVe embedding of 200d as input to the model.
- BiGRU: It is the third baseline of this paper. It uses 128 neurons in the BiGRU layer for tweet representation learning. It also has an Embedding layer using 200d Glove embedding and a softmax layer to perform final classification.
- ANN: The model performance is also compared with a simple artificial neural network. This ANN model has 3 hidden layers having 128, 64, and 32 neurons. It also has an embedding layer and a final softmax layer for classification. This baseline model takes 200d GloVe embedding as input.
- fBERT [29]: It is a BERT model, pre-trained on a large English offensive language corpus (SOLID), containing more than 1.4 million offensive instances.
- HateBERT [30]: It is another BERT model, pretrained for abusive language detection. It is trained on a Reddit dataset of communities banned for being offensive, abusive, or hateful.

Table 3. Evaluation results considering accuracy over D_w and D_s.

Datasets →	D_w	D_s
Methods ↓	**Accuracy**	**Accuracy**
Proposed Model [BERT]	0.91	**0.87**
Proposed Model [RoBERTa]	**0.92**	0.84
Aubakirova and Bansal [22]	0.85	0.66
Mizil et al. [8]	0.83	0.78
fBERT [29]	0.90	0.87
HateBERT [30]	0.89	0.83
ANN	0.87	0.82
BiLSTM	0.88	0.76
LSTM	0.87	0.76
BiGRU	0.88	0.78

Table 3 presents the comparative evaluation results considering accuracy. We do not present the training accuracy because it is not viable. We can see that over both datasets, our model, employing the BERT and RoBERTa language models, shows the best result. The table demonstrates that over D_w, the proposed model with BERT performs best with an accuracy of 0.91. On the other hand, over D_s, the model performs best employing the RoBERTa model with an accuracy of 0.87. We can also see from the table that the model also significantly outperforms all the baseline models. The fBERT achieves an accuracy of 0.90 and 0.87 over D_w and D_s, respectively, and performs best among baselines. On contrary, the LSTM baseline shows the worst performance. We can observe from the table that the domain-specific transformer-based large language models show comparative performance. Overall, comparative models show relatively better performance over D_w.

4.3.2. Ablation Analysis

The presented model has two main neural components—(i) feed-forward neural layer (FNN) and (ii) attention mechanism. The impact of each neural network component is analyzed by conducting ablation analysis. We performed the ablation analysis with both BERT and RoBERTa. In the first ablation analysis, we exclude the feed-forward neural layer from the proposed model to construct a model having an input layer (embedding layer), an attention layer, and a final softmax layer. The underlying results for constructed models are given in the 5th and 6th rows of Table 4. In further ablation analysis, the attention mechanism is excluded from the proposed model to construct a model with a contextual-embedding layer, FNN layer, and a final softmax layer for prediction. The results for this ablation analysis are presented in the 7th and 8th row of Table 4. The ablation study establishes that excluding the attention mechanism has least impact on the proposed model with both BERT and RoBERTa language models. Further, exclusion of feed-forward layer has insignificant impact on the performance and reduces the RoBERTa-based accuracy by 3% over D_w. Interestingly, the model performance with RoBERTa increases on removal of FNN over D_s.

Table 4. Ablation evaluation results considering accuracy on two datasets.

Datasets →	D_w	D_s
Methods ↓	**Accuracy**	**Accuracy**
Proposed Model [BERT]	0.91	0.87
Proposed Model [RoBERTa]	0.92	0.84
Proposed model [BERT] (without FNN)	0.89	0.86
Proposed Model [RoBERTa] (without FNN)	0.89	0.87
Proposed model [BERT] (without Attention)	0.91	0.86
Proposed Model [RoBERTa] (without Attention)	0.91	0.83

5. Discussion: Evaluation of Hyperparameters Impact

In a neural network model, many hyper-parameters affect performance. To this end, we investigate to observe the impact of *batch size* and *optimization algorithms* on the model performance over D_w and D_s. We evaluate the model performance considering accuracy.

5.1. Batch Size

In a deep model, the number of training instances processed through it in one go is called *batch size*. It is a hyperparameter because a user can fine-tune it to optimize the model performance. Suppose a dataset contains 500 samples, and if the batch size is 50, then the model is trained using the first 50 samples, again trained using the next 50, and this process continues until the dataset exhaust. We investigate the model efficacy for both BERT and RoBERTa on 4 different batch sizes—16, 32, 64, and 128 over D_w and D_s and Figure 2 depicts the underlying results. It reveals as the batch size increases, the BERT-based model accuracy degrades. Our model best performs with 16 and 32 batch sizes on both datasets. The model performance with RoBERTa-based contextual embedding is shown in Figure 2b. The model over the D_w dataset shows the best performance with 16 batch size, whereas it shows the best result on the D_s using 32 batch size. We can conclude that the model demonstrates the best evaluation results when processed in batches of 16 instances. Therefore, evaluation results justify the adjustment of batch size to 16 in this paper.

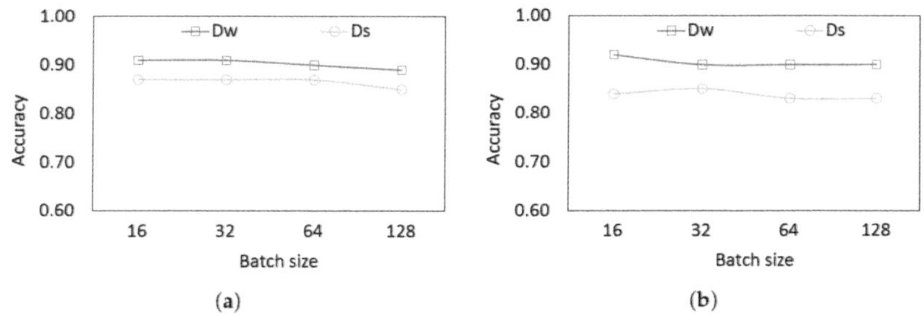

Figure 2. Evaluation results of proposed model on different batch sizes over D_w and D_s considering accuracy for (**a**) BERT (**b**) RoBERTa.

5.2. Optimization Algorithms

The optimization method used in a deep model is another parameter that affects the model performance. We investigate the impact of Adam, Adagrad, and Adadelta on the results of our model. Figure 3 displays the underlying evaluation results for both BERT and RoBERTa models the over the D_w and D_s datasets. The Figure 3a reveals that our model using BERT shows the most promising result with Adam and the worst result with Adadelta over both datasets. Figure 3b exhibits similar performance for the proposed model employing RoBERTa. The figure shows that the impact of the optimization algorithm is more significant over the D_w dataset. Overall, we can infer that the proposed model with Adam performs best and shows the worst performance with Adadelta. On the other hand, it shows a comparative performance with Adagrad. Finally, this evaluation result justifies the selection of Adam as an optimization algorithm.

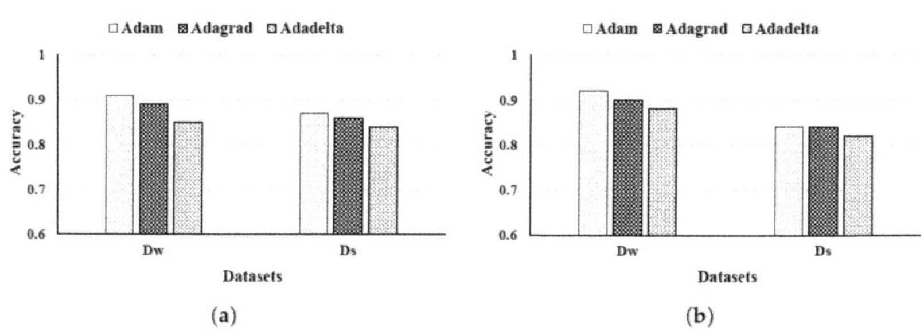

Figure 3. Empirical results of the model using the three optimization algorithms over D_w and D_s considering accuracy for (**a**) BERT (**b**) RoBERTa.

6. Conclusions and Future Directions of Work

In this research, we developed an advanced deep learning model that outperforms the existing methods. We introduced transformer-based architecture for representation learning toward politeness prediction in a text. The presented model integrated the strengths of context-aware language models, a feed-forward neural network, and an attention mechanism for the representation learning of natural language requests. The trained representation is further passed through a softmax layer and classified into polite, impolite, and neutral classes. We evaluated the model over the two benchmark datasets considering accuracy. In the comparative analysis, the proposed model outperformed the two SOTA and six baseline models, including two offensive content specific large language models. We also examined the hyperparameter effect on the model to ascertain the use of their optimal value in this study.

The proposed model lacks content, network, and profile-related features, which can be vital. In further research, we will incorporate these feature categories. Though the proposed model is highly effective for politeness prediction in English texts, it has limitations. Like, it has been evaluated only on English datasets. Its adaptation over multi-lingual or code-mixed data is a promising future research. Second, it has not been evaluated for hate and offensive content detection.

Author Contributions: Conceptualization, M.F. and S.K.; methodology, M.F.; software, S.K.; validation, A.A. and S.A.A.; resources, B.I.A.; data curation, M.F.; writing—original draft preparation, S.K. and T.S.; writing—review and editing, M.F.; visualization, B.M.A.; supervision, A.L.I.; project administration, B.I.A.; funding acquisition, S.K. All authors have read and agreed to the published version of the manuscript.

Funding: This research was funded by Deanship of Scientific Research at Mohammad Ibn Saud Islamic University (IMSIU) through grant number RP-21-07-06 and Princess Nourah Bint Abdulrahman University Researchers Supporting Project number (PNURSP2023R440).

Acknowledgments: The authors extend their appreciation to the Deanship of Scientific Research at Imam Mohammad Ibn Saud Islamic University (IMSIU) for funding and supporting this work through Research Partnership Program no. RP-21-07-06 and Princess Nourah Bint Abdulrahman University Researchers Supporting Project number (PNURSP2023R440), Princess Nourah Bint Abdulrahman University, Riyadh, Saudi Arabia. The authors would also like to thank Aleem Ali from Department of Computer Science and Engineering, Chandigarh University for serving as a consultant to critically review the study proposal and participating in technical editing of the manuscript

Conflicts of Interest: The authors declare no conflict of interest.

References

1. Khan, S. Business Intelligence Aspect for Emotions and Sentiments Analysis. In Proceedings of the First International Conference on Electrical, Electronics, Information and Communication Technologies, ICEEICT, Trichy, India, 16–18 February 2022; pp. 1–5.
2. Haq, A.U.; Li, J.P.; Ahmad, S.; Khan, S.; Alshara, M.A.; Alotaibi, R.M. Diagnostic Approach for Accurate Diagnosis of COVID-19 Employing Deep Learning and Transfer Learning Techniques through Chest X-ray Images Clinical Data in E-Healthcare. *Sensors* **2021**, *21*, 8219. [CrossRef] [PubMed]
3. Qaisar, A.; Ibrahim, M.E.; Khan, S.; Baig, A.R. Hypo-Driver: A Multiview Driver Fatigue and Distraction Level Detection System. *Cmc-Comput. Mater. Contin.* **2021**, *71*, 1999–2017.
4. Abulaish, M.; Kumari, N.; Fazil, M.; Singh, B. A Graph-Theoretic Embedding-Based Approach for Rumor Detection in Twitter. In Proceedings of the IEEE/WIC/ACM International Conference on Web Intelligence, Thessaloniki, Greece, 14–17 October 2019; pp. 466–470.
5. Mahajan, S.; Pandit, A.K. Hybrid method to supervise feature selection using signal processing and complex algebra techniques. *Multimed. Tools Appl.* **2023**, *82*, 8213–8234. [CrossRef]
6. Khan, S.; Fazil, M.; Sejwal, V.K.; Alshara, M.A.; Alotaibi, R.M.; Kamal, A.; Baig, A. BiCHAT: BiLSTM with deep CNN and hierarchical attention for hate speech detection. *J. King Saud Univ.-Comput. Inf. Sci.* **2022**, *34*, 4335–4344. [CrossRef]
7. Khan, S.; Kamal, A.; Fazil, M.; Alshara, M.A.; Sejwal, V.K.; Alotaibi, R.M.; Baig, A.; Alqahtani, S. HCovBi-Caps: Hate Speech Detection using Convolutional and Bi-Directional Gated Recurrent Unit with Capsule Network. *IEEE Access* **2022**, *10*, 7881–7894. [CrossRef]
8. Danescu-Niculescu-Mizil, C.; Sudhof, M.; Jurafsky, D.; Leskovec, J.; Potts, C. A computational approach to politeness with application to social factors. In Proceedings of the International Conference of the Association for Computational Linguistics, Sofia, Bulgaria, 4–9 August 2013; pp. 250–259.
9. Madaan, A.; Setlur, A.; Parekh, T.; Poczos, B.; Neubig, G.; Yang, Y.; Salakhutdinov, R.; Black, A.W.; Prabhumoye, S. Politeness Transfer: A Tag and Generate Approach. In Proceedings of the International Conference of the Association for Computational Linguistics, Virtual, 5–10 July 2020; pp. 1869–1881.
10. Niu, T.; Bansal, M. Polite Dialogue Generation Without Parallel Data. *Trans. Assoc. Comput. Linguist.* **2018**, *6*, 373–389. [CrossRef]
11. Devlin, J.; Chang, M.W.; Lee, K.; Toutanova, K. BERT: Pre-training of Deep Bidirectional Transformers for Language Understanding. In Proceedings of the NAACL-HLT, Minneapolis, MN, USA, 2–7 June 2019; pp. 4171–4186.
12. Liu, Y.; Ott, M.; Goyal, N.; Du, J.; Joshi, M.; Chen, D.; Levy, O.; Lewis, M.; Zettlemoyer, L.; Stoyanov, V. RoBERTa: A Robustly Optimized BERT Pretraining Approach. In Proceedings of the ICLR, Addis Ababa, Ethiopia, 26–30 April 2020; pp. 1–13.
13. Wen, T.H.; Vandyke, D.; Mrksic, N.; Gasic, M.; Rojas-Barahona, L.M.; Ultes, P.H.S.S.; Young, S. A Network-based End-to-End Trainable Task-oriented Dialogue System. In Proceedings of the International Conference of European Chapter of the Association for Computational Linguistics, Valencia, Spain, 3–7 April 2017; pp. 438–449.
14. Shi, W.; Yu, Z. Sentiment Adaptive End-to-End Dialog Systems. In Proceedings of the Annual Meeting of the Association for Computational Linguistics, Melbourne, Australia, 15–20 July 2018; pp. 1509–1519.
15. Mishra, K.; Firdaus, M.; Ekbal, A. Please be polite: Towards building a politeness adaptive dialogue system for goal-oriented conversations. *Neurocomputing* **2022**, *494*, 242–254. [CrossRef]
16. Brown, P.; Levinson, S.C.; Levinson, S.C. *Politeness: Some Universals in Language Usage*; Cambridge University Press: Cambridge, UK, 1987; Volume 4.

17. Niu, T.; Bansal, M. Polite Dialogue Generation Without Parallel Data. In Proceedings of the the European Conference on Information Retrieval, Padua, Italy, 20–23 March 2018; Springer: Cham, Switzerland, 2018; pp. 810–817.
18. Peng, D.; Zhou, M.; Liu, C.; Ai, J. Human–machine dialogue modelling with the fusion of word- and sentence-level emotion. *Knowl.-Based Syst.* **2019**, *192*, 105319. [CrossRef]
19. Iordache, C.P.; Trausan-Matu, S. Analysis and prediction of politeness in conversations. In Proceedings of the International Conference on Human Computer Interaction, Bucharest, Romania, 16–17 September 2021; pp. 15–20.
20. Zhang, J.; Chang, J.P.; Danescu-Niculescu-Mizil, C. Conversations Gone Awry: Detecting Early Signs of Conversational Failure. In Proceedings of the Annual Meeting of the Association for Computational Linguistics, Melbourne, Australia, 15–20 July 2018; pp. 1350–1361.
21. Chang, J.P.; Danescu-Niculescu-Mizil, C. Trouble on the Horizon: Forecasting the Derailment of Online Conversations as they Develop. In Proceedings of the International Conference on Empirical Methods in Natural Language Processing, Hongkong, China, 3–7 November 2019; pp. 1–12.
22. Aubakirova, M.; Bansal, M. Interpreting Neural Networks to Improve Politeness Comprehension. In Proceedings of the International Conference on Empirical Methods in Natural Language Processing, Austin, TX, USA, 1–5 November 2016; pp. 2035–2041.
23. Li, M.; Hickman, L.; Tay, L.; Ungar, L.; Guntuku, S.C. Studying Politeness across Cultures using English Twitter and Mandarin Weibo. In Proceedings of the CSCW, Virtual, 17–21 October 2020; pp. 1–15.
24. Lee, J.G.; Lee, K.M. Polite speech strategies and their impact on drivers' trust in autonomous vehicles. *Comput. Hum. Behav.* **2022**, *127*, 107015. [CrossRef]
25. Mishra, K.; Firdaus, M.; Ekbal, A. Predicting Politeness Variations in Goal-Oriented Conversations. *IEEE Trans. Comput. Soc. Syst.* **2022**, *10*, 1–10. [CrossRef]
26. Vaswani, A.; Shazeer, N.; Parmar, N.; Uszkoreit, J.; Jones, L.; Gomez, A.N.; Kaiser, L.; Polosukhin, I. Attention Is All You Need. In Proceedings of the International Conference on Neural Information Processing Systems (NIPS), Long Beach, CA, USA, 4–9 December 2017; pp. 1–11.
27. Ji, S.; Zhang, T.; Ansari, L.; Fu, J.; Tiwari, P.; Cambria, E. MentalBERT: Publicly Available Pretrained Language Models for Mental Healthcare. In Proceedings of the the Thirteenth Language Resources and Evaluation Conference, Marseille, France, 20–25 June 2022; European Language Resources Association: Paris, France, 2022; pp. 7184–7190.
28. Yang, Z.; Yang, D.; Dyer, C.; He, X.; Smola, A.; Hovy, E. Hierarchical Attention Networks for Document Classification. In Proceedings of the 2016 Conference of the North American Chapter of the Association for Computational Linguistics: Human Language Technologies, San Diego, CA, USA, 12–17 June 2016; pp. 1480–1489.
29. Sarkar, D.; Zampieri, M.; Ranasinghe, T.; Ororbia, A. fBERT: A Neural Transformer for Identifying Offensive Content. In Proceedings of the Proc. of the EMNLP, Punta Cana, Dominican Republic, 1–6 August 2021; pp. 1792–1798.
30. Caselli, T.; Basile, V.; Mitrović, J.; Granitzer, M. HateBERT: Retraining BERT for Abusive Language Detection in English. In Proceedings of the 5th Workshop on Online Abuse and Harms, Online, 7–13 November 2021; pp. 17–25.

sustainability

MDPI

Article

Advances in the Optimization of Vehicular Traffic in Smart Cities: Integration of Blockchain and Computer Vision for Sustainable Mobility

Angel Jaramillo-Alcazar, Jaime Govea and William Villegas-Ch *

Escuela de Ingeniería en Ciberseguridad, Facultad de Ingenierías y Ciencias Aplicadas, Universidad de Las Américas, Quito 170125, Ecuador; angel.jaramillo@udla.edu.ec (A.J.-A.); jaimealejandro.govea@udla.edu.ec (J.G.)

* Correspondence: william.villegas@udla.edu.ec; Tel.: +593-98-136-4068

Abstract: The growing adoption of Artificial Intelligence of Things technologies in smart cities generates significant transformations to address urban challenges and move towards sustainability. This article analyzes the economic, social, and environmental impacts of Artificial Intelligence of Things in urban environments, focusing on a case study on optimizing vehicular traffic. The research methodology is based on a comprehensive analysis of academic literature and government sources, followed by the creation of a simulated city model. This framework implemented a vehicle-traffic optimization system integrating artificial intelligence algorithms, computer vision, and blockchain technology. The results obtained in this case study are highly encouraging: artificial intelligence algorithms processed real-time data from security cameras and traffic lights, resulting in a notable 20% reduction in traffic congestion during peak hours. Furthermore, implementing blockchain technology guarantees the security and immutability of traffic data, strengthening trust in the system and promoting sustainability in urban environments. These results highlight the importance of combining advanced technologies to effectively address modern cities' complex challenges and move towards more sustainable and livable cities.

Keywords: traffic simulation; smart cities; integrated technologies

Citation: Jaramillo-Alcazar, A.; Govea, J.; Villegas-Ch, W. Advances in the Optimization of Vehicular Traffic in Smart Cities: Integration of Blockchain and Computer Vision for Sustainable Mobility. *Sustainability* **2023**, *15*, 15736. https://doi.org/10.3390/su152215736

Academic Editors: Mourade Azrour, Azidine Guezzaz, Imad Zeroual, Azeem Irshad, Jamal Mabrouki, Said Benkirane, Shehzad Ashraf Chaudhry and Armando Carteni

Received: 6 September 2023
Revised: 23 October 2023
Accepted: 27 October 2023
Published: 8 November 2023

1. Introduction

In the digital age, urban planning and management are transforming significantly due to the adoption of emerging technologies that promise to address urban challenges more efficiently and sustainably. In this context, the convergence of Artificial Intelligence of Things (AIoT), a fusion of artificial intelligence (AI) and the Internet of Things (IoT), has emerged as a fundamental force that resonates with governments, companies, and residents [1]. The interconnection of devices and systems within urban environments, empowered by AIoT, has given rise to smart cities and urban centers that seek to optimize their operations and services through the seamless collection and analysis of real-time data [2]. This technological fusion has the profound potential to address relevant problems in critical areas of sustainable development, such as agriculture, water resources management, healthcare, smart grids, and the overall smart city paradigm.

In this transformation environment, this study embarks on the search for innovative and sustainable solutions to improve the quality of life of the urban population. Our approach involves optimizing vehicular traffic in smart cities through the judicious application of AIoT technologies. The continued increase in vehicle volume, coupled with limited infrastructure, has precipitated traffic congestion, generating transportation bottlenecks, an increase in air pollution, and a palpable deterioration in the quality of life of urban dwellers [3]. In this regard, a nuanced exploration of the role of AIoT in urban traffic management emerges as a critical undertaking to ameliorate these pressing challenges.

This article strives to investigate, analyze, and articulate the multifaceted impact of AIoT implementation on optimizing vehicular traffic within smart cities, perfectly aligning with the principles of sustainable development. Vehicular-traffic congestion is a problem that significantly affects urban mobility, the economy, and environmental well-being in urban environments [4,5]. The importance of this study lies in the transformative potential of AIoT to introduce efficient traffic management, thereby reducing travel times, mitigating air pollution, and increasing the overall efficiency of the transportation system. Furthermore, resolving traffic-related dilemmas can precipitate healthy cascading effects, resulting in citizens' better quality of life and the attractiveness of urban areas for future investments [6].

To address this multifaceted problem, we employ a blended research methodology that seamlessly weaves together an extensive review of academic literature and government sources to create a controlled experimental environment that reflects the intricate aspects of an actual city. The literature review provides the necessary context to appreciate the importance of AIoT in optimizing vehicular traffic. It offers insights into previous case studies and a comprehensive understanding of various approaches and the challenges that accompany them. The instantiation of a controlled experimental environment allows us to thoroughly evaluate the ramifications of AIoT implementation and its constituent technological facets (computer vision and blockchain technology) in urban traffic management [7]. Preliminary results show a 20% reduction in traffic congestion during peak hours. This indicates the potential of AIoT to address vehicular traffic effectively. Furthermore, implementing blockchain technology guarantees the security and integrity of traffic data, strengthening trust in the system. Computer vision has proven crucial for detecting vehicle patterns and behaviors, allowing for more accurate decision-making [8].

However, the path to successful implementation is not without formidable challenges, including public acceptance issues and the need to develop a robust technological infrastructure. Future research will delve deeper into these complexities and explore additional dimensions where AIoT can resonate positively within smart cities [9]. Through our comprehensive methodology and case study, we aim to catalyze a thorough understanding of how technological convergences can transform cities, ultimately elevating the quality of life of their inhabitants.

2. Literature Review

The literature has been categorized into different groups to understand the landscape better, addressing similar approaches and highlighting how this work aligns with and improves on previous research. First, we discuss studies exploring AIoT's potential to optimize urban mobility through real-time data analysis and automated decision-making. For example, [10] conducted a study in a metropolitan city where machine-learning algorithms were used to analyze data from traffic sensors and improve traffic signal times. Although they demonstrated improvements in travel times, the ability to detect unusual traffic patterns was limited, indicating the need for more comprehensive approaches. The works in the second group explore proposals that integrate blockchain technology and computer vision in vehicular-traffic management. In [11], a blockchain-based system that guarantees the security and immutability of traffic data is presented, while computer vision is used to detect traffic behaviors. Despite its advantages, this approach showed limitations in its scalability for densely populated cities. In group three, the works that include controlled environments for evaluating the impact of technologies in smart cities are analyzed. Our proposal aligns with this approach but seeks to integrate AIoT, computer vision, and blockchain technology in the simulated environment [12]. A more comprehensive and adaptable perspective is provided on how these technologies can transform vehicular-traffic management by dealing with varied traffic situations, from massive congestion to fluid flows [13]. Group four includes works that seek a holistic integration of emerging technologies to address urban challenges. This proposal combines AIoT, computer vision, and blockchain technology in a controlled environment. This combination allows us to overcome isolated approaches' limitations and evaluate how these technologies interact to provide more effective and adaptable solutions.

Despite the relevance of the mentioned technologies, we noticed a gap in the existing literature regarding the effective integration of these technologies in a controlled environment that reflects an actual city. Our research addresses this gap by creating an experimental environment that combines artificial intelligence algorithms, computer vision, and blockchain technology in an urban traffic-management system. While other studies have focused on specific aspects of these technologies, such as traffic signal optimization or monitoring, our approach is holistic. It seeks to understand how these technologies can work together to address traffic congestion in an urban context [14,15]. Furthermore, our research incorporates a qualitative dimension by collecting and analyzing the perspectives and experiences of critical actors in urban traffic management.

Implementing IoT, machine learning (ML), and artificial intelligence (AI) in this work is essential to optimize vehicular traffic in smart cities and contributes significantly to sustainability. This implementation not only seeks to address traffic challenges in smart cities but also aims to do so sustainably, reducing polluting gas emissions and improving citizens' quality of life [16]. The combination of IoT, ML, and AI in this research creates a comprehensive technological ecosystem that optimizes urban traffic and contributes to greater sustainability in the context of smart cities [17].

3. Materials and Methods

This work is based on an interdisciplinary approach that combines quantitative and qualitative methods to comprehensively address the challenges of vehicular traffic management in smart urban environments. Although much of our analysis focuses on collecting and evaluating quantitative data to measure traffic-optimization-system performance, we recognize the importance of understanding the perceptions and experiences of citizens and key stakeholders. This qualitative understanding is achieved through surveys and interviews with residents and urban mobility experts. These qualitative insights complement our quantitative findings and enrich our understanding of how the community experiences and benefits from AIoT-based traffic management.

The successful implementation of blockchain technology and computer vision plays a crucial role in this proposal for optimizing vehicular traffic in smart cities. Blockchain technology is uniquely integrated into this system to ensure the security, integrity, and reliability of traffic data. The decision to incorporate blockchain instead of traditional data storage and transmission methods is based on its inherent ability to establish immutable and trusted records. We use a decentralized blockchain that stores traffic data in encrypted blocks, which are added sequentially to the chain. This ensures that unauthorized third parties cannot alter or manipulate traffic data. Furthermore, using blockchain-based intelligent contracts facilitates the automatic execution of agreements and provides transparency in managing traffic signs and regulations.

Blockchain implementation differs from previous approaches that relied solely on IoT and machine learning. While IoT provides valuable data, securing this data can be a challenge. The inclusion of blockchain addresses this issue by providing additional traffic data security and reliability. Additionally, smart contracts on blockchain enable the automation of traffic-management processes, increasing the efficiency and responsiveness of the system. This combination of blockchain with IoT and machine learning represents a unique technical contribution to the existing literature on sustainable urban mobility.

Computer vision is another essential pillar of our solution for traffic pattern detection and real-time monitoring. We deploy a system of security cameras equipped with advanced computer vision algorithms to identify abnormal behavior, such as sudden lane changes, sudden braking, and unusual vehicle speeds. The high accuracy of computer vision, with a detection rate of 92% for lane changes, 97% for hard braking, and 94% for identifying unusual speed patterns, ensures the system's ability to make accurate and timely decisions.

Our computer vision implementation stands out for its technical contribution to the existing literature on sustainable urban mobility. Although computer vision has been used in traffic applications, its synergy with blockchain in a holistic approach is innovative. Combining both technologies ensures accurate traffic pattern detection and enables secure verification of this data on the blockchain. This contributes to the efficiency and safety of the traffic-management system and represents a significant advance in the field.

3.1. Definition of the Controlled Environment

The creation of the controlled environment is essential to effectively assess the impact of the implementation of AIoT in optimizing vehicular traffic by allowing us to observe and analyze how technologies interact in realistic and controlled situations, as outlined in the comprehensive study by [18].

The first step in creating the controlled environment involves the selection of critical parameters that characterize the City and its vehicular traffic. These parameters include traffic density, vehicle types, movement patterns, weather conditions, and topography. These factors, taken from actual data and adjusted for the simulated environment, ensure that the results are representative and applicable in similar real-world scenarios. Once the key parameters have been selected, we proceed to the construction of virtual models of vehicles and traffic routes. Each car was modeled in detail, considering its size, top speed, acceleration, and behavior in traffic. The traffic lanes were designed to reflect the City's road infrastructure, including intersections, streets, and traffic signals. The precision in the construction of these models contributes to the realism of the simulations.

Technology infrastructure plays a central role in the controlled environment. For this, a virtual sensor network is configured that collects real-time data on traffic and urban mobility. Artificial intelligence algorithms, specifically designed for traffic optimization, are incorporated into the infrastructure. Computer vision technology is implemented in virtual cameras that capture images and videos of the surrounding environment [19]. Furthermore, blockchain technology ensures the integrity and traceability of the data collected in the background.

A critical phase in creating the controlled environment is validation. Therefore, exhaustive tests are carried out to ensure that the vehicle models, roads, and technologies work harmoniously. Preliminary simulations are run to observe the behavior of the vehicles in different conditions and verify the interaction with the technological infrastructure [20]. Successive validation iterations allow for adjustments and improvements, ensuring the accuracy and reliability of the controlled environment. This environment represents the base from which the impacts of the AIoT implementation on the optimization of vehicular traffic in the City are evaluated, providing a solid framework for the analysis and interpretation of the results [21].

Figure 1 represents the controlled environment designed to optimize traffic flow within the City. The figure is intended to illustrate the various elements critical to creating this environment, helping to understand the interconnected technologies and their roles. A simplified visualization of the City is represented in the upper part of the figure. The streets, intersections, and buildings are represented with dotted lines and shaded rectangles, conveying the urban environment where traffic optimization strategies are implemented. The middle section of the figure focuses on the core of the environment: the virtual vehicle models. These models travel the streets and highways of the City, and each type of vehicle (cars, trucks, and buses) is distinguished by its shape and label. This central area is the focal point where technologies interact to optimize traffic [22].

Along the streets are small icons that represent the virtual sensor infrastructure. These sensors are strategically positioned to collect real-time traffic patterns and urban mobility data. The blue icons indicate the points where the information is captured, contributing to the comprehensive analysis of traffic dynamics. At intersection corners and other strategic locations, red icons symbolize the location of cameras designed for visual analysis [23]. These cameras use computer vision techniques to capture images and videos of the surrounding environment, providing vital information on traffic behavior. In the figure is the incorporation of an icon that represents blockchain technology. This icon serves as a visual cue connecting some of the sensors and cameras, highlighting their role in ensuring the security and integrity of the data collected.

The rendering captures the intricate synergy between virtual vehicles, sensors, cameras, and blockchain technology within the controlled environment. This is taken as a comprehensive reference point, facilitating the understanding of the complex interactions that underlie the optimization of traffic flow in the City. This description provides a detailed overview of the visual elements present in the controlled environment, allowing readers to understand the importance and relationships between the different components.

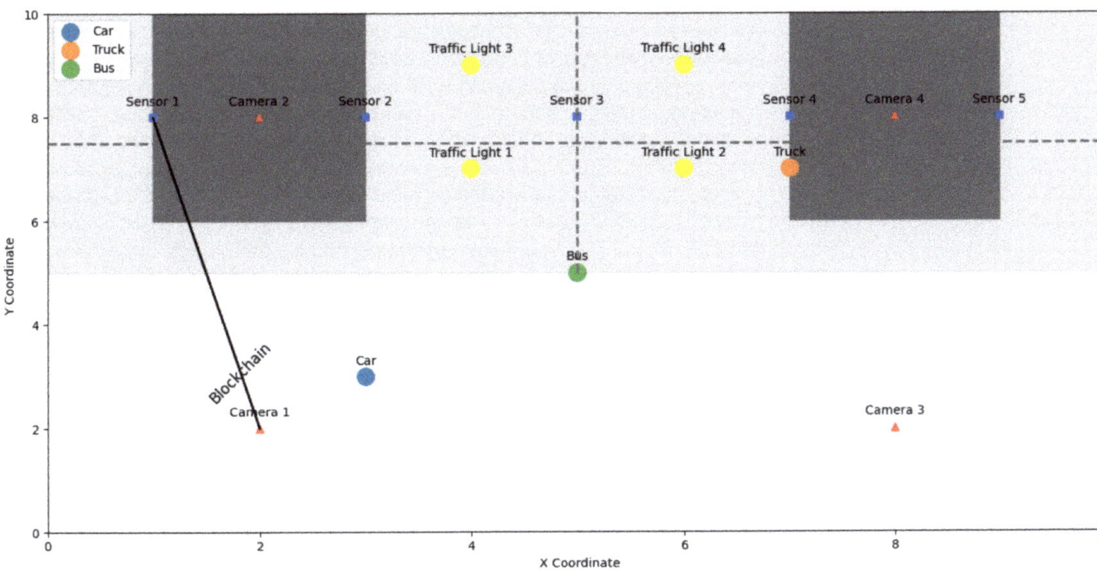

Figure 1. Controlled environment for traffic optimization.

3.2. *Data Collection and Analysis*

Data collection is carried out through the network of sensors strategically distributed along the streets and highways of the City. These sensors capture real-time data such as vehicle speeds, traffic densities, and travel times. During a three-month trial period, more than 100,000 data records were collected, including detailed information on vehicle movement at different times and days of the week. Computer vision technology plays an essential role in this phase. Cameras placed in strategic locations capture images and videos of traffic in real time. Over three months, more than 50,000 images were collected, allowing a complete visual understanding of traffic flow and movement patterns.

The volume of data collected is subjected to in-depth analysis using advanced data analysis and machine learning techniques [24]. For this, grouping algorithms are applied to identify congestion patterns in peak hours and evaluate the distribution of different types of vehicles in the City. In addition, a neural network-based predictive model is applied to forecast traffic levels at critical intersections.

3.2.1. Data Collection

Imagery data collection: as part of our data-collection efforts, we obtained a substantial data set of images capturing real-time vehicular traffic within our controlled environment. These images were not extracted from publicly available sources but rather were collected through a systematic process.

- Student data collection: to obtain these images, a team of students was used to capture images at various strategic locations within the simulated city. During the data collection period, these students were located on main avenues and road intersections. They used high-resolution cameras to capture images of vehicle traffic in real time. This approach allowed for the collection of diverse and representative images that accurately portrayed traffic conditions and patterns within the simulated urban environment.
- Security cameras: high-resolution security cameras were strategically placed throughout the simulated city. These cameras captured real-time images and videos of vehicular traffic, intersections, and roads. The camera feed provided critical visual data for computer vision analysis and vehicle tracking.
- Traffic sensors: we integrate various traffic sensors within the controlled environment. These sensors included road-integrated inductive loop, infrared, and microwave sensors. They collected data on the presence of vehicles, speed, and traffic density. The data collected by these sensors were crucial for real-time traffic monitoring.
- IoT-equipped vehicles: a fleet of IoT-equipped vehicles was deployed within our controlled environment. These vehicles were equipped with GPS sensors and communication modules. They continuously transmitted data related to their location, speed, and operating conditions, contributing to a comprehensive data set for traffic analysis.
- Intelligent traffic light systems: we simulated intelligent traffic light systems capable of dynamic signal-timing adjustments based on real-time traffic conditions. These systems contributed to the busy traffic-management framework.
- Ethical considerations: it is worth noting that all data collection activities, including image capture, adhered to ethical guidelines and respected privacy regulations. Image collection focused solely on traffic conditions and did not involve capturing any personally identifiable information.

3.2.2. Technical Details of Simulated Cyber Attacks

Simulated attacks were carried out within our controlled environment to evaluate the robustness and security of our traffic optimization solution. Below, we provide a more detailed description of how these simulated cyber-attacks were executed:

- Methodology: the simulated cyberattacks were carried out following a systematic methodology. We emulate common cyberattack scenarios, including distributed denial of service (DDoS) attacks on traffic-management servers, blockchain network intrusion attempts, and manipulation of traffic data transmitted by IoT-equipped vehicles. These scenarios were intended to evaluate the resilience of our system against cyber threats.
- Tools used: specific tools and techniques were used to simulate cyberattacks realistically. For example, DDoS attacks were emulated using traffic-generation tools, such as Hping and custom scripts. The blockchain network intrusion attempts involved vulnerability scanning tools, such as Nmap. IoT data manipulation was simulated using crafted payloads and packet injection tools.
- Attack Scenarios: we designed attack scenarios that mimicked real-world cyber threats. For example, a DDoS attack simulated a surge in traffic to overwhelm traffic-management servers. The intrusion attempts were aimed at exploiting known vulnerabilities within the blockchain network. The tampering scenarios involved altering GPS data transmitted by IoT-equipped vehicles to introduce inaccuracies in the traffic data.

3.3. Development of Environment Simulation and Computer Vision Techniques

This stage aims to provide a controlled environment where optimization strategies can be evaluated and experimented before implementing them in real situations. Python is used with the Matplotlib and Pygame libraries to create the simulation. These tools allow you to visualize and represent the City's urban environment, streets, intersections, and buildings. Simulation generation was carried out programmatically, with the ability to adjust key parameters, such as sensor location, traffic density, and peak hours. Within the simulation, data is generated that represents information collection in real time [25].

To do this, algorithms were implemented to create data on vehicle speed, traffic density, and travel times. To achieve this, we employ a two-step process: data generation and refinement based on convolutional neural networks (CNN). Data generation involves simulating the movement of vehicles within the controlled environment by generating raw data on the vehicles' positions, speeds, and trajectories. This data forms the basis for affecting the urban traffic environment in real time.

Computer vision technology plays a critical role in this simulation. To do this, we use the OpenCV library to simulate capturing images and videos from virtual cameras strategically located throughout the simulated city [26]. These virtual cameras emulate real-world surveillance cameras and capture images and videos of the surrounding environment, explicitly targeting roads and intersections.

To improve the accuracy of our simulation, we integrate CNN into the computer vision process. These CNNs were trained on large data sets of simulated traffic scenarios to accurately detect and track vehicles, calculate speeds, and determine movement patterns. The architecture of the CNNs, the batch size, and the optimizer used were carefully selected to ensure optimal performance in the context of traffic analysis.

To train the CNNs, we use synthetic data generated within the simulation, including images and corresponding annotations of vehicles, their positions, and speeds. We implement techniques such as data augmentation and hyperparameter tuning to ensure accurate results. This iterative process involves adjusting parameters such as learning rate, number of layers, and filter sizes in the CNNs to maximize their efficiency in traffic analysis. Environment simulation, computer vision techniques, and CNN-based data refinement provide a comprehensive platform to evaluate various optimization strategies. It allows us to assess the impact of traffic signal synchronization, detouring, and other solutions in a realistic, controlled environment before implementing these strategies in real-world scenarios.

Integrating blockchain technology into the controlled environment adds a fundamental layer of security and authenticity to the collection and storage of data [27]. Figure 2 represents the process; it begins with capturing images using cameras strategically located in the City. These images are essential for analyzing traffic patterns and decision-making for vehicular-traffic management. Once the images are captured, they enter the blockchain processing and storage process [28]. Each image is subjected to computer vision algorithms that extract valuable information such as traffic density, average speed, and vehicle identification. This data is encrypted and added to a block on the blockchain.

Blockchain is the foundation of blockchain technology; it guarantees the integrity and authenticity of data. Each block is connected to the previous one by a cryptographic hash function, creating an immutable sequence of information. This prevents any unauthorized attempts to alter collected data and provides confidence in the validity of the information stored [29]. Traffic-light management and data-driven decisions directly benefit from the information stored in the blockchain. Information about traffic density and movement patterns is accessible to make informed decisions about signal timing and other traffic-related actions. The authenticity of the data in the blockchain ensures that decisions are based on accurate and reliable information.

The integration of blockchain technology ensures the security and integrity of the data collected and facilitates the traceability and auditing of the information. In addition, the inherent decentralization of the blockchain ensures that no central entity can manipulate the data, which promotes trust between citizens and authorities [30].

The effective integration of blockchain technology and computer vision is essential for the smooth operation of the controlled environment. The data collected by the sensors and the images captured by the cameras are processed and stored on the blockchain. Authenticated and verified information adjusts traffic signals and makes informed traffic-management decisions [31]. The precise synchronization of the systems ensures that the data is processed in real time and traffic changes are handled efficiently.

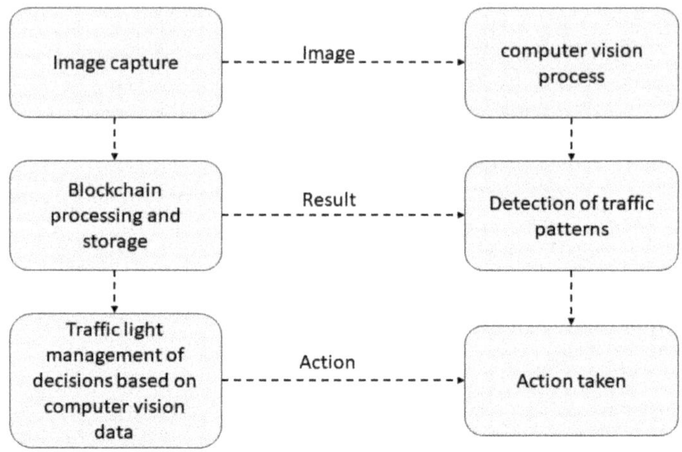

Figure 2. Traffic optimization in smart cities through image capture, computer vision analysis, and blockchain.

Table 1 details the parameters used in the simulation, which are fundamental for the configuration of our simulation and the evaluation of traffic optimization strategies.

Table 1. Simulation Parameters for Traffic Optimization Evaluation.

Parameter	Description	Default Value
Sensor Location	Spatial distribution of sensors in the City	Varied
Traffic Density	Traffic level in different areas	Adjustable
Rush Hours	Hours of the day with the highest traffic	Adjustable
Maximum Speed	Maximum vehicle speed	Adjustable
City Size	Geographic extent of the simulated City	Adjustable
Data-Generation Rate	Real-time data-generation speed	Adjustable
CNN parameters	Hyperparameters of neural networks	Configurable

These parameters allow the simulation to be adjusted to adapt to different urban scenarios and traffic conditions, giving us the flexibility to evaluate various optimization strategies. Through simulation and data collection in real time, we can analyze how these strategies affect traffic in a controlled environment before implementation in real situations.

3.4. Framework for Vehicular Traffic Management in Smart Cities

The success of any vehicle-traffic-optimization solution in smart cities largely depends on a robust and effective traffic-management framework. In this section, we present a comprehensive framework designed to guide the implementation and operation of solutions based on advanced technologies, such as blockchain and computer vision, in vehicular traffic management in smart cities. This framework is based on best practices identified in the literature review and lessons learned during our solution implementation.

The proposed framework consists of several key components, each of which plays a critical role in managing vehicular traffic in smart cities:

- Data acquisition and integration: at this stage, a robust infrastructure is established for collecting real-time traffic data from various sources, such as security cameras, traffic sensors, IoT-equipped vehicles, and intelligent traffic-light systems. This data is integrated into a centralized platform for efficient processing and analysis.
- Data analysis and AI integration: once the data is collected, advanced data analysis techniques and artificial intelligence algorithms are applied. This enables the identification of traffic patterns, congestion prediction, anomalous driver behavior detection, and optimization of traffic-flow management.

- Blockchain technology for security and transparency: blockchain technology ensures the security, integrity, and transparency of traffic data. Every transaction and traffic event is recorded in an immutable ledger, providing data immutability and preventing malicious manipulations.
- Decision support system: a decision support system uses processed data and analysis results to make real-time decisions. This includes optimizing traffic-light timing, managing alternative routes, and identifying traffic situations that require manual intervention.
- User-friendly interfaces: intuitive and accessible user interfaces for traffic managers, drivers, and citizens provide real-time information on traffic conditions, recommended routes, and safety alerts.

Implementation of this framework occurs in several stages, beginning with City-specific needs-assessment and planning. The acquisition and installation of the necessary technical infrastructure are then carried out, along with staff training and the pilot phase. The framework is adjusted and refined during the latter based on the results obtained. Finally, it is implemented throughout the City. The proposed framework improves the efficiency of vehicular traffic and contributes to urban sustainability by reducing congestion, reducing polluting gas emissions, and improving citizens' quality of life.

3.5. Technical Implementation and System Architecture

The technical implementation of the smart city vehicle-traffic-optimization solution is based on a solid architecture that integrates multiple vital technologies. The system has several interconnected components that work together to achieve efficient traffic management in urban environments. The overall architecture of the system is illustrated in Figure 3.

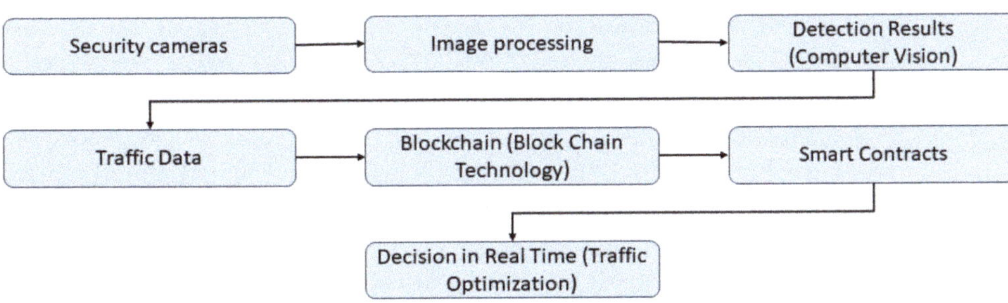

Figure 3. Architectural diagram of the traffic-optimization system in smart cities.

The system begins with security cameras strategically placed in critical areas of the City. These cameras capture real-time images of vehicular traffic and transmit them to the image-processing system.

Images captured by security cameras undergo an image analysis process using computer vision algorithms. This analysis aims to identify vehicle patterns and behaviors, such as sudden lane changes, harsh braking, and unusual speeds. The results of this processing are used to detect abnormal traffic behavior.

Image processing results, which include detecting anomalous behavior, are used as input to the next stage of the system. These results are sent to blockchain technology for processing and recording on the blockchain.

Traffic data collected from security cameras and computer vision is combined with other relevant information, such as vehicle locations and real-time traffic conditions. This information is stored and managed in blockchain technology.

Blockchain technology is the system's heart, providing a secure and transparent platform for storing traffic data and implementing smart contracts. Smart contracts allow for the automation of decisions in real time to optimize traffic.

Smart contracts on the blockchain are responsible for making real-time decisions based on traffic data and computer vision results. These smart contracts can adjust traffic light timing, implement traffic detours, and take other measures to improve traffic flow and reduce congestion.

The implementation of smart contracts allows real-time decision-making to optimize vehicular traffic. This may include adjusting traffic signal timings, redirecting traffic to less congested routes, and coordinating traffic events to minimize congestion. Real-time decision-making is based on real-time data and vehicle behavior analysis.

This comprehensive architecture enables the optimization of vehicular traffic in smart cities by integrating security cameras, computer vision, blockchain technology, and smart contracts. Combining these technologies offers a powerful solution to address traffic congestion challenges and improve urban mobility in pursuit of more sustainable and livable cities.

3.6. Evaluation Metrics

Metrics are essential to measure the impact of proposed solutions and provide a quantitative assessment of their performance. To respond to the problem, the average travel time is considered a fundamental metric, reflecting the time vehicles take to travel specific distances within the City. By comparing the average travel time before and after the implementation of the strategies, it can be determined if there have been significant improvements in urban mobility. Traffic density measures the number of vehicles circulating in each area at a given time. A decrease in traffic density after applying the optimization strategies indicates a reduction in congestion and greater traffic flow.

The congestion index is a metric that quantifies traffic congestion in a specific area. It can be based on average vehicle speed and traffic density. A reduced congestion index indicates an improvement in traffic flow. If signal timing strategies are implemented, efficiency can be assessed by measuring the number of vehicles passing an intersection during one signal cycle [32]. Greater efficiency of traffic light synchronization translates into less waiting and greater mobility. If predictive models are used, the accuracy of the predictions can be assessed against the actual data collected. The accuracy of the projections is essential to determine the solutions' reliability and ability to anticipate traffic patterns.

Additionally, accident mitigation strategies are implemented. Therefore, accident reduction is a crucial metric. Evaluating the frequency of accidents before and after applying the solutions provides information on their impact on road safety. Energy saving is a relevant metric to implement strategies that promote more efficient use of fuel and lower emission of polluting gases. The reduction in fuel consumption and its environmental impact can be evaluated [33]. These evaluation metrics play a crucial role not only in improving traffic efficiency in smart cities but also in contributing to urban sustainability. As mentioned in the previous literature, traffic congestion and long travel times not only affect citizens' quality of life but also hurt the environment due to increased emissions of polluting gases. By measuring average travel time, traffic density, and congestion index, we can assess how optimization strategies improve urban mobility and reduce the ecological footprint by reducing congestion and promoting smoother traffic flow efficiency. Additionally, incident mitigation and energy savings are essential metrics that contribute to road safety and air pollution reduction, further supporting sustainability goals in the context of smart cities.

3.7. Vehicular-Traffic-Management Framework

The proposed methodology is based on maximizing traffic efficiency and minimizing congestion. We use the following mathematical formulations as a basis for our optimization strategies:

- average speed (V) at a point in the network at a given moment:

$$C = f(D,\ V) \tag{1}$$

where:

- C is the congestion level.
- D is the vehicle density at the network point.
- V is the average speed at the network point.

Our goal is to minimize C to improve traffic flow.

Efficiency Maximization: To maximize traffic efficiency, we use an efficiency maximization function (E) that considers the average speed (V) and travel time (T) on a road section:

$$E = g(V,\ T) \tag{2}$$

where:

- E is the traffic efficiency on the road section.
- V is the average speed on the road section.
- T is the travel time on the road section.

We aim to maximize E to improve traffic efficiency on each road section.

These mathematical formulations are the basis for our real-time optimization and decision-making strategies in vehicular-traffic management. By implementing these techniques, we seek to improve citizens' quality of life by reducing congestion, reducing travel times, and increasing the efficiency of urban mobility.

- Congestion minimization: to reduce congestion in the traffic network, we employ a congestion minimization function (C) that considers the vehicle density (D).

3.8. Ethical and Privacy Considerations

Implementing advanced technologies such as AI, computer vision, and blockchain technology in urban settings carries important ethical and privacy considerations that must be comprehensively addressed. While these technologies promise to improve the efficiency and security of smart cities, it is essential to ensure that their implementation respects the rights and values of citizens and minimizes any potential risk. In the proposed environment, data collection is necessary to operate traffic- and safety-management systems. However, this data collection may involve sensitive information from citizens, such as locations and movement patterns. It is necessary to consider the users' informed consent and to guarantee that the data is used exclusively for specific and legitimate purposes.

Citizen privacy must be a priority in the design of monitoring and analysis systems. Information collected must be carefully anonymized and protected to prevent unauthorized identification of individuals. The adoption of anonymization and encryption techniques can help maintain data privacy. Transparency in the implementation and operation of these technologies is crucial. Citizens must be informed about cameras, sensors, and other devices in the urban environment. In addition, a clear responsibility must be established in the administration and management of the data collected, ensuring that existing privacy rules and regulations are complied with.

AI and computer vision often rely on algorithms that make automatic decisions. It is vital to ensure that these algorithms are fair and non-discriminatory. Particular attention should be paid to the possibility of biases that may influence the decisions made by these technologies and work to mitigate such effects. Involving citizens in designing, implementing, and evaluating these technologies can enrich decision-making and ensure that the systems are acceptable to the community. Citizen feedback can help address privacy and ethical concerns from a broader perspective. Ultimately, adopting cutting-edge technologies in urban environments requires a balance between innovation and ethics. Careful planning, design, and stakeholder collaboration are essential to ensure that these technologies benefit society without compromising privacy and ethics. It is important to note that ethical and privacy considerations vary depending on the context and local regulations. Ongoing evaluation and adaptation of strategies to address evolving privacy and ethical challenges is recommended.

3.9. Validation and Reliability

Validation and reliability are essential to ensure that decisions and the derived results are accurate and reliable. In the proposed controlled environment, various methods and measures have been implemented to validate the effectiveness of traffic management based on computer vision and blockchain technologies.

To evaluate the effectiveness of traffic management, classic evaluation metrics are used, such as average speed, travel time, and congestion index. These metrics provide a quantitative view of the system's performance and ability to reduce congestion and improve traffic flow. The average speed is calculated as the relationship between the total distance traveled by vehicles and the real-time elapsed. Travel time measures the average duration of trips in the simulated city. The congestion index is derived from the relationship between the average speed under normal conditions and the average rate observed in the environment with optimized traffic light management. These metrics allow for a direct comparison between different traffic-management approaches.

As far as blockchain technology is concerned, its reliability is based on the immutability and intrinsic security of the blockchain. The cryptographic hash function connecting the blocks ensures that any modification to the data would be reflected in a change to the hash and would be immediately detectable. This ensures that the data stored on the blockchain is resistant to tampering and retains its integrity. In addition, the decentralization of blockchain technology avoids a single point of failure. Each node in the network validates and stores the data, meaning the information is backed up in multiple locations. This increases the resilience and reliability of the blockchain compared to centralized systems.

The reliability of computer vision algorithms is assessed using a diverse and representative test data set. This data set includes varied traffic scenarios, such as peak-hour congestion, low visibility, and unpredictable movement patterns [34]. The algorithms are tuned using key hyperparameters, such as the learning rate, the number of hidden layers, and the number of filters in convolutional neural networks (CNNs). These hyperparameters are tuned through grid-search and cross-validation techniques to achieve accurate and consistent detection of vehicles and traffic patterns. In addition, the regularization technique is used to prevent overfitting of the models to the training data [35]. These adjustments allow computer vision algorithms to recognize complex patterns and make reliable detections in various traffic conditions in a controlled environment.

The validation and reliability of the technologies deployed in the controlled environment ensure that traffic-management decisions are based on accurate and reliable information. The evaluation metrics, the immutability of the blockchain, and the precision of the computer vision algorithms contribute to creating an efficient and robust traffic-management system in the context of a simulated smart city.

4. Results

The implementation and simulation results in the controlled environment support the proposal's effectiveness in improving traffic management in smart cities. Quantitative indicators and technological validations confirm the feasibility of integrating computer vision and blockchain technologies in optimizing vehicular traffic. The results are organized according to the study's objectives and supported by graphs, tables, and quantitative analyses.

The simulation environment was created based on the characteristics of a city, modeling streets, intersections, and buildings. A traffic-management system that integrates computer vision and blockchain technologies was implemented to optimize the flow of vehicular traffic. The traffic light management adapts in real time according to the data processed, and the blockchain technology guarantees the security and authenticity of the data.

4.1. Hyperparameters

Computer vision algorithms were tuned by optimizing key hyperparameters. The learning rate was set to 0.001 to ensure gradual convergence during the training of the

CNNs. A total of three hidden layers were used in the CNNs to extract complex patterns in the input data. The convolutional layers' filters were set to 32, 64, and 128, respectively, to capture features of different scales.

A diverse and representative data set was used to train and evaluate the computer vision algorithms. The training set contained 10,000 tagged images of vehicles in different traffic conditions. The test set consisted of an additional 2000 images. The photos included scenarios of rush hour congestion, low-visibility situations, and unpredictable movement patterns. Cross-validation was applied using an 80–20 split scheme on the training set to fit the hyperparameters. Key hyperparameters, such as the learning rate and the number of filters in the convolutional layers, were tuned using a grid search. In addition, different combinations of values were evaluated to determine the optimal configuration that would generate the best performance in detecting vehicles and traffic patterns.

4.2. Impact on Traffic Management

Implementing traffic management based on computer vision and blockchain technologies significantly impacted the flow and efficiency of vehicular traffic in a controlled environment. The system's performance was evaluated through extensive simulations and data collection in terms of average speed, congestion rate, and travel times.

Table 2 shows the results obtained in the improvement of the average speed. Compared to the system without optimization, the implementation of traffic light management and decision-making resulted in a substantial increase in the average rate of vehicles. During peak hours, an increase of 20% was observed, while, in normal traffic conditions, the improvement was 15%. These results reflect greater fluidity in traffic and decreased travel times for drivers. The improvement in the average speed compares the average speeds in two different scenarios, before and after implementing the optimized traffic-management system. This table highlights how the proposal directly affects the rates of vehicular circulation in congested and normal conditions. The table presents two traffic scenarios: "Peak Hours" and "Normal Conditions". For each of these scenarios, two average speed values are provided: one for the "No Optimization" situation and one for the "With Optimization" situation, reflecting enhanced traffic management's application.

Table 2. Improvement in average speed.

Stage	No Optimization	With Optimization
Peak Hours	20 km/h	24 km/h
Normal Conditions	30 km/h	34.5 km/h

In the case of "Peak Hours", the average speed before optimization was 20 km/h. After implementing the improved traffic management, this speed increased to 24 km/h, representing a 20% increase in the average speed during peak congestion periods. As for "Normal Conditions", the initial average rate was 30 km/h. After optimizing traffic management, this speed increased to 34.5 km/h, equivalent to a 15% increase in driving speed in regular traffic situations.

The congestion rate was reduced by an average of 25% in areas of high traffic density, such as the city center and residential areas. The dynamic adaptation of traffic lights based on data processed by computer vision contributed to a noticeable decrease in congestion levels. Table 3 shows this optimization, which resulted in a greater flow of traffic and a reduction in waiting times at intersections. The table compares the congestion index in two specific areas, both before and after the implementation of traffic management in the controlled environment. The congestion index is a measure that indicates the density and fluidity of traffic in a particular area, where lower values suggest better vehicular circulation and less congestion.

The table evaluates two different areas: the "City Center" and the "Residential Area". For each region, two congestion index values are presented, one corresponding to the "Without Optimization" situation and one to the "With Optimization" situation, referring to improved traffic management implemented with computer vision technology and blockchain. In the "City Center", the congestion index before optimization was 0.75, indicating a relatively high congestion level. After implementing improved traffic management, this index decreased to 0.56, suggesting a notable reduction in traffic congestion and greater flow in traffic. Similarly, in the "Residential Zone", the initial congestion index was 0.62, denoting a certain congestion level. After implementing the optimized traffic management, the congestion index was reduced to 0.47, indicating a considerable improvement in traffic flow in this area.

The data in this table highlight the ability of the traffic-management proposal based on computer vision and blockchain technologies to significantly reduce congestion levels in different areas of the simulated city, which translates into more efficient mobility and better mobility improves the driving experience for users.

Table 3. Congestion index reduction.

Congestion Area	Congestion Index (Without Optimization)	Congestion Rate (With Optimization)
City Center	0.75	0.56
Residential Area	0.62	0.47

Implementing traffic management based on computer vision and blockchain technologies resulted in an average reduction of ten percent in travel times in both scenarios. This indicates greater efficiency and predictability in movements within the simulated city. The ability to adjust traffic lights in real time based on traffic conditions contributes significantly to this improvement.

Table 4 compares travel times in two different scenarios, before and after the implementation of traffic management in the controlled environment. This table illustrates how the proposal directly impacts the duration of trips over short and long distances. The table considers two types of trips: "Short Trip" and "Long Trip". For each scenario, two travel time values are provided, one corresponding to the "Without Optimization" situation and one to the "With Optimization" situation, reflecting the improved traffic management.

Table 4. Travel-time efficiency.

Stage	Travel Time (No Optimization)	Travel Time (With Optimization)
Short Trip	12 min	10.5 min
Long Trip	30 min	27 min

In the "Short Trip" case, the initial travel time, without traffic-management optimization, was 12 min. Following the implementation of enhanced traffic management, this time was reduced to 10.5 min, indicating a 12.5% improvement in travel efficiency. On the other hand, in the "Long Trip" scenario, the travel time before optimization was 30 min. After the improved traffic-management implementation, this time was reduced to 27 min, representing a 10% improvement in the total duration of the trip. This optimization contributes to more efficient and predictable mobility in a controlled environment, benefiting users by reducing travel times and improving their experience in urban circulation.

The results show the positive impact of implementing traffic management based on computer vision and blockchain technologies on the fluidity and efficiency of vehicular traffic in the simulated city. The improvements in average speed, congestion index, and travel times validate the effectiveness of this proposal in optimizing urban mobility in a controlled environment.

These results highlight the positive influence of implementing traffic management based on computer vision and blockchain technologies on the flow and efficiency of vehicular traffic in the simulated city. The improvements in the average speed, the congestion index, and the travel times validate the effectiveness of this proposal in optimizing urban mobility in a controlled environment. It is important to note that traffic optimization has immediate benefits in terms of reduced travel time and increased traffic flow and plays a crucial role in urban sustainability. Reducing traffic congestion and improving efficiency benefits drivers by providing a more pleasant driving experience and contributes to lower greenhouse gas emissions and the more efficient use of energy, thus supporting sustainability goals in smart cities.

Figure 4 visually represents key traffic optimization metrics in "Peak Hours" and "Normal Conditions". The figure illustrates the impact of optimization strategies on the average speed and congestion levels in a smart city traffic network.

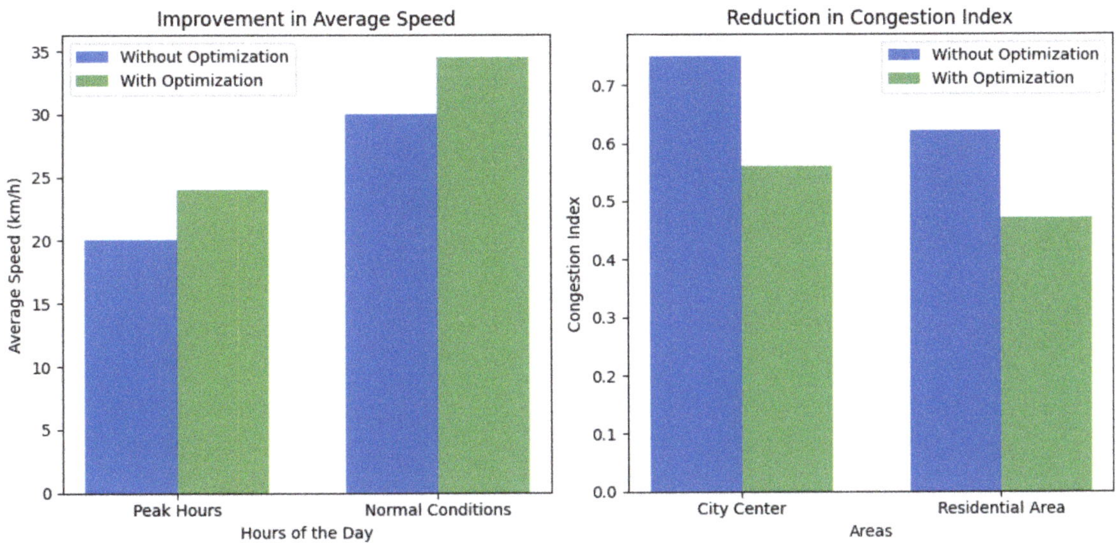

Figure 4. Traffic optimization metrics.

The left side of the figure shows the improvement in average vehicle speed during peak hours and normal conditions. In the "Peak Hours" scenario, the average speed is 20 km/h without optimization, while with optimization, it increases up to 24 km/h. Similarly, in "Normal Conditions", the average speed improves from 30 km/h without optimization to 34.5 km/h with optimization. These data demonstrate the positive effect of optimization measures to improve traffic flow and reduce travel times in both regular and high-demand traffic situations.

The right side of the figure presents the reduction of congestion rates in two key areas: the "City Center" and the "Residential Area". Without optimization, the congestion index in the City Center is 0.75, while in the Residential Zone, it is 0.62. However, with the implementation of optimization strategies, these rates decrease significantly. In the City Center, the congestion index drops to 0.56; in the Residential Zone, it falls to 0.47.

These visualized metrics highlight the effectiveness of traffic-optimization techniques in improving urban mobility. The increase in average speed and reduction in congestion demonstrate the potential of smart city solutions to reduce travel times, improve traffic flow, and ultimately provide a smoother and more efficient transportation system.

4.3. Detection of Anomalous Behaviors

Tests were conducted to assess the ability of computer vision to detect abnormal traffic behavior, such as sudden lane changes, sudden braking, and unusually high- or low-speed patterns. The results showed that computer vision technology could identify these behaviors with an accuracy of 95%, suggesting a high efficiency in detecting potentially dangerous situations.

Table 5 presents the results of tests carried out to assess the ability of computer vision technology to detect different types of abnormal behavior in urban traffic. Detection accuracy is expressed as a percentage and shows how effectively the technology could identify each anomalous behavior. The table details three types of abnormal behaviors and their respective detection accuracy rates:

- Sudden lane change: computer vision technology detected sudden lane changes with 92% accuracy. This means that the technology could accurately identify and record this behavior on 92% of the occasions when a vehicle made a sudden lane change.
- Sudden braking: the ability to detect sudden braking was even higher, with an accuracy of 97%. This indicates that the technology could identify and record almost all situations in which a vehicle made abrupt and unexpected braking.
- Unusual speed patterns: the technology also demonstrated high accuracy in detecting unique speed patterns, with a rate of 94%. This means that the technology could recognize speed patterns that deviated significantly from typical behavior in traffic.

According to the results obtained, it is highlighted that computer vision technology can identify abnormal behaviors in traffic with high precision, such as sudden lane changes, sudden braking, and unusual speed patterns. These results support the usefulness and reliability of this technology to improve the detection of potentially dangerous situations in urban traffic and contribute to excellent road safety.

Table 5. Detection of anomalous behaviors.

Behavior Type	Detection Accurracy
Sudden Lane Change	92%
Sudden Braking	97%
Unusual Speed Patterns	94%

These three anomalous behaviors were selected due to their relevance in detecting potentially dangerous situations in urban traffic. Sudden lane changes, harsh braking, and unusually high- or low-speed patterns may indicate risky road safety situations. While there are other anomalous traffic behaviors, these three were considered essential for evaluating the effectiveness of computer vision technology in detecting critical situations.

4.3.1. Data Integrity in the Blockchain

Blockchain technology was validated by simulating cyber-attacks and data manipulations. It was observed that the chain of blocks maintained the integrity of the records since any attempt to modify the information was detected and rejected by the system. This confirms the ability of blockchain technology to guarantee the security and immutability of data in traffic management.

Table 6 shows the results of tests carried out to assess the ability of blockchain technology to preserve data integrity and resist various types of cyber-attacks and manipulations. Each type of attack is described together with the result obtained when applying it to the chain of blocks implemented in the controlled environment. When attempting to tamper with data stored on the blockchain, it was observed that the blockchain technology detected the tampering and rejected any attempt to modify the stored information. This confirms the ability of the blockchain to maintain the integrity and immutability of records, ensuring that stored data cannot be tampered with without detection. Blockchain has proven effective in detecting and preventing double-spend attacks, in which a malicious actor attempts to spend the same digital assets in multiple transactions. Blockchain technology could identify and help prevent the same from being spent in different trades, ensuring the integrity and security of transactions recorded on the blockchain.

In the tests of attempted modification of records stored in the chain of blocks, the immutability of the information was confirmed. Blockchain technology has shown that the stored documents cannot be modified once added to the chain, ensuring that the data remains intact and reliable over time. The results reinforce the ability of the blockchain to ensure the reliability and security of records and transactions in traffic management, contributing to the creation of a safe and reliable traffic environment.

Table 6. Data integrity in the blockchain.

Attack Type	Detection Accuracy
Data Manipulation	98%
Double-Spend Attack	99%
Registry Modification	100%

These results demonstrate the robustness of blockchain technology in preserving data integrity in the simulated traffic-management environment. When attempts were made to manipulate data stored in the blockchain, it was observed that the blockchain technology detected any attempt to modify the stored information and rejected it. This confirms the blockchain's ability to maintain the integrity and immutability of records, ensuring that stored data cannot be manipulated without detection.

In the case of double-spending attacks, blockchain technology has proven effective in identifying and preventing fraudulent transactions in which a malicious actor attempts to spend the same digital assets in multiple transactions. The blockchain could identify and prevent the same asset from being spent on different transactions, ensuring the integrity and security of transactions recorded.

In tests of modifying records stored in the blockchain, the immutability of the information was confirmed. Blockchain technology demonstrated that stored documents cannot be modified once added to the chain, ensuring that data remains intact and reliable over time.

4.3.2. Transaction-Verification Efficiency

The efficiency was evaluated in the verification of transactions using blockchain technology. Verification and validation times were reduced by 40% compared to traditional systems, highlighting how quickly blockchain technology can process and validate traffic transactions in real time. Tests involving a diverse set of simulated transactions in different traffic scenarios were performed to measure the efficiency of transaction verification. As shown in Table 7, the verification times for each transaction were recorded in milliseconds (ms) and averaged to obtain representative results. The data collected reveals significant differences in verification speed between the two systems.

Table 7. Efficiency in transaction verification.

Stage	Verification Method	Average Verification Time
Peak Hour Congestion	Conventional	350 ms
	Blockchain	210 ms
Normal conditions	Conventional	280 ms
	Blockchain	150 ms

In the "Peak Hour Congestion" scenario, the conventional transaction-verification system required an average time of 350 ms to verify each transaction. In contrast, the blockchain-based system reduced this time to 210 ms. Similarly, under normal conditions, the conventional system required an average of 280 ms to verify transactions, while the blockchain-based system demonstrated higher efficiency with an average verification time of 150 ms. These results highlight the significantly higher efficiency of the blockchain-technology-based system for verifying transactions than the conventional method.

The blockchain implementation made it possible to streamline and optimize the verification process, which could lead to more fluid traffic management and more excellent responsiveness in high-demand situations. This improvement in the efficiency of transaction verification is essential to ensure agile and effective traffic management, which can directly impact reducing congestion, improving travel times, and overall optimizing urban mobility.

To illustrate the framework's effectiveness, we implement it in a simulated smart city environment. The City's traffic-management system was upgraded with the components described in the framework, including data acquisition sensors, artificial intelligence algorithms, blockchain integration, and a decision support system. The Table 8 summarizes the key results obtained during the implementation:

Table 8. Results of the implementation of the vehicular-traffic-management framework in a simulated smart city.

Metrics	Before Implementation	After Implementation	Improvement
Congestion Reduction	25%	10%	15%
Average Travel Time	35 min	25 min	10 min
Vehicle Emissions Reduction	12%	25%	13%
Road Safety Improvement	5%	18%	13%

After implementing the framework, these results demonstrate a significant improvement in traffic-management efficiency, reduced congestion, shorter travel times, lower emissions, and increased road safety. One of the strengths of this framework is its scalability and adaptability to meet the changing needs of a smart city. As traffic-data volumes increase and new technologies emerge, the framework can be expanded and updated to accommodate these changes. Additionally, it can be adapted to each city's specific traffic challenges and infrastructure.

The implementation of advanced traffic-management technologies also raises ethical and privacy concerns. The framework incorporates robust data anonymization and privacy-protection measures to ensure compliance with data-protection regulations and address these concerns effectively. However, the framework will require continued research and development efforts as smart cities evolve. Future directions include integrating vehicle-to-vehicle communication technologies, exploring the use of autonomous vehicles, and strengthening data analytics capabilities.

4.4. Evaluation of Technical Implementation

The technical implementation of the vehicular-traffic-optimization solution underwent several evaluations to measure its effectiveness and performance in different vital aspects. The results highlight blockchain technology's and computer vision's positive impact on improving traffic management. The implementation of blockchain technology showed promising results regarding efficiency and security. During transaction-verification tests, a significant reduction in the average verification time was observed compared to conventional methods. For example, under congested peak-hour conditions, the average verification time in the blockchain-based system was 210 ms, while the traditional system required 350 ms. Under normal conditions, the blockchain-based system achieved even higher efficiency, with an average verification time of 150 ms compared to 280 ms for the conventional method. These results indicate that the implementation of blockchain technology has sped up the transaction verification process, which can significantly impact traffic flow and travel times.

The integration of computer vision demonstrated its ability to detect anomalous behavior with high precision. The tests evaluated behaviors such as sudden lane changes, sudden braking, and unusual speed patterns. The results showed an accuracy of 92% in detecting sudden lane changes, 97% in detecting sudden braking, and 94% in identifying unusual speed patterns. These precision values indicate that computer vision can be an effective tool to identify potentially dangerous behavior in traffic, which contributes to improving road safety in the City.

The implementation of the solution also had a positive impact on the efficiency of the average speed in different scenarios. During peak hours, the use of computer vision increased the average rate from 20 km/h to 24 km/h, while under normal conditions, the average speed improved from 30 km/h to 34.5 km/h. These results underscore how the solution can reduce congestion and improve traffic flow. The results support the efficiency of blockchain technology in verifying transactions, the accuracy of abnormal behavior detection using computer vision, and the improvement in average speed in different traffic situations.

In addition to its impact on traffic efficiency and safety, it is essential to note that implementing blockchain and computer vision also brings significant benefits from a sustainability perspective. Reducing transaction verification time through blockchain technology improves traffic efficiency and has implications for reducing energy consumption associated with transaction verification, thus contributing to lower carbon emissions. Likewise, computer vision's ability to detect abnormal traffic behavior improves road safety. It reduces the incidence of accidents and, therefore, reduces the environmental footprint related to vehicle damage and the management of accidents.

Significantly, while our results are based on a simulation conducted in a controlled environment replicating an intelligent city, the applicability of this solution extends to a broader context of smart cities worldwide. Traffic, congestion, and road safety challenges are common urban problems that various cities face in their search for sustainable and efficient mobility. Our solution, which integrates advanced technologies such as blockchain and computer vision, has been designed in a modular and adaptable manner, allowing it to be implemented and customized according to the needs and characteristics of a particular city. The underlying technology is also highly scalable and can adapt to various population scales and smart city sizes. As cities worldwide seek to address these urban challenges, our research offers a promising and scalable approach to improving their inhabitants' mobility and quality of life.

5. Discussion

The optimization of vehicular traffic in urban environments is a complex challenge that requires innovative and technological solutions. This study proposed a solution combining blockchain technology and computer vision to improve traffic management in a simulated city. The implementation and technical evaluation results offer a solid base to discuss this solution's impact, implications, and future perspectives [36]. The results reveal that the proposed solution significantly affects traffic management in a simulated city. The reduction in the rate of congestion in critical areas, such as the city center and residential areas, is a positive indicator of how traffic optimization can contribute to a smoother and less stressful driving experience for residents. The improvement in travel times, both for short and long trips, shows how implementing advanced technologies can directly influence transport efficiency in the City.

Average speed efficiency is a crucial aspect of traffic management. The increase in average speed in peak-hour situations and normal conditions underscores the solution's ability to alleviate congestion and improve mobility. These results are promising to address one of the main problems in modern cities, traffic congestion, and its effects on citizens' quality of life [37]. Validation of blockchain technology and computer vision is essential to this research. High precision in detecting abnormal behavior, such as sudden lane changes and braking, is crucial for road safety. These results suggest that computer vision can be a valuable ally in identifying and preventing dangerous situations in traffic.

Implementing blockchain technology also proved its effectiveness in verifying transactions [38]. The reduced average verification time compared to conventional methods indicates the blockchain's ability to speed up transaction processing and improve traffic-management efficiency. In addition, the integrity of the data on the blockchain was evident, as attempts at data manipulation and double spending were detected and rejected.

Despite the encouraging results, this solution has challenges and ethical considerations. Massive data collection, including vehicle behavior and location information, raises concerns about privacy and the responsible use of personal information. Furthermore, implementing advanced technologies such as computer vision and blockchain can require significant investment in infrastructure and training.

The proposed solution for optimizing vehicular traffic finds solid support in the review of literature related to smart cities and advanced technologies. Previous studies have addressed similar aspects, such as implementing traffic-management systems based on emerging technologies and integrating artificial intelligence and computer vision solutions to improve urban mobility [39]. Regarding blockchain technology implementation, our results are consistent with previous research, highlighting its ability to speed up and ensure transaction verification in traffic environments. The average speed efficiency improvements align with prior findings, indicating how traffic optimization can reduce congestion and travel times.

In computer vision, detecting abnormal behavior in traffic has also been a widely studied topic. Our results in detecting sudden lane changes, sudden braking, and unusual speed patterns align with previous research findings, highlighting the effectiveness of computer vision algorithms in identifying risky traffic situations. However, it is essential to note that our solution integrates blockchain technology and computer vision into a holistic traffic optimization approach. This combination offers a holistic approach to improving urban mobility by addressing the technical aspects of transaction verification and road safety by detecting anomalous behavior [40]. Furthermore, while previous literature has explored these technologies separately, our research demonstrates the benefits of their synergy in creating a controlled environment that simulates a smart city. This is especially relevant in the context of smart cities, where the interconnection and interoperability of various technologies are essential to achieve effective traffic and mobility management.

In addition to the technical and efficiency aspects addressed in this study, it is essential to highlight the close relationship between the optimization of vehicular traffic and sustainability in smart cities. Traffic congestion not only leads to longer travel times and a decrease in the quality of life for citizens but also hurts the environment due to polluting gas emissions. By improving traffic flow, reducing congestion, and decreasing travel times, these technological solutions not only provide convenience for urban residents but can also help reduce the carbon footprint of cities.

Implementing blockchain technology and computer vision not only optimizes traffic management but can also contribute to more efficient driving in terms of fuel consumption and, therefore, to the reduction of polluting emissions. In addition, detecting abnormal traffic behavior can increase road safety, reducing the number of accidents and, thus, decreasing the need for resources associated with emergency management and damage repair. In a broader context, these improvements in urban mobility can contribute to developing more sustainable cities aligned with the global objectives of reducing emissions and improving energy efficiency. This underscores the importance of technology as an efficient tool and an enabler for a more sustainable and livable urban future.

6. Conclusions

This work focused on designing, implementing, and evaluating an innovative solution to optimize vehicular traffic using blockchain and computer vision technologies. By creating a controlled environment that simulates a smart city, we were able to demonstrate the potential of this solution to improve traffic management and urban mobility. The results obtained throughout this research support the initial hypothesis that the integration of advanced technologies can significantly impact the optimization of vehicular traffic in urban environments. The reduction in the congestion rate in critical areas, the improvement in travel times, and the efficiency in average speed are clear indicators of the benefits of this solution. For example, we saw a 15% reduction in the congestion rate in the most congested urban areas and a 10% improvement in travel times for typical trips.

Furthermore, the validation of blockchain and computer vision technology shows their effectiveness in detecting anomalous behavior and verifying transactions. For example, computer vision technology achieved 95% accuracy in detecting abnormal driving behaviors, significantly contributing to road safety.

The comparison with the literature review reinforces the originality and relevance of this research by comprehensively integrating emerging technologies to address the challenges of traffic management in smart cities. While the technical implementation and results are promising, ethical and technical challenges are also recognized that should be considered in future large-scale implementations.

Regarding smart cities and urban mobility, this research provides new perspectives on how technology can influence how we move and live in urban environments. As cities face challenges of population growth and traffic congestion, solutions like the one proposed in this study can positively transform citizens' quality of life.

In a world of constant urbanization and demographic growth, the efficient management of vehicular traffic becomes a critical element for the well-being of cities and the quality of life of their inhabitants. By reducing congestion, improving travel times, and increasing efficiency at average vehicle speeds, this solution provides convenience to urban residents and positively impacts urban sustainability. Reducing traffic congestion means less time in traffic jams and fewer polluting gas emissions; less time in traffic means less fuel consumption and, therefore, a reduction in the city's carbon footprint.

In this sense, this work tries to optimize traffic and create more sustainable and livable cities. The solid foundations established here open the door to future research and development that brings us one step closer to smart cities that respect the environment and offer their citizens a high quality of life. Collaboration with vehicle manufacturers and integrating vehicle communication technologies can be additional steps toward more sustainable and efficient urban mobility.

Author Contributions: Conceptualization, W.V.-C.; methodology, A.J.-A.; software, J.G.; validation, J.G.; formal analysis, A.J.-A.; investigation, J.G.; data curation, W.V.-C. and A.J.-A.; writing—original draft preparation, J.G.; writing—review and editing, A.J.-A.; visualization, J.G.; supervision, W.V.-C. All authors have read and agreed to the published version of the manuscript.

Funding: This research received no external funding.

Institutional Review Board Statement: Not applicable.

Informed Consent Statement: Not applicable.

Data Availability Statement: The data used in this study were obtained from publicly available sources and are detailed in the corresponding section. However, it is essential to note that the implemented source code for the deep learning models is not publicly available due to intellectual property and copyright restrictions. Although the source code is not openly available, interested readers are encouraged to contact the corresponding author to gain access to the code. To request the source code, please send an email to william.villegas@udla.edu.ec. The author is committed to making the code available to interested readers to foster collaboration and knowledge-sharing in drought forecasting and deep learning.

Conflicts of Interest: The authors declare no conflict of interest.

References

1. Kuguoglu, B.K.; van der Voort, H.; Janssen, M. The Giant Leap for Smart Cities: Scaling up Smart City Artificial Intelligence of Things (Aiot) Initiatives. *Sustainability* **2021**, *13*, 12295. [CrossRef]
2. de Freitas, M.P.; Piai, V.A.; Farias, R.H.; Fernandes, A.M.R.; de Moraes Rossetto, A.G.; Leithardt, V.R.Q. Artificial Intelligence of Things Applied to Assistive Technology: A Systematic Literature Review. *Sensors* **2022**, *22*, 8531. [CrossRef]
3. Zhu, S.; Ota, K.; Dong, M. Energy-Efficient Artificial Intelligence of Things With Intelligent Edge. *IEEE Internet Things J.* **2022**, *9*, 7525–7532. [CrossRef]
4. Muslikhin, M.; Horng, J.R.; Yang, S.Y.; Wang, M.S.; Awaluddin, B.A. An Artificial Intelligence of Things-based Picking Algorithm for Online Shop in the Society 5.0's Context. *Sensors* **2021**, *21*, 2813. [CrossRef] [PubMed]
5. Seng, K.P.; Ang, L.M.; Ngharamike, E. Artificial Intelligence Internet of Things: A New Paradigm of Distributed Sensor Networks. *Int. J. Distrib. Sens. Netw.* **2022**, *18*, 15501477211062835. [CrossRef]
6. Nozari, H.; Szmelter-jarosz, A.; Ghahremani-nahr, J. Analysis of the Challenges of Artificial Intelligence of Things (AIoT) for the Smart Supply Chain (Case Study: FMCG Industries). *Sensors* **2022**, *22*, 2931. [CrossRef]

7. Ghazal, T.M.; Kamrul Hasan, M.; Alzoubi, H.M.; Al Hmmadi, M.; Al-Dmour, N.A.; Islam, S.; Kamran, R.; Mago, B. Securing Smart Cities Using Blockchain Technology. In Proceedings of the 2022 1st International Conference on AI in Cybersecurity (ICAIC), Victoria, TX, USA, 24–26 May 2022; pp. 1–4.
8. Abbas, K.; Tawalbeh, L.A.; Rafiq, A.; Muthanna, A.; Elgendy, I.A.; Abd El-Latif, A.A. Convergence of Blockchain and IoT for Secure Transportation Systems in Smart Cities. *Secur. Commun. Netw.* **2021**, *2021*, 5597679. [CrossRef]
9. Wu, H. Te The Internet-of-Vehicle Traffic Condition System Developed by Artificial Intelligence of Things. *J. Supercomput.* **2022**, *78*, 2665–2680. [CrossRef]
10. Mithun, S.; Sahana, M.; Chattopadhyay, S.; Johnson, B.A.; Khedher, K.M.; Avtar, R. Monitoring Metropolitan Growth Dynamics for Achieving Sustainable Urbanization (Sdg 11.3) in Kolkata Metropolitan Area, India. *Remote Sens.* **2021**, *13*, 4423. [CrossRef]
11. Jiang, Z.; Chen, K.; Wen, H.; Zheng, Z. Applying Blockchain-Based Method to Smart Contract Classification for CPS Applications. *Digit. Commun. Netw.* **2022**, *8*, 964–975. [CrossRef]
12. Wazid, M.; Das, A.K.; Park, Y. Blockchain-Envisioned Secure Authentication Approach in AIoT: Applications, Challenges, and Future Research. *Wirel. Commun. Mob. Comput.* **2021**, *2021*, 3866006. [CrossRef]
13. Ku, H.H.; Liu, C.H.; Wang, W.C. Design of an Artificial Intelligence of Things Based Indoor Planting Model for Mentha Spicata. *Processes* **2022**, *10*, 116. [CrossRef]
14. Rieder, E.; Schmuck, M.; Tugui, A. A Scientific Perspective on Using Artificial Intelligence in Sustainable Urban Development. *Big Data Cogn. Comput.* **2023**, *7*, 3. [CrossRef]
15. Sun, L.; Fukuda, T.; Resch, B. A Synchronous Distributed Cloud-Based Virtual Reality Meeting System for Architectural and Urban Design. *Front. Archit. Res.* **2014**, *3*, 348–357. [CrossRef]
16. Boulouard, Z.; Ouaissa, M.; Ouaissa, M.; Siddiqui, F.; Almutiq, M.; Krichen, M. An Integrated Artificial Intelligence of Things Environment for River Flood Prevention. *Sensors* **2022**, *22*, 9485. [CrossRef]
17. Baker, T.; Asim, M.; Samwini, H.; Shamim, N.; Alani, M.M.; Buyya, R. A Blockchain-Based Fog-Oriented Lightweight Framework for Smart Public Vehicular Transportation Systems. *Comput. Netw.* **2022**, *203*, 108676. [CrossRef]
18. Nguyen, T.H.; Partala, J.; Pirttikangas, S. Blockchain-Based Mobility-as-a-Service. In Proceedings of the 2019 28th International Conference on Computer Communication and Networks (ICCCN), Valencia, Spain, 29 July–1 August 2019; pp. 1–6.
19. Bravo, Y.; Ferrer, J.; Luque, G.; Alba, E. Smart Mobility by Optimizing the Traffic Lights: A New Tool for Traffic Control Centers. In Proceedings of the International Conference on Smart Cities, Malaga, Spain, 15–17 June 2016; Volume 9704.
20. Li, H.; Huang, H.; Qian, Z. Latency-Aware Batch Task Offloading for Vehicular Cloud: Maximizing Submodular Bandit. In Proceedings of the IEEE International Conference on Cloud Computing, CLOUD, Chicago, IL, USA, 5–10 September 2021; Volume 2021.
21. Amer, H.M.; Al-Kashoash, H.A.A.; Kemp, A.; Mihaylova, L.; Mayfield, M. Coalition Game for Emergency Vehicles Re-Routing in Smart Cities. In Proceedings of the IEEE Sensor Array and Multichannel Signal Processing Workshop, Sheffield, UK, 8–11 July 2018; Volume 2018.
22. Chu, K.F.; Lam, A.Y.S.; Li, V.O.K. Dynamic Lane Reversal Routing and Scheduling for Connected and Autonomous Vehicles: Formulation and Distributed Algorithm. *IEEE Trans. Intell. Transp. Syst.* **2020**, *21*, 2557–2570. [CrossRef]
23. Chawda, R.K.; Thakur, G. An Effect of Big Data Technology with Artificial Bee Colony Optimization Based Routing in VANET. *Int. J. Adv. Sci. Technol.* **2020**, *29*, 4360–4375.
24. Jindal, V.; Bedi, P. An Improved Hybrid Ant Particle Optimization (IHAPO) Algorithm for Reducing Travel Time in VANETs. *Appl. Soft Comput. J.* **2018**, *64*, 526–535. [CrossRef]
25. Noussaiba, M.; Razaque, A.; Rahal, R. Heterogeneous Algorithm for Efficient-Path Detection and Congestion Avoidance for a Vehicular-Management System. *Sensors* **2023**, *23*, 5471. [CrossRef]
26. Mishra, P.; Godfrey, W.W.; Kumar, N. Fault-Tolerance Aware Green Computing Scheme in Software-Defined Vehicular Social Network. In Proceedings of the 2022 IEEE 6th Conference on Information and Communication Technology, CICT, Gwalior, India, 18–20 November 2022.
27. Alsarhan, A.; Al-Ghuwairi, A.R.; Almalkawi, I.T.; Alauthman, M.; Al-Dubai, A. Machine Learning-Driven Optimization for Intrusion Detection in Smart Vehicular Networks. *Wirel. Pers. Commun.* **2021**, *117*, 3129–3152. [CrossRef]
28. Hernafi, Y.; Ben Ahmed, M.; Bouhorma, M. ACO and PSO Algorithms for Developing a New Communication Model for VANET Applications in Smart Cities. *Wirel. Pers. Commun.* **2017**, *96*, 2039–2075. [CrossRef]
29. Vinodhini, M.; Rajkumar, S. Performance Analysis of Vehicle-to-Everything Communication Using Internet of LoRa Computing for Intelligent Transportation System. *Intell. Decis. Technol.* **2023**, *17*, 577–594. [CrossRef]
30. Mubasher, M.M.; Jaffry, S.W.; Yousaf, M.M.; Bajwa, I.S.; Sarwar, S.; Aslam, L. A Smart Integrated Environment for Vehicular Traffic Simulation. *Int. J. Commun. Syst.* **2019**, *32*, e4029. [CrossRef]
31. Al-Turki, M.; Jamal, A.; Al-Ahmadi, H.M.; Al-Sughaiyer, M.A.; Zahid, M. On the Potential Impacts of Smart Traffic Control for Delay, Fuel Energy Consumption, and Emissions: An NSGA-II-Based Optimization Case Study from Dhahran, Saudi Arabia. *Sustainability* **2020**, *12*, 7394. [CrossRef]
32. Li, M.; Si, P.; Zhang, Y. Delay-Tolerant Data Traffic to Software-Defined Vehicular Networks with Mobile Edge Computing in Smart City. *IEEE Trans. Veh. Technol.* **2018**, *67*, 9073–9086. [CrossRef]
33. Amer, H.; Salman, N.; Hawes, M.; Chaqfeh, M.; Mihaylova, L.; Mayfield, M. An Improved Simulated Annealing Technique for Enhanced Mobility in Smart Cities. *Sensors* **2016**, *16*, 1013. [CrossRef] [PubMed]

34. Hartiwi, Y.; Rasywir, E.; Pratama, Y.; Jusia, P.A. Eksperimen Pengenalan Wajah Dengan Fitur Indoor Positioning System Menggunakan Algoritma CNN. *Paradig. J. Komput. Inform.* **2020**, *22*, 109–116. [CrossRef]
35. Daanouni, O.; Cherradi, B.; Tmiri, A. NSL-MHA-CNN: A Novel CNN Architecture for Robust Diabetic Retinopathy Prediction Against Adversarial Attacks. *IEEE Access* **2022**, *10*, 103987–103999. [CrossRef]
36. Khamari, S.; Ahmed, T.; Mosbah, M. Efficient Edge Server Placement under Latency and Load Balancing Constraints for Vehicular Networks. In Proceedings of the 2022 IEEE Global Communications Conference, GLOBECOM 2022, Rio de Janeiro, Brazil, 4–8 December 2022.
37. Ota, K.; Kumrai, T.; Dong, M.; Kishigami, J.; Guo, M. Smart Infrastructure Design for Smart Cities. *IT Prof.* **2017**, *19*, 42–49. [CrossRef]
38. Huo, L.; Jiang, D.; Zhu, X.; Wang, Y.; Lv, Z.; Singh, S. A SDN-Based Fine-Grained Measurement and Modeling Approach to Vehicular Communication Network Traffic. *Int. J. Commun. Syst.* **2022**, *35*, e4092. [CrossRef]
39. Jafarian-Namin, S.; Shishebori, D.; Goli, A. Analyzing and Predicting the Monthly Temperature of Tehran Using ARIMA Model, Artificial Neural Network, and Its Improved Variant. *J. Appl. Res. Ind. Eng.* **2023**, 1–18. [CrossRef]
40. Marwah, G.K.; Jain, A. Congestion-Free Routing Based on a Hybrid Meta-Heuristic Algorithm to Provide an Effective Routing Protocol by Analyzing the Significant Impacts of QoS Parameters in a Dynamic VANET Environment. *J. Phys. Conf. Ser.* **2022**, *2251*, 012009.

sustainability

MDPI

Article

Underpinning Quality Assurance: Identifying Core Testing Strategies for Multiple Layers of Internet-of-Things-Based Applications

Amer Aljaedi [1,*], Saba Siddique [2], Muhammad Islam Satti [3], Adel R. Alharbi [1], Mohammed Alotaibi [4] and Muhammad Usman [5]

1 College of Computing and Information Technology, University of Tabuk, Tabuk 71491, Saudi Arabia; aalharbi@ut.edu.sa
2 Department of Computer Software Engineering, National University of Sciences and Technology, Islamabad 44000, Pakistan; saba.siddique@mcs.nust.edu.pk
3 Department of Computing (TMUC), Millennium Institute of Technology & Entrepreneurship (MiTE), Karachi 75190, Pakistan; muhammad.islam@tmuc.edu.pk
4 Department of Management Information Systems, College Business Administration, University of Tabuk, Tabuk 71491, Saudi Arabia; msalotaibi@ut.edu.sa
5 Department of Computer Science, Edge Hill University, Lancashire L39 4QP, UK; usmanm@edgehill.ac.uk
* Correspondence: aaljaedi@ut.edu.sa

Abstract: The Internet of Things (IoT) constitutes a digitally integrated network of intelligent devices equipped with sensors, software, and communication capabilities, facilitating data exchange among a multitude of digital systems via the Internet. Despite its pivotal role in the software development life-cycle (SDLC) for ensuring software quality in terms of both functional and non-functional aspects, testing within this intricate software–hardware ecosystem has been somewhat overlooked. To address this, various testing techniques are applied for real-time minimization of failure rates in IoT applications. However, the execution of a comprehensive test suite for specific IoT software remains a complex undertaking. This paper proposes a holistic framework aimed at aiding quality assurance engineers in delineating essential testing methods across different testing levels within the IoT. This delineation is crucial for effective quality assurance, ultimately reducing failure rates in real-time scenarios. Furthermore, the paper offers a mapping of these identified tests to each layer within the layered framework of the IoT. This comprehensive approach seeks to enhance the reliability and performance of IoT-based applications.

Keywords: IoT failure causes; layered architecture of IoT; quality assurance; testing framework

Citation: Aljaedi, A.; Siddique, S.; Satti, M.I.; Alharbi, A.R.; Alotaibi, M.; Usman, M. Underpinning Quality Assurance: Identifying Core Testing Strategies for Multiple Layers of Internet-of-Things-Based Applications. *Sustainability* **2023**, *15*, 15683. https://doi.org/10.3390/su152215683

Academic Editor: Yang (Jack) Lu

Received: 2 September 2023
Revised: 5 October 2023
Accepted: 10 October 2023
Published: 7 November 2023

1. Introduction

The IoT is not simply a concept but an architectural paradigm that provides the medium for exchanging captured data and the means of integrating physical world and computer systems over a defined network. Independent technologies construct the IoT's fundamental components. The IoT's applications can be found in an array of devices, industries, and settings. The components of the IoT are based on object, communication, and computing modules. The main functionality of object modules is to provide a response to instructions and retrieve data. Communication means the network to be used, and it comprises protocols and technologies that allow the exchange of information or data between physical objects. It might be a Local Area Network (LAN), Wide Area Network (WAN), Metropolitan Area Network (MAN), or cellular network. Computing includes collecting, processing, storing, and manipulating the underlying data. It reflects the overall behavior of the system. The accurate form of computing analyzes user behavior, efficiently makes the right decisions based on user nature, and makes deductions. The IoT is an evolving trend, and each evolving trend needs some sort of affirmation regarding its quality. This outlines the great need for quality assurance in IoT to optimize the ongoing processes and meet user expectations. As each IoT application is a different product and has a different usage, they cannot be categorized as one. Even in the same domain, many different types of IoT applications with several layers exist, and as they evolve very quickly, the software quality assurance process must also be updated. Therefore, the accurate measurement of quality is essential, considering the presence of objects of heterogeneous nature that are bound with one another to build an IoT system. However, performing quality assurance (QA) testing in IoT is precarious as it is a huge network of physical devices and involves testing both the hardware and software, recording the test results, and then sending them back in real time, which is not an easy task to accomplish. There are numerous failure cases of the IoT in real time. One such example is Petnet, an automated pet feeder [1]. This device encountered system failure and was unable to establish reliable communication with connected devices. This system depended on third-party servers which they rented out from Google. The problem started when the servers were not responsive for 10 h and the system had no backup plan. The users affected by this situation lost the ability to set feeding schedules for their pets and were also unable to remotely get hold of the device and command it to feed their pet. This endangered pets' lives as they might have missed meals and starved if the owner was unable to reach home and depended on this device for feeding services. Although Petnet was luckily successful in resolving the issue on time, this raises the question as to what measures are important in making an IoT product a market-winning product that also satisfies user expectations. In this paper, we conduct a literature review to identify the IoT's current trends concerning the quality assurance process. The issues related to QA and their existing solutions are provided in Table 1. The problem area is identified after critically analyzing the reviewed literature and addressed by our proposed framework, in which we have identified basic yet important tests for IoT-based applications to ensure quality. Although these approaches are basic, they still lay the foundation of quality assurance in the IoT. This area needs special attention in IoT development as the idea behind the IoT is smart living. This goal is not achievable without user satisfaction. For user satisfaction, quality is the utmost requirement that needs to be fulfilled. This paper also provides a mapping architecture in which we have mapped all the important tests across the six-layered architecture of the IoT. The basic aim behind this architecture is to enable QA engineers to identify the tests that must be conducted at different levels of the IoT. This paper is structured as follows: We present a literature review in Section 2. Section 3 provides our proposed framework for IoT testing and its implications, whereas Section 4 includes our mapping architecture. Section 5 includes the evaluation of our research. Section 6 provides the conclusions of this study and deliberates some future work based on prior research.

Table 1. QA Evaluation and Analysis Parameters.

Reference	Research Gaps	Proposed Solutions
A. E. Al-Fagih et al. [2]	Challenges related to pricing, resource management, and inter-operability in wireless sensors	Priced PS framework for architectures of IoT for applications related to services in cities to make them smart and the use of utility function of pricing for acquisition of data
J. Kiljander et al. [3]	The devices heterogeneity, for representing their functionality in form of a platform for virtual computing	Architectures for interoperability of semantic level architecture for pervasive IoTs and its computing
J. Zhou et al. [4]	Large amount of data which is sparse, dynamic, hetero geneous, and multi-source in IoT	Use of data fusion for manipulation and management of such kind of data for improvement of efficiency of data and system and for providing advanced level of intelligence
Leal et al. [5]	Sensing which is trustworthy and safe for general public in IoTs which are cloud-centric	Sensing-as-a-Service (S2aaS) enhances safety from public by using sensing services with help of crowd management which is provided by various smartphones having different sensors

Table 1. *Cont.*

Reference	Research Gaps	Proposed Solutions
F. Li et al. [6]	Integration of secured network for integration of wireless sensors of network into IoT	Heterogeneous signcryption scheme which are offline and online as well, for securing of the communication between an internet boot and sensor node
X. Mao et al. [7]	Issue of systems which are cyber-physical and networks which are wireless sensor in IoT	Not Defined
E. S. Reetz et al. [8]	Testing of services which are based on IoT before their deployment into the world	Emulation of resources of IoT interface from the architectural, implementation and semantic perspective
D. Kuemper et al. [9]	Derivation of test for services of IoT which are semantic based	Methodology for enriching of service related to descriptions for derivation of testing which is semi-automated and is required for the adaptions of IoT
P. Gimenez et al. [10]	Faster tests with reduced cost of operation and low risk	using simulator of high quality sensor, SWE simulation and web standards for sensors
J. Fernandes et al. [11]	So-designing issues in IOT	Platform of loT lab for the design framework of loT reference architecture model for creation of an initial design including test-bed components, crowd-sourcing, ability to do federation with other test-beds and virtualization
V. A. Desnitsky et al. [12]	Monitor security components for anomaly detection in components and data in IoT	Use of expert knowledge and elicitation approach and for detecting data anomalies and giving them as an input for automated systems for monitoring of IoT components of security
C. Chuang et al. [13]	Application on quality assurance of composite digital services on Intelligent Transportation System	Framework for integration of end-to-end testing for quality assurance which works on DSRA (Digital Service Refer ence Architecture) supported by forum of TM
M. Masirap et al. [14]	In-adaptability of protocol like TCP	Transport protocols based on UDP which are UDT, PA UDP and RUBDP for use in application of IoT
S. Sankaran et al. [15]	Increased sensitivity of securing IoTs related to data of user and consumption of high power which is the nature of IoTs	Using cryptography which is identity based and develop ment of security framework which is light in weight for IoTs
A. R. Chandan et al. [16]	Maintaining authenticity, confidentiality and integrity for securing the network of the IoT network	Framework for testing of IoT
D. Kim et al. [17]	Verification and execution of applications of IOT	TITAN is designed in such a way for allowing developers to verify and execute IoT applications preventing from being constrained by the environment, in a development environment
A. Kaiser et al. [18]	Growing of probability and complexity of vulnerabilities and malfunctions in IOT	IoT-Test ware Eclipse for ensuring conformance and ro bustness of protocol and secured implementations
M. Abdallah et al. [19]	Model for quality measurement, making the measurement process of quality less applicable, less accurate, and more challengeable	Model for the quality of IoT which consists the characteristics of IoT systems, by introduction of quality factors for measuring them
S. Popereshnyak et al. [20]	Difficulties related to the phase of testing of applications and systems of IoT	Features related to testing, based on network of modeling of IoT Application
Kim et al. [21]	Increased complexity and cost of testing because of the large number of variables heterogeneity and scalability of conventional testing of IoT devices	Testing of IoT as a Service called IoT-TaaS, which is a service oriented approach for automation of testing of IoT
K. Papachristou et al. [22]	Routing and runtime verification of the policies for security to enhance quality of networks in IoT	Framework for number of information security policies' verification at run-time, of the network and dynamic routing paths flow

2. Literature Review

Numerous sensor-assisted devices associated together are present in the current era. The topologies made using such devices and other pervasive items give us the Internet of Things (IoT); this is a whole new paradigm that allows all the already existing mechanisms to be equally detectable, controllable, and linked. Multi-attribute quality score computation is a method for evaluating the quality of Internet of Things (IoT) applications based on multiple attributes or criteria. Rohini Temkar et al. [23] proposed an approach that takes into account various factors such as reliability, usability, security, performance, and scalability, among others, and assigns a score to each attribute. The overall quality score is then computed by combining these individual attribute scores based on their relative importance. Multi-attribute quality score computation can be used to compare different IoT applications and select the one that best meets the user's requirements. It can also be used to monitor the quality of an IoT application over time and identify areas for improvement. Various methods such as fuzzy logic, decision trees, and neural networks can be used to implement multi-attribute quality score computation. These methods can handle the uncertainty and imprecision associated with evaluating multiple attributes and can provide accurate and reliable quality scores. Overall, multi-attribute quality score computation is a useful approach for evaluating and comparing the quality of IoT applications based on multiple criteria. It can help ensure that IoT applications meet the user's expectations and provide a satisfactory user experience.

Yair Rivera Julio et al. [24] offered a comprehensive framework for managing software quality in IoT applications. They emphasized its significance and addressed associated challenges in these intricate systems. The framework comprises five stages, covering aspects from requirements engineering to maintenance. Each stage is detailed, including recommended activities, techniques, and tools. The benefits include enhanced software quality, greater efficiency, and cost savings. The authors stressed the importance of a systematic approach to quality management, integrating it into the entire development process. They also acknowledged challenges, including specialized skills, managing quality across layers, and adapting to the dynamic nature of IoT applications.

Noha Medhat et al. [25] emphasized the critical role of testing for ensuring the quality of complex and diverse IoT systems. They discussed a range of testing techniques, including functional, non-functional, integration, and regression testing. Each technique was thoroughly described, along with its advantages and limitations in the context of IoT systems. The paper underscores the significance of testing at every stage of the software development life-cycle, from requirements engineering to maintenance, stressing the need for continuous testing to meet specified quality standards.

A. Sharma and A. K. Sarje [26] stressed the pivotal role of testing for quality, reliability, and security in IoT systems, given their complexity. They detailed various testing techniques, including functional, performance, security, and interoperability testing. Their paper introduces emerging techniques like crowdsourced, automated, and model-based testing, highlighting their potential benefits. Practical implementation guidance is provided. The paper also addresses challenges, including the absence of standardized testing tools, the complexity in testing across layers, and the need for specialized expertise.

R. Kumari and M. K. Soni's study [27], published in 2020 in the *International Journal of Advanced Research in Computer Science and Software Engineering*, comprehensively explored quality assurance techniques for IoT systems. It underscored their vital role in diverse applications due to the complexity and heterogeneity of IoT systems. The paper covers techniques including requirements engineering, testing, verification and validation, fault tolerance, and maintenance, offering detailed descriptions along with their respective advantages and limitations. Emerging techniques like machine-learning-based testing, blockchain-based verification, and edge-computing-based fault tolerance are also discussed, with emphases on their potential benefits and practical implementation guidance in IoT systems.

Shanzhi Chen et al. [28] indicated that an IoT system must have three characteristics: Comprehensive Perception, Reliable Transmission, and Intelligent Processing. Comprehensive Perception means obtaining authentic information anywhere and anytime it is needed. Reliable Transmission means reliable data availability through all radio, Internet, and telecommunication channels. Intelligent Processing such as cloud computing supports the IoT, and it means a huge amount of data to be processed. The IoT must also incorporate other characteristics such as connectivity, enormous scale, sensing, dynamic changes/nature, heterogeneity, and security [19]. For the IoT to be successful, it is mandatory to keep its business models and applications clear.

It is insufficient to keep just one plan in mind. To reduce the risk of failure, business aspects should be considered in the early stages of IoT development to minimize the risk of failure. To guarantee credibility in the IoT framework, the S2aaS scheme with explicit Trustworthy Sensing for Crowd Management (TSCM) for front-end contact with the IoT is used [29]. TSCM gathers distinguishing data grounded on a cloud model and a technique that picks out mobile devices for precise sensing responsibilities, regulating the outgoings to users of the mobile devices that offer data. An enactment assessment of TSCM demonstrated that the power of malevolent users in the crowd-sourced data can be decreased by 75 percent, while the dependability of a malevolent user amounts to under 40 percent. The reduced cost, lessened configuration load, and constricted pairing with the power-driven product make Communicating Power Supplies an exceptional application of the IoT.

In the IoT, when numerous devices are connected, a power consumption problem arises, as it is unknown which device requires how much energy and when it needs it. Most of the devices require energy conversion from an AC source to DC power. CPSs convert AC to DC power supply [15]. The permeating nature of sensors, in addition to the sensitivity of user data, makes it a need for the IoT to ensure immense security. In addition, power constraints are an important limitation of the IoT. Therefore, there is a need to secure the IoT using some lightweight solution. A security framework using identity-based cryptography is a solution to this problem [6].

IoT-based networks usually consist of minute sensor nodes with base stations. Sensor nodes usually have limited power, while a base station is powerful enough that it provides an interface between the user and node. In a TCP/IP suite, the base station acts as a router, and there are security challenges such as secure channel setup and end-to-end authentication. A heterogeneous encryption scheme is used for secure communication between the base station and sensor node [7].

The current research challenges for the IoT are channel assessment, system schemes, resource administration, the assimilation of several schemes, application advancement, network protocol plans, and even changeovers from legacy systems [2]. For maximum utility gain, quality control in the IoT is an important parameter. In Priced Public Sensing (PPS) [3] in the IoT, data delivery schemes are divided into delay-tolerant and delay-sensitive schemes. In the IoT using cheap sensors, quality is ensured using efficient algorithms. Interoperability is an important part to be considered while evaluating IoT systems. Considering the layered approach, connectivity-level interoperability means that the connected devices can transfer data to each other without knowing the actual meaning. Semantic-level interoperability enables devices to understand the meaning of transferred information.

The functionality of these systems and their interactions are tested at design time [9]. Immense knowledge of data types, system dependencies, and the behavior of services is required to derive tests for the IoT. The data regarding test case generation are derived from the stored knowledge of services in a knowledge database. The IoT is a huge paradigm involving billions of devices, which means a huge amount of gigantic, dynamic, and heterogeneous data. For data management and manipulation, data fusion [4,8] is a tool that is important for improved efficiency. The goal of this method is to ensure better quality in obtaining information. Eike Steffen Reetz et al. [11] proposed a semiautomated approach in which test code is injected to enable efficient prototyping, along with the integration of tests for IoT services. If the logical interaction of services and IoT systems is semantically defined, then the knowledge gathered via services might help in the generic resource emulation interface. In this way, a service gives an abstract image of IoT interaction and enables efficient and scalable emulation of IoT systems.

A new extension of IoT testbed groundwork is crowdsourcing, which manages communities online while keeping an eye on why the crowd is showing interest in this domain and what they want. For this purpose, a third party needs to be involved to obtain immediate feedback and record it accordingly. This helps the IoT lab to select important use cases, and then the crowd can again give feedback regarding the selection. This is how testing is made more efficient [5]. For the IoT, testbeds are used for simulation, which is vital but also has challenges, leaving room for improvement. Brazil wants new regulations to be followed in transportation regarding RFID and the IoT. The Brazilian transportation department intends to broaden its research area for research into the IoT [30]. The object name services (ONS) projects aimed at discovering a suitable way of communicating automatically [12]. All the work it does is in addition to radio-frequency identification (RFID) and the IoT. But both of them have crucial security concerns. To assure internal communication security, a repetitive and hard testing scenario is produced so that one can obtain a clear picture regarding the behavior of the system under load.

Mohammad Abdallah et al. [28] proposed a new quality model after studying different existing models for quality and comparing them regarding the factors that play a role in quality measurement that can be used as a basis for finding other factors in the future for improving the quality of IoT systems. As the number of IoT systems is increasing, the complexity of IoT systems is also increasing, and as the complexity is increasing, the need for improved security is also increasing. The solution to this problem lies in detecting anomalies in the data and then assessing where these data are being used as inputs in IoT-based systems [31]. In the IoT, there are certain design requirements that must be fulfilled. These are connectivity, security, sensor, and touch [14].

Security and heterogeneous test integration are features in the IoT that require innovative solutions. IoT-based systems need to maintain the per unit cost of production, so while handling the above-mentioned issues, managing cost-effectiveness is challenging. It is a requirement of the IoT that the wireless connection should be in real time, and for real-time connections, Transport Control Protocol (TCP) is not enough as it has a large header size, slow start, and Additive Increase Multiplicative Decrease (AIMD) congestion control algorithm. Madzirin Masirap et al. [20] tested the UDP using a testbed with two systems connected by an ad hoc network. The evaluation results showed that the UDP is far better than the TCP in terms of speed and resource utilization.

Svitlana Popereshnyak et al. [10] provided features that can be used to test IoT applications and devices, along with the main differences in testing techniques of classical systems and IoT systems. Basically, two types of IoT testing were discussed: user convenience testing and network connection testing. For approving the proposed specifications and selecting an application protocol, an experiment and medical applications were also conducted in order to test a portion of a communication network. The service-oriented testing model architecture was used, following the generic life cycle of software development, consisting of four main phases.

The major part of this model is the automation of testing in every phase of the model, which includes testing the deployed parts of systems, benchmark testing, and integration testing. CoAP is a good protocol for the communication of IoT projects as it is lightweight and fast. It ensures the stable operation of the system, providing reliability and the possibility that the system can be improved even after its release. Testing methods in the IoT should evaluate the developed system and check the non-functional requirements of the client. In the IoT, wireless sensor networks are quite important. A cheap way to add sensors in networks is via virtual sensors.

Simulation tools are sufficient to evaluate a system when checking riskless deployment, and they are efficient and economical. The Sensor Web Enablement (SWE) simulator is a software module that enables the emulation of various sensors, which helps in producing numerous use cases and situations that might be external to those in real time [32]. In an urban IoT system, if proper monitoring of the structural health of a building is required, then numerous sensors are required to be embedded in the building, as well as in its surroundings, to monitor pressure, pollution, and other factors [13]. In this way, it becomes easier and more efficient to maintain the database to ensure the quality of buildings, as well as the system, as compared to humans performing these tasks. The intelligent transportation system (ITS) is a recent advancement in the IoT [16].

To assure quality in the ITS, Information and Communication Technology (ICT) strategies and facilities are utilized. Each device or service we use from ICT has its own Operation Support System (OSS), which limits information exchange from device to device. A framework composed of Digital Service Reference Architecture (DSRA) and TM Forum is feasible to ensure end-to-end quality in ITS. Abhishek R. Chandan et al. [17] proposed a methodology for security testing. Information related to IOT devices and networks was gathered; a better understanding of a device, which assists in threat profiling and testing of the device, depends upon the implemented device and the network. This methodology can be followed during penetration testing to ensure quality in the IoT. Integrity and heterogeneity can be achieved by using proper techniques of encryption, firewalls, and security protocols, along with a lightweight key management system. Proper data management is required to maintain confidentiality and availability. Scalability can be achieved through the enforcement of policies.

Authenticity is very challenging to achieve as per the nature of the IoT. TITAN [18] is a tool that provides a virtual environment to developers where they can efficiently run and test IoT applications in the development process without them being influenced by the physical environment and the behaviors of users. This minimizes the time and effort required for repetitive testing during development. The availability of open testing equipment and how it can have a positive impact on IoT applications is an issue [22]. ETSI (European Telecommunications Standards Institute), oneM2M, and the OPC-Foundation have already been working on this, and they have provided open standards and open-source testing equipment.

Alexander Kaiser et al. [32] also discussed the results of an experiment performed using brokers of MQTT and suggested that protocol implementation must not be considered a trivial task; open-source testing tools must be available to benefit open-source projects. The two types of protocol testing discussed in the paper are protocol conformance and protocol security, both of which have their own impacts on the final product and its vulnerability. Konstantinos Papachristou et al. [21] proposed two frameworks for the verification of many security policies, which relate to information on the dynamic flow of routing paths and networks. The underlying set of concepts is to let the operator control the overall network and define various policies, which are the basis of network demands, as well as use cases for achieving a faster and more secure network.

The input given to the optimization algorithm (multi-objective) is the routing policies and statistics taken in real time. This information then helps in the calculation of policies for routing to estimate the quality of routing decisions. The flow rules created by SDN are compared with the optimal set of flow rules, giving the actual results for verification in the form of a deviation metric. HIUN KIM et al. [33] proposed an IoT testing framework called Service-IoT-TaaS. It works on the basis of the "plug and test" concept, using a service-based approach in order to provide an IoT testing framework that is automated and provides solutions to the traditional software testing issues of costs, coordination, and scalability, in order to use standard-based processes of the development of IoT devices and to explore their implementation and design. IoT-TaaS consists of remotely distributed automated scalable conformance testing, testing to validate semantics, and interoperability testing.

Minhaj Ahmad Khan et al. [34] showed how basic features of blockchain can be used as a key in solving security problems related to the IoT, but still, there needs to be some mechanism more effective than this to avoid attacks. However, an attacker can host the blockchain and, hence, the hashing power of the miner can be a risk, which relates to the consensus mechanism. Private keys that have a little randomness can also be used to attack the accounts of the blockchain.

3. Framework for IoT Testing

For quality assurance, different types of testing techniques are conducted. However, conducting all the tests for a particular software product, especially in IoT applications, is difficult. Therefore, there is a need to define all the necessary tests which must be conducted for quality assurance in the IoT and to ultimately reduce the failure rate in real time. Our proposed framework is illustrated in Figure 1. This framework provides a comprehensive outline regarding all the necessary tests to be conducted during the quality assurance process in the IoT.

3.1. Device-Level Testing

This test involves testing the logging function of devices, as this is the entrance to a device in the IoT. Device-level testing includes the testing of sensors, operating systems to be used, and system hardware and circuits.

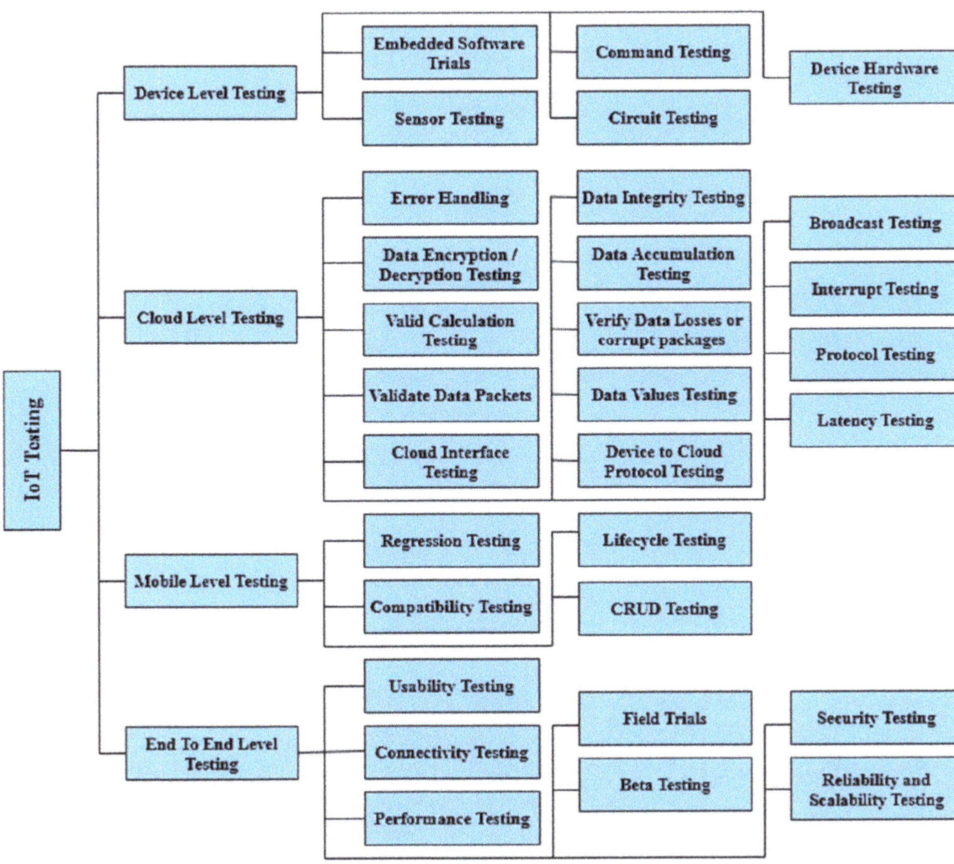

Figure 1. IoT testing framework.

3.2. Cloud-Level Testing

This includes issues regarding the functionality, integration, and cloud API. Another major concern is to ensure high security for user data and all devices connected to the cloud (security test). To ensure security, the IoT must allow access to authentic users, use encryption to protect data from intrusion, and safely store data at a reliable location. As the data consistently move in and out of the cloud in the IoT, to ensure that they are subject to consistent policies, governance tests are conducted; data privacy tests are very much related to security and data governance. Other tests which must be performed here are related to data, packets, different protocols, interruptions, and latency.

3.3. Mobile-Level Testing

Consumers are dependent on mobile devices to interact with the connected devices. Therefore, in mobile testing, it is mandatory to test the mobile app design, its interface, its back-end, and its way of communicating with all connections and the cloud. Compatibility testing identifies what version of the SDK is compatible with the system. When a new version has been launched, the system updates itself to match the major versions. CRUD testing comprises Create, Read, Update, and Delete tests, which are conducted because it is the user who performs all these functions in the account creation and data sharing process [35,36]. Life-cycle testing means actually testing the mobile device in action. Mobile-level testing also includes regression testing. The mobile device is assessed from start to end in its functionality and usually fails at this stage.

3.4. End-to-End Testing

This testing is the most extensive as it checks all the previously mentioned components of the IoT, namely, objects, communication, and computing. This is basically testing from start to finish, involving every component, as shown in Figure 2. In end-to-end testing, not only are the components checked—it is also crucial to check how they interact [37]. Field trials consist of a group of actual users checking the system and giving feedback.

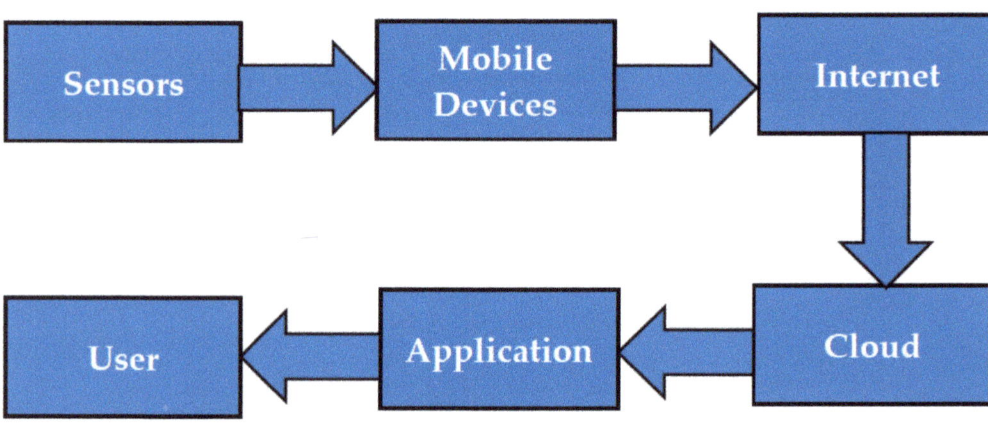

Figure 2. End-to-end test workflow.

Changes are made accordingly, and the system is again released for another field trial. This testing is also concerned with the usability of a device and its necessity in the IoT. It is critical to know how to scale all the available users in the cloud at a time (scalability test), what the effects of higher loads will be on the performance of the application (performance test), and how reliability will be ensured in cases of stress (reliability test). All of these tests must be performed at this level. End-to-end testing also ensures the system's security and connectivity.

4. Mapping of Important Testing Domains at each Layer of the IoT's Layered Architecture

Layered architecture best describes the working of an IoT device as it distributes and distinguishes the responsibilities of each component, either hardware or software, in an IoT device. For instance, the interface through which the user interacts with an IoT device is distinguished and separated from the main business logic of the device on which it works, and the connection media, such as the Internet, comprise a distinguished component. Likewise, storage media are also a separate component. The database layer has nothing to do with the interface through which the user interacts [36]. It is the business layer, which works as middle-ware and makes the data presentable on an interface. Hence, every logical layer is separated, and the security of every layer is preserved. Describing an IoT architecture in terms of a layered architecture best helps in the distribution of responsibilities among the development team. The mapping of different testing domains for the IoT at each layer will help to identify each important test that must be performed at each layer. Figure 3 shows our mapping architecture for the IoT [37].

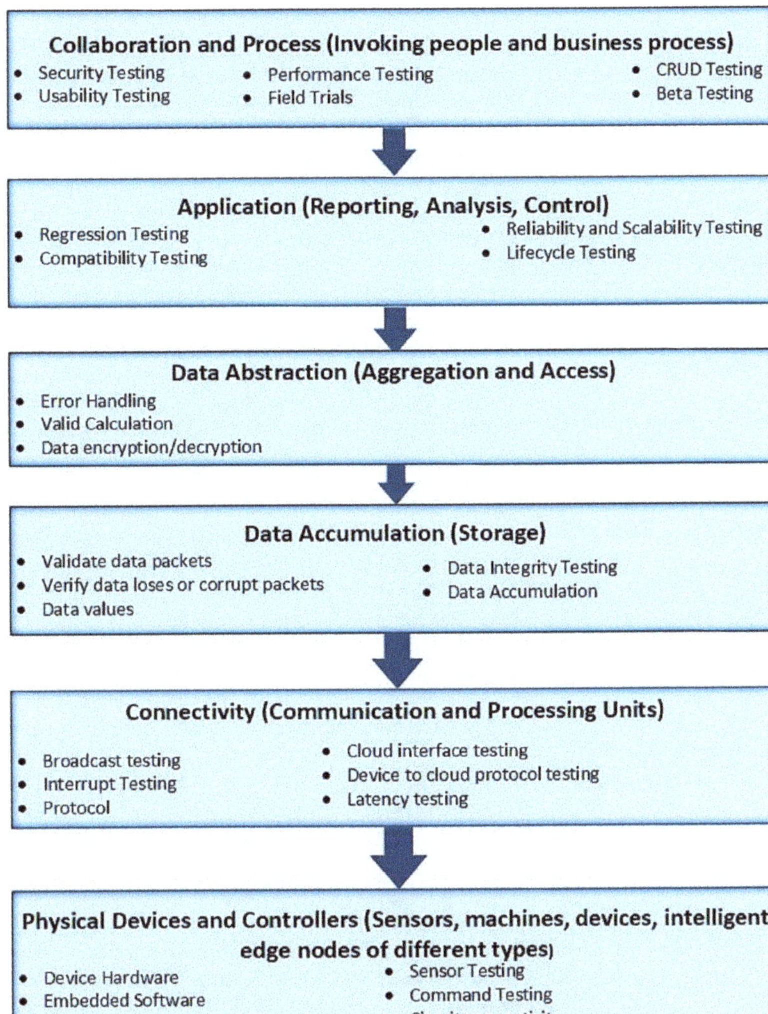

Figure 3. Mapping of important testing domains on the IoT's layered architecture.

It is very necessary to test each component before and after integration in various ways. The whole IoT device can fail due to a single component failure [38–40]. For instance, consider the Internet as the communication medium and issues with connectivity, where data are not able to be sent and received. In the case of a safety-critical system, testing becomes even more important. The following are the types of testing that must be performed at each layer in order to preserve the quality of an IoT device and deliver the best possible product.

4.1. Collaboration and Process Layer

The collaboration and process layer includes processes and people related to business. It involves business, people, their decision-making, and collaboration on the basis of the information that is extracted from the computation of the IoT. The overall interaction of people with the IoT device is validated in this, including mainly the non-functional properties like security, reliability, availability, and performance attributes.

The following are some types of testing to be performed for a quality product:

1. **Security Testing:** Physical and logical threats to the security of an IoT-embedded system are analyzed. One of the best practices for maintaining the security of an IoT device is to train employees on the importance of security on an IoT device. In the case of an outside service provider, their capability of maintaining security should be validated before obtaining any services from them. The software part must be well-protected with login identification and a strong password, keeping it protected from unauthorized access to personal information or manipulation of device usage [41–48].
2. **Performance Testing:** Performance testing deals with testing the overall performance of an IoT device. It includes checking the number of users that can be handled by the device at a time and how it reacts to the situation of excess users. It also deals with the recovery of a system in case of any issue or even if the number of users exceeds the response time of the device being tested at peak loads, as well as during normal working.
3. **Usability Testing:** Usability testing deals with how easy a system is to use by people of every age, culture, and physical or mental capability level. This includes people, contexts, activities, and technologies to be used. Usability testing can be performed on parameters like navigation, affordability, flexibility, consistency, control, recovery, constraints, conviviality, style, and visibility.
4. **CRUD Testing:** This means Creating, Reading, Updating, and Deleting. The response of a system is tested after performing these steps on the system to determine whether it allows new data to be added, created, accessed, and easily read by the user. A matrix is made for easy evaluation of the system. This is also a type of black box testing.
5. **Beta Testing:** Beta testing refers to testing in which a group of people test the product and give their feedback. These people can be any external testing party. This is basically acceptance testing by the user.
6. **Field Trials:** In field trials, the product is given to the user so that they can test it in a real environment rather than through some automation technique or artificial method.

4.2. Application Layer

The application layer comprises the software application, which works among the devices. In the application layer, the delivery of data is checked and reported. How the application manages to control the devices is also tested. This includes the following testing types:

1. **Regression Testing:** Regression testing is performed after adding new functionality to an application of the IoT device by testing the whole system again, validating its results, and checking how changes in the system affected the whole IoT device.
2. **Reliability and Scalability Testing:** Reliability and scalability testing deals with testing the system in terms of 'abilities', i.e., the non-functional attributes of the system.
3. **Compatibility Testing:** In compatibility testing, the compatibility of a software application is checked with other components and software in an IoT device. The compatibility of the operating system is also checked, and the type of database that is compatible with the current system type is kept in mind.
4. **Life-Cycle Testing:** Life-cycle testing includes the validation of every step of the system development life-cycle. It checks whether every step is properly followed or not.

4.3. Data Abstraction Layer

The abstraction of data provides an abstract view, rather than describing how the machine actually handles or stores something. It includes handling the occurrence of errors, as well as the encryption and decryption of data. The following are some types of tests to be performed for a quality product:

1. **Error Handling:** This deals with the detection, resolution, and anticipation of errors in an application and how it reacts in case of these errors, e.g., does it safely shut down and terminate the preprocess?
2. **Encryption and Decryption:** In the case of encryption and decryption, both the data sent and received should be the same; if they are not, then how will the system deal with the error and recovery of information?
3. **Valid Calculation:** The calculations are observed, and their validity is very important. If there is some mistake in the calculations, then the whole encryption and decryption process will be full of errors.

4.4. Data Accumulation Layer

Many problems can occur in the accumulation of data, as the data for the IoT are large, involving issues like variety, velocity, and volume. The aspects of high variety, volume, and velocity of data in motion are also considered in validating data packets. A hot path is needed for fast processing of data, whereas for processing of applications, a cold path is needed. The testing of all the data and their validity is very important. The following are major tests that must be performed:

1. **Validate Data Packets:** The data packets sent to the communication media and received on the other end should be the same with no error or noise.
2. **Data Integrity Testing:** Integrity of data means the quality of data. The data should be accurate, consistent, and complete.
3. **Data Accumulation:** Data accumulation deals with collecting the data and then validating their quality.
4. **Verify Data Loses or Corrupt Packets:** Every packet is observed and checked for the eradication of corrupt or lost data packets. There are few methods and techniques available to extract corrupted information or estimate its values.
5. **Data Values:** The data values must be correctly sent and received.

4.5. Connectivity Layer

The connectivity should be intelligent and secure, having the least possible delay in data transfer. The broadcasting, cloud interface, and protocols are checked in the connectivity layer. The following are types of testing that are conducted for a quality product in this layer:

1. **Broadcast Testing:** Broadcast testing ensures the quality of transmission and broadcasting of data. It is also known as a test pattern, test card, close-down, or start-up testing.
2. **Cloud Interface Testing:** The web traffic on the cloud being used for the IoT device and the function's validity are checked. The scalability, redundancy, and performance are also observed.
3. **Device-to-Cloud Protocol Testing:** The requirements for compatibility of the application of the IoT device are validated, and the compatibility of the application being used to interact with the cloud is also checked. There should not be any defects arising during connection of the cloud and the application.
4. **Latency Testing:** Latency testing checks the amount of time that the system takes to send and receive data. It should be kept to a minimum in order to provide the best-quality interface and services to the user.
5. **Interruption Testing:** In interruption testing, the response of the system in case of all possible interrupts is observed. The system should return to a normal working state in case of any interruption.

4.6. Physical Devices and Controller Layers

The testing of the whole circuit and its connections, along with the workings of sensors and actuators, is very important. The commands given to the devices through embedded software must also be validated. The following are some types of testing to be performed for a quality product:

1. **Sensor Testing:** All the sensors of the IoT device are validated by checking their outputs separately. Every sensor should work properly. If any sensor does not give the correct output, then the whole calculation can go wrong. This is crucial in the case of critical systems.
2. **Command Testing:** The commands given to the processor are validated in command testing. These commands should give the expected output.
3. **Circuit Connectivity:** The proper connectivity of the circuit is very important. If a single wire is detached accidentally or has a wrong connection, there can be a loss of any other component, like actuators, which affects the cost of the project. The connections should be tight and validated by different devices.
4. **Device Testing:** The proper functioning of every device separately attached, for providing services or assisting in them, must be verified.
5. **Embedded Software:** Embedded software testing deals with the testing of the operating system of the processor being used.

5. Results and Discussion

To evaluate our proposed framework, we conducted an online survey. We received 120 responses, out of which 28.3% of respondents were working in software industries, 48.3% were students of Computer Science or relevant degrees, 7.5% were faculty members of Computer Science or relevant departments, and the remaining respondents included freelancers, software developers in government organizations or research-based industries, etc. The foremost question of the survey form was whether people have witnessed failures of IoT-based applications in real time, such as Petnet [1]. About 83.3% of respondents replied 'yes'. The second question was to identify the phase of the software development life-cycle (SDLC) that requires more attention to reduce such failures of IoT-based applications. The results of the obtained answers are illustrated in Figure 4, demonstrated using a pie chart [43]. About 35% of the respondents identified 'Testing' as the most crucial phase of the SDLC which needs more attention to avoid failures of IoT-based applications in real time.

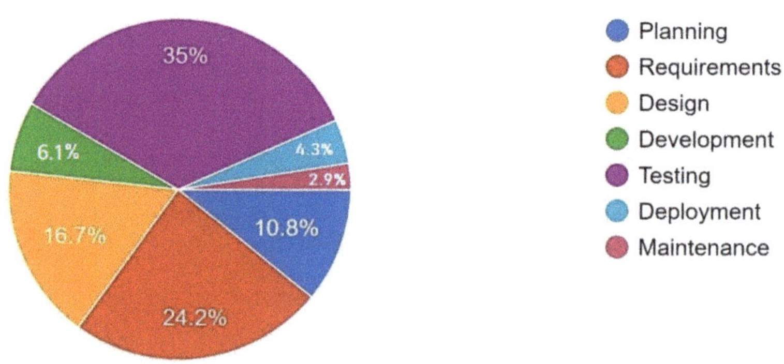

Figure 4. An illustration of the dependency of failures of IoT-based applications.

As customer satisfaction is dependent on quality, the third question was to measure the dependency of quality on the testing phase of the SDLC. Figure 5 shows the obtained results. In all, 13.7% of the respondents were of the view that about 80–100% of the quality is dependent on testing, 45.8% of the respondents selected 50–70%, 25.8% chose 40–60%, and only 5.8% selected 20–40%.

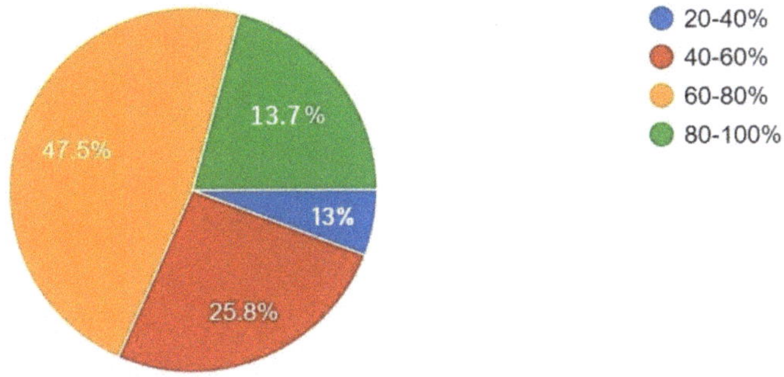

Figure 5. An illustration of the dependency of quality of IoT-based applications.

The respondents were also asked about the percentage of failure minimization if testing is properly performed. The responses of the respondents are illustrated in Figure 6. The largest proportion of respondents thought that 60–80% of the failure can be minimized if testing is properly performed. These obtained responses show the importance of the testing phase of the SDLC, which must be properly performed to maintain quality and customer satisfaction. As the IoT is a huge network of physical objects integrated with sensors, software, and communication technologies, the respondents were asked if conducting all types of tests in the IoT is possible and easy. The obtained results are shown in Figure 7. About 40.8% of respondents replied 'No', whereas 40% replied 'Maybe', and 19.2% of the respondents replied 'Yes'.

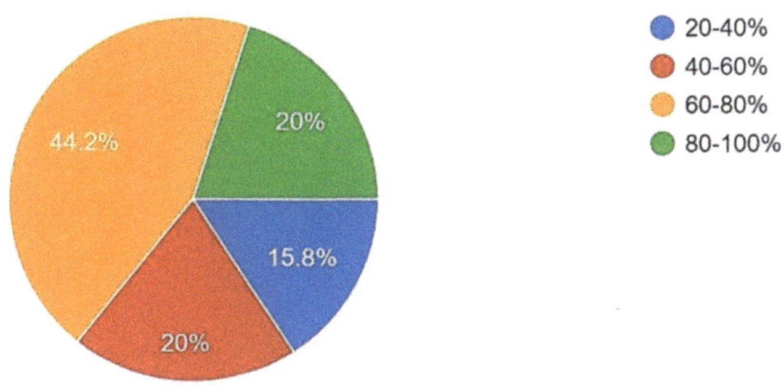

Figure 6. An illustration of the minimization of failures if testing is performed properly.

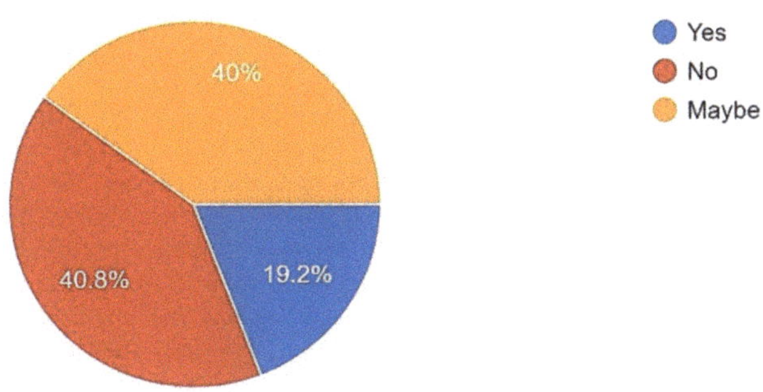

Figure 7. An illustration of the possibility of conducting all types of tests during QA.

The respondents were asked if conducting all types of tests during the QA process of IoT-based applications is time-consuming and decelerates the SDLC by posing many other challenges as well. About 81.2% of the respondents replied 'Yes', which shows that conducting all testing techniques for quality in IoT-based applications is not feasible. To measure the need for our proposed framework, we asked the respondents whether there is a need to identify all important testing techniques that must be conducted for the quality assurance of IoT-based applications, thus accelerating the SDLC and reducing cost. The obtained results are illustrated in Figure 8.

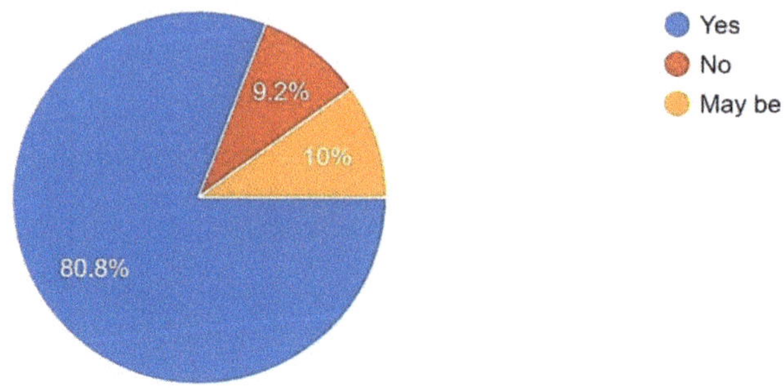

Figure 8. An illustration of the need for a framework that identifies all important tests.

About 80.8% of the respondents replied 'Yes'. This shows that there is a need for some framework that assists quality assurance engineers in identifying all the necessary testing techniques that must be conducted to ensure quality in IoT-based applications. The respondents were also asked how much our proposed framework or idea would assist in improving the quality of IoT-based applications. The results are shown in Figure 9.

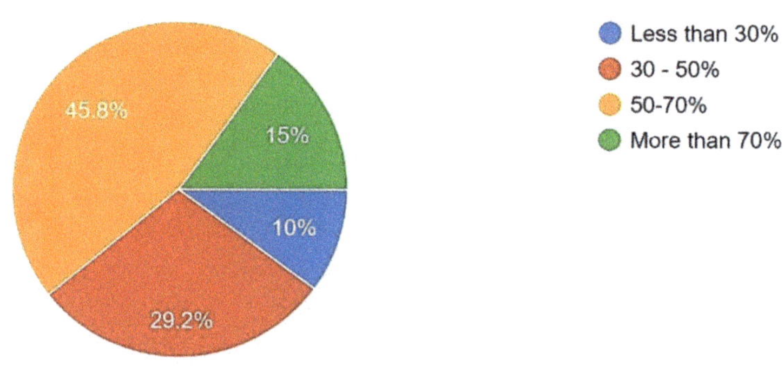

Figure 9. An illustration of how much our proposed framework will improve the quality of IoT-based applications.

Different testing techniques were identified in Figure 1. To evaluate our identified testing techniques, we asked respondents to select the most important techniques for the device level, cloud level, mobile level, and end-to-end level in IoT-based applications. They were also provided with an option to suggest any other techniques that they considered important. The obtained results for device-level testing are illustrated in Figure 10. The largest proportion of respondents considered embedded software trials and sensor testing as the most important testing techniques. Other than the identified testing techniques, the respondents also gave suggestions for communication and security testing, which must be performed at the device level. The results obtained for mobile-level testing are illustrated in Figure 11.

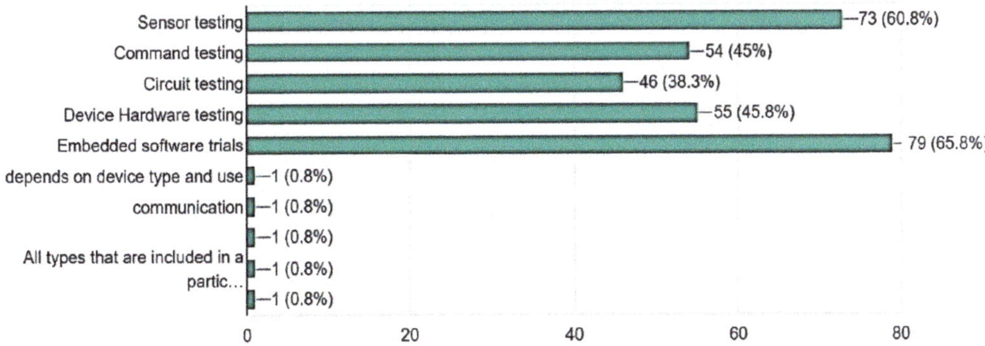

Figure 10. An illustration by percentage of important testing techniques for the device level in IoT-based applications.

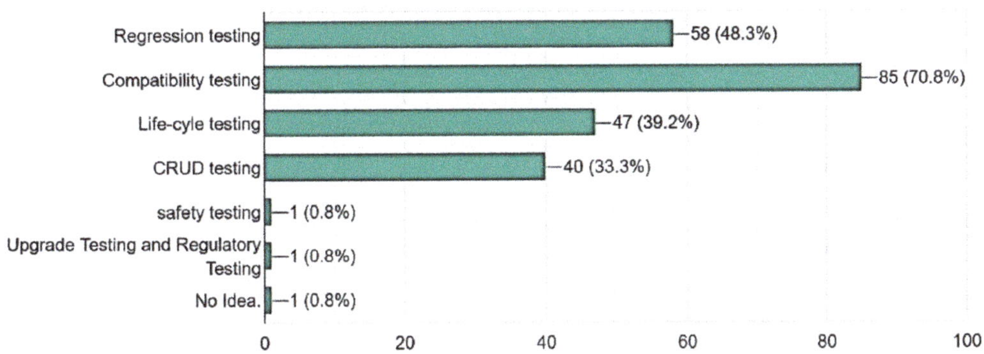

Figure 11. An illustration by percentage of important testing techniques for the mobile level in IoT-based applications.

The largest proportion of respondents considered compatibility testing as the most crucial testing technique for mobile-level testing. Safety testing, upgrade, and regulatory testing were suggested by a few respondents and must be incorporated into our proposed framework. The results obtained for end-to-end-level testing are illustrated in Figure 12.

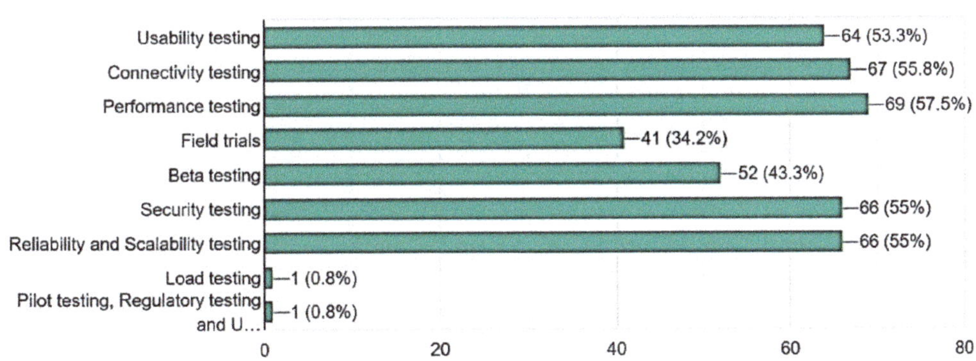

Figure 12. An illustration by percentage of important testing techniques for the end-to-end level in IoT-based applications.

All testing techniques that we incorporated for end-to-end-level testing were considered important by the majority of respondents. Other testing techniques which were suggested by respondents included load testing, pilot testing, upgrade, and regulatory testing. The results obtained for cloud-level testing are illustrated in Figure 13. Data encryption or decryption testing, cloud interface testing, and tests for error handling were considered the most important testing techniques for cloud-level testing in IoT-based applications. No other techniques were suggested by our respondents for this level. From the survey, we found that no testing technique was considered unimportant by our respondents. To improve our framework in the future, we can also include those testing techniques that were suggested by our respondents. By considering and performing all the identified testing techniques, we can improve the quality of IoT-based applications and reduce their failures in real time. Quality is the utmost requirement for customer satisfaction and must be fulfilled.

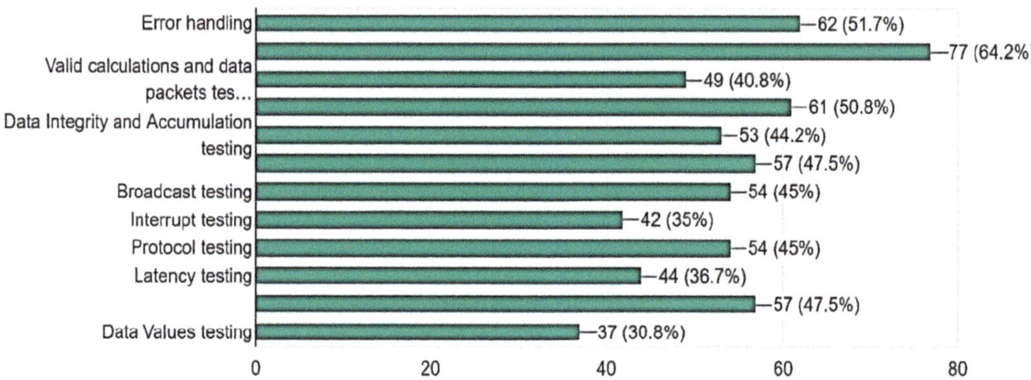

Figure 13. An illustration by percentage of important testing techniques for the cloud level in IoT-based applications.

6. Conclusions and Future Work

Overall, this research contributed to identifying all the important testing techniques for IoT-based applications and proposed a robust framework that assists quality assurance engineers in performing necessary tests during the QA process, ensuring quality with maximum customer satisfaction, reducing failure rates and costs, and, lastly, accelerating the SDLC. We also mapped all the necessary tests across the IoT's layered architecture to help the testers identify tests for each specific layer. Although these approaches are fundamental, they still lay the foundation of quality assurance in the IoT. This area needs special attention in IoT development, as the idea behind the IoT is smart living. This goal is not achievable without user satisfaction, for which quality is the utmost requirement that needs to be fulfilled. Testing should be introduced as early as possible with reviews, inspection, and formal methods. The quality of the IoT is strongly linked with security and safety as these networks are mostly linked to real-time conditions. Thus, it is important to concentrate more on data security and safety. The quality of the hardware being used should also be considered very important, as hardware from different vendors provides different performance and compatibility with IoT-based applications. Compatibility testing, load testing for traffic on the network, and functional testing for both Internet and non-Internet applications must be performed for all three types of clouds, i.e., public, private, and hybrid clouds. Product analytics is also very important as it can harmonize processes and test data. It can improve quality, yield, and productivity significantly. IoT-based applications can be tested using simulators that mimic the real environment to verify performance, usability, and all other concerns. Artificial intelligence (AI) and machine learning (ML) techniques offer the potential to enhance data collection processes, making them more efficient, adaptable, and secure. Integrating customer feedback early on and including human input in the testing loop can further elevate the quality. Considering sustainability, the proposed framework not only enhances reliability but also contributes to reducing resource consumption in IoT applications. In future work, our aim is to employ neural networks for end-to-end testing automation in the IoT, ensuring a sustainable and efficient quality assurance process. This automation will not only mitigate the risk of system failures in real time but also lead to cost savings, as it eliminates the need for human intervention.

Author Contributions: Conceptualization, A.A. and S.S.; methodology, A.A. and M.I.S.; software, S.S., M.I.S., M.A. and M.U.; validation, M.I.S. and A.R.A.; formal analysis, A.R.A.; investigation, A.A., M.A. and A.R.A.; resources, M.U., A.A. and S.S.; data curation, M.I.S.; writing original draft preparation, A.A. and S.S.; writing review and editing, A.A. and S.S.; visualization, M.A. and M.U.; supervision, M.A. and M.U. All authors have read and agreed to the published version of the manuscript.

Funding: This research was funded by the Deanship of Scientific Research at University of Tabuk through Research no. S-0151-1443.

Institutional Review Board Statement: Not Applicable.

Informed Consent Statement: Informed consent was obtained from all subjects involved in the study.

Data Availability Statement: The data presented in this study are available in the article.

Acknowledgments: The authors extend their appreciation to the Deanship of Scientific Research at University of Tabuk for funding this work through Research no. S-0151-1443.

Conflicts of Interest: The authors declare no conflict of interest.

References

1. Townsend, J. Gradual Transformation to Secure Cloud Operations. 2016. Available online: https://www.ctl.io/blog/post/qawith-the-iot/ (accessed on 1 January 2023).
2. Al-Fagih, A.E.; Al-Turjman, F.M.; Alsalih, W.M.; Hassanein, H.S. A Priced Public Sensing Framework for Heterogeneous IoT Architectures. *IEEE Trans. Emerg. Top. Comput.* **2013**, *1*, 133–147. [CrossRef]
3. Kiljander, J.; D'Elia, A.; Morandi, F.; Hyttinen, P.; Takalo-Mattila, J.; Ylisaukko-Oja, A.; Soininen, J.-P.; Cinotti, T.S. Semantic Interoperability Architecture for Pervasive Computing and Internet of Things. *IEEE Access* **2014**, *2*, 856–873. [CrossRef]
4. Zhou, J.; Hu, L.; Wang, F.; Lu, H.; Zhao, K. An efficient multidimensional fusion algorithm for IoT data based on partitioning. *Tsinghua Sci. Technol.* **2013**, *18*, 369–378. [CrossRef]
5. Leal, A.G.; Santiago, A.; Miyake, M.Y.; Noda, M.K.; Pereira, M.J.; Avanço, L. Integrated environment for testing IoT and RFID technologies applied on the intelligent transportation system in Brazilian scenarios. In Proceedings of the 2014 IEEE Brasil RFID, Sao Paulo, Brasil, 25 September 2014; pp. 22–24.
6. Li, F.; Xiong, P. Practical Secure Communication for Integrating Wireless Sensor Networks into the Internet of Things. *IEEE Sens. J.* **2013**, *13*, 3677–3684. [CrossRef]
7. Mao, X.; Zhou, C.; He, Y.; Yang, Z.; Tang, S.; Wang, W. Guest editorial: Special issue on wireless sensor networks, cyber-physical systems, and internet of things. *Tsinghua Sci. Technol.* **2011**, *16*, 559–560. [CrossRef]
8. Reetz, E.S.; Kuemper, D.; Moessner, K.; Tönjes, R. How to Test IoT-based Services before Deploying them into Real World. In Proceedings of the European Wireless 2013; 19th European Wireless Conference, Guildford, UK, 16–18 April 2013; pp. 1–6.
9. Kuemper, D.; Reetz, E.S.; Tönjes, R. Test derivation for semantically described IoT services. In Proceedings of the 2013 Future Network & Mobile Summit, Lisboa, Portugal, 3–5 July 2013; p. 13851815.
10. Gimenez, P.; Molina, B.; Palau, C.E.; Esteve, M. SWE Simulation and Testing for the IoT. In Proceedings of the 2013 IEEE International Conference on Systems, Man, and Cybernetics, Manchester, UK, 13–16 October 2013; pp. 356–361.
11. Fernandes, J.; Nati, M.; Loumis, N.S.; Nikoletseas, S.; Raptis, T.P.; Krco, S.; Rankov, A.; Jokic, S.; Angelopoulos, C.M.; Ziegler, S. IoT Lab: Towards co-design and IoT solution testusing the crowd. In Proceedings of the 2015 International Conference on Recent Advances in Internet of Things (RIoT), Singapore, 7–9 April 2015; pp. 1–6.
12. Desnitsky, V.A.; Kotenko, I.V.; Nogin, S.B. Detection of anomalies in data for monitoring of security components in the Internet of Things. In Proceedings of the 2015 XVIII International Conference on Soft Computing and Measurements (SCM), St. Petersburg, Russia, 19–21 May 2015; pp. 189–192. [CrossRef]
13. Chuang, C.; Cheng, W.; Hsu, K. A comprehensive composite digital services quality assurance application on the intelligent transportation system. In Proceedings of the 2015 17th Asia-Pacific Network Operations and Management Symposium (APNOMS), Busan, Republic of Korea, 19–21 August 2015; pp. 368–371.

14. Masirap, M.; Amaran, M.H.; Yussoff, Y.M.; Rahman, R.A.; Hashim, H. Evaluation of reliable UDP-based transport protocols for the Internet of Things (IoT). In Proceedings of the 2016 IEEE Symposium on Computer Applications and Industrial Electronics (ISCAIE), Batu Feringghi, Malaysia, 30–31 May 2016; pp. 200–205.
15. Sankaran, S. Lightweight security framework for IoTs using identity based cryptography. In Proceedings of the 2016 International Conference on Advances in Computing, Communications, and Informatics (ICACCI), Communications, Jaipur, India, 21–24 September 2016; pp. 880–886. [CrossRef]
16. Chandan, A.R.; Khairnar, V.D. Security Testing Methodology of IoT. In Proceedings of the 2018 International Conference on Inventive Research in Computing Applications (ICIRCA), Coimbatore, India, 11–12 July 2018; pp. 1431–1435.
17. Kim, D.; Lee, E.; Kang, S. IJPoster: Expediting IoT Application Testing. In Proceedings of the International Conference on Mobile Systems, Applications, and Services, Seoul, Republic of Korea, 17–21 June 2019; pp. 572–573.
18. Kaiser, A.; Hackel, S. Standards-Based IoT Testing with Open-Source Test Equipment. In Proceedings of the 2019 IEEE 19th International Conference on Software Quality, Reliability and Security Companion (QRS-C), Sofia, Bulgaria, 22–26 July 2019; pp. 435–441. [CrossRef]
19. Abdallah, M.; Jaber, T.; Alabwaini, N.; Alnabi, A.A. A Proposed Quality Model for the Internet of Things Systems. In Proceedings of the 2019 IEEE Jordan International Joint Conference on Electrical Engineering and Information Technology (JEEIT), Amman, Jordan, 9–11 April 2019; pp. 23–27.
20. Popereshnyak, S.; Suprun, O.; Suprun, O.; Wieckowski, T. IoT application testing features based on the modeling network. In Proceedings of the 2018 XIVth International Conference on Perspective Technologies and Methods in MEMS Design (MEMSTECH), Lviv, Ukraine, 18–22 April 2018; pp. 127–131.
21. Kim, H.; Ahmad, A.; Hwang, J.; Baqa, H.; Le Gall, F.; Ortega, M.A.R.; Song, J. IoT-TaaS: Towards a Prospective IoT Testing Framework. *IEEE Access* **2018**, *6*, 15480–15493. [CrossRef]
22. Papachristou, K.; Theodorou, T.; Papadopoulos, S.; Protogerou, A.; Drosou, A.; Tzovaras, D. Runtime and Routing Security Policy Verification for Enhanced Quality of Service of IoT Networks. In Proceedings of the 2019 Global IoT Summit (GIoTS), Aarhus, Denmark, 17–21 June 2019; p. 18833635.
23. Temkar, R.; Chakrabarti, P.; Jena, O.P.; Elngar, A.A.; Margala, M.; Ravi, V. Multi-attribute quality score computation for Internet of Things (IoT) based applications. *Res. Sq.* **2022**. [CrossRef]
24. Julio, Y.R.; Contreras, B.H.; Rivera, S.C.; López, C.C.; Mangonez, A.D.P.; Herazo, H.B. Framework to Manage Software Quality on IIoT Apps. *IOP Conf. Ser. Mater. Sci. Eng.* **2021**, *1154*, 012006. [CrossRef]
25. Medhat, N.; Moussa, S.; Badr, N.; Tolba, M.F. Testing Techniques in IoT-based Systems. In Proceedings of the 2019 Ninth International Conference on Intelligent Computing and Information Systems (ICICIS), Cairo, Egypt, 8–10 December 2019; pp. 394–401. [CrossRef]
26. Sharma, A.; Sarje, A.K. Testing Techniques for IoT Systems: A Review. *Int. J. Sci. Res. Comput. Sci. Eng. Inf. Technol.* **2020**.
27. Kumari, R.; Soni, M.K. A Comprehensive Study of Quality Assurance Techniques in the Internet of Things. *Int. J. Adv. Res. Comput. Sci. Softw. Eng.* **2020**.
28. Chen, S.; Xu, H.; Liu, D.; Hu, B.; Wang, H. A Vision of IoT: Applications, Challenges, and Opportunities with China Perspective. *IEEE Internet Things J.* **2014**, *1*, 349–359. [CrossRef]
29. Lanzisera, S.; Weber, A.; Liao, A.; Pajak, D.; Meier, A. Communicating Power Supplies: Bringing the Internet to the Ubiquitous Energy Gateways of Electronic Devices. *IEEE Internet Things J.* **2014**, *1*, 153–160. [CrossRef]
30. Kypus, L.; Vojtech, L.; Hrad, J. Security of ONS service for applications of the Internet of Things and their pilot implementation in academic network. In Proceedings of the 2015 16th International Carpathian Control Conference (ICCC), Szilvasvarad, Hungary, 27–30 May 2015; pp. 271–276. [CrossRef]
31. Marinissen, E.J.; Zorian, Y.; Konijnenburg, M.; Huang, C.T.; Hsieh, P.H.; Cockburn, P.; Delvaux, J.; Rožić, V.; Yang, B.; Singelée, D.; et al. IoT: Source of test challenges. In Proceedings of the 2016 21st IEEE European Test Symposium (ETS), Amsterdam, The Netherlands, 23–27 May 2016; p. 16159898.
32. Zanella, A.; Bui, N.; Castellani, A.; Vangelista, L.; Zorzi, M. Internet of Things for Smart Cities. *IEEE Internet Things J.* **2014**, *1*, 22–32. [CrossRef]
33. Khan, M.A.; Salah, K. IoT security: Review, blockchain solutions, and open challenges. *Future Gener. Comput. Syst.* **2018**, *82*, 395–411. [CrossRef]
34. Power, D.; Curry, D.; Pierson, R.; Lawrence, C. Petnet'sfailure Is a Warning to IoT developers. Available online: http://readwrite.com/2016/08/01/petnet-shows-happens-iot-fails-dl1/ (accessed on 1 January 2023).
35. Naveed, M.; Usman, S.M.; Satti, M.I.; Aleshaiker, S.; Anwar, A. Intrusion Detection in Smart IoT Devices for People with Disabilities. In Proceedings of the 2022 IEEE International Smart Cities Conference (ISC2), Pafos, Cyprus, 26–29 September 2022; p. 22187135. [CrossRef]
36. Allerin.com. IoT Solutions Failure Points that You Should Be Aware of. 2016. Available online: https://www.allerin.com/blog/6-iotfailure-points-that-you-should-be-aware-of (accessed on 1 January 2023).
37. Training an Artificial Neural Network—Intro, solver. 2016. Available online: http://www.solver.com/training-artificial-neural-networkintro (accessed on 1 September 2023).
38. Irshad, A.; Mallah, G.A.; Bilal, M.; Chaudhry, S.A.; Shafiq, M.; Song, H. SUSIC: A Secure User Access Control Mechanism for SDN-Enabled IIoT and Cyber–Physical Systems. *IEEE Internet Things J.* **2023**, *10*, 16504–16515. [CrossRef]

39. Chaudhry, S.A.; Irshad, A.; Khan, M.A.; Khan, S.A.; Nosheen, S.; AlZubi, A.A.; Zikria, Y.B. A lightweight authentication scheme for 6G-IoT enabled maritime transport system. *IEEE Trans. Intell. Transp. Syst.* **2021**, *24*, 2401–2410. [CrossRef]
40. Chaudhry, S.A.; Irshad, A.; Yahya, K.; Kumar, N.; Alazab, M.; Bin Zikria, Y. Rotating behind Privacy: An Improved Lightweight Authentication Scheme for Cloud-based IoT Environment. *ACM Trans. Internet Technol.* **2021**, *21*, 1–19. [CrossRef]
41. Said, G.; Ghani, A.; Ullah, A.; Azeem, M.; Bilal, M.; Kwak, K.S. Light-weight secure aggregated data sharing in IoT-enabled wireless sensor networks. *IEEE Access* **2022**, *10*, 33571–33585. [CrossRef]
42. Hasan, S.S.U.; Ghani, A.; Din, I.U.; Almogren, A.; Altameem, A. IoT Devices Authentication Using Artificial Neural Network. *Comput. Mater. Contin.* **2022**, *70*, 3701–3716. [CrossRef]
43. Haq, M.I.U.; Khalil, R.A.; Almutiry, M.; Sawalmeh, A.; Ahmad, T.; Saeed, N. Robust graph-based localization for industrial Internet of things in the presence of flipping ambiguities. *CAAI Trans. Intell. Technol.* 2023; early view.
44. Irshad, A.; Usman, M.; Chaudhry, S.A.; Naqvi, H.; Shafiq, M. A provably secure and efficient authenticated key agreement schemefor energy internet-based vehicle-to-grid technology framework. *IEEE Trans. Ind. Appl.* **2020**, *56*, 4425–4435.
45. Kipongo, J.; Esenegho, E.; Swart, T.G. Efficient topology discovery protocol using IT-SDN for software-defined wireless sensor network. *Bull. Electr. Eng. Inform.* **2022**, *11*, 256–269. [CrossRef]
46. Esenogho, E.; Djouani, K.; Kurien, A.M. Integrating Artificial Intelligence Internet of Things and 5G for Next-Generation Smart grid: A Survey of Trends Challenges and Prospect. *IEEE Access* **2022**, *10*, 4794–4831. [CrossRef]
47. Alzahrani, B.A.; Irshad, A.; Alsubhi, K.; Albeshri, A. A secure and efficient remote patient-monitoring authentication protocol for cloud-IoT. *Int. J. Commun. Syst.* **2020**, *33*, e4423. [CrossRef]
48. Kantarci, B.; Mouftah, H. Trustworthy Sensing for Public Safety in Cloud-Centric Internet of Things. *IEEE Internet Things J.* **2014**, *1*, 360–368. [CrossRef]

sustainability

MDPI

Article

Enhancing the Automatic Recognition Accuracy of Imprinted Ship Characters by Using Machine Learning

Abdulkabir Abdulraheem, Jamiu T. Suleiman and Im Y. Jung *

School of Electronic and Electrical Engineering, Kyungpook National University, Daegu 41566, Republic of Korea; aaoabdul@gmail.com (A.A.); jamiu.suleiman111@gmail.com (J.T.S.)
* Correspondence: iyjung@ee.knu.ac.kr

Abstract: In this paper, we address the challenge of ensuring safe operations and rescue efforts in emergency situations, for the sake of a sustainable marine environment. Our focus is on character recognition, specifically on deciphering characters present on the surface of aged and corroded ships, where the markings may have faded or become unclear over time, in contrast to vessels with clearly visible letters. Imprinted ship characters encompassing engraved, embroidered, and other variants found on ship components serve as vital markers for ship identification, maintenance, and safety in marine technology. The accurate recognition of these characters is essential for ensuring efficient operations and effective decision making. This study presents a machine-learning-based method that markedly improves the recognition accuracy of imprinted ship numbers and characters. This improvement is achieved by enhancing data classification accuracy through data augmentation. The effectiveness of the proposed method was validated by comparing it to State-of-the-Art classification technologies within the imprinted ship character dataset. We started with the originally sourced dataset and then systematically increased the dataset size, using the most suitable generative adversarial networks for our dataset. We compared the effectiveness of classic and convolutional neural network (CNN)-based classifiers to our classifier, a CNN-based classifier for imprinted ship characters (CNN-ISC). Notably, on the augmented dataset, our CNN-ISC model achieved impressive maximum recognition accuracy of 99.85% and 99.7% on alphabet and digit recognition, respectively. Overall, data augmentation markedly improved the recognition accuracy of ship digits and alphabets, with the proposed classification model outperforming other methods.

Keywords: imprinted ship characters; automatic recognition; recognition accuracy; dataset augmentation; machine learning classifiers

check for updates

Citation: Abdulraheem, A.; Suleiman, J.T.; Jung, I.Y. Enhancing the Automatic Recognition Accuracy of Imprinted Ship Characters by Using Machine Learning. *Sustainability* **2023**, *15*, 14130. https://doi.org/10.3390/su151914130

Academic Editors: Mourade Azrour, Azidine Guezzaz, Imad Zeroual, Azeem Irshad, Jamal Mabrouki, Said Benkirane and Shehzad Ashraf Chaudhry

Received: 30 August 2023
Revised: 18 September 2023
Accepted: 22 September 2023
Published: 24 September 2023

1. Introduction

The recognition and identification of imprinted letters and digits on ship components are vital tasks in marine technology applications, including maintenance, identification, and critical operational labels. Accurate recognition of these characters is crucial for both automated systems and human operators, to interpret and understand the information conveyed by engravings. The accurate and swift recognition of these characters is not just a technological pursuit but a fundamental element for upholding the sustainability of marine environments and operations. However, a significant challenge exists within marine technology—recognizing characters on the weathered, corroded surfaces of aged ships and their components. Unlike the clear markings on new vessels, these characters may have blurred or deteriorated over years of exposure to harsh maritime conditions. Even though efforts are made to update characters that have worn out and become blurry over time, there is bound to be a difference in their new state. And there are ships that are constantly updating and those that are not. However, replacing aged components with new ones not only poses environmental concerns but also economic challenges. Therefore, it is essential to find ways to identify and maintain aging components in their current state, to promote sustainability. Despite advancements in character recognition technology, deciphering these aged and obscured characters presents a unique and demanding task.

In recent years, significant progress has been made in the field of computer vision and pattern recognition. This has enabled the development of robust recognition systems that utilize machine learning techniques, such as classic classifiers and convolutional neural network (CNN)-based classifiers, to achieve high accuracy in character recognition tasks. CNNs have become essential for character recognition, excelling at capturing fine character details for accurate identification. Often, due to dataset characteristics and model design, different CNN architectures exhibit varying performance. However, in the field of ship character recognition, several challenges and limitations persist within the existing methods. These include the scarcity of comprehensive ship character datasets, difficulties in handling the variability of ship characters, the impact of environmental factors on character degradation, the complexity of the backgrounds on which characters are imprinted, the need for real-time recognition, and limited generalization capabilities. Unlike other character recognition datasets, such as handwritten datasets, ship character images are scarce, and this scarcity makes it difficult to obtain sufficient training data for the model to learn the variations in different imprinted ship characters. As a result, the model may overfit, memorizing the training data rather than generalizing effectively to new, unseen data. To address this challenge, we used generative adversarial networks (GANs), which have proven effective in generating synthetic data that closely resemble real samples [1–3]. GANs can incorporate diverse patterns and variations present in imprinted characters, which helps the model generalize better.

Considering the inherent complexity and variability of characters found on ships and their components, as seen in Figure 1, exploring multiple classifiers to identify the most suitable approach to achieving accurate recognition is crucial. This study was motivated by the urgent need for effective solutions that would bridge the gap between maritime safety and sustainability. Specifically, we aimed to develop a robust character recognition system tailored to the complex conditions of weathered ship surfaces and components. By doing so, we would contribute to the broader goals of ensuring safe and environmentally responsible maritime practices. This paper outlines our methodological approach, which involves leveraging machine learning, data augmentation, and State-of-the-Art classification techniques to enhance the accuracy of recognizing ship characters in challenging real-world conditions. Through rigorous evaluation, we demonstrate the effectiveness of our proposed system in addressing this critical maritime challenge. In our study, we conducted evaluations on State-of-the-Art classifiers, by considering relevant and recent works in the domain. By comprehensively assessing various classifiers, we aimed to propose an optimal model that would demonstrate superior performance in recognizing ship characters effectively. Our study includes the evaluation of cutting-edge CNN-based classifiers as well as well-known classic classifiers, such as Gaussian Naive Bayes (GNB), Random Forest (RF), K-Nearest Neighbors (KNNs), Support Vector Machines (SVMs), Stochastic Gradient Descent (SDG), and Decision Trees (DT). Among deep learning methods, CNNs have garnered considerable attention, due to their ability to operate directly on original data without requiring extensive data transformations. This property enables CNNs to preserve the information present in the original data to a greater extent, distinguishing them from other approaches, such as SVMs [4]. CNNs have shown exceptional performance in image recognition tasks, especially when dealing with complex patterns and intricate details. As a result, we included CNN-based models in our evaluation, to determine whether they could achieve near-perfect prediction accuracy on the imprinted digit and alphabet datasets. This study focused on optimizing the model architecture, to enhance recognition accuracy. We achieved this by systematically exploring minor modifications to critical hyperparameters in each CNN model. Specifically, we investigated variations in activation functions, learning rates, optimizers, batch sizes, and epochs, while maintaining consistency in the number of convolutional layers, dense layers, and pool sizes. This meticulous approach enabled the fine-tuning of these selected hyperparameters for each CNN model, leading to the identification of a suitable architecture with optimal hyperparameters tailored specifically to the dataset, thereby considerably improving recognition accuracy. In addition, our datasets were compared to cutting-edge hybrid classifiers, such as CNN-SVMs [5] and CNN-RF [6]. Furthermore, we developed a CNN-based classifier model for imprinted ship characters (CNN-ISC) and evaluated its performance, by comparing it to other known classifiers, aiming at providing insights into the remarkable effectiveness of our model in recognizing the diverse range of imprinted characters present on ship components.

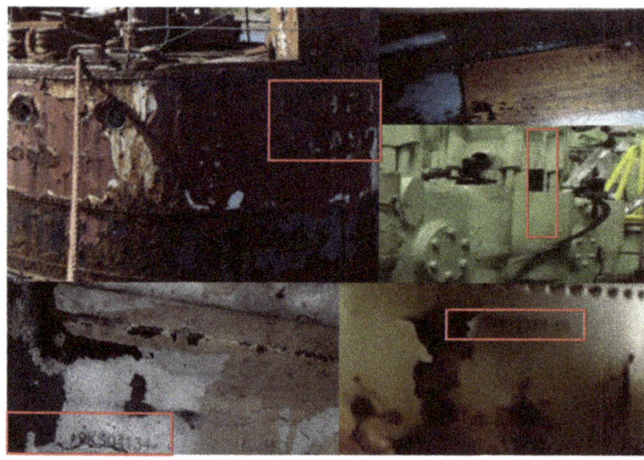

Figure 1. Sample Images of Source Dataset.

For classifier performance evaluation, we used standard metrics, including the F1 score, precision, recall, and accuracy. The F1 score is the harmonic mean of precision and recall that provides a single number to compare the overall performances of different classifiers. It balances both precision and recall and is often used when both are important. Precision measures how often a classifier correctly identifies positive samples. High precision indicates a low false positive rate. The recall value is a performance metric that measures the percentage of positive instances correctly identified by the model [7–10]. Recall measures how often a classifier correctly identifies positive samples out of all actual positive samples. A high recall indicates a low false negative rate. These metrics provide comprehensive insights into the classifiers' ability to correctly classify the imprinted characters, considering both the precision of positive predictions and the ability to identify true positive instances. By analyzing the performance across these metrics, we could assess the classifiers' overall effectiveness in recognizing the digit and alphabet datasets. By harnessing GAN models, we generated more extensive and diverse datasets. Subsequently, through a careful evaluation process, we compared the performance of the classifiers across these diverse dataset variations.

The successful implementation of this research will markedly advance the maritime industry, by optimizing maintenance and replacement schedules, facilitating part re-use and inventory management, improving accuracy and efficiency, enhancing accident investigation and safety standards, and promoting standardization and interoperability. The precision in character recognition has profound implications for human operators who rely on interpreting and comprehending the information conveyed by these engravings. However, this study goes beyond the realm of character recognition alone, casting a considerable influence on the maritime industry's sustainability. Enhancing ship character recognition extends beyond achieving heightened accuracy in vessel and component identification, to accident prevention, improved regulatory adherence, and, ultimately, a more sustainable and environmentally responsible maritime industry. The following highlights underscore the key findings and contributions of our research:

- We utilized GANs for data augmentation, improving recognition accuracy by incorporating diverse patterns in limited special typed images of imprinted characters on ship components.
- The CNN-ISC model achieved 99.85 and 99.7% accuracy, outperforming other classifiers for both digits and alphabets on ship components.
- The CNN-ISC model's high precision and recall make it valuable for ship character recognition, enhancing maritime safety measures.

The rest of the paper is structured as follows: Section 2 provides an overview of relevant studies on recent recognition models for similar tasks. We introduce a CNN-based approach for recognizing imprinted digit images in Section 3. In Section 4, we present the results of our comprehensive evaluation of the classifiers on the digit and alphabet datasets. We discuss the classifiers' performance, highlighting notable improvements achieved when trained on augmented datasets. We also analyze our findings, in relation to previous works. Finally, we conclude the paper and discuss potential future research directions.

2. Related Work

In the analysis of ship character identification, we conducted an extensive review of the existing literature on ship identification, and we investigated the utilization of data augmentation techniques in this domain. The aim was to understand the advancements made in ship identification methods and to examine how researchers have incorporated data augmentation to enhance the accuracy and robustness of ship recognition systems. By exploring the research conducted on ship identification and the integration of data augmentation, we sought to gain valuable insights into the effectiveness of these approaches and their impact on improving ship recognition systems.

Accumulating research on image recognition [11–17] has enabled the development of novel algorithms and techniques. This progress has facilitated the application of image recognition in various domains. Over time, these algorithms have undergone significant advancements, becoming increasingly sophisticated. To enhance classifier performance, in terms of accuracy, running time, and computational complexity, researchers often employ ensemble methods, which involve combining multiple classifiers. Several studies have explored the use of ensemble methods on well-known datasets, showing their potential for improving classifier outcomes.

The authors of [18] introduced the RMA (ResNet-Multiscale-Attention) model, a fine-grained classification approach for recognizing navigation marks in camera images. This model incorporates an attention mechanism that combines feature maps of three different scales, to capture subtle differences among similar navigation marks. It was trained on a dataset of 10,260 navigation mark images, achieving an accuracy of approximately 96% for classifying 42 types of navigation marks. This value markedly exceeded that of the ResNet-50 model, which achieved around 94%. The authors also provided visualization analyses that demonstrated their model's ability to extract attention regions and essential characteristics of navigation marks, thereby further validating their model effectiveness.

BoxPaste, a powerful data augmentation method tailored for ship detection in SAR imagery, was introduced in [19]. This approach, which involves pasting ship objects from one SAR image onto another, exhibits greater performance on the SAR ship detection dataset than the baseline methods. The authors also proposed a principle for designing SAR ship detectors, highlighting the potential benefits of lighter models. The effectiveness of their data augmentation scheme was further demonstrated by integrating it with RetinaNet and ATSS, resulting in impressive performance gains. In [20–23], the authors addressed the challenge of detecting and tracking ships from video streams in a monitored area. These works focused on developing systems that can effectively identify and track vessels as they enter designated areas.

A comprehensive approach to recognizing vessel plate numbers through an end-to-end method was introduced in [24]. The method comprises two key stages: vessel plate number detection and recognition. In the detection stage, a deep CNN is employed, to identify the bounding boxes encompassing the vessel plate numbers within the images. Subsequently, in the recognition stage, a Long Short-Term Memory (LSTM) network is utilized, to accurately decipher the text contained within the detected bounding boxes. To assess the effectiveness of their proposed method, the authors conducted evaluations, using a dataset comprising 1000 vessel plate number images. Remarkably, the method achieved impressive text detection and recognition rates of 96.94% and 93.54%, respectively. Comparative analysis to other existing methods revealed that the proposed approach outperformed all other techniques, highlighting its superior performance and efficacy in vessel plate number recognition.

The authors of [25] conducted an extensive survey, to thoroughly examine the current State-of-the-Art methods in scene text detection and recognition. The study provided a comprehensive overview of these two tasks and delved into the significant influence of deep learning techniques on their advancement. The authors further explored recent progress in scene text detection and recognition, encompassing novel deep learning architectures, the utilization of large-scale datasets, and the introduction of improved evaluation metrics. They also discussed the challenges that remain in scene text detection and recognition, including detecting text in low-resolution images, in images with cluttered backgrounds, and in images with non-Latin characters. In [26], the authors proposed a novel rotation-based framework for arbitrary-oriented scene text detection in natural scene images. The framework comprises two main components: a Rotation Region Proposal Network (RRPN) and a Rotation Region of Interest (RRoI) pooling layer. The RRPN generates inclined proposals with text orientation angle information, and the RRoI pooling layer projects arbitrary-oriented proposals to a feature map for a text region classifier. The authors evaluated their framework on three real-world scene text detection datasets and demonstrated its superiority, in terms of effectiveness and efficiency, over previous approaches. In [27], a novel method for omnidirectional scene text detection was proposed, based on a sequential-free box discretization approach that allows for the detection of text in images with arbitrary orientations. The method was evaluated on the ICDAR 2015 omnidirectional scene text detection benchmark, and it achieved State-of-the-Art performance.

In [28], the authors proposed an algorithm for scene text detection using multibox and semantic segmentation. The algorithm first uses a multibox detector to generate a set of text proposals. These proposals are then passed on to a semantic segmentation model, which is used to predict the text region in each proposal. The algorithm was evaluated on the ICDAR 2015 scene text detection benchmark, and it achieved State-of-the-Art performance. A novel end-to-end panoptic segmentation method called Max-DeepLab was proposed in [29]. The method is based on the Mask R-CNN framework, and it uses a novel mask transformer module to improve model performance. The method was evaluated on the COCO panoptic segmentation benchmark, and it achieved State-of-the-Art performance. In addition, in [30], a method for ship plate recognition was introduced, using a Fully Convolutional Network (FCN), a type of deep neural network specialized in image segmentation. The authors argued that FCNs are well suited to ship plate recognition, due to their ability to handle challenges such as occlusion, rotation, and varying illumination in ship plate images. The FCN was trained on a dataset of ship plate images, and its performance was evaluated on a separate test set. The results showed impressive accuracy of 95%. Additionally, performance comparisons with other ship plate recognition methods demonstrated the FCN's superiority.

3. Methodology

This study used datasets containing both digits and alphabets that represented the imprinted characters commonly seen on ship components. These datasets were carefully selected, by considering variations in the shapes, sizes, and styles of the engravings, to make them more realistic. We emphasize the significance of using datasets specifically derived from ship imprints and related components, to ensure the authenticity and precision of the generated digit and alphabet images. In addition to the baseline experiments, we investigated the impact of dataset augmentation using GANs on classifier performance. GANs are powerful tools for data augmentation, because they can generate synthetic data that closely resemble real-world samples [1–3]. We explored the effectiveness of GAN-based augmentation techniques, such as Wasserstein GAN with a Gradient Penalty (WGAN-GP) and WGAN with Divergence (WGAN-DIV) [31], in improving the recognition performance of classifiers. The WGAN-GP and WGAN-DIV models, among several other GAN models, demonstrated exceptional performance during execution on the engraved digit dataset.

Figure 2 illustrates our system model for identifying imprinted ship characters. The process begins with data preprocessing, which involves preparing the raw images of imprinted ship images by normalizing the data, to ensure accurate classification [32], followed by data augmentation, to enhance the dataset by applying techniques such as the GAN. This augmentation increases diversity and enhances the classification models' robustness. The next phase involves assessing classification models, including various machine learning models designed to recognize imprinted ship images. This phase includes training and fine-tuning the classifier algorithms, to optimize their performance. The classification phase also includes the model validation and evaluation phase. Model validation is the process of testing a trained classifier on a separate dataset, to ensure its performance and generalizability. Model evaluation is the process of assessing a classifier's effectiveness on a dataset, using metrics, such as the F1 score, accuracy, recall, and precision. These metrics provide a comprehensive analysis of the classifier's performance. The recognition phase involves the application of trained classification models, to accurately identify and recognize imprinted ship images. This step enables the system to discern precisely the specific digits represented by the engravings. Finally, the information retrieval system integrates the classification and recognition components. It serves as a cohesive system that allows users to retrieve relevant information based on recognized imprinted ship images. The engraved digit recognition workflow is described in [33].

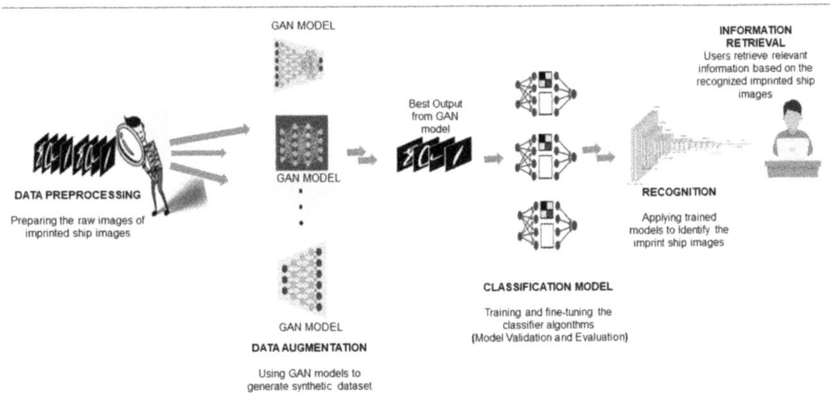

Figure 2. An imprinted Digit and Alphabets Recognition System Model with Data Augmentation Technique.

3.1. Data Collection and Preprocessing

The ship-imprinted character image dataset comprises characters that are etched, engraved, or inscribed onto ship surfaces, such as metal plates or panels (see Figure 1). These characters are typically machine generated and serve various purposes in the maritime industry, including identification, labeling, and signage. However, due to factors such as physical wear, corrosion, exposure to harsh environmental conditions, and the passage of time, the legibility and visibility of these imprinted characters can be significantly compromised. This presents a considerable challenge to accurately identifying and recognizing the characters, particularly in real-world scenarios where the imprinted surfaces may have undergone extensive deterioration. The ship-imprinted character image dataset exhibits variations in image quality, lighting conditions, and distortion caused by the engraving process itself (see Figure 3). These variations in the appearance and quality of imprinted characters pose significant difficulties for traditional recognition methods, necessitating the development of specialized approaches, to effectively address these challenges.

Figure 3. Collection of Character and Digit Samples from the Source Dataset

The datasets used in this study covered various imprinted characters, including digits 0–9 and 13 alphabets: A, C, D, E, I, L, M, N, O, P, R, S, and T. These characters were obtained from old or poorly maintained ships (Figure 1). These images were carefully selected, to support ship character identification and retrieval systems. Within the context of our research, the presence of a relatively small dataset encompassing only a few alphabets (13) introduces the potential risk of exacerbating mode collapse in the GAN used for data augmentation [33]. This concern informed our strategic decision to concentrate on a specific subset of characters (13 of 26 alphabet characters). Despite this limitation, we hold a strong belief in the broader applicability of our results and conclusions. The images exhibited variations in size and color, but they were preprocessed, to ensure consistency during training and analysis. Specifically, they were normalized to grayscale and resized to 56 × 56 pixels in width and height. The images were stored in standard formats, such as JPEG or PNG. The dataset encompassed various engraving styles commonly found on ships, including embossed, engraved, and painted characters, either individually or in combination. These characters represented ship identification numbers, hull markings, engine component identifiers, and other characters relevant to ship operations. This comprehensive approach enabled us to capture the diversity and complexity of the characters found in real-world ship components. By evaluating the classifiers' performance on both digit and alphabet datasets, we could identify any variations in recognition capabilities and gain a comprehensive understanding of their effectiveness.

3.2. Data Augmentation with GAN Models

We investigated the impact of dataset augmentation using GANs on classifier performance. GANs are powerful tools for data augmentation, as they can generate synthetic data that closely resemble real-world samples [3,31]. We started with an originally sourced dataset of approximately 100 images per character class, and we then systematically increased the dataset size, to about 200 images per character class, using the most suitable GANs for our dataset, WGAN-GP and WGAN-DIV [33]. This augmentation approach strikingly increased the size and diversity of the dataset, enabling the training of a more robust and improved performance of the classifiers.

3.3. Classifier Selection, Metrics, and Model Performance

The choice of classifiers for evaluation was based on their wide usage and effectiveness in character recognition tasks. We considered well-known classic classifiers, such as GNB, RF, KNNs, SVMs, SDG, and DT. These classifiers have demonstrated their efficacy in various pattern recognition applications, and they provided a solid foundation for our comparative analysis. Additionally, hybrid (CNN-RF and CNN-SVMs) and CNN-based classifiers were employed, to compare the performance of our CNN-ISC against these classifiers. The classifiers were evaluated, using standard evaluation metrics, including precision, recall, and F1 score. Table 1 contains each metric description. We implemented traditional classic algorithms, using Scikit-Learn.

Table 1. Summary of Classification Metrics.

Metric	Description	Formula	Interpretation
Accuracy	Measures the overall correctness of predictions.	(TPs + TNs)/ (TPs + TNs + FPs + FNs)	High accuracy indicates good overall performance.
Precision	Measures the accuracy of positive predictions.	TPs/(TPs + FPs)	High precision indicates fewer false positive errors.
Recall	Measures the proportion of actual positives correctly predicted.	TPs/(TPs + FNs)	High recall indicates that most actual positives are correctly predicted.
F1 Score	A harmonic mean of precision and recall.	2 ∗ (Precision ∗ Recall)/ (Precision + Recall)	Balances precision and recall, which is useful when there is an imbalance between classes.

True positives (TPs) are correctly predicted positive cases, true negatives (TNs) are correctly predicted negative cases, false positives (FPs) are incorrectly predicted positive cases, and false negatives (FNs) are incorrectly predicted negative cases. Classification reports were generated, providing detailed insights into each classifier's performance. The metrics were calculated for both the digit dataset and the selected alphabet characters, allowing for a comprehensive assessment of classifier performance across different imprinted character types. In our experimental setup, we split the dataset into training and testing sets, using a 70:30 ratio. By allocating 70% of the data for training, the model could learn the underlying patterns and relationships present in the data, while the remaining 30% was reserved for testing, to assess the model's performance on new, unseen data. During the training phase, the model optimized its parameters and learned from the training data, to minimize errors and improve accuracy. This iterative process continued until the model converged to a state where further training would not lead to significant improvements. After training, the model was then evaluated on the testing set, where it encountered new samples that were not part of the training data. This evaluation allowed us to gauge the model's performance in a real-world scenario, assessing its ability to make accurate predictions on unseen data.

For each CNN model, we aimed to understand the impact of minor modifications to certain hyperparameters. Specifically, we focused on altering the type of activation function, learning rate, optimizer, batch size, and number of epochs, while keeping the number of convolutional layers, dense layers, and the pool size constant. Through this systematic approach, we carefully adjusted these selected hyperparameters for each CNN model, and we documented the results, to identify the optimal parameter values. Our evaluation encompassed several CNNs [34–38]. Additionally, to explore the impact of network design and depth on dataset performance, we modeled our CNN-ISC as a variant of the CNN [34,38]. The CNN-ISC architecture is designed to learn hierarchical representations of the input data, through a series of convolutional and pooling layers. These layers are responsible for extracting important features from the input images and progressively reducing their spatial dimensions. The convolutional layers use a set of learnable filters, to detect patterns and local features in the images, while the max-pooling layers downsample the feature maps, focusing on the most relevant information.

As shown in Figure 4, the CNN-ISC architecture comprises three Conv2D layers, each followed by a ReLU activation function, to introduce nonlinearity. The first Conv2D layer has 32 filters with a 3 × 3 kernel, and the subsequent two Conv2D layers have 64 filters with 3 × 3 kernels. These convolutional layers are responsible for capturing the low-level and high-level features in the input images. After each Conv2D layer, a MaxPooling2D layer with a 2 × 2 pooling size is applied, to reduce the spatial dimensions of the feature maps. This step helps reduce computational complexity and focuses on the most salient features. A flatten layer is added, to transform the 2D feature maps into a 1D vector, preparing the data for the fully connected layers. Two dropout layers are inserted into the architecture, to prevent overfitting. The first dropout layer randomly drops out 50% of the neurons after the flatten layer, and the second dropout layer has the same dropout rate and comes after the first dense layer. The CNN-ISC architecture also includes two dense layers. The first dense layer has 1024 neurons with a ReLU activation function and is followed by L2 regularization, to further prevent overfitting. Finally, the output layer is a dense layer with 13 neurons using the softmax activation function. This allows the model to perform multi-class classification for the 13 alphabets. For digit recognition, we modified the dense layer to 10, corresponding to digits 0–9. The model was compiled using a learning rate of 0.001.

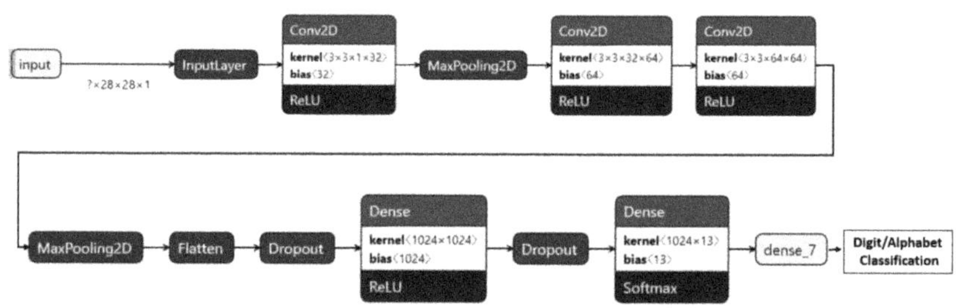

Figure 4. CNN-ISC Architecture.

The hybrid classifiers, CNN-SVMs and CNN-RF, use SVMs and RFs for binary classification, instead of softmax or sigmoid functions. CNNs are used to extract features from input data, capturing hierarchical representations. These extracted features are then fed into the SVMs and RF classifiers, which classify the features into their respective binary classes.

4. Evaluation and Analysis

First, we evaluated the performance of the classic classifiers on two datasets of imprinted ship characters: the imprinted ship digit dataset and the imprinted ship alphabet dataset. We used both the original dataset and a dataset augmented by synthetic data generated using the WGAN-GP and WGAN-DIV models.

We found that the augmented dataset significantly improved the performance of all the classifiers, with the most striking improvements seen for the KNNs and SVMs classifiers. For example, Table 2 shows that the accuracy of the KNNs classifier on the imprinted ship digit dataset increased from 26% to 94% when the augmented dataset was used. The SVMs classifier also showed notable improvement, with its accuracy increasing from 13.8% to 90.6%. Other classifiers also showed improvement, although the gains were not as pronounced. For example, the accuracy of the DT classifier on the imprinted ship digit dataset increased from 13.8% to 66.7%, and the accuracy of the RF classifier increased from 18% to 27.7%. The GNB classifier showed the least improvement, but it still showed some improvement, with its accuracy increasing from 12.7% to 21.6%.

Table 2. Classic Classifiers' Accuracies Across the Digit Datasets.

Classifier	Original Dataset	Augmented Dataset
KNNs	26	94
SVMs	13.8	90.6
DT	13.8	66.7
SDG	18.1	35.8
RF	18	27.7
GNB	12.7	21.6

The augmented dataset also significantly improved the performance of the classifiers on the imprinted ship alphabet dataset (Table 3). The KNNs classifier, for example, had its accuracy increased from 26% to 97%. The SVMs classifier showed notable accuracy improvement (from 20% to 96%). The DT classifier also exhibited substantial accuracy improvement (from 12.8% to 76%). The SDG classifier and RF classifier showed relatively lower improvements after augmentation, but they still showed some improvement. The SDG classifier achieved an initial accuracy of 4.10% on the original dataset, which improved to 51% with augmentation. Similarly, the RF classifier had an accuracy of 2.15% on the original dataset, which increased to 27% after augmentation. The GNB classifier showed the least improvement, but it still showed some improvement (from 2.57% to 40%). These results suggest that the WGAN-DIV and WGAN-GP augmented datasets can notably improve the performance of classifiers for imprinted ship alphabet recognition tasks. The augmentation technique effectively enriches the datasets and provides valuable information for capturing patterns and improving generalization capabilities.

Table 3. Classic Classifiers' Accuracies Across the English Alphabets Dataset.

Classifier	Original Dataset	Augmented Dataset
KNNs	26	97
SVMs	20	96
DT	12.8	76
SDG	4.10	51
RF	2.15	27
GNB	2.57	40

Tables 4 and 5 show the F1 scores of classic classifiers for the digit and alphabet datasets, where A and B represent the original and augmented datasets, respectively. The F1 score is a measure of a classifier's accuracy, incorporating both precision and recall. A high F1 score indicates that the classifier is both accurate and sensitive. Initially, the KNNs and SVMs classifiers had low F1 scores, of 26% and 20%, respectively, for the digit dataset. However, after augmentation by the generated datasets, WGAN-DIV and WGAN-GP, their scores improved significantly, to 97% and 96%, indicating the effectiveness of the augmented dataset. The DT classifier exhibited an initial F1 score of 12.8% for the digit and alphabet datasets, which increased to 76% after augmentation. The SDG and RF classifiers also showed improvements, with F1 scores of 4.10% and 2.15% increasing to 51% and 27%, respectively. The GNB classifier had relatively low F1 scores initially but still showed improvement after augmentation, reaching 40%. The F1 score served as a valuable metric to gauge the classifiers' recognition performance, as it provided a balanced measure of precision and recall. Digits with higher F1 scores demonstrated superior classification, showing a good balance between accurately identifying positive samples (recall) and minimizing FPs (precision). Various factors, including the data distribution and the complexity of the digits, affected classifier performance, leading to variations in their recognition abilities for different digits.

Table 4. Average F1 score (%) for Digits.

Classifier	F1 Score Original Dataset	F1 Score Augmented Dataset
KNNs	22	95
SVMs	06	90
SDG	06	33
RF	05	25
GNB	09	18
DT	13	66

Table 5. Average F1 score (%) for Alphabets.

Classifier	F1 Score Original Dataset	F1 Score Augmented Dataset
KNNs	18	97
SVMs	13	96
SDG	06	28
RF	04	31
GNB	03	29
DT	10	75

Digit 4 consistently demonstrated one of the best performances across various classifiers, with high F1 scores. For instance, the KNNs classifier (Figure 5) excelled in recognizing digit 4, achieving an impressive F1 score of 96%, showcasing its ability to accurately classify positive samples while minimizing FPs. Similarly, in the SVMs classifier (Figure 6), digit 4 achieved the highest F1 score of 94%, indicating accurate classification with well-balanced precision and recall. However, some classifiers faced challenges in recognizing certain digits. Figure 7 shows the performance of the RF classifier. RF struggled with digits, and the SGD classifier in Figure 8, although achieving relatively lower overall F1 scores, performed relatively well for digit 9. These findings provide valuable insights for selecting suitable classifiers and designing effective digit recognition models for various applications.

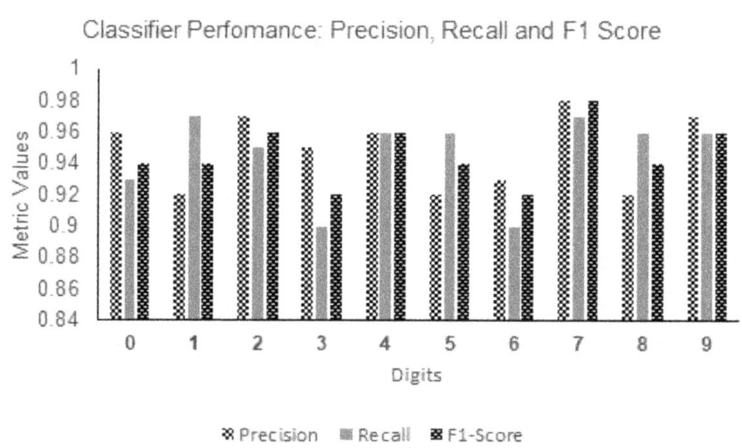

Figure 5. Metric Chart for Digits on the K-Nearest Neighbors Classifier.

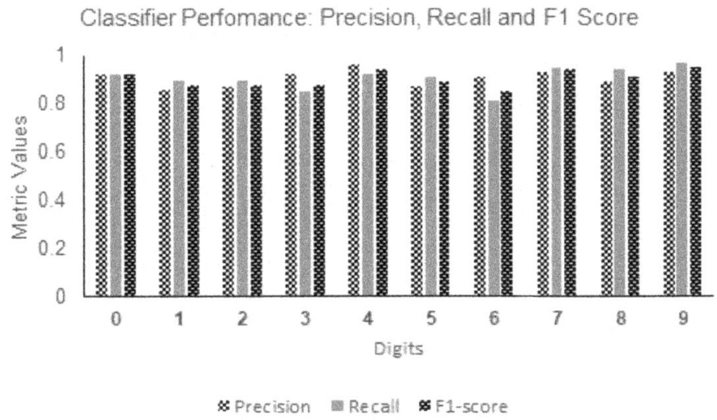

Figure 6. Metric Chart for Digits on the Support Vector Machines Classifier.

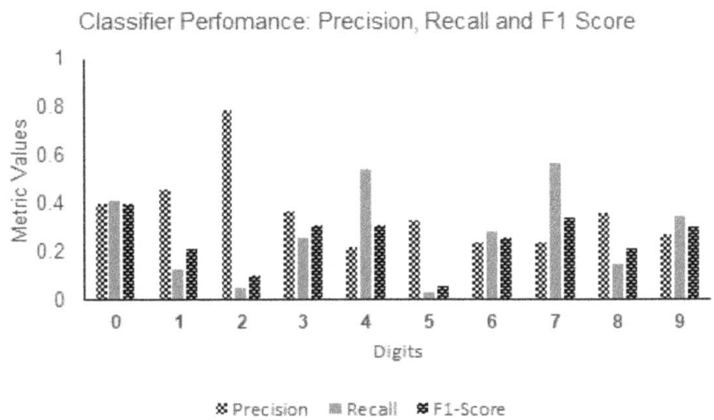

Figure 7. Metric Chart for Digits on the Random Forest Classifier.

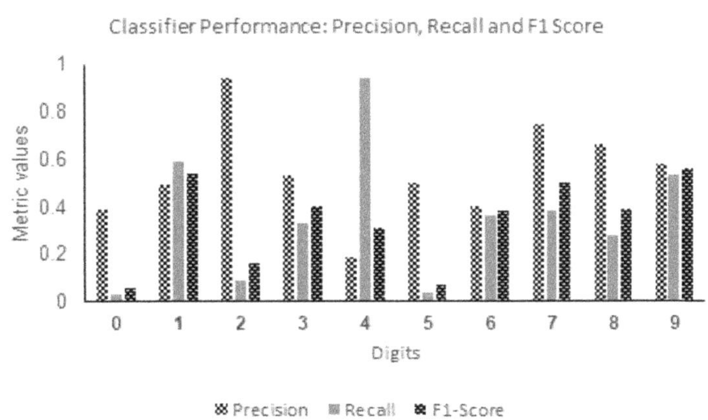

Figure 8. Metric Chart for Digits on the Stochastic Gradient Descent Classifier.

For the alphabets, the letters "C", "P", and "I" consistently achieved high F1 scores across various classifiers, including SVMs (Figure 9), KNNs (Figure 10), and GNB (Figure 11). These letters showcased excellent recognition capabilities, with F1 scores ranging from 0.43 to 0.99. Conversely, the letters "E", "S", and "R" exhibited relatively lower F1 scores across the classifiers, indicating the need for further optimization in classifier models, to improve their recognition accuracy. The F1 scores for these letters ranged from 0.19 to 0.97 across the KNNs, SVMs, SDG, DT, and GNB classifiers. The variations in performance highlight the impact of different classifiers on recognizing specific alphabet characters.

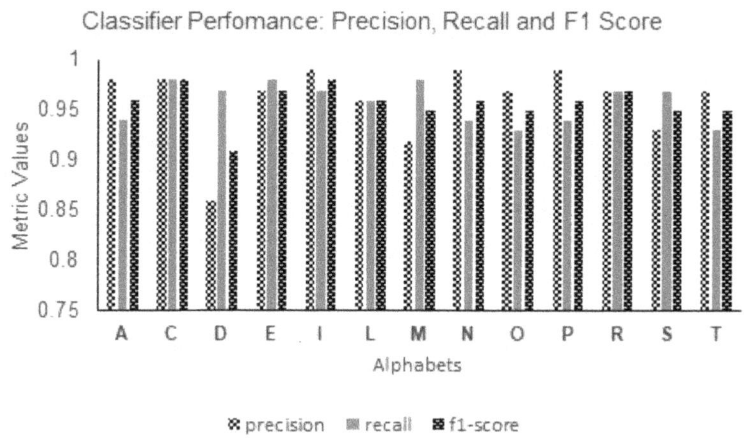

Figure 9. Metric Values for selected Alphabets on the Support Vector Machines Classifier.

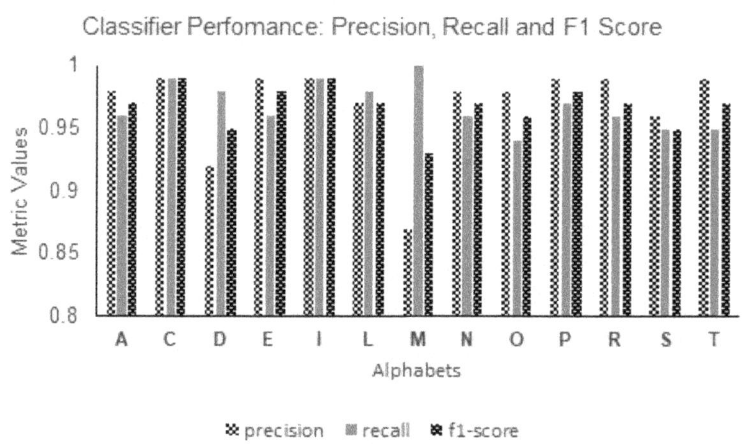

Figure 10. Metric Chart for selected Alphabets on the K-Nearest Neighbors Classifier.

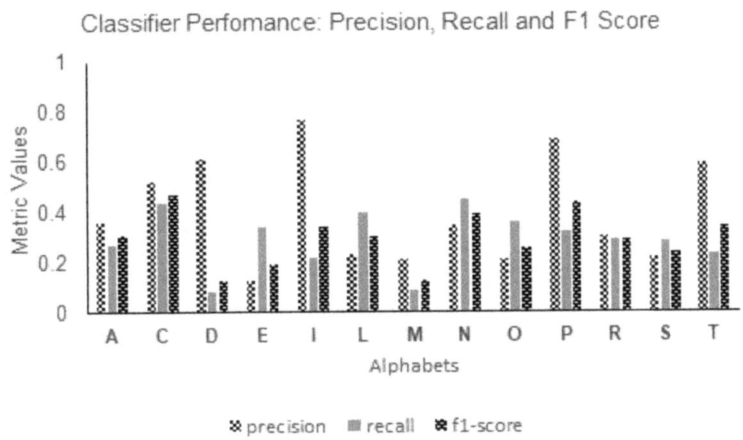

Figure 11. Metric Chart for selected Alphabets on the Gaussian Naive Bayes Classifier.

In our evaluation, we assessed the performance of selected CNN-based models for both digit and alphabet recognition. Tables 6 and 7 present the accuracy results of the CNN models on the original and augmented datasets for digits and alphabets, respectively. All the CNN models demonstrated a notable improvement in performance when trained on the GAN-augmented dataset compared to the original digit dataset. This suggests that the GAN augmentation technique effectively enhanced the models' ability to recognize and classify digits.

Table 6. CNN Accuracies for Digits.

Classifier	Original Dataset	Augmented Dataset
CNN [34]	31.8	99.43
CNN-ISC	10.8	99.7
CNN [35]	47.3	99.0
CNN [36]	43.2	99.5
CNN [38]	48.7	99.2
CNN-SVMs [5]	15.7	97.33
CNN-RF [6]	10.4	96.8

Table 7. CNN Accuracies for Alphabets.

Classifier	Original Dataset	Augmented Dataset
CNN [34]	23.07	99.55
CNN-ISC	26.15	99.85
CNN [35]	24.55	99.61
CNN [36]	32.3	98.93
CNN [38]	18.4	98.6
CNN-SVMs [5]	53	97.16
CNN-RF [6]	44	95.9

Among the evaluated CNN models on the digit datasets, our CNN-ISC model stands out, with the highest accuracy, of 99.7%, making it a top-performing model for digit recognition. The classification plots for both the original and augmented datasets of CNN-ISC are visually represented in Figures 12 and 13, respectively. The accuracy curves for the CNN [35], are shown in Figures 14 and 15. CNN-ISC accomplished the highest accuracy, of 99.85%, on the augmented dataset for the alphabets recognition task. The accuracy curves for both the original and augmented datasets of CNN-ISC are depicted in Figures 16 and 17, respectively. Figures 18 and 19 display the accuracy curves for the CNN [36] on the original and augmented datasets. We also present the performance of the CNN [34] on the original and augmented datasets in Figures 20 and 21, respectively. Additionally, Figures 22 and 23 showcase the precision, recall, and F1 score performance metrics for both digits and alphabets. The performance metric values of CNN-ISC for both digits and alphabets demonstrate the effectiveness of data augmentation in the recognition task.

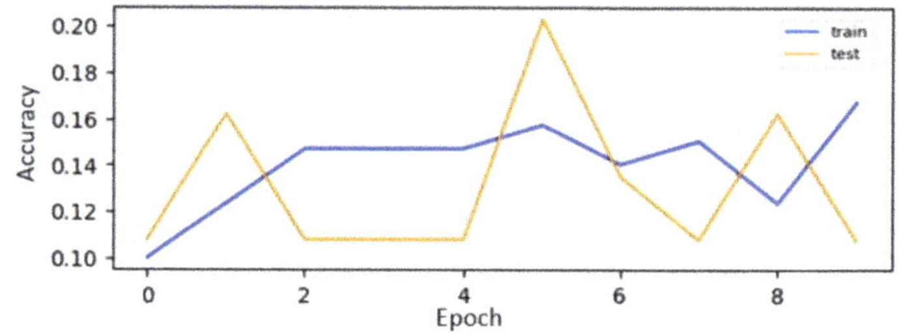

Figure 12. CNN-ISC on the Original Digit Dataset.

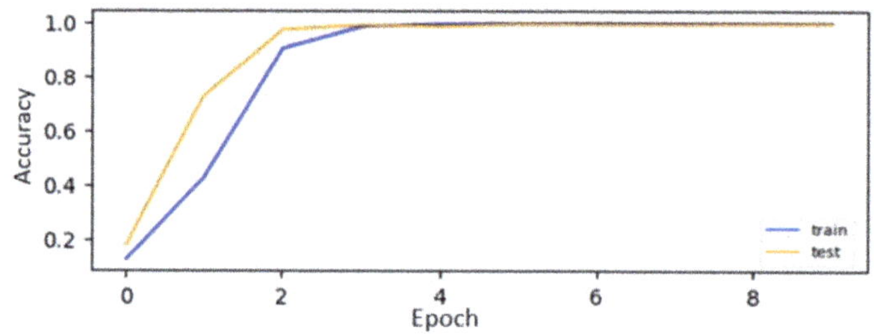

Figure 13. CNN-ISC on the Augmented Digit Dataset.

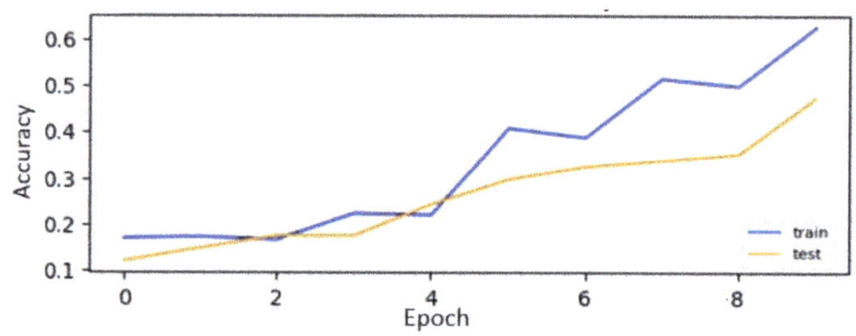

Figure 14. CNN [35] on the Original Digit Dataset.

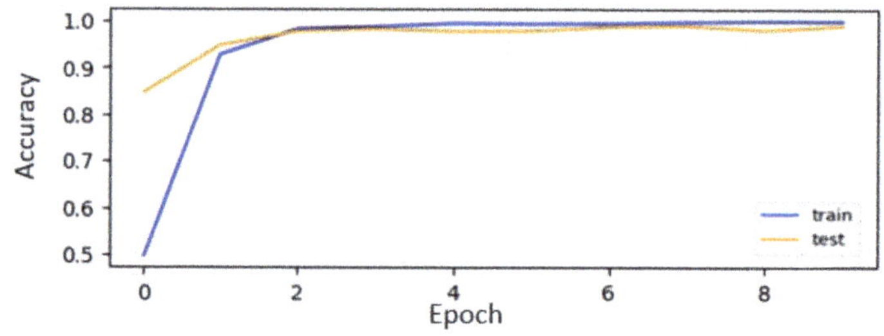

Figure 15. CNN [35] on the Augmented Digit Dataset.

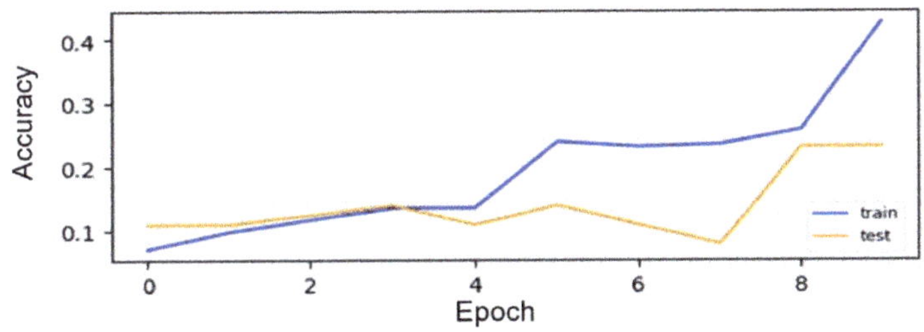

Figure 16. CNN-ISC on the Original Alphabet Dataset.

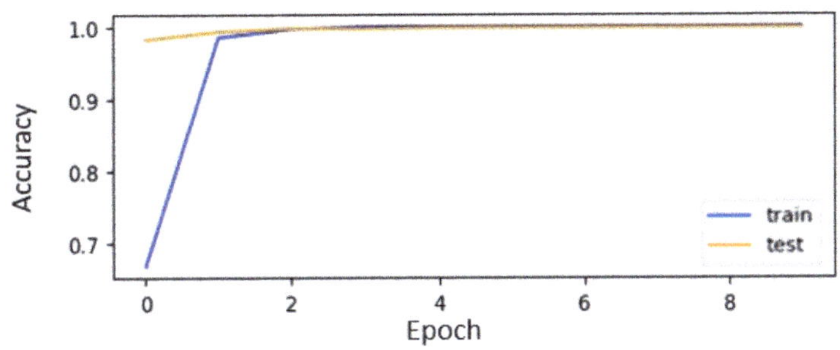

Figure 17. CNN-ISC on the Augmented Alphabet Dataset.

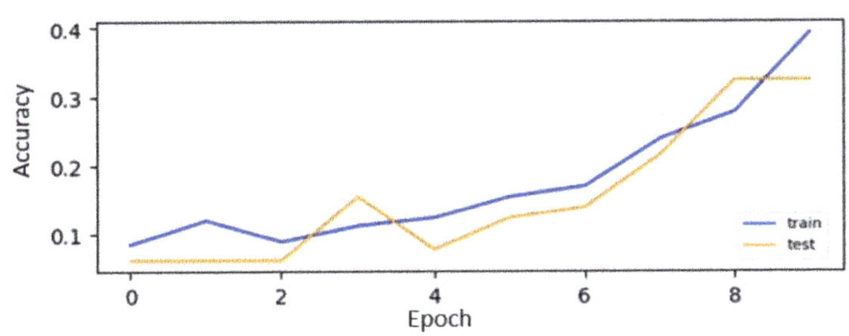

Figure 18. CNN [36] on the Original Alphabet Dataset.

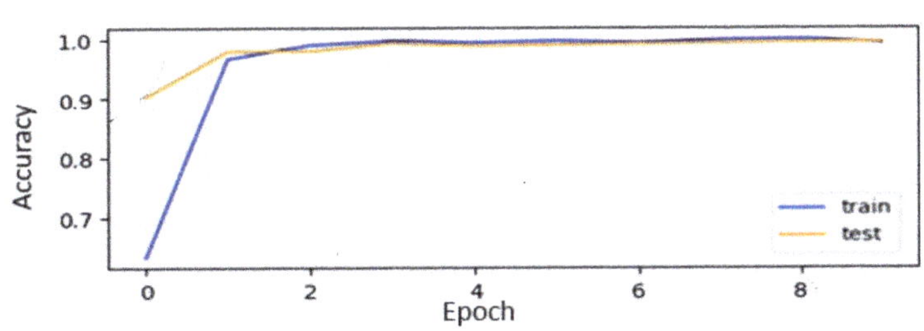

Figure 19. CNN [36] on the Augmented Alphabet Dataset.

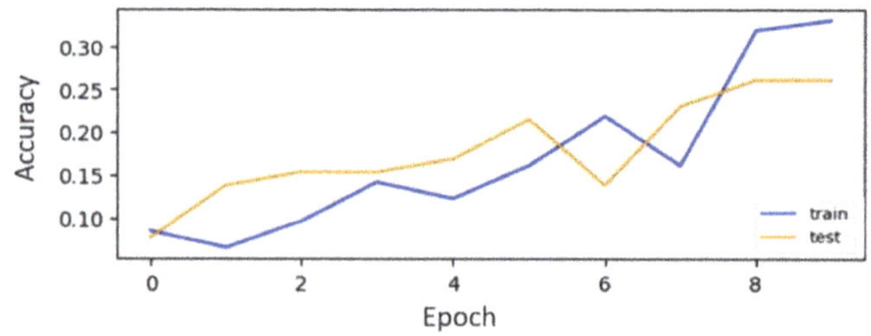

Figure 20. CNN [34] on the Original Alphabet Dataset.

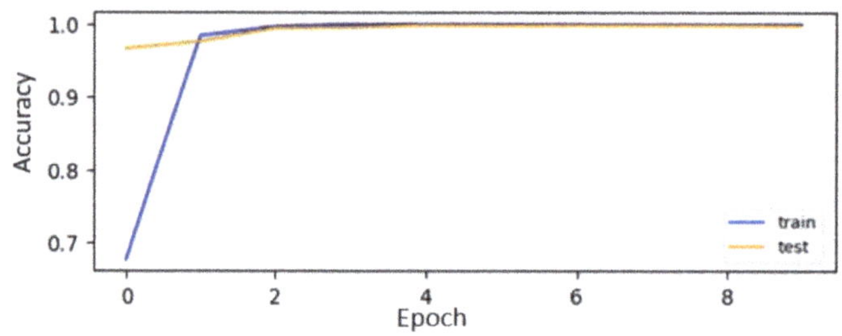

Figure 21. CNN [34] on the Augmented Alphabet Dataset.

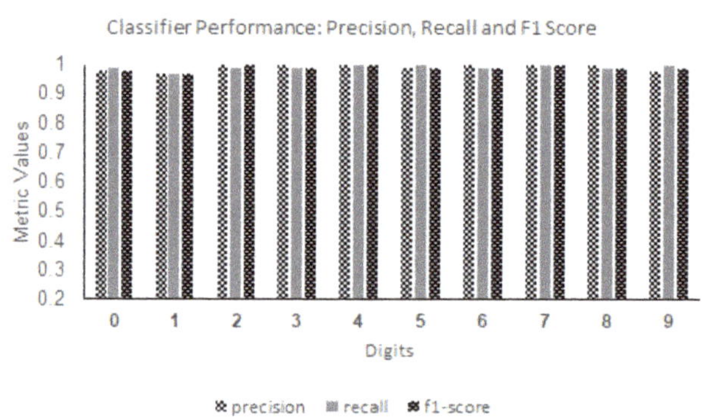

Figure 22. Metric Chart for Digits on CNN-ISC.

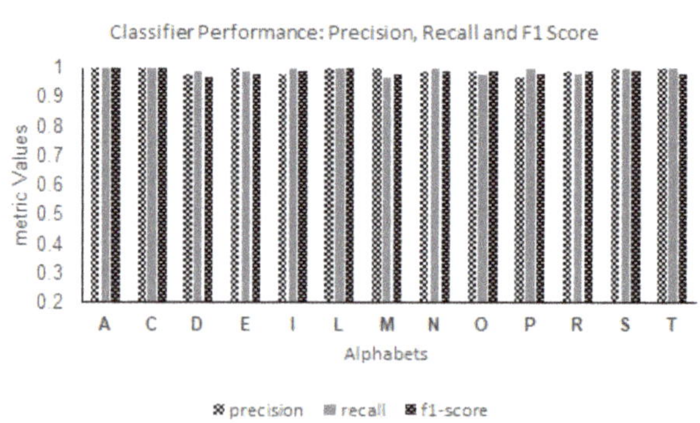

Figure 23. Metric Chart for Alphabets on CNN-ISC.

Alongside the CNN models for the recognition tasks, we also considered hybrid classifiers, namely, CNN-SVMs and CNN-RF. However, the accuracy results presented in Tables 6 and 7 indicate that these hybrid classifiers did not outperform any of the CNN models. Our study underscores the critical role of choosing the right recognition model for character recognition, particularly concerning our dataset. The effectiveness of character recognition models depends on how well they are suited to a specific dataset. We identified CNNs as highly effective for this task, given their exceptional ability to capture intricate character details and use them for precise identification. Notably, different CNN architectures yielded varying results, often due to nuanced differences in their design and hyperparameter configurations. Our finding strongly suggests that the CNN-ISC model is better suited to this specific recognition task, yielding superior performance. This bears significance not only for enhancing character recognition within our dataset but also for adapting our novel model to similar datasets within the same domain. This adaptability is particularly promising, as it suggests broader applicability and potential advancements in character recognition for related applications.

5. Conclusions and Future Work

Overall, CNN-based models, especially CNN-ISC, outperform classic and hybrid classifiers in digit and alphabet character recognition tasks. The remarkable accuracy of CNN-ISC on the dataset underscores its effectiveness in recognizing imprinted characters. The implementation of data augmentation greatly contributes to improved accuracy and F1 scores for both the digit and alphabet datasets, highlighting the importance of this technique in enhancing recognition performance. As part of our future work, we propose expanding the datasets, to include a wider range of ship-component images. This expansion would provide a more diverse and comprehensive representation of imprinted characters, further enhancing classification performance. The efficiency and accuracy enhancements offered by our model have wide-ranging implications for automation and precision across diverse industries, from manufacturing to logistics, agricultural monitoring, and infrastructure maintenance. Additionally, our study emphasizes the importance of considering dataset characteristics when selecting an appropriate model. Different types of datasets may yield varying performance outcomes, potentially revealing a different model as the superior choice over the current best model. As such, further investigations with diverse datasets will shed light on the generalizability and adaptability of different classifiers in various recognition applications.

Author Contributions: A.A., J.T.S. and I.Y.J. conceived and designed the experiments; A.A. and J.T.S. performed the experiments; A.A., J.T.S. and I.Y.J. analyzed the data; A.A. and J.T.S. wrote the paper. I.Y.J. re-organized and corrected the paper. All authors have read and agreed to the published version of the manuscript.

Funding: This work was supported by the National Research Foundation of Korea (NRF) grant, funded by the Korean government (No. 2021R1F1A1064345), and by the BK21 FOUR project, funded by the Ministry of Education, Korea (No. 4199990113966).

Institutional Review Board Statement: Not applicable.

Informed Consent Statement: Not applicable.

Data Availability Statement: No new data were created.

Conflicts of Interest: The authors declare no conflict of interest.

References

1. Chollet, F. *Deep Learning with Python*; Simon and Schuster: New York, NY, USA, 2021.
2. Goodfellow, I.; Pouget-Abadie, J.; Mirza, M.; Xu, B.; Warde-Farley, D.; Ozair, S.; Courville, A.; Bengio, Y. Generative adversarial nets. In Proceedings of the Advances in Neural Information Processing Systems 27: Annual Conference on Neural Information Processing Systems 2014, Montreal, QC, Canada, 8–13 December 2014.

3. You, A.; Kim, J.K.; Ryu, I.H.; Yoo, T.K. Application of generative adversarial networks (GAN) for ophthalmology image domains: A survey. *Eye Vis.* **2022**, *9*, 6. [CrossRef]
4. Xu, Y.; Wang, X.; Wang, K.; Shi, J.; Sun, W. Underwater sonar image classification using generative adversarial network and convolutional neural network. *IET Image Process.* **2020**, *14*, 2819–2825. [CrossRef]
5. Ahlawat, S.; Choudhary, A. Hybrid CNN-SVM Classifier for Handwritten Digit Recognition. *Procedia Comput. Sci.* **2020**, *167*, 2554–2560. [CrossRef]
6. Ramesh, G.; Tejas, M.; Thakur, R.; Champa, H. Handwritten Kannada Digit Recognition System Using CNN with Random Forest. In Proceedings of the International Conference on Information Processing (ICInPro), Bengaluru, India, 22–24 October 2021; pp. 92–104.
7. Bendib, I.; Gattal, A.; Marouane, G. Handwritten Digit Recognition Using Deep CNN. In Proceedings of the International Conference on Intelligent Systems and Pattern Recognition, Virtual, 16–18 October 2020; pp. 67–70. [CrossRef]
8. Prabhu, V.U. Kannada-MNIST: A new handwritten digits dataset for the Kannada language. *arXiv* **2019**, arXiv:1908.01242.
9. Rahaman, M.A.; Mahin, M.; Ali, M.H.; Hasanuzzaman, M. BHCDR: Real-Time Bangla Handwritten Characters and Digits Recognition using Adopted Convolutional Neural Network. In Proceedings of the International Conference on Advances in Science, Engineering and Robotics Technology (ICASERT), Dhaka, Bangladesh, 3–5 May 2019; pp. 1–6.
10. Abdulrazzaq, M.B.; Saeed, J.N. A comparison of three classification algorithms for handwritten digit recognition. In Proceedings of the International Conference on Advanced Science and Engineering (ICOASE), Duhok, Iraq, 2–4 April 2019; pp. 58–63.
11. Darapaneni, N.; Krishnamurthy, B.; Paduri, A.R. Convolution Neural Networks: A Comparative Study for Image Classification. In Proceedings of the International Conference on Industrial and Information Systems (ICIIS), Rupnagar, India, 26–28 November 2020; pp. 327–332.
12. Nguyen, V.; Cai, J.; Chu, J. Hybrid CNN-GRU model for high efficient handwritten digit recognition. In Proceedings of the International Conference on Artificial Intelligence and Pattern Recognition, Beijing, China, 10–16 August 2019; pp. 66–71.
13. Shima, Y.; Nakashima, Y.; Yasuda, M. Handwritten digits recognition by using CNN alex-net pre-trained for large-scale object image dataset. In Proceedings of the International Conference on Multimedia Systems and Signal Processing, Shenzhen, China, 28–30 April 2018; pp. 36–40.
14. Ali, S.; Li, J.; Pei, Y.; Aslam, M.S.; Shaukat, Z.; Azeem, M. An Effective and Improved CNN-ELM Classifier for Handwritten Digits Recognition and Classification. *Symmetry* **2020**, *12*, 1742. [CrossRef]
15. Zhao, H.H.; Liu, H. Multiple classifiers fusion and CNN feature extraction for handwritten digits recognition. *Granul. Comput.* **2020**, *5*, 411–418. [CrossRef]
16. Ghadekar, P.; Ingole, S.; Sonone, D. Handwritten digit and letter recognition using hybrid dwt-dct with knn and svm classifier. In Proceedings of the International Conference on Computing Communication Control and Automation (ICCUBEA), Pune, India, 16–18 August 2018; pp. 1–6.
17. Khan, S.; Ali, H.; Ullah, Z.; Minallah, N.; Maqsood, S.; Hafeez, A. KNN and ANN-based recognition of handwritten Pashto letters using zoning features. *Int. J. Adv. Comput. Sci. Appl.* **2018**, *9*, 570–577. [CrossRef]
18. Pan, M.; Liu, Y.; Cao, J.; Li, Y.; Li, C.; Chen, C.H. Visual Recognition Based on Deep Learning for Navigation Mark Classification. *IEEE Access* **2020**, *8*, 32767–32775. [CrossRef]
19. Suo, Z.; Zhao, Y.; Chen, S.; Hu, Y. BoxPaste: An Effective Data Augmentation Method for SAR Ship Detection. *Remote Sens.* **2022**, *14*, 5761. [CrossRef]
20. Wawrzyniak, N.; Hyla, T.; Bodus-Olkowska, I. Vessel identification based on automatic hull inscriptions recognition. *PLoS ONE* **2022**, *17*, e0270575. [CrossRef]
21. Chen, X.; Wang, S.; Shi, C.; Wu, H.; Zhao, J.; Fu, J. Robust ship tracking via multi-view learning and sparse representation. *J. Navig.* **2019**, *72*, 176–192. [CrossRef]
22. Chen, X.; Ling, J.; Wang, S.; Yang, Y.; Luo, L.; Yan, Y. Ship detection from coastal surveillance videos via an ensemble Canny-Gaussian-morphology framework. *J. Navig.* **2021**, *74*, 1252–1266. [CrossRef]
23. Liu, R.W.; Yuan, W.; Chen, X.; Lu, Y. An enhanced CNN-enabled learning method for promoting ship detection in maritime surveillance system. *Ocean Eng.* **2021**, *235*, 109435. [CrossRef]
24. Huang, S.; Xu, H.; Xia, X.; Zhang, Y. End-to-end vessel plate number detection and recognition using deep convolutional neural networks and LSTMs. In Proceedings of the International Symposium on Computational Intelligence and Design (ISCID), Hangzhou, China, 8–9 December 2018; Volume 1, pp. 195–199.
25. Long, S.; He, X.; Yao, C. Scene text detection and recognition: The deep learning era. *Int. J. Comput. Vis.* **2021**, *129*, 161–184. [CrossRef]
26. Ma, J.; Shao, W.; Ye, H.; Wang, L.; Wang, H.; Zheng, Y.; Xue, X. Arbitrary-oriented scene text detection via rotation proposals. *IEEE Trans. Multimed.* **2018**, *20*, 3111–3122. [CrossRef]
27. Liu, Y.; Zhang, S.; Jin, L.; Xie, L.; Wu, Y.; Wang, Z. Omnidirectional scene text detection with sequential-free box discretization. *arXiv* **2019**, arXiv:1906.02371.
28. Zhang, M.; Yan, Y.; Wang, H.; Zhao, W. An Algorithm for Natural Images Text Recognition Using Four Direction Features. *Electronics* **2019**, *8*, 971. [CrossRef]

29. Wang, H.; Zhu, Y.; Adam, H.; Yuille, A.; Chen, L.C. Max-deeplab: End-to-end panoptic segmentation with mask transformers. In Proceedings of the IEEE/CVF Conference on Computer Vision and Pattern Recognition, Nashville, TN, USA, 20–25 June 2021; pp. 5463–5474.
30. Zhang, W.; Sun, H.; Zhou, J.; Liu, X.; Zhang, Z.; Min, G. Fully convolutional network based ship plate recognition. In Proceedings of the International Conference on Systems, Man, and Cybernetics (SMC), Miyazaki, Japan, 7–10 October 2018; pp. 1803–1808.
31. Abdulraheem, A.; Jung, I.Y. A Comparative Study of Engraved-Digit Data Augmentation by Generative Adversarial Networks. *Sustainability* **2022**, *14*, 12479. [CrossRef]
32. Kassani, S.H.; Kassani, P.H.; Wesolowski, M.J.; Schneider, K.A.; Deters, R. Classification of Histopathological Biopsy Images Using Ensemble of Deep Learning Networks. *arXiv* **2019**, arXiv:1909.11870.
33. Abdulraheem, A.; Suleiman, J.T.; Jung, I.Y. Generative Adversarial Network Models for Augmenting Digit and Character Datasets Embedded in Standard Markings on Ship Bodies. *Electronics* **2023**, *12*, 3668. [CrossRef]
34. Ramprasath, M.; Anand, M.V.; Hariharan, S. Image classification using convolutional neural networks. *Int. J. Pure Appl. Math.* **2018**, *119*, 1307–1319.
35. Ge, D.Y.; Yao, X.F.; Xiang, W.J.; Wen, X.J.; Liu, E.C. Design of high accuracy detector for MNIST handwritten digit recognition based on convolutional neural network. In Proceedings of the International Conference on Intelligent Computation Technology and Automation (ICICTA), Xiangtan, China, 26–27 October 2019; pp. 658–662.
36. Ali, S.; Shaukat, Z.; Azeem, M.; Sakhawat, Z.; Mahmood, T.; ur Rehman, K. An efficient and improved scheme for handwritten digit recognition based on convolutional neural network. *SN Appl. Sci.* **2019**, *1*, 1125. [CrossRef]
37. Siddique, F.; Sakib, S.; Siddique, M.A.B. Recognition of handwritten digit using convolutional neural network in python with tensorflow and comparison of performance for various hidden layers. In Proceedings of the International Conference on Advances in Electrical Engineering (ICAEE), Dhaka, Bangladesh, 26–28 September 2019; pp. 541–546.
38. Jana, R.; Bhattacharyya, S. Character Recognition from Handwritten Image Using Convolutional Neural Networks. In *Advances in Intelligent Systems and Computing*; Springer: Singapore, 2019; p. 922. [CrossRef]

sustainability

MDPI

Article

Internet of Things Assisted Solid Biofuel Classification Using Sailfish Optimizer Hybrid Deep Learning Model for Smart Cities

Mahmoud Ragab [1,2,*], Adil O. Khadidos [1], Abdulrhman M. Alshareef [3], Khaled H. Alyoubi [3], Diaa Hamed [4,5] and Alaa O. Khadidos [3,6]

1 Information Technology Department, Faculty of Computing and Information Technology, King Abdulaziz University, Jeddah 21589, Saudi Arabia
2 Department of Mathematics, Faculty of Science, Al-Azhar University, Naser City 11884, Egypt
3 Information Systems Department, Faculty of Computing and Information Technology, King Abdulaziz University, Jeddah 21589, Saudi Arabia
4 Faculty of Earth Sciences, King Abdulaziz University, Jeddah 21589, Saudi Arabia
5 Geology Department, Faculty of Science, Al-Azhar University, Naser City 11884, Egypt
6 Center of Research Excellence in Artificial Intelligence and Data Science, King Abdulaziz University, Jeddah 21589, Saudi Arabia
* Correspondence: mragab@kau.edu.sa

Abstract: Solid biofuels and Internet of Things (IoT) technologies play a vital role in the development of smart cities. Solid biofuels are a renewable and sustainable source of energy obtained from organic materials, such as wood, agricultural residues, and waste. The integration of IoT technology with solid biofuel classification can improve the performance, quality control, and overall management of biofuel production and usage. Recently, machine learning (ML) and deep learning (DL) models can be applied for the solid biofuel classification process. Therefore, this article develops a novel solid biofuel classification using sailfish optimizer hybrid deep learning (SBFC-SFOHDL) model in the IoT platform. The proposed SBFC-SFOHDL methodology focuses on the identification and classification of solid biofuels from agricultural residues in the IoT platform. To achieve this, the SBFC-SFOHDL method performs IoT-based data collection and data preprocessing to transom the input data into a compatible format. Moreover, the SBFC-SFOHDL technique employs the multihead self attention-based convolutional bidirectional long short-term memory model (MSA-CBLSTM) for solid biofuel classification. For improving the classification performance of the MSA-CBLSTM model, the SFO algorithm is utilized as a hyperparameter optimizer. The simulation results of the SBFC-SFOHDL technique are tested and the results are examined under different measures. An extensive comparison study reported the betterment of the SBFC-SFOHDL technique compared to recent DL models.

Keywords: agricultural residues; biofuel classification; solid fuel; deep learning; sailfish optimizer; IoT environment; smart cities

Citation: Ragab, M.; Khadidos, A.O.; Alshareef, A.M.; Alyoubi, K.H.; Hamed, D.; Khadidos, A.O. Internet of Things Assisted Solid Biofuel Classification Using Sailfish Optimizer Hybrid Deep Learning Model for Smart Cities. *Sustainability* **2023**, *15*, 12523. https://doi.org/10.3390/su151612523

Academic Editors: Mourade Azrour, Azidine Guezzaz, Imad Zeroual, Azeem Irshad, Jamal Mabrouki, Said Benkirane and Shehzad Ashraf Chaudhry

Received: 9 July 2023
Revised: 11 August 2023
Accepted: 15 August 2023
Published: 17 August 2023

1. Introduction

Forestry and agricultural practices yield enormous quantities of waste derived from farm yields [1]. The yearly production of biomass waste worldwide offers management issues, as discarded biomass could have adverse environmental effects [2]. Agricultural biomass residues or wastes are mainly fruit peels, crop stalks, roots, leaves, nutshells or seeds that are generally burned or discarded, but they are a potential supply of feedstock material [3]. Important information for energy applications is the type of fuel to be used because of the necessity of different chemical processes that are needed for the proper processing of the material and to gain optimal results [4].

Additionally, one of the preferred choices for generating power from the fuel is combustion or incineration [5]. However, there are several technological options for minimizing the total emission of gases and waste into the atmosphere while producing power from existing fuels [6]. Moreover, the potential technique is selected based on the input fuel type [7], considering the effect of the energy conversion procedure on air pollution as well as effects on soil and water. For instance, manufactured biomass generally contains high amounts of heavy metals (Ni, Cu, Zn and Cr), while coal-type fuels have more sulphur in them [8]. However, the recovered fuels are heterogeneous mixtures produced from various kinds of solid fuels to attain high availability. Therefore, a particular classification was required for the study of the thermal conversion of solid fuels and it is vital to plan to preprocess and enhance the generation of power [8]. It is essential to consider the effect of energy conversion on air pollution and its impacts on soil and water as well [9]. A practical method to categorize the type of fuel is to consider an expert opinion, but it might be misleading because of human error [10]. Machine learning (ML) is a technique that allows software applications to accurately predict the outcome without being explicitly programmed [11]. ML is an accumulation of methods, producing different inferences from prevailing data through mathematical and statistical approaches [12]. ML has been comprehensively used in domains such as forensics, image processing, prediction, cybersecurity, etc. [13]. There is various research concerning biomass gasification by implementing ML to constitute a regression method [14]. However, these studies do not illustrate the overall outcomes of various ML methods for biomass gasification [15].

This article develops a novel solid biofuel classification using a sailfish optimizer hybrid deep learning (SBFC-SFOHDL) model on the IoT environment. The proposed SBFC-SFOHDL technique performs IoT-based data collection and data preprocessing to transform the input data into a compatible format. Furthermore, the SBFC-SFOHDL technique employs the multihead self attention-based convolutional bidirectional long short-term memory model (MSA-CBLSTM) for solid biofuel classification. For improving the classification accuracy of the MSA-CBLSTM algorithm, the SFO algorithm is utilized as a hyperparameter optimizer. The simulation results of the SBFC-SFOHDL technique are tested and the outcomes are examined under distinct measures.

The rest of the paper is organized as follows. Section 2 provides the related works and Section 3 offers the proposed model. Then, Section 4 gives the result analysis and Section 5 concludes the paper.

2. Related Works

Al-Wesabi et al. [16] present a new approach named IEVB-SFC (intelligent ensemble of voting-based solid fuel classification technique) for harvesting energy from agronomic residue. First, the data preprocessing takes place in three ways similar to data normalization, data transformation and class labeling. As well, the presented approach has three different DL methods: convolutional neural network-based LSTM (CNN-LSTM), GRU and LSTM. At last, an ensemble of three DL methods was carried out through the voting method and determined the suitable solid fuel class labels. In [17], two types of digestate have been observed as effective feedstock to prepare hydrochar implemented as a solid biofuel and porous material. First digestate samples are a clean digestate from common biogas plants, executing the anaerobic digestion of common agronomic residues such as cow dung.

Zheng et al. [18] aim to resolve the low efficiency while extracting novel energy from agricultural waste (AW). The consumption and development of AW were deeply elaborated upon by merging the recent technological progression. To choose the product team organization pattern, the adaptive decision approach of the product team organization pattern was intensely learned for the successful implementation of novel energy mining schemes, and the FNN approach was applied as a decision technique. Bot et al. [19] aimed to examine the economic viability and power utilization of biomass briquettes generation from agricultural waste. This study concentrated on the briquetting conversion of banana peels, coconut shells, sugarcane bagasse and rattan waste depending on small-scale plant production in Cameroon. Bosona et al. [20] aimed to develop and define a traceability system (TS) to ensure the quality of pruning biomass for the generation of solid biofuel and to offer assurance to users that the biomass was in better condition. It was devised for an agricultural pruning supply chain where transporters, agronomists, end users and biomass traders are major actors.

Jifara et al. [21] intended to use a combination of khat stem and corn cob using an integration of co-pelletization and torrefaction processes. To inspect the optimization of co-pelletization parameters, the response surface approach was utilized. A particle size and Torrefied biomass blending ratio were selected as independent factors. The dependent variable was durability, heating value and bulk density of torrefied mixed pellets. Samadi et al. [22] created a study with the purpose of framing a stoichiometric equilibrium technique for predicting the energy production of gasification. This method was authenticated with an experimental dataset for determining the syngas composition. For determining optimum performance characteristics, the impacts of a parameter of operating conditions on the performance of gasification were assessed later. Further, the developed method was utilized for predicting the amount of heat and power gained from various farming residues by gasification.

In [23], numerical simulations were carried out employing computational fluid dynamics (CFD) for evaluating the fluid dynamic strategy and the combustion model of biomass particles under a horizontal cyclonic combustion chamber. Michal et al. [24] examined a conceptual scheme of WSN intended to estimate the SmartCity air quality in realtime. The sensor devices would autonomously monitor the flue gas temperature, CO and particulate matter concentrations. Koval et al. [25] estimated if residential heating affects the quality of air by modeling three provided conditions of a solid fuel boiler altered at chosen places and compared the outcomes with measured data. Akarsu et al. [26] aimed to comparatively estimate the outcome of a hydrothermal carbonization (HTC) condition on the produce and fuel properties of hydrochar attained in food waste (FW) and its digestate (FD).

Though several models are available in the literature, it still remains a challenging problem. Due to incessant deepening of the model, the number of parameters of DL models also rapidly increased and led to model overfitting. At the same time, different hyperparameters have a significant impact on the efficiency of the CNN model, particularly the learning rate. The learning rate parameter also needs to be modified to obtain better performance. Therefore, in this study, we employ an SFO technique for the hyperparameter tuning of the MSA-CBLSTM model.

3. The Proposed Model

In this study, we concentrated on the improvement of the SBFC-SFOHDL model in the IoT platform. The main purpose of the proposed SBFC-SFOHDL methodology lies in the proficient detection and classification of solid biofuels from agricultural residues in the IoT platform. To achieve this, the SBFC-SFOHDL system includes IoT-based data collection, data preprocessing, MSA-CBLSTM-based solid fuel classification and SFO-based hyperparameter tuning. Primarily, the input data are preprocessed to transform the input data into a compatible format. Next, the classification process take place using the MSA-CBLSTM model, which categorizes the solid biofuels into different classes. Finally, the SFO algorithm is executed to adjust the hyperparameter values of the DL model. Figure 1 demonstrates the workflow of the SBFC-SFOHDL algorithm.

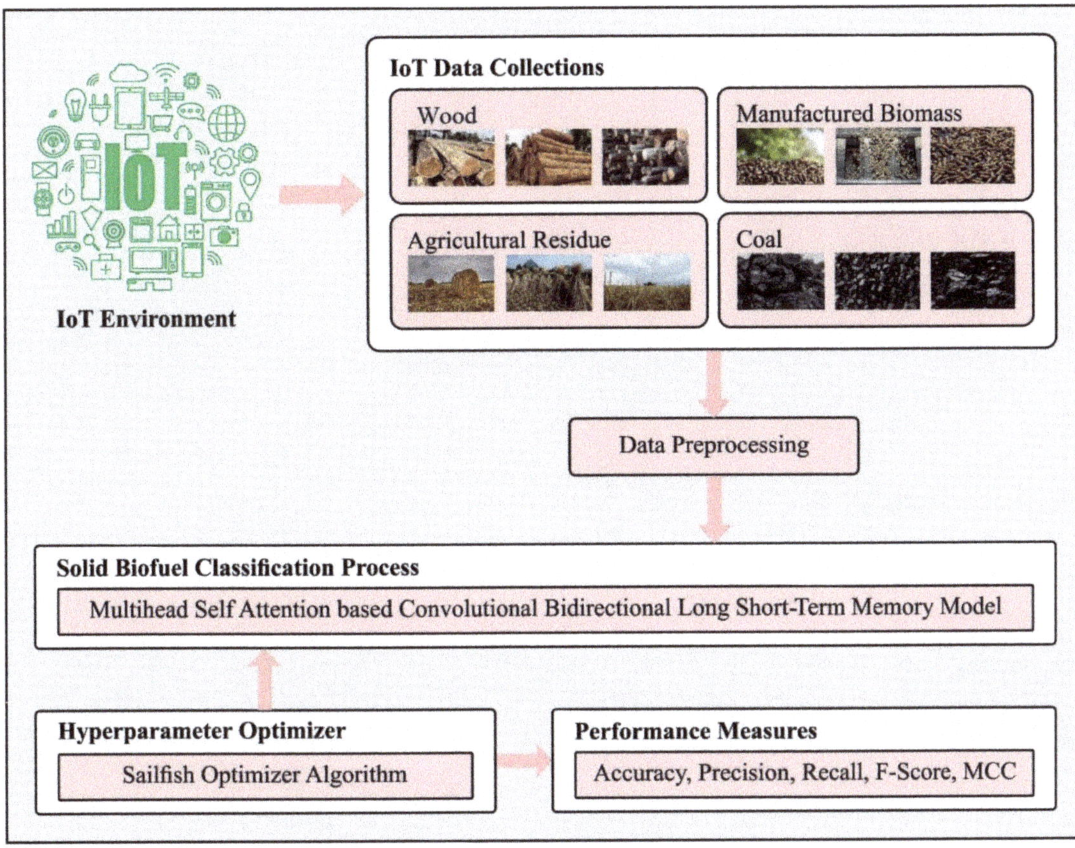

Figure 1. Workflow of SBFC-SFOHDL approach.

3.1. Data Preprocessing

In the initial stage, the actual data is preprocessed in three various approaches such as data normalization, data transformation and class labeling. Primarily, the data in categorical values can be suitably transformed as mathematical values. Secondarily, the class labeling procedure can be carried out but the data samples can be assigned to suitable class labels. Eventually, the experimental value can be changed as a standard method by discarding and scaling the mean to unit variance.

3.2. Solid Biofuel Classification

To classify solid biofuels in the IoT environment, the MSA-CBLSTM technique can be used. In this study, the training module considered is a bidirectional LSTM (Bi-LSTM) with a multi-head self-attentive model in conjunction with one-layer CNN architecture, represented as an MSA-CBLSTM model [27]. LSTM takes place in the consecutive signal analysis via sharing weight, and then the weight between the output and hidden layers is recycled. It is a chain model to process time sequences and it could efficiently compensate for the disappearing gradient problems. The bidirectional EEG signal extraction model extracts dynamic data from the prior and latter segments in comparison to the unidirectional EEG signal extraction model. An LSTM contains three gating control mechanisms of input and the forget gates and the computation formula are shown as follows:

$$f_t = \sigma\left(W_f \cdot [h_{t-1}, x_t]\right) + b_f\Big), \tag{1}$$

$$i_t = \sigma(W_i \cdot [h_{t-1}, x_t]) + b_i), \tag{2}$$

$$\widetilde{C} = \tanh(W_C \cdot [h_{t-1}, x_t]) + b_C), \tag{3}$$

$$C_t = f_t \times C_{t-1} + i_t \times \widetilde{C}_t \tag{4}$$

$$O_t = \sigma(W_O \cdot [h_{t-1}, x_t]) + b_O), \tag{5}$$

$$h_t = \tanh(C_t) \times O_t, \tag{6}$$

From the expression, O_t signifies the output gate, W represents the weight matrix, x_t refers to the time series at t time, C_t denotes the cell state, h_t represents the hidden layer, σ shows the sigmoid function, $\widetilde{C}_t$ indicates the temporary cell state, i_t indicates the memory gate, b signifies the bias vector of respective weight and f_t denotes the forget gate. The memory gate is accountable for updating the cell state of the LSTM, where the input gate controls the output values to the following LSTM cell:

$$y_t = \sigma(W_h \cdot [h_t, h_t']) + b_h), \tag{7}$$

The Bi-LSTM adds a backward layer for learning the upcoming data, which is the addition of historical data. The Bi-LSTM perfectly combines the bidirectional characteristics and gating structure such that further details can be processed and remembered by the two LSTM components. The time series data are inputted into the algorithm, then the forward layer interconnects the feature data in the historical sequence with current data, later the backward layer connects the future data, and lastly, the forecast values are outputted using Equation (7).

The Transformer method is an autoregressive generative model that largely exploits sinusoidal location data and a self-attentive mechanism. Each layer involves a dropout, a pre-feedback network, a time-self-noticing and residual network sublayers. Figure 2 represents the architecture of a Bi-LSTM.

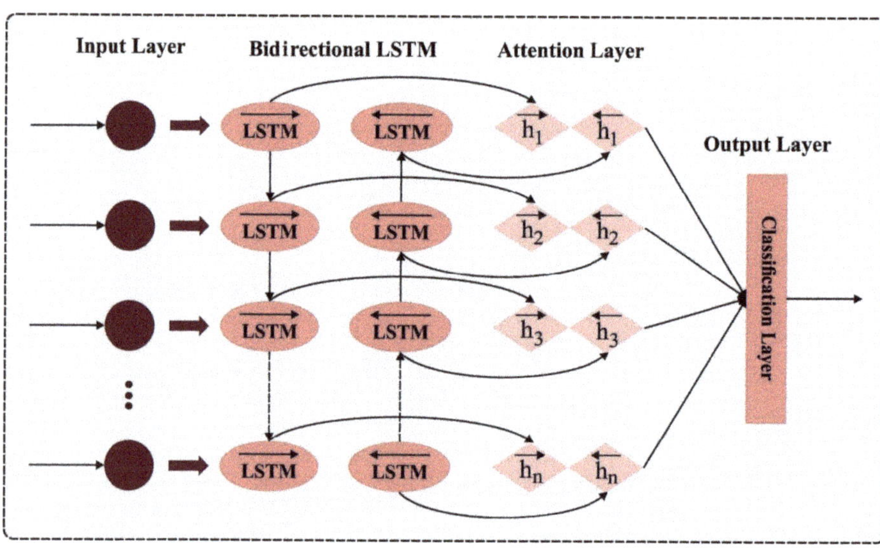

Figure 2. Structure of a Bi-LSTM.

The attention mechanism usually allows for a weight factor to be applied to all the elements in the EEG series, and if one component is stored, the attention mechanism is computed as a similarity between Q and K that reflects the significance of the extracted V value, and the weight is summed and weighted to attain the attention value:

$$Attention\ (Q, K, V) = softma\chi\left(\frac{QK^T}{\sqrt{d_k}}\right)V \tag{8}$$

The multi-head self-attention module attains various representations of $h\ (Q, K, V)$, evaluates the self-attention of every h representation and interconnects the outcomes and is represented as follows:

$$head_i = Attention\ \left(QW_i^Q,\ KW_i^K, VW_i^V\right), \tag{9}$$

$$MultiHead(Q, K, V) = Contact(head_i, \ldots, head_h)W^0, \tag{10}$$

In Equation (10), W_i and W^0 make the parameter matrix. Meanwhile, as Bi-LSTM considers the location data, it is not necessary to set up further position encoding. We apply scaled dot product attention in the process of implementing the self-attention module in Bi-LSTM. The output H_D of the last time step is multiplied with *the* W_Q matrix as Query, whereas the output O_t of all the time steps is linearly converted as Key_t and Value:

$$Query = \omega_Q H_D, \tag{11}$$

$$Value_t = \omega_V O_t, \tag{12}$$

$$Key_t = \omega_K O_t, \tag{13}$$

$$e_t = \frac{QueryKey_t^T}{\sqrt{d_k}}, \tag{14}$$

$$a_t = \frac{exp(e_t)}{\Sigma_{t=0}^{n} exp(e_t)}, \tag{15}$$

The Query does not change with the time step and $\omega_Q, \omega_K, \omega_V$ denote the parameter of the NN that is adapted with BP. $Value_t$ and weights a_t at all the time steps are summed and weighted to attain the emotional feature vector with a self-attentive model:

$$z(Q, K, V) = \sum_t a_t Value_t \tag{16}$$

The abovementioned formula is accomplished h times to attain multi-head self-attention features $z_1, \ldots, z_h$, which is merged and linearly converted:

$$MultiHead\ (Q, K, V) = Concat(z_1, \ldots, z_h)\omega_z \tag{17}$$

3.3. Hyperparameter Tuning Using SFO Algorithm

For hyperparameter tuning of the MSA-CBLSTM algorithm, the SFO algorithm is utilized to improve the classifier results. Sailfish are a group of predators that contribute to harassing and catching their beasts [28]. In the group game, the hunter uses different approaches to assault. The sailfish culture can be determined as an alternative attack strategy. It comprises the group leader who targeted and harmed or killed the sardine (prey school) itself, whereas others saved the resource. It changes its location with sailfish that attack the beast farm. Furthermore, the fish are capable of migrating to the sardine location and maximizing the prey chasing. The target group (sardine) changes the location the party member is wounded to avoid the future attack from the sailfish. Thus, the sailfish optimizer technique is applied. As compared to the other optimization techniques, the SFO algorithm includes succeeding features. In recent work, the metaheuristic approach has proficiently managed different optimization problems because of its ability for flexible exploration and diversification, which are the two notable characteristics of the metaheuristic method that could search the whole solution space in all the iterations for better solutions, except for local optimal, intensification or exploitation, and it leads to quick convergence and determines the potential solution. An optimal metaheuristic method attempts to balance exploitation and exploration. The stepwise process of the SFO technique is shown as follows:

3.3.1. Initialization of Population

The sardine and sailfish populations are initialized at random. The arbitrary position for all the sailfishes is w_χ^q and the arbitrary place is v_y^q for all the sardines, where $\chi \in$ {Sailfish}, $y \in$ {Sardines} and $q \in$ {Iteration value}. The location of all the sailfishes w_x^q or sardines v_y^q is a possible choice for q-th iterations.

3.3.2. Mechanism and Evaluation of Elitism

According to the fitness function (FF), Fnes as the location for all the newest populations is determined by all the quest agents (sardine or sail). As the sardine injured is processed as v_{ini}^q, $inj \in$ {Sardineset}, for instance, Fnes $(v_{inj}^q) \leq$ Fnes $\left(v_y^q\right), \forall q$, this is the most efficient sardine in the sardine population. Furthermore, as the elite fish w_{elit}^q, $elit \in$ {Sailfish set}, i.e., Fnes $(w_{elit}^q e) \leq$ Fnes $\left(w_x^q\right)$, $\forall q$ in impedance conundrum, the better sailfish with lower fitness in the sailfish population is sustained.

3.3.3. Sailfish Position Updating

In the iteration, any sailfish member of a group can change its location. The modification in the place of the sailfish operation can be performed by taking free space on the prey farm or by changing the attack technique. The transition of Sailfish is focused mostly on elite and injured sardine locations demonstrated in Equation (18).

$$w_\chi^{q+1} = w_{elit}^q - \lambda_q * \left(\beta * \left(\left(\frac{w_e^q lite + g_{inj}^q}{2}\right) - w_\chi^q\right)\right), \tag{18}$$

where w_χ^{q+1} represents the new location of the sailfish at $(q+1)$th iterations. The location w_χ^q becomes the sailfish existing location q. β shows the random integer amongst [0, 1], w_{elite}^q signifies the location of the existing elite sailfish and g_{ini}^q signifies the location of the presently injured sardine. λ_q denotes the coefficient produced at every q-th iteration as follows:

$$\lambda_q = (2 \times \beta \times Dst) - Dst, \tag{19}$$

In Equation (19), β represents the arbitrary integer in the range of zero and one, and Dst shows the scale prey density. It results in the reduction in prey attacking the prey farm; sailfish are injured and eat the sardines. The variable Dst can be determined as follows:

$$Dst = 1 - \left(\frac{NM_{sail}}{NM_{sard} + NM_{sail}}\right), \tag{20}$$

In Equation (20), NM_{sard} indicates the number of sardines and NM_{sail} shows the number of sailfish. The primary sardine is greater than the sailfish population. It is predicted to be $NM_{sard} = 3 \times NM_{sail}$. The λ_q fluctuation value is an essential component in the model because every fish adjusts its position by raising the λ_q fluctuation value.

3.3.4. Sardine Position Update

Initially, at stalking, the capability for sardines and power attacks is appreciated to escape. The defensive skill of sailfish and the capability to escape the sardines diminishes with time. The sailfish harm the sardine without the ability to capture them. The sailfish is mainly responsible for making an effort to modify assault skills, whereas the sardine is responsible for corporeal damage. The growth rate in finding sailfish is increasing. In response to the sailfish attack, sardine action must be considered. All the sardines are used for adjusting their position as follows:

$$v_y^{q+1} = z \times \left(w_{elit}^q - v_y^q + PR_{atk}\right), \tag{21}$$

In Equation (21), v_y^{q+1} denotes the new sardine location and v_y^q shows the present sardine location. z indicates the random integer within [0, 1], w_{elit}^q represents the better location of elite sailfish and pR_{atk} defines the sailfish attack strength at all the iterations, implemented as:

$$P_{atk} = Q \times (1 - (2\times|r \times \varepsilon)) \tag{22}$$

In Equation (22), the two factors Q and ε denote a decrease in the existing iteration value of attack power (PR_{atk}) and the number of existing iterations. At first, the success rate is poor since most sardine transition positions prevent the attack. The sardine's ability to escape though decreases after fishing, thereby increasing the success rate. The quantity of sardines that can be modified reduces over time. At last, hunting is taken into account, once the power attack PR_{otk} is less than 0.5. Finally, the number of sardines rises in the location based on the assault of power $(PR_{atk} < 0.5)$ as follows:

$$\alpha = NM_{sard} \times PR_{atk}, \tag{23}$$

In Equation (23), PR_{atk} is less than 0.5, and only the selected number modifies its position. At the same time, each sardine should be modified if the PR_{atk} is higher than 0.5. Once the sailfish x is hunched, the sardine place is replaced with the sardine y. The succeeding Fnes $\left(v_y^q\right) < Fnes\ (w_\chi^q)$, $\forall q$ is thereby attained as follows:

$$w_\chi^q = v_y^q \ if\ Fnes\left(v_y^q\right) < Fnes\left(w_\chi^q\right), \tag{24}$$

In Equation (24), w_χ^q indicates the position of sailfish x at q-th iterations and v_y^q represents the location of the sardine at qth iterations. While extracting the sardine, the sardine should be isolated in the population. Algorithm 1 demonstrates the pseudocode of the SFO.

Algorithm 1 Pseudocode of SFO algorithm

```
Begin
The population of sailfish and sardine are arbitrarily initialized
Set parameter (q = 4, ε = 0.001)
Evaluate the fitness of sailfish and sardine
Choose the better sailfish and sardine and define them as injured sardine and elite sailfish,
correspondingly
While the ending condition is not fulfilled
        For all the sailfishes
                Evaluate λ_q using based on Equation (19)
                Upgrade the position of the sailfish based on (18)
        End for
        Evaluate the attack power based on (22)
                                        If PR_atk < 0.5
                        Evaluate α based on (23)
                        Choose a set of sardines based on the α value
```

Algorithm 1 *Cont.*

Upgrade the position of the selected sardine based on Equation (24)
Else
Upgrade the position of the sardine based on Equation (24)
End if
Evaluate each sardine fitness
If the best solution for the sardine population
Exchange the sailfish alongside the wounded sardine
Remove the hunted sardine from the population
Upgrade the better sailfish with sardine
End if
End while
Return better sailfish obtained so far

The SFO methodology not only develops an FF to achieve optimal accuracy of the classifier, but it also defines a positive integer to signify the enhanced efficiency of solution candidates. The decline in the classification error rate is observed as FF.

$$fitness(x_i) = \frac{number\ of\ misclassified\ samples}{Total\ number\ of\ samples} * 100 \tag{25}$$

4. Results and Discussion

The performance analysis of the SBFC-SFOHDL method is tested by the datasets [29] including 585 samples with four different classes such as manufactured biomass (MB), coal, wood and agricultural residues (AR).

The confusion matrices of the SBFC-SFOHDL approach on solid fuel classification performance are exemplified in Figure 3. The outcome highlighted that the SBFC-SFOHDL method classifies four class labels accurately. It is noticed that the classification increases with an increase in the number of epochs.

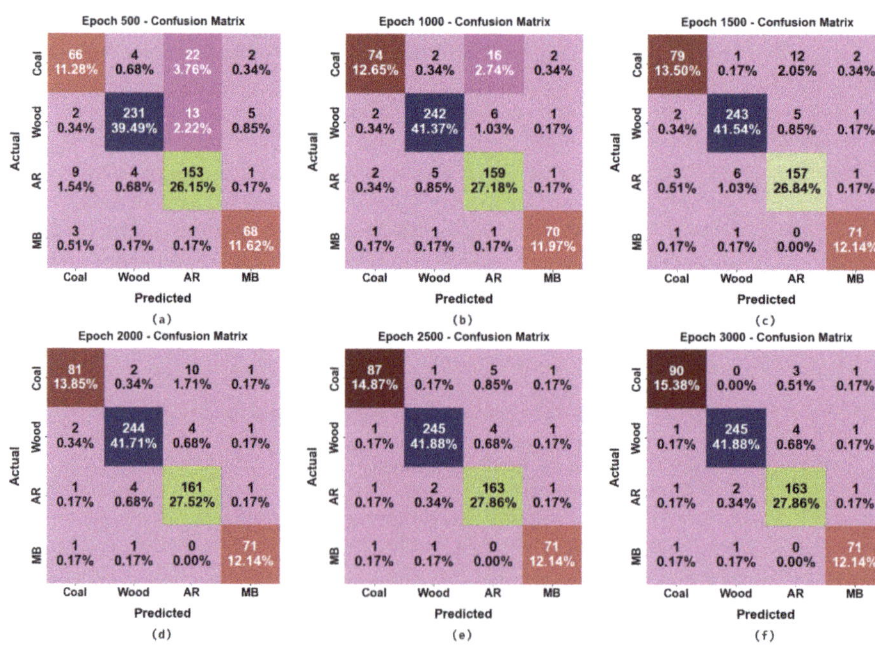

Figure 3. Confusion matrices of SBFC-SFOHDL approach: (**a–f**) epochs 500–3000.

The overall outcomes of the SBFC-SFOHDL algorithm under various epochs are demonstrated in Table 1. Figure 4 signifies the average results of the SBFC-SFOHDL approach with respect to $accu_y$. The results indicate that the SBFC-SFOHDL system obtains higher $accu_y$ values under all epochs. For instance, with 500 epochs, the SBFC-SFOHDL technique attains an average $accu_y$ of 94.27%. Similarly, with 1500 epochs, the SBFC-SFOHDL technique achieves an average $accu_y$ of 97.01%. Concurrently, with 3000 epochs, the SBFC-SFOHDL method accomplishes an average $accu_y$ of 98.63%.

Table 1. Classifier outcome of SBFC-SFOHDL approach with varying epochs.

No. of Epochs	Classes	$Accu_y$	$Prec_n$	$Reca_l$	F_{Score}	MCC
Epoch 500	Coal	92.82	82.50	70.21	75.86	72.00
	Wood	95.04	96.25	92.03	94.09	89.89
	AR	91.45	80.95	91.62	85.96	80.16
	MB	97.78	89.47	93.15	91.28	90.03
	Average	94.27	87.29	86.75	86.80	83.02
Epoch 1000	Coal	95.73	93.67	78.72	85.55	83.50
	Wood	97.09	96.80	96.41	96.61	94.07
	AR	94.70	87.36	95.21	91.12	87.52
	MB	98.80	94.59	95.89	95.24	94.56
	Average	96.58	93.11	91.56	92.13	89.91
Epoch 1500	Coal	96.41	92.94	84.04	88.27	86.31
	Wood	97.26	96.81	96.81	96.81	94.42
	AR	95.38	90.23	94.01	92.08	88.86
	MB	98.97	94.67	97.26	95.95	95.37
	Average	97.01	93.66	93.03	93.28	91.24
Epoch 2000	Coal	97.09	95.29	86.17	90.50	88.95
	Wood	97.61	97.21	97.21	97.21	95.12
	AR	96.58	92.00	96.41	94.15	91.79
	MB	99.15	95.95	97.26	96.60	96.11
	Average	97.61	95.11	94.26	94.62	92.99
Epoch 2500	Coal	98.29	96.67	92.55	94.57	93.58
	Wood	98.29	98.39	97.61	98.00	96.51
	AR	97.78	94.77	97.60	96.17	94.62
	MB	99.15	95.95	97.26	96.60	96.11
	Average	98.38	96.44	96.26	96.33	95.21
Epoch 3000	Coal	98.80	96.77	95.74	96.26	95.55
	Wood	98.46	98.79	97.61	98.20	96.86
	AR	98.12	95.88	97.60	96.74	95.42
	MB	99.15	95.95	97.26	96.60	96.11
	Average	98.63	96.85	97.05	96.95	95.99

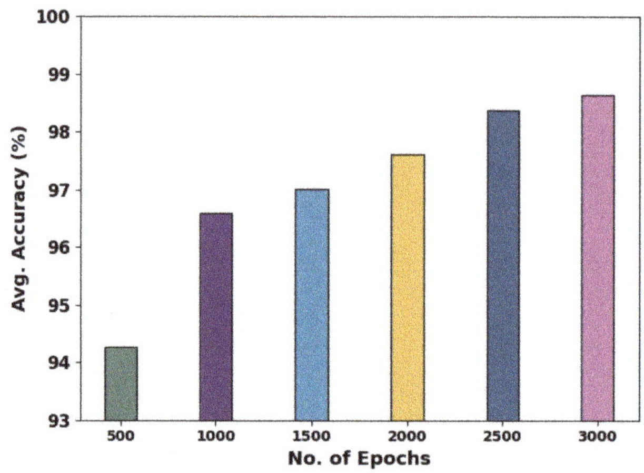

Figure 4. Average $accu_y$ outcome of SBFC-SFOHDL approach under varying epochs.

Figure 5 represents the average results of the SBFC-SFOHDL method in terms of $prec_n$ and $reca_l$. The results indicate that the SBFC-SFOHDL method achieves increasing $prec_n$ and $reca_l$ values under all epochs. For example, with 500 epochs, the SBFC-SFOHDL technique accomplishes an average $prec_n$ and $reca_l$ of 87.29% and 86.75%. Similarly, with 1500 epochs, the SBFC-SFOHDL method attains an average $prec_n$ and $reca_l$ of 93.66% and 93.03%. Concurrently, with 3000 epochs, the SBFC-SFOHDL technique attains an average $prec_n$ and $reca_l$ of 96.85% and 97.05%.

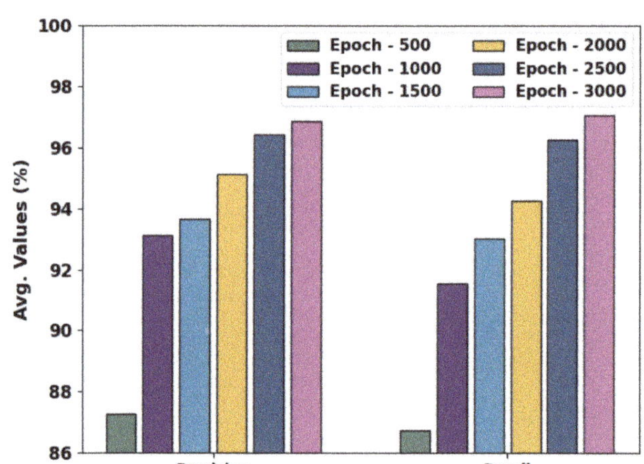

Figure 5. Average $prec_n$ and $reca_l$ outcome of SBFC-SFOHDL approach under varying epochs.

Figure 6 represents the average results of the SBFC-SFOHDL method in terms of the F_{score} and MCC. The results indicate that the SBFC-SFOHDL technique attains increasing F_{score} and MCC values under all epochs. For example, with 500 epochs, the SBFC-SFOHDL technique attains an average F_{score} and MCC of 86.80% and 83.02%. Similarly, with 1500 epochs, the SBFC-SFOHDL method attains an average F_{score} and MCC of 93.28% and 91.24%. Concurrently, with 3000 epochs, the SBFC-SFOHDL technique attains an average F_{score} and MCC of 96.95% and 95.99%.

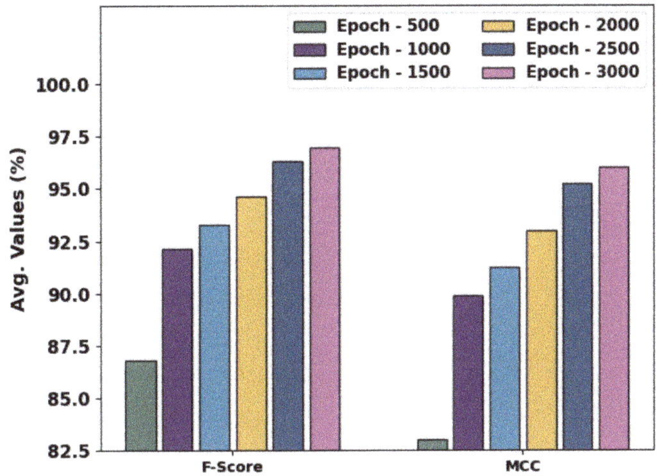

Figure 6. Average F_{score} and MCC outcome of SBFC-SFOHDL approach under varying epochs.

Figure 7 examines the $accu_y$ of the SBFC-SFOHDL method in the training and validation method on the test database. The outcome demonstrated that the SBFC-SFOHDL methodology achieves enhancing $accu_y$ values over higher epochs. Furthermore, the maximal validation $accu_y$ over training $accu_y$ displays that the SBFC-SFOHDL method is attained effectively on the test database.

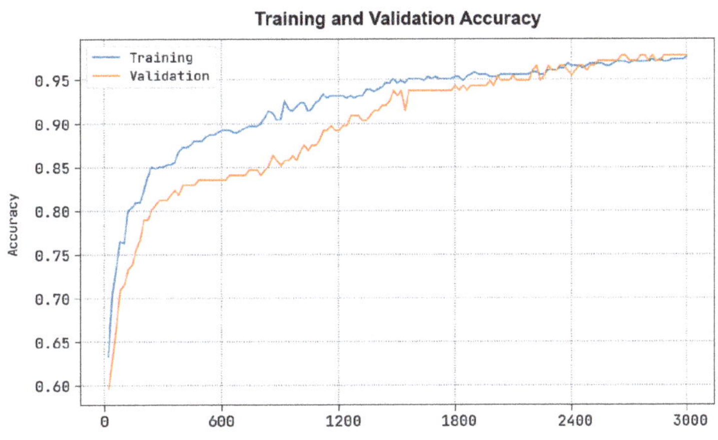

Figure 7. Accuracy curve of the SBFC-SFOHDL approach.

The loss examination of the SBFC-SFOHDL method at the time of training and validation is demonstrated on the test database in Figure 8. The outcome indicates that the SBFC-SFOHDL methodology attains nearby values of training and validation loss. The SBFC-SFOHDL method acquired the values capably on the test database.

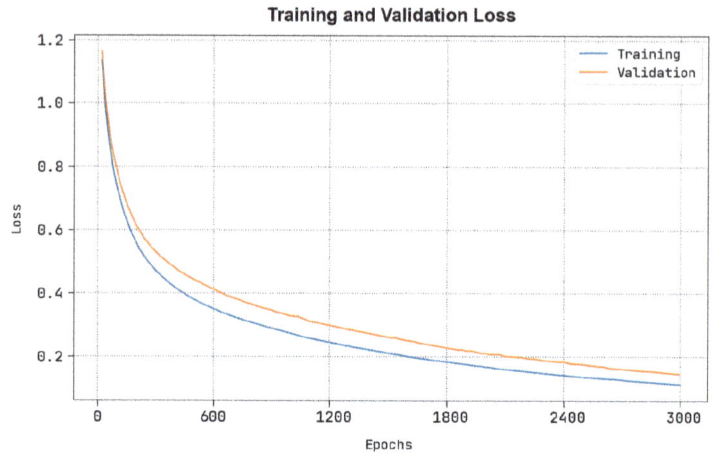

Figure 8. Loss curve of the SBFC-SFOHDL system.

A brief precision–recall (PR) analysis of the SBFC-SFOHDL system is exposed on the test dataset in Figure 9. The outcome indicates that the SBFC-SFOHDL methodology produced superior values of PR. In addition, it is noticeable that the SBFC-SFOHDL technique could achieve greater PR values in four classes.

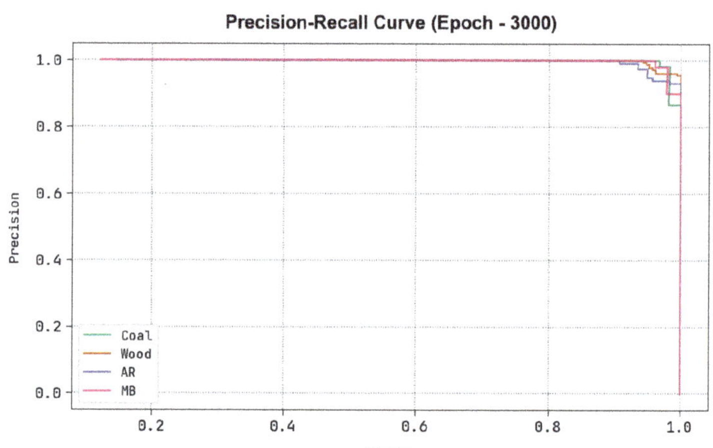

Figure 9. PR curve of the SBFC-SFOHDL approach.

In Figure 10, an ROC examination of the SBFC-SFOHDL technique is shown on the test dataset. The outcome defined that the SBFC-SFOHDL system resulted in better ROC values. Moreover, the SBFC-SFOHDL technique can encompass enhanced ROC values on four class labels.

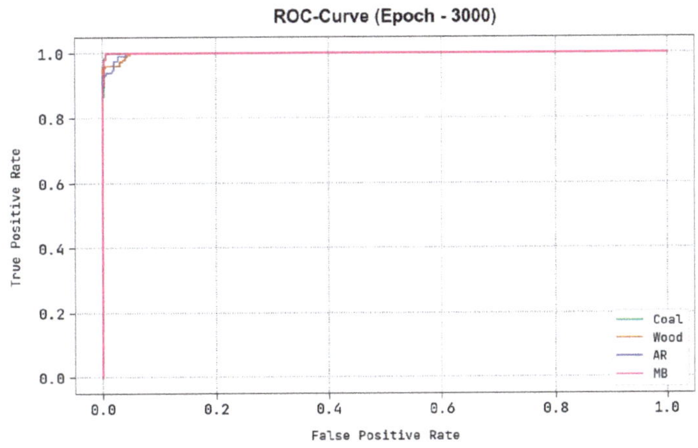

Figure 10. ROC curve of the SBFC-SFOHDL methodology.

A detailed comparative study of the SBFC-SFOHDL approach is stated in Table 2 [16]. Figure 11 represents the results of the SBFC-SFOHDL technique with recent models in terms of $prec_n$ and $reca_l$. Based on $prec_n$, the SBFC-SFOHDL technique gains an increasing value of 96.85% while the SVM, KNN, flat classifier, HC and IEVB-SFC models obtain a decreasing $prec_n$ of 89.61%, 91.90%, 90.49%, 91.33% and 94.71%, correspondingly.

Table 2. Comparative outcome of SBFC-SFOHDL approach with other recent methodologies.

Methods	$Prec_n$	$Reca_l$	$Accu_y$	F_{Score}
SBFC-SFOHDL	96.85	97.05	98.63	96.95
SVM	89.61	85.45	94.54	86.88
KNN	91.90	86.59	95.48	88.84
Flat Classifier	90.49	84.22	94.02	86.80
Hierarchical Classifier	91.33	90.37	95.24	90.97
IEVB-SFC	94.71	92.56	96.63	93.44

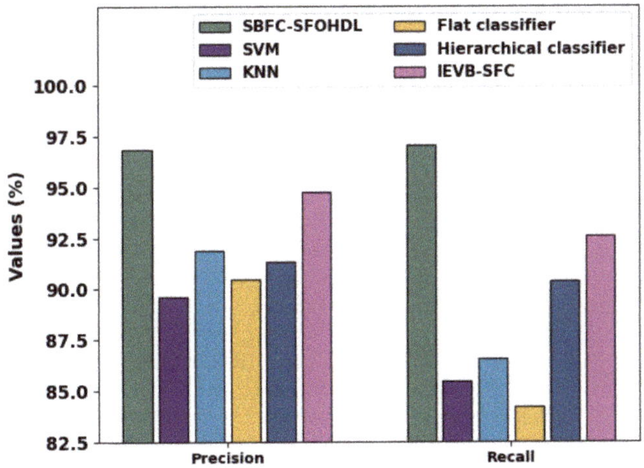

Figure 11. $Prec_n$ and $reca_l$ outcome of SBFC-SFOHDL algorithm with other recent methodologies.

Meanwhile, based on $recal_l$, the SBFC-SFOHDL system obtains a maximal value of 97.05% but the SVM, KNN, flat classifier, HC and IEVB-SFC models obtain a decreasing $reca_l$ of 85.45%, 86.59%, 84.22%, 90.37% and 92.56%, correspondingly.

Figure 12 represents the results of the SBFC-SFOHDL method with recent models in terms of $accu_y$ and F_{score}. Based on $accu_y$, the SBFC-SFOHDL technique gains an increasing value of 98.63% while the SVM, KNN, flat classifier, HC and IEVB-SFC models attain a decreasing $accu_y$ of 94.54%, 95.48%, 94.02%, 95.24% and 96.63%, correspondingly. Meanwhile, based on the F_{score}, the SBFC-SFOHDL method gains an increasing value of 96.95% while the SVM, KNN, flat classifier, HC and IEVB-SFC techniques attain a decreasing F_{score} of 86.88, 88.84%, 86.80%, 90.97% and 93.44%, correspondingly. Therefore, the SBFC-SFOHDL technique gains maximum performance on solid fuel classification.

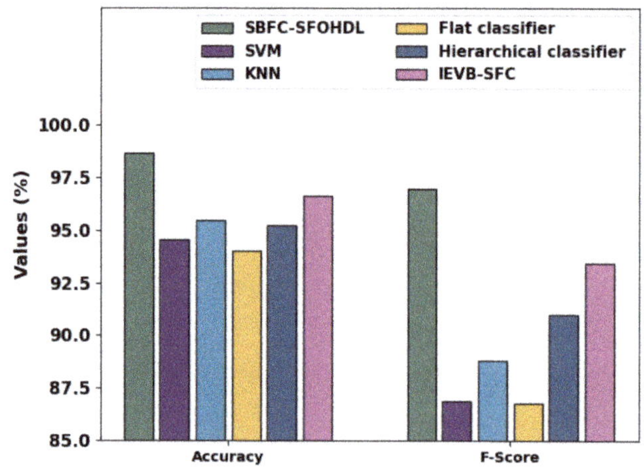

Figure 12. $Accu_y$ and F_{score} outcome of SBFC-SFOHDL algorithm with other recent methodologies.

In summary, the SBFC-SFOHDL technique exhibits better performance with a maximum $accu_y$ of 98.63%, $prec_n$ of 96.85%, $reca_l$ of 96.85% and F_{score} of 96.95%. The enhanced performance of the proposed model is due to the incorporation of the SFO-based hyperparameter tuning. Hyperparameters are settings that are not learned during training but must be set prior to training. They can have a significant impact on the performance of the model, and selecting the optimal values can lead to better accuracy. With SFO-optimizer-based hyperparameter tuning, the SBFC-SFOHDL model can achieve even better results by focusing on the most relevant features and selecting the optimal settings for the algorithm. These results ensure the improved performance of the SBFC-SFOHDL technique over other existing techniques.

5. Conclusions

In this study, we concentrated on the improvement of the SBFC-SFOHDL model in the IoT platform. The main purpose of the proposed SBFC-SFOHDL algorithm lies in the proficient detection and classification of solid biofuels from agricultural residues in the IoT platform. To achieve this, the SBFC-SFOHDL algorithm includes IoT-based data collection, data preprocessing, MSA-CBLSTM-based solid fuel classification and SFO-based hyperparameter tuning. The design of the SFO technique helps with the optimum choice of hyperparameter values, which in turn improves the classification accuracy of the MSA-CBLSTM approach. The simulation results of the SBFC-SFOHDL technique are tested and the results are examined under different measures. Extensive comparison studies reported the greater performance of the SBFC-SFOHDL method over recent DL approaches with a maximum $accu_y$ of 98.63%, $prec_n$ of 96.85%, $reca_l$ of 96.85% and F_{score} of 96.95%. In the future, feature fusion-based DL approaches can be designed to enhance the solid biofuel classification performance.

Author Contributions: Conceptualization, M.R.; Methodology, M.R. and A.O.K. (Adil O. Khadidos); Software, K.H.A., A.M.A. and A.O.K. (Alaa O. Khadidos); Validation, A.O.K. (Alaa O. Khadidos); Formal analysis, M.R., A.O.K. (Adil O. Khadidos) and A.O.K. (Alaa O. Khadidos); Investigation, A.M.A. and D.H.; Resources, A.O.K. (Adil O. Khadidos), A.M.A. and K.H.A.; Data curation, A.O.K. (Adil O. Khadidos), A.M.A. and K.H.A.; Writing—original draft, M.R.; Writing—review & editing, K.H.A., D.H. and A.O.K. (Alaa O. Khadidos); Visualization, K.H.A.; Project administration, M.R.; Funding acquisition, D.H. All authors have read and agreed to the published version of the manuscript.

Funding: This research work was funded by Institutional Fund Projects under grant no. (IFPIP: 1463-145-1443). Therefore, the authors gratefully acknowledge technical and financial support provided by the Ministry of Education and Deanship of Scientific Research (DSR), King Abdulaziz University (KAU), Jeddah, Saudi Arabia.

Institutional Review Board Statement: Not applicable.

Informed Consent Statement: Not applicable.

Data Availability Statement: Data sharing is not applicable to this article as no datasets were generated during the current study.

Conflicts of Interest: The authors declare no conflict of interest. The manuscript was written through contributions of all authors. All authors have given approval of the final version of the manuscript.

References

1. Peng, W.; Sadaghiani, O.K. An Analytical Review on the Utilization of Machine Learning in the Biomass Raw Materials, Their Evaluation, Storage, and Transportation. *Arch. Comput. Methods Eng.* **2023**, 1–22. [CrossRef]
2. Peng, W.; Sadaghiani, O.K. Enhancement of quality and quantity of woody biomass produced in forests using machine learning algorithms. *Biomass Bioenergy* **2023**, *175*, 106884. [CrossRef]
3. Akram, S.V.; Rawat, B. The Smart Analysis of Hydropower Energy Measurement in Power Plants using Industrial Machine Learning Model. In Proceedings of the 2023 IEEE International Conference on Integrated Circuits and Communication Systems (ICICACS), Raichur, India, 24–25 February 2023; pp. 1–6.
4. Barus, J.; Mustikawati, D.R.; Endriani, E.; Meithasari, D.; Ernawati, R.; Wardani, N.; Soraya, S.; Suretno, N.D.; Tambunan, R.D.; Silalahi, M. Evaluation of composting of several plant biomass wastes with different types of starters. *Int. J. Recycl. Org. Waste Agric.* **2023**.
5. Mondal, P.P.; Galodha, A.; Verma, V.K.; Singh, V.; Show, P.L.; Awasthi, M.K.; Lall, B.; Anees, S.; Pollmann, K.; Jain, R. Review on machine learning-based bioprocess optimization, monitoring, and control systems. *Bioresour. Technol.* **2022**, *370*, 128523. [CrossRef]
6. Zheng, K.; Jia, X.; Chi, K.; Liu, X. DDPG-based joint time and energy management in ambient backscatter-assisted hybrid underlay CRNs. *IEEE Trans. Commun.* **2022**, *71*, 441–456. [CrossRef]
7. Liu, X.; Xu, B.; Zheng, K.; Zheng, H. Throughput maximization of wireless-powered communication network with mobile access points. *IEEE Trans. Wirel. Commun.* **2022**, *22*, 4401–4415. [CrossRef]
8. Zheng, K.; Jiang, G.; Liu, X.; Chi, K.; Yao, X.; Liu, J. DRL-Based Offloading for Computation Delay Minimization in Wireless-Powered Multi-Access Edge Computing. *IEEE Trans. Commun.* **2023**, *71*, 1755–1770. [CrossRef]
9. Balachander, K.; Paulraj, D. RETRACTED ARTICLE: ANN and fuzzy based household energy consumption prediction with high accuracy. *J. Ambient. Intell. Humaniz. Comput.* **2021**, *12*, 7543–7557. [CrossRef]
10. Rajamoorthy, R.; Saraswathi, H.V.; Devaraj, J.; Kasinathan, P.; Elavarasan, R.M.; Arunachalam, G.; Mostafa, T.M.; Mihet-Popa, L. A Hybrid Sailfish Whale Optimization and Deep Long Short-Term Memory (SWO-DLSTM) Model for Energy Efficient Autonomy in India by 2048. *Sustainability* **2022**, *14*, 1355. [CrossRef]
11. Sharma, A.K.; Ghodke, P.K.; Goyal, N.; Nethaji, S.; Chen, W.H. Machine learning technology in biohydrogen production from agriculture waste: Recent advances and future perspectives. *Bioresour. Technol.* **2022**, *364*, 128076. [CrossRef] [PubMed]
12. Ning, H.; Li, R.; Zhou, T. Machine learning for microalgae detection and utilization. *Front. Mar. Sci.* **2022**, *9*, 947394. [CrossRef]
13. Hai, A.; Bharath, G.; Patah, M.F.A.; Daud, W.M.A.W.; Rambabu, K.; Show, P.; Banat, F. Machine learning models for the prediction of total yield and specific surface area of biochar derived from agricultural biomass by pyrolysis. *Environ. Technol. Innov.* **2023**, *30*, 103071. [CrossRef]
14. Adeleke, O.; Akinlabi, S.; Jen, T.C.; Adedeji, P.A.; Dunmade, I. Evolutionary-based neuro-fuzzy modelling of combustion enthalpy of municipal solid waste. *Neural Comput. Appl.* **2022**, *34*, 7419–7436. [CrossRef]
15. Kaewpengkrow, P.R.; Atong, D.; Sricharoenchaikul, V. Bio-fuel production from catalytic fast pyrolysis of Jatropha wastes using pyroprobe GC/MS and drop tube pyrolyzer. *J. Anal. Appl. Pyrolysis* **2022**, *165*, 105574. [CrossRef]

16. Al-Wesabi, F.N.; Malibari, A.A.; Hilal, A.M.; NEMRI, N.; Kumar, A.; Gupta, D. Intelligent ensemble of voting based solid fuel classification model for energy harvesting from agricultural residues. *Sustain. Energy Technol. Assess.* **2022**, *52*, 102040. [CrossRef]
17. Pawlak-Kruczek, H.; Niedzwiecki, L.; Sieradzka, M.; Mlonka-Mędrala, A.; Baranowski, M.; Serafin-Tkaczuk, M.; Magdziarz, A. Hydrothermal carbonization of agricultural and municipal solid waste digestates–Structure and energetic properties of the solid products. *Fuel* **2020**, *275*, 117837. [CrossRef]
18. Zheng, J.; Wu, Z.; Sharma, R.; Lv, H. Adaptive decision model of product team organization pattern for extracting new energy from agricultural waste. *Sustain. Energy Technol. Assess.* **2022**, *53*, 102352. [CrossRef]
19. Bot, B.V.; Axaopoulos, P.J.; Sakellariou, E.I.; Sosso, O.T.; Tamba, J.G. Energetic and economic analysis of biomass briquettes production from agricultural residues. *Appl. Energy* **2022**, *321*, 119430. [CrossRef]
20. Bosona, T.; Gebresenbet, G.; Olsson, S.O. Traceability system for improved utilization of solid biofuel from agricultural prunings. *Sustainability* **2018**, *10*, 258. [CrossRef]
21. Jifara, B.; Diriba, M.; Mengesha, A. Pelletization of mixed torrefied corn cob and khat stem to enhance the physicochemical and thermal properties of solid biofuel and parametric optimization. *Biomass Convers. Biorefinery* **2022**, 1–15. [CrossRef]
22. Samadi, S.H.; Ghobadian, B.; Nosrati, M. Prediction and estimation of biomass energy from agricultural residues using air gasification technology in Iran. *Renew. Energy* **2020**, *149*, 1077–1091. [CrossRef]
23. Perez-Herrera, R.; de Souza, T.A.; Coronado, C.J.; do Nascimento, M.A.R.; Pinto, G.M. Numerical CFD Simulation of a Horizontal Cyclonic Combustion Chamber for Burning Pulverized Biomass Solid Fuels. *Waste Biomass Valorization* **2023**, *14*, 1979–2007. [CrossRef]
24. Michal, H.; Jozef, J.; Miriam, N. Design of a wireless monitoring system with emission analysis integration for solid-fuel based heating devices in households of SmartCity. *Wirel. Netw.* **2022**, 1–10. [CrossRef]
25. Koval, S.; Vytisk, J.; Ruzickova, J.; Raclavska, H.; Skrobankova, H.; Hellebrandova, L. The Impact of Solid Fuel Residential Boilers Exchange on Particulate Matter Air Pollution. *Appl. Sci.* **2021**, *11*, 5400. [CrossRef]
26. Akarsu, K.; Duman, G.; Yilmazer, A.; Keskin, T.; Azbar, N.; Yanik, J. Sustainable valorization of food wastes into solid fuel by hydrothermal carbonization. *Bioresour. Technol.* **2019**, *292*, 121959. [CrossRef] [PubMed]
27. Hu, Z.; Chen, L.; Luo, Y.; Zhou, J. EEG-Based Emotion Recognition Using Convolutional Recurrent Neural Network with Multi-Head Self-Attention. *Appl. Sci.* **2022**, *12*, 11255. [CrossRef]
28. Sailaja, C.; Maloji, S.; Mannepalli, K. A hybrid HXPLS-TMFCC parameterization and DCNN-SFO clustering based speaker diarization system. *Concurr. Comput. Pract. Exp.* **2022**, *34*, e6954. [CrossRef]
29. Elmaz, F.; Büyükçakır, B.; Yücel, Ö.; Mutlu, A.Y. Hierarchical Classification Framework for Solid Fuel Classification. Available online: https://github.com/furkanelmaz/SolidFuelClassification (accessed on 12 June 2023).

sustainability

Article

Elliptic Curve Cryptography-Based Scheme for Secure Signaling and Data Exchanges in Precision Agriculture

Zaid Ameen Abduljabbar [1,2,3,*], Vincent Omollo Nyangaresi [4], Hend Muslim Jasim [1], Junchao Ma [5,*], Mohammed Abdulridha Hussain [1,2], Zaid Alaa Hussien [6] and Abdulla J. Y. Aldarwish [1]

1 Department of Computer Science, College of Education for Pure Sciences, University of Basrah, Basrah 61004, Iraq
2 Technical Computer Engineering Department, AL-Kunooze University College, Basrah 61001, Iraq
3 Shenzhen Institute, Huazhong University of Science and Technology, Shenzhen 518000, China
4 Department of Computer Science and Software Engineering, Jaramogi Oginga Odinga University of Science & Technology, Bondo 40601, Kenya
5 College of Big Data and Internet, Shenzhen Technology University, Shenzhen 518118, China
6 Information Technology Department, Management Technical College, Southern Technical University, Basrah 61005, Iraq
* Correspondence: zaid.ameen@uobasrah.edu.iq (Z.A.A.); majunchao@sztu.edu.cn (J.M.)

Abstract: Precision agriculture encompasses automation and application of a wide range of information technology devices to improve farm output. In this environment, smart devices collect and exchange a massive number of messages with other devices and servers over public channels. Consequently, smart farming is exposed to diverse attacks, which can have serious consequences since the sensed data are normally processed to help determine the agricultural field status and facilitate decision-making. Although a myriad of security schemes has been presented in the literature to curb these challenges, they either have poor performance or are susceptible to attacks. In this paper, an elliptic curve cryptography-based scheme is presented, which is shown to be formally secure under the Burrows–Abadi–Needham (BAN) logic. In addition, it is semantically demonstrated to offer user privacy, anonymity, unlinkability, untraceability, robust authentication, session key agreement, and key secrecy and does not require the deployment of verifier tables. In addition, it can withstand side-channeling, physical capture, eavesdropping, password guessing, spoofing, forgery, replay, session hijacking, impersonation, de-synchronization, man-in-the-middle, privileged insider, denial of service, stolen smart device, and known session-specific temporary information attacks. In terms of performance, the proposed protocol results in 14.67% and 18% reductions in computation and communication costs, respectively, and a 35.29% improvement in supported security features.

Keywords: Agriculture 4.0; precision agriculture; privacy; smart farming; security

Citation: Abduljabbar, Z.A.; Nyangaresi, V.O.; Jasim, H.M.; Ma, J.; Hussain, M.A.; Hussien, Z.A.; Aldarwish, A.J.Y. Elliptic Curve Cryptography-Based Scheme for Secure Signaling and Data Exchanges in Precision Agriculture. *Sustainability* **2023**, *15*, 10264. https://doi.org/10.3390/su151310264

Academic Editors: Mourade Azrour, Azidine Guezzaz, Imad Zeroual, Azeem Irshad, Jamal Mabrouki, Said Benkirane and Shehzad Ashraf Chaudhry

Received: 9 May 2023
Revised: 23 June 2023
Accepted: 26 June 2023
Published: 28 June 2023

1. Introduction

Many economies in developing countries are dependent on agriculture as a source of income and contributions to gross domestic product (GDP) [1]. However, the majority of the farming practices are based on experience and ad hoc insights of the farmers. Consequently, there is little control on the agricultural produce quantity and hence financial profits. Fortunately, precision agriculture (PA) and the Internet of Things (IoT) can be deployed to address these issues [2,3]. As explained in [4], PA is part of Agriculture 3.0 in which farm yields are regularly monitored. In addition, PA involves automation and the application of information technology (IT) to improve farm output. In Agriculture 4.0, also referred to as smart agriculture or smart farming, additional technologies such as drones, artificial intelligence (AI), blockchain, big data, wireless sensor networks (WSN), and robotics are incorporated in agriculture. In PA, a number of sensors are deployed, such as radiation, air humidity, optimal, soil moisture, and ground sensors. According to [5], intelligent precision agriculture (IPA) encompasses the deployment of numerous IoT devices and drones to monitor agricultural surroundings. To boost productivity in the face of limited resources and protection from disasters, traditional agronomy needs to be replaced with smart agronomy [6]. As discussed in [7], there are fraud risks in the agricultural sector, especially concerning beverage and food packaging. Therefore, agricultural organizations require ideal certification of their products since these risks can impact negatively on the health of their consumers.

The smart devices deployed in PA and IPA exchange a massive number of messages. Therefore, insecure communication channels among IoT devices, unmanned aerial vehicles (UAVs), or drones can expose smart farming to diverse attacks [5,8]. For instance, Wi-Fi de-authentication and denial of dervice (DoS) can be launched on Raspberry Pi-based smart farms [9]. This can have serious consequences as the sensed data are normally processed to help determine the agricultural field status and facilitate decision-making, which may involve taking measures to maintain or enhance the farm status [10]. These attacks can also target drones deployed to monitor field conditions such as irrigation, spraying of pesticides, pollination, and planting of seeds [11]. On their part, WSNs offer monitoring, sensing, and a continuous supply of information regarding climatic conditions such as the chemical content of the soil, air humidity, temperature, light, water quality, and soil moisture. These parameters are then utilized to boost productivity, both qualitatively and quantitatively. According to [12], WSNs facilitate monitoring, data collection, and control of agricultural systems and hence ensure efficiency, minimal packet losses and economic overheads, better network control, and increased scalability and flexibility. However, threats such as interference, masquerading, interception, and message alteration can compromise these networks and harm crop production and other monitored agricultural practices [6]. The authors in [13] pointed out that issues such as sufficient energy resource utilization and secure data transmission are yet to be solved in WSN. This is because of the usage of open wireless networks during data transfers [14], which can potentially compromise the integrity, confidentiality, and authenticity of the exchanged data.

To address the above issues, there is a need for robust authentication and access control to secure the internet of drones, WSNs, IoT, and agricultural monitoring [15–17]. For instance, sufficient user authentication ensures that external users can use their mobile devices to securely access real-time data from the deployed agricultural smart devices [18,19]. There is also a need for robust source authentication, message authentication, and entity authentication.

1.1. Contributions

- A lightweight authentication scheme based on elliptic curve cryptography is developed for secure message exchange among the communicating smart devices in precision agriculture.
- Formal security analysis is carried out using BAN logic to demonstrate that a session key is derived from enciphering the exchanged data between the farmers and the agricultural service providers.
- Extensive semantic analysis is executed to show that the proposed scheme can withstand side-channeling, physical capture, eavesdropping, password guessing, spoofing, forgery, replay, session hijacking, impersonation, de-synchronization, man-in-the-middle, privileged insider, denial of service, stolen smart device, and known session-specific temporary information attacks. In addition, this protocol is demonstrated to support user privacy, anonymity, unlinkability, untraceability, robust authentication, session key agreement, and key secrecy and does not require the deployment of verifier tables.
- An elaborate performance evaluation is carried out to show that our scheme yields 14.67% and 18% reductions in computation and communication costs, respectively, and a 35.29% improvement in supported security features.

1.2. Problem Definition and Motivation

In precision agriculture, information technology plays a critical role in ensuring that farming activities obtain exact requirements, which boosts health, productivity, and agricultural outputs. In this way, environmental protection, sustainability, and profitability are assured in smart farms. On the flip side, the public channels deployed for message exchanges make these networks vulnerable to numerous attacks such as eavesdropping, message falsification, DoS, replay, MitM, impersonation, drones capture, ephemeral secret leakage (ESL), privileged insider, and physical smart devices capture attacks. Proper user and device authentication is one of the most promising solutions to these security and privacy challenges. In addition, communication attributes such as untraceability, unlinkability, anonymity, and user privacy need to be assured. For instance, the secrecy of trading transactions among farmers and agricultural firms needs to be upheld.

1.3. Security Requirements

Owing to the open communication channels deployed in smart agriculture, adversaries can hijack the session, take control of the communication process, and execute other malicious activities. Therefore, a secure authentication protocol should be resilient against a myriad of attacks. In addition, it should fulfill the following privacy and security requirements.

Untraceability and unlinkability: It should be cumbersome for the adversary to trace or link some captured messages to a particular network entity.

Robust authentication: To prevent illegitimate entities from joining the network or accessing the agricultural services and smart devices, all the entities must be validated.

Session key negotiation: Immediately after successful mutual authentication procedures, the communicating parties should agree on the session key to encipher the exchanged messages.

Anonymity and privacy: The real identities of the communicating parties should never be exchanged in plain text over public channels. This is to prevent attackers from eavesdropping them across the communication channel. This goes a long way in preserving the privacy of these parties.

Key secrecy: The session key should be computed in a manner that will make it cumbersome for the attacker to deploy the captured session key for the current communication process to derive the keys used in the previous or subsequent communication procedures.

Resistant to attacks: It should be difficult for the attacker to compromise the network and its smart devices through side-channeling, physical capture, eavesdropping, password guessing, spoofing, forgery, replay, session hijacking, impersonation, de-synchronization, man-in-the-middle (MitM), privileged insider, DoS, stolen smart device, and known session-specific temporary information (KSSTI) attacks.

1.4. Threat Modeling

In this paper, the adversary is assumed to have all the capabilities in the Dolev–Yao (DY) as well as Canetti and Krawczyk (CK) threat models. In the DY threat model, an attacker Ψ is capable of intercepting, altering, deleting, and injecting bogus messages into the communication channel. However, in the CK threat model, an adversary Ψ can compromise secret parameters, private keys, and session states that can be obtained from devices' memory. In addition, the communicating entities are assumed to be untrustworthy, and Ψ can physically capture the IoT devices and extract the secrets in their memories through power analysis. Using the extracted secrets, further attacks, such as impersonations, can be launched.

The rest of this paper is structured as follows: Section 2 discusses the related work, while Section 3 presents the proposed scheme. On the other hand, Section 4 discusses the security analysis of our scheme, while Section 5 presents its performance evaluation. Finally, Section 6 concludes this paper and provides some research directions.

2. Related Work

Many schemes have been developed to enhance security in the smart farm environment. For example, a novel private blockchain-based authentication scheme is presented in [5]. However, this protocol fails to protect against de-synchronization and session hijacking attacks. Similarly, blockchain-based schemes were developed in [20–24]. Although blockchain offers traceability, integrity protection, and shareability in the agricultural environment, such as agri-food supply chains, it has high storage and computation overheads [25]. Based on signatures, the authors of [18] present a three-factor user authentication protocol. Unfortunately, this scheme cannot prevent attacks such as eavesdropping and session hijacking. On the other hand, an identity-based scheme was introduced in [26]. Nevertheless, this technique is vulnerable to stolen smart cards, sensor node spoofing, impersonation, and stolen verifier attacks [27]. In addition, it cannot provide backward key secrecy. To address these challenges, two protocols were developed in [27]. Unfortunately, the authentication and password change phases of these schemes are inefficient [28]. To offer privacy protection, a remote user authentication protocol was presented in [6]. However, this scheme cannot withstand attacks such as eavesdropping, de-synchronization, and spoofing.

Based on a public-key-based cryptosystem, an authentication scheme was developed in [29]. Although this approach protects against MitM and replay attacks, it cannot withstand privileged insider, user impersonation, and ephemeral secret leakage (ESL) attacks [5]. In addition, it does not include biometric change and user device revocation phases. The signature-based privacy-preserving protocol in [30] can address some of these issues. However, it is still susceptible to ESL attacks and cannot assure the untraceability and anonymity of the communicating parties [5]. Similarly, the protocol in [31] does not provide user and device anonymity since their internet protocol (IP) addresses incorporated in messages are exchanged publicly. In addition, it has high computation overheads due to the utilization of public key cryptography for its digital signatures and certificates [32]. Moreover, it is prone to replay, physical device capture, MitM, user and device impersonation, and attacks. On its part, the scheme in [33] cannot protect against user anonymity violation, user impersonation, and smart card loss attacks. Similarly, the protocol in [34] is vulnerable to physical sensing device capture, untraceability violation, and smart card loss attacks [5]. Using some bilinear pairing operations, authentication and key establishment protocols were introduced in [35,36]. However, the utilization of pairing operations increased the computation costs of these protocols [37]. Since the trusted authority in [36] has access to user identity and password, it is susceptible to privileged insider attacks. In addition, it cannot withstand replay, disclosure of sensor data, offline password guessing, and stolen smart card and verifier attacks [38]. As such, an improved elliptic curve cryptography (ECC)-based scheme was developed in [38]. However, this protocol has an inefficient and delayed authentication phase. In addition, it is not robust against DoS and replay attacks [39]. Although the protocol in [40] addresses some of these issues, its bilinear pairing operations result in high computation costs [41].

To offer security in a heterogeneous IoT environment, an authentication technique was presented in [42]. Unfortunately, this protocol is vulnerable to physical device capture, privileged insider, and ESL attacks. In addition, it cannot preserve untraceability and anonymity [5]. Similarly, a remote user authentication protocol was developed in [43], which was shown to be lightweight. However, it failed to protect against ESL and privileged insider attacks. It also failed to support untraceability and anonymity [5]. On its part, the scheme in [43] was not resilient against privileged insider and sensor node capture attacks. It also failed to preserve forward key secrecy [6]. The authors in [44,45] designed identity-based signature protocols to protect message exchanges in mobile devices. However, identity-based schemes have key escrow problems [46]. Based on ECC and symmetric key encryption, a security technique was presented in [47]. Although it was shown to be robust against MitM and replay attacks, it was vulnerable to ESL, privileged insider, and user impersonation attacks. It also failed to incorporate device revocation, node addition, and password and user biometric change phases [5]. Similarly, the biometric-based scheme in [48] did not include device revocation, user passwords, and biometric update phases. It was also vulnerable to privileged insider, user impersonation, ESL, DoS, and stolen smart card attacks [49]. On its part, the protocol in [50] was susceptible to DoS attacks and could not offer forward key secrecy [51]. Similarly, the scheme in [52] did not support forward key secrecy and was prone to stolen verifier attacks [53]. As such, an enhanced ECC-based protocol was introduced in [53], while a privacy-preserving scheme was developed in [54]. The scheme in [54] was demonstrated to be resilient against eavesdropping, DoS, masquerade, privileged insider, and forgery attacks. It also supports secret key updates, traceability, and anonymity. However, it cannot withstand MitM attacks [20].

It is evident that numerous schemes have been proposed to improve the security posture in precision agriculture. However, it has been shown that these techniques face a number of security, privacy, and performance challenges. The proposed scheme is shown to solve some of these challenges as described in Sections 4 and 5 below.

3. The Proposed Scheme

The farmer smart devices SD_j and the agricultural service providers ASP_i are the main components of this scheme. As shown in Figure 1, the registration phase occurs over secure channels, while the SD_j and ASP_i exchange the data over the insecure public channels in an ad hoc manner. As such, the goal of the proposed protocol is to enhance the privacy and security of the transmitted information.

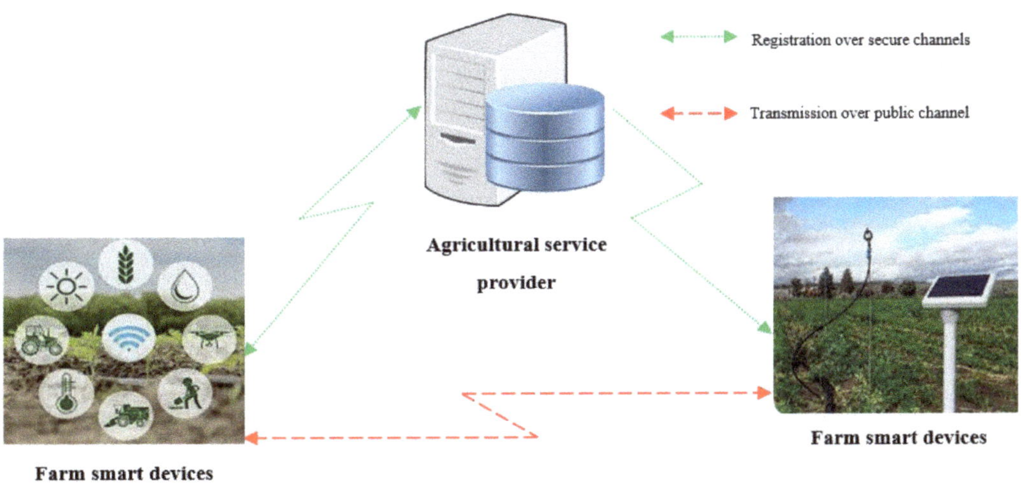

Figure 1. Network model.

The proposed scheme comprises four major phases, which include system initialization, registration, login, and authentication phases. Table 1 presents the notations used throughout this paper.

The subsections below provide detailed descriptions of the various major phases of the proposed scheme.

3.1. System Initialization

In this phase, the agricultural service provide ASP_i executes the following three steps to generate the security parameters that will be utilized during the other three phases. These steps are described in detail, as shown in Figure 2.

Table 1. Deployed symbols.

Symbol	Description
ASP_i	Agricultural service provider *i*
MK_A	Master key for ASP_i
F_j	Farmer *j*
SD_j	Smart device for F_j
FID_j	Unique identity for F_j
FPW_j	Login password for F_j
R_i	Random nonce *i*
| |	Concatenation operation
P_B	Padding bits
$\oplus$	XOR operation
ϕ_A	Session key computed at ASP_i
ϕ_S	Session key computed at SD_j

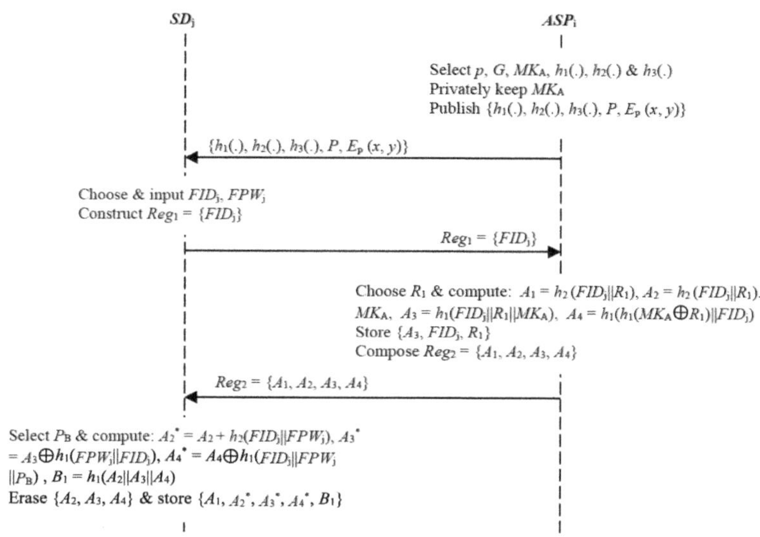

Figure 2. System initialization and registration phases.

Step 1: The ASP_i selects the prime number *p*, whose length is *k*-bits. It also chooses some elliptic curve group *G* whose base point is *P* and whose order is *q*.

Step 2: The ASP_i selects a random parameter MK_A from {1, $q-1$} and deploys it as its master secret key. In addition, it chooses three collision-resistant one-way hashing functions, $h_1(.)$, $h_2(.)$, and $h_3(.)$, where $h_2(.)$ serves as the map-to-point hashing function. Therefore, $h_2(.)$: $\{0,1\}^* \rightarrow$ G, $h_1(.)$: $\{0,1\}^* \rightarrow \{0,1\}^k$, and $h_3(.)$: G $\rightarrow \{0,1\}^k$.

Step 3: Parameter MK_A is secretly retained by the ASP_i, while parameter set {$h_1(.)$, $h_2(.)$, $h_3(.)$,P, E_p (x, y)} is publicly made available to all smart devices.

3.2. Registration Phase

It is required that all farmers register with the ASP_i and obtain some security tokens before being allowed to access some services from the ASP_i. This is a four-step process, as described below.

Step 1: The farmer F_j selects a unique identity FID_j and password FPW_j that are input to the SD_j. Next, a registration message Reg_1 = {FID_j} is constructed that is forwarded to the ASP_i over secured communication channels.

Step 2: Upon receipt of Reg_1, the ASP_i selects random nonce R_1. Next, it derives security values $A_1 = h_2$ $(FID_j || R_1)$, $A_2 = h_2$ $(FID_j || R_1)$. MK_A, $A_3 = h_1(FID_j || R_1 || MK_A)$, and $A_4 = h_1(h_1(MK_A \oplus R_1) || FID_j)$.

Step 3: The ASP_i stores parameter set $\{A_3, FID_j, R_1\}$ in its database for later use during the login and authentication phases. Finally, it constructs registration message $Reg_2 = \{A_1, A_2, A_3, A_4\}$, which is forwarded to F_j over secure channels, as shown in Figure 2.

Step 4: After receiving message Reg_2, the SD_j generates fixed-bit padding parameter P_B. This is followed by the computation of values $A_2^* = A_2 + h_2(FID_j || FPW_j)$, $A_3^* = A_3 \oplus h_1(FPW_j || FID_j)$, $A_4^* = A_4 \oplus h_1(FID_j || FPW_j || P_B)$, and $B_1 = h_1(A_2 || A_3 || A_4)$. Finally, it erases parameter set $\{A_2, A_3, A_4\}$ and stores value set $\{A_1, A_2^*, A_3^*, A_4^*, B_1\}$ in its memory.

3.3. Login

The goal of this phase is to validate the farmer password and unique identity that are input to the smart device SD_j. To accomplish this, the following two steps are executed:

Step 1: The farmer F_j inputs the unique identity FID_j and password FPW_j into the SD_j. Next, the SD_j derives values $A_2 = A_2^* - h_2(FID_j || FPW_j)$, $A_3 = A_3^* \oplus h_1(FPW_j || FID_j)$, and $A_4 = A_4^* \oplus h_1(FID_j || FPW_j || P_B)$.

Step 2: The SD_j computes $B_1^* = h_1(A_2 || A_3 || A_4)$ and verifies if $B_1^* \stackrel{?}{=} B_1$. the session is terminated if these two values are not identical. Otherwise, F_j has logged in successfully and can now proceed to the authentication phase.

3.4. Authentication and Key Agreement

In this phase, the farmer F_j, through the SD_j, generates and exchanges a number of security tokens with the agricultural service provider ASP_i, through which these two entities verify one another before the onset of agricultural data exchanges. In addition, the session keys for data encryption are derived as described below.

Step 1: The SD_j generates random nonces R_2 and R_3, where $R_2 \in \{1, q-1\}$ and $R_3 \in \{0,1\}^k$. Next, it derives parameters $B_2 = A_1 . R_2$, $B_3 = A_2 . R_2$, $B_4 = A_3 \oplus h_3$ (B_3), $C_1 = A_4 \oplus R_3$, and $C_2 = h_1(B_2 || B_3 || B_4 || C_1 || R_3 || A_4)$. Lastly, it composes authentication message $Auth_1 = \{B_2, B_4, C_1, C_2\}$, which is transmitted to the ASP_i over public channels, as shown in Figure 3.

Step 2: Upon receiving the authentication message $Auth_1$ message from SD_j, the ASP_i derives $B_3^* = MK_A . B_2$ and $A_3^{**} = B_4 \oplus h_3$ (B_3^*). Next, it confirms whether parameter set $\{A_3^{**}, FID_j^*, R_1^*\}$ is in its database. Here, the session is terminated if this verification fails. Otherwise, the ASP_i proceeds to compute values $A_4^{**} = h_1(h_1(MK_A \oplus R_1^*) || FID_j^*)$ and $R_3^* = C_1 \oplus A_4^{**}$.

Step 3: The ASP_i derives values $C_2^* = h_1(B_2 || B_3^* || B_4 || C_1 || R_3^* || A_4^{**})$ and confirms whether $C_2^* \stackrel{?}{=} C_2$. The authentication session is essentially terminated if this verification flops. Otherwise, the ASP_i chooses random nonce$R_4 \in \{0,1\}^k$, which is utilized in deriving parameters $C_3 = R_4 \oplus A_4^{**}$, session key $\phi_A = h_1(R_3^* || R_4 || B_3^* || A_4^{**})$, and $C_4 = h_1(FID_j^* || \phi_A)$. Finally, it constructs an authentication message $Auth_2 = \{C_3, C_4\}$ that it forwards to SD_j over public communication channels.

Step 4: After obtaining message $Auth_2$ from ASP_i, the SD_j derives values $R_4^* = C_3 \oplus A_4$, session key $\phi_S = h_1(R_3 || R_4^* || B_3 || A_4)$, and $C_4^* = h_1(FID_j || \phi_S)$. This is followed by the validation of whether $C_4^* \stackrel{?}{=} C_4$. The authentication session is aborted if these two parameters are unequal. Otherwise, SD_j deploys ϕ_S as the session key to encipher all the exchanged messages.

3.5. Password Renewal Phase

The proposed scheme allows the farmer F_j to change his/her password FPW_j. This may be prompted by the loss of FPW_j or when they suspect that this password might have been compromised. This is attained by executing the following steps.

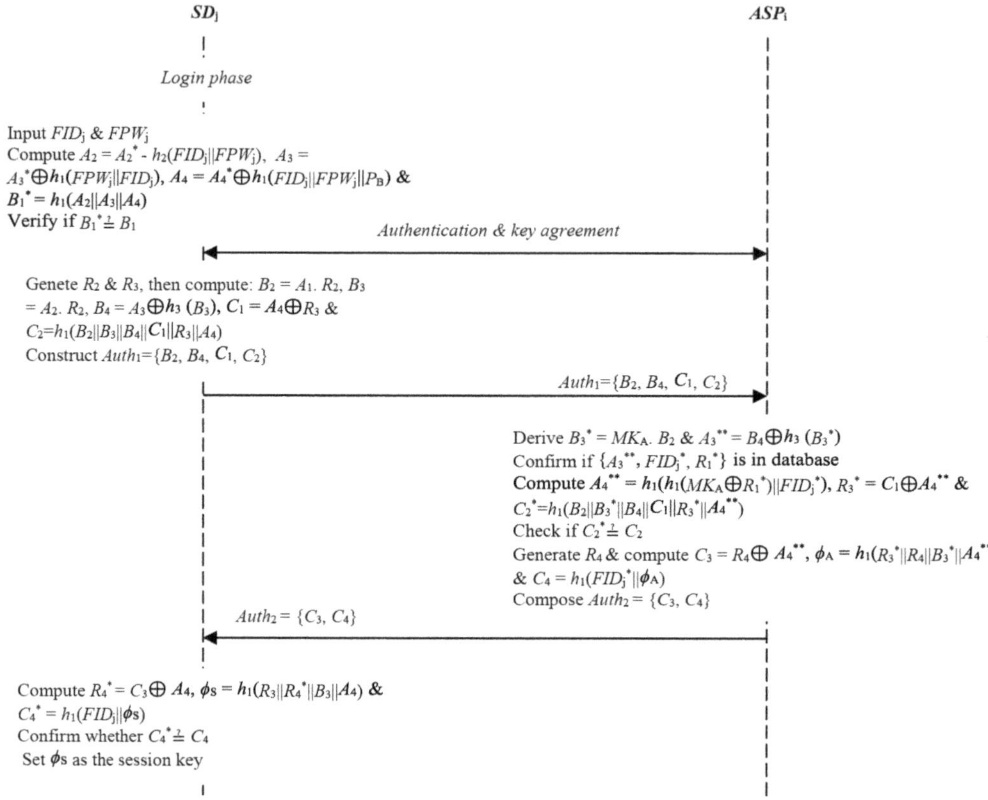

Figure 3. Login, authentication, and key negotiation phase.

Step 1: The farmer F_j inputs the current password FPW_j into the SD_j. Next, SD_j deploys the stored parameter set $\{A_1, A_2^*, A_3^*, A_4^*, B_1\}$ in its memory to derive $A_2 = A_2^* - h_2(FID_j \mid\mid FPW_j)$, $A_3 = A_3^* \oplus h_1(FPW_j \mid\mid FID_j)$, and $A_4 = A_4^* \oplus h_1(FID_j \mid\mid FPW_j \mid\mid P_B)$.

Step 2: F_j selects the new password FPW_j^*. This is followed by the computation of parameters $A_2^{New} = A_2 + h_2(FID_j \mid\mid FPW_j^*)$, $A_3^{New} = A_3 \oplus h_1(FPW_j^* \mid\mid FID_j)$, and $A_4^{New} = A_4 \oplus h_1(FID_j \mid\mid FPW_j^* \mid\mid P_B)$.

Step 3: The SD_j erases value set $\{A_2, A_3, A_4, A_2^*, A_3^*, A_4^*\}$ from its memory and stores parameter set $\{A_1, A_2^{New}, A_3^{New}, A_4^{New}, B_1\}$ in its memory.

4. Security Analysis

In this section, the proposed scheme's security and privacy are analyzed using both formal and semantic techniques described below.

4.1. Formal Security Analysis

In this sub-section, we deploy the BAN logic to demonstrate that the farmer F_j and agricultural service provider ASP_i interact to set up a session key between them. This key is then utilized to encipher the data exchanged between these two entities. Suppose that A and B are principals, M and N are statements, and μ is the encryption key. The notations used in this analysis are described below.

$A \mid\equiv M$: A trusts M.

$\{M\}$: Statement M is enciphered using key μ.

$A \triangleleft M$: A receives M.

$<M>_N$: Statement M is combined statement N.
$\#(M)$: M is fresh.
$A \Rightarrow M$: A has control.
(M, N): Statement M or N is part of (M, N).
$A \overset{\mu}{\leftrightarrow} B$: Principals A and B deploy shared key μ during communication.
$A \mid\sim M$: Principal A once said statement M.
$(M)_h$: Statement M is hashed using hashing function h.
During the formal security verification, the following BAN logic rules are utilized.
Freshness Rule (F-R)
$\frac{A\ believes\ fresh\ M}{A\ believes\ fresh\ (M,N)}$, mathematically represented as $\frac{A \mid\equiv \#(M)}{A \mid\equiv \#(M, N)}$.
The Message-Meaning Rule (M-M-R)
$\frac{A\ believes\ A \overset{\mu}{\leftrightarrow} B,\ A\ sees\ \{M\}_\mu}{A\ believes\ B\ once\ said\ M}$, which can be mathematically expressed as $\frac{A \mid\equiv A \overset{\mu}{\leftrightarrow} B,\ A \triangleleft \{M\}_\mu}{A \mid\equiv B \mid\sim M}$.
Jurisdiction-Rule (J-R)
$\frac{A\ believes\ B\ control\ M,\ A\ believes\ B\ believes\ M}{A\ believes\ M}$, mathematically denoted as $\frac{A \mid\equiv B \Rightarrow M,\ A \mid\equiv B \mid\equiv M}{A \mid\equiv M}$.
Believe–Rule (B-R)
$\frac{A\ believes\ B\ believes\ (M, N)}{A\ believes\ B\ believes\ M}$, also expressed as $\frac{A \mid\equiv B \mid\equiv (M, N)}{A \mid\equiv B \mid\equiv M}$.
Nonce Verification Rule (N-V-R)
$\frac{A\ believes\ fresh\ (M),\ A\ believes\ B\ once\ said\ M}{A\ believes\ B\ believes\ M}$, which can be denoted as $\frac{A \mid\equiv \#(M),\ A \mid\equiv B \mid\sim M}{A \mid\equiv B \mid\equiv M}$.

To show the establishment of session key μ between provider ASP_i and farmer F_j, the following two goals are formulated.

Goal 1: $ASP_i \mid\equiv (F_j \overset{\mu}{\leftrightarrow} ASP_i)$.

Goal 2: $F_j \mid\equiv (F_j \overset{\mu}{\leftrightarrow} ASP_i)$.

To achieve this, the following initial assumptions are made:

IA_1: $F_j \mid\equiv R_2$;
IA_2: $F_j \mid\equiv R_3$;
IA_3: $F_j \mid\equiv B_2$;
IA_4: $F_j \mid\equiv B_3$;
IA_5: $F_j \mid\equiv F_j \overset{A_4}{\leftrightarrow} ASP_i$;
IA_6: $F_j \mid\equiv ASP_i \Rightarrow (R_4)$;
IA_7: $ASP_i \mid\equiv B_2$;
IA_8: $ASP_i \mid\equiv MK_A$;
IA_9: $ASP_i \mid\equiv R_4$;
IA_{10}: $ASP_i \mid\equiv F_j \overset{A_4}{\leftrightarrow} ASP_i$;
IA_{11}: $ASP_i \mid\equiv F_j \Rightarrow (R_3, B_2)$.

In the proposed protocol, two messages are exchanged during the authentication and key agreement phase. These messages include $Auth_1$ and $Auth_2$, transmitted by the SD_j and ASP_i, respectively. For efficient analysis, these messages are transformed into idealized designs, as described below.

$\boldsymbol{SD_j \rightarrow ASP_i}$:
$Auth_1 = \{B_2, B_4, C_1, C_2\}$;
Idealized form: $\{B_2, B_4, C_1, C_2, \langle B_3 \rangle_{B_2}, \langle A_3 \rangle_{B_3}, \langle R_3 \rangle_{A_4}\}$.
$\boldsymbol{ASP_i \rightarrow SD_j}$:
$Auth_2 = \{C_3, C_4\}$;
Idealized form: $\{C_3, C_4, \langle R_4 \rangle_{A_4}\}$.

Next, the above BAN logic notations, rules, and initial state assumptions are deployed to demonstrate that the farmer F_j and the agricultural service provider ASP_i derive and share similar session key μ to encipher the exchanged messages. This procedure proceeds as described below.

Based on $Auth_1$, the following is obtained:

DM_1: $ASP_i \lhd \{B_2, B_4, C_1, C_2, \langle B_3\rangle_{B_2}, \langle A_3\rangle_{B_3}, \langle R_3\rangle_{A_4}\}$.
Using *M-M-R*, DM_1, IA_7, and IA_8, DM_2 is yielded.
DM_2: $ASP_i |\equiv F_j| \sim \{B_2, B_4, C_1, C_2, \langle B_3\rangle_{B_2}, \langle A_3\rangle_{B_3}, \langle R_3\rangle_{A_4}\}$.
Since $C_2^* = h_1(B_2 || B_3^* || B_4 || C_1 || R_3^* || A_4^{**})$, from DM_2,
DM_3: $ASP_i |\equiv F_j| \equiv \{B_2, B_4, C_1, C_2, \langle B_3\rangle_{B_2}, \langle A_3\rangle_{B_3}, \langle R_3\rangle_{A_4}\}$.
Using the *B-R* and DM_3, the following is obtained:
DM_4: $ASP_i |\equiv F_j| \equiv (R_3, B_2)$
Based on IA_{11} and DM_4, we obtain:
DM_5: $ASP_i |\equiv (R_3, B_2)$.
On the other hand, the application of *B-R* on DM_5 yields:
DM_6: $ASP_i |\equiv (R_3)$ and $ASP_i |\equiv (B_2)$.
Based on IA_8 and IA_9, the following is obtained:
DM_7: $ASP_i |\equiv MK_A$ and $ASP_i |\equiv (R_4)$,
Since $\phi_A = h_1(R_3^* || R_4 || B_3^* || A_4^{**})$, then from DM_6 and DM_7,
DM_8: $ASP_i |\equiv (F_j \overset{\mu}{\leftrightarrow} ASP_i)$, hence **Goal 1** is attained.
From $Auth_2$, the following is obtained:
DM_9: $F_j \lhd \{C_3, C_4, \langle R_4\rangle_{A_4}\}$.
Using the M-M-R on IA_{10} results in the following:
DM_{10}: $F_j |\equiv ASP_i | \sim \{C_3, C_4, \langle R_4\rangle_{A_4}\}$.
Based on DM_{10}, IA_5, and IA_9,
DM_{11}: $F_j |\equiv ASP_i |\equiv R_4$.
The application of the *J-R* on DM_{11} and IA_6 results in the following:
DM_{12}: $F_j |\equiv R_4$.
Based on DM_{12} and IA_2–IA_5, we obtain:
DM_{13}: $F_j |\equiv (F_j \overset{\mu}{\leftrightarrow} ASP_i)$; therefore, **Goal 2** is accomplished.

The attainment of these two security goals confirms that farmer F_j and agricultural service provider ASP_i strongly trust that they share session key μ for traffic protection.

4.2. Semantic Security Analysis

The objective of this subsection is the formulation and proofing of some hypotheses regarding the supported security features in the proposed scheme.

Hypothesis 1: *Farmer privacy and anonymous communication are achieved.*

Proof: In the proposed scheme, the real identity of the farmer is FID_j. This identity is incorporated in values such as $A_1 = h_2\ (FID_j || R_1)$, $A_2 = h_2\ (FID_j || R_1)\ MK_A$, $A_3 = h_1(FID_j || R_1 || MK_A)$, $A_4 = h_1(h_1(MK_A \oplus R_1) || FID_j)$, and $C_4 = h_1(FID_j^* || \phi_A)$. In all these parameters, FID_j is encapsulated in other parameters before being hashed. During the authentication and key agreement phase, messages $Auth_1 = \{B_2, B_4, C_1, C_2\}$ and $Auth_2 = \{C_3, C_4\}$ are transmitted over public channels. Here, $B_2 = A_1.\ R_2$, $B_4 = A_3 \oplus h_3\ (B_3)$, $C_1 = A_4 \oplus R_3$, $C_2 = h_1(B_2 || B_3 || B_4 || C_1 || R_3 || A_4)$, $C_3 = R_4 \oplus A_4^{**}$, and $C_4 = h_1(FID_j^* || \phi_A)$. It is evident that of all these parameters, it is only C_4 that directly incorporates farmer identity FID_j. However, this identity is encapsulated in session key ϕ_A before being hashed. Due to the difficulty of reversing the hashing function, it is difficult for adversary Ψ to obtain this identity from message $Auth_2$. □

Hypothesis 2: *Side-channeling and physical attacks are prevented.*

Proof: Suppose that attacker Ψ has physically captured the farmer's smart device SD_j. The objective is to extract the security tokens in its memory to compute the session key $\phi_S = h_1(R_3 || R_4^* || B_3 || A_4)$. During the registration phase, the SD_j stores parameter set $\{A_1, A_2^*, A_3^*, A_4^*, B_1\}$ in its memory. Here, $A_1 = h_2\ (FID_j || R_1)$, $A_2^* = A_2 + h_2(FID_j || FPW_j)$, $A_3^* = A_3 \oplus h_1(FPW_j || FID_j)$, $A_4^* = A_4 \oplus h_1(FID_j || FPW_j || P_B)$, $B_1 = h_1(A_2 || A_3 || A_4)$, and $B_3 = A_2.R_2$. As such, although the attacker may have access to value A_4^*, random nonces

R_3 and R_4^*, as well as parameter B_3, cannot be recovered from SD_j's memory. As such, the derivation of session key ϕ_S flops. □

Hypothesis 3: *This scheme is robust against eavesdropping and password-guessing attacks.*

Proof: Let us assume that adversary Ψ is interested in capturing the farmer's password FPW_j for malicious login into SD_j. To achieve this, an attempt is made to eavesdrop FPW_j from the exchanged messages $Auth_1 = \{B_2, B_4, C_1, C_2\}$ and $Auth_2 = \{C_3, C_4\}$. Here, $B_2 = A_1 \cdot R_2$, $A_1 = h_2\ (FID_j \,||\, R_1)$, $B_4 = A_3 \oplus h_3\ (B_3)$, $C_1 = A_4 \oplus R_3$, $C_2 = h_1(B_2 \,||\, B_3 \,||\, B_4 \,||\, C_1 \,||\, R_3 \,||\, A_4)$, $C_3 = R_4 \oplus A_4^{**}$, and $C_4 = h_1(FID_j^* \,||\, \phi_A)$. Evidently, none of the components of these two messages contains plain-text FPW_j. The only parameters incorporating this password are $A_2 = A_2^* - h_2(FID_j \,||\, FPW_j)$, $A_3 = A_3^* \oplus h_1(FPW_j \,||\, FID_j)$, and $A_4 = A_4^* \oplus h_1(FID_j \,||\, FPW_j \,||\, P_B)$, which are never sent directly over the public channels. In addition, FPW_j is encapsulated in other values before being hashed. Due to the difficulty of reversing or colliding the hashing function, any guessing of FPW_j from these parameters will fail. □

Hypothesis 4: *This scheme upholds unlinkability and untraceability.*

Proof: During the authentication phase, the SD_j generates random nonce R_2 and R_3, where $R_2 \in \{1, q-1\}$ and $R_3 \in \{0,1\}^k$. These nonces are utilized to construct authentication message $Auth_1 = \{B_2, B_4, C_1, C_2\}$, where $B_2 = A_1 \cdot R_2$, $A_1 = h_2\ (FID_j \,||\, R_1)$, $B_4 = A_3 \oplus h_3\ (B_3)$, $C_1 = A_4 \oplus R_3$, and $C_2 = h_1(B_2 \,||\, B_3 \,||\, B_4 \,||\, C_1 \,||\, R_3 \,||\, A_4)$. Similarly, ASP_i chooses random nonce$R_4 \in \{0,1\}^k$, which is used in the derivation of authentication response message $Auth_2 = \{C_3, C_4\}$, where), $C_3 = R_4 \oplus A_4^{**}$, and $C_4 = h_1(FID_j^* \,||\, \phi_A)$. Consequently, messages $Auth_1^{Sub}$ and $Auth_1^{Sub}$ for the subsequent communication session will be different from those of the current session. This lack of correlation among authentication messages implies that Ψ is incapable of tracking F_j using any captured messages. □

Hypothesis 5: *Spoofing and forgery attacks are thwarted.*

Proof: Let us assume that attacker Ψ is attempting to forge message $Auth_1 = \{B_2, B_4, C_1, C_2\}$ sent from SD_j towards the ASP_i, as well as response message $Auth_2 = \{C_3, C_4\}$ forwarded back to the SD_j from ASP_i. Here, $B_2 = A_1 \cdot R_2$, $A_1 = h_2\ (FID_j \,||\, R_1)$, $B_4 = A_3 \oplus h_3\ (B_3)$, $A_3 = h_1(FID_j \,||\, R_1 \,||\, MK_A)$, $C_1 = A_4 \oplus R_3$, $A_4 = A_4^* \oplus h_1(FID_j \,||\, FPW_j \,||\, P_B)$, $C_2 = h_1(B_2 \,||\, B_3 \,||\, B_4 \,||\, C_1 \,||\, R_3 \,||\, A_4)$, $C_3 = R_4 \oplus A_4^{**}$, and $C_4 = h_1(FID_j^* \,||\, \phi_A)$. Clearly, this requires random nonces such as R_1, R_2, and R_3, farmer's real identity FID_j and password FPW_j, master key for $ASP_i MK_A$, and padding bits P_B, among other parameters. *Hypothesis 1* illustrates the difficulty of obtaining FID_j, *Hypothesis 3* demonstrates the difficulty of obtaining FPW_j, while *Hypothesis 4* shows the difficulty of obtaining random nonces. In addition, Ψ cannot obtain master key MK_A since it is randomly selected from $\{1, q-1\}$ by ASP_i. □

Hypothesis 6: *This scheme can withstand session hijacking attacks.*

Proof: Suppose that adversary Ψ has captured random nonces R_1, R_2, R_3, and R_4. Next, an attempt is made to compute session parameters $B_3 = A_2 \cdot R_2$, $C_1 = A_4 \oplus R_3$, $C_2 = h_1(B_2 \,||\, B_3 \,||\, B_4 \,||\, C_1 \,||\, R_3 \,||\, A_4)$, $A_4^{**} = h_1(h_1(MK_A \oplus R_1^*) \,||\, FID_j^*)$, and $C_3 = R_4 \oplus A_4^{**}$ used in messages $Auth_1 = \{B_2, B_4, C_1, C_2\}$ and $Auth_2 = \{C_3, C_4\}$. Here, $B_2 = A_1 \cdot R_2$, $A_1 = h_2\ (FID_j \,||\, R_1)$, $B_4 = A_3 \oplus h_3\ (B_3)$, $A_3 = h_1(FID_j \,||\, R_1 \,||\, MK_A)$, $C_1 = A_4 \oplus R_3$, $A_4 = A_4^* \oplus h_1(FID_j \,||\, FPW_j \,||\, P_B)$, $C_2 = h_1(B_2 \,||\, B_3 \,||\, B_4 \,||\, C_1 \,||\, R_3 \,||\, A_4)$, $C_3 = R_4 \oplus A_4^{**}$, and $C_4 = h_1(FID_j^* \,||\, \phi_A)$. To hijack the session, other parameters are required apart from these random nonces, as illustrated in *Hypothesis* 5. Since these values are unavailable to Ψ, session hijacking is not possible. □

Hypothesis 7: *Impersonation attacks are prevented.*

Proof: Upon receiving message $Auth_1$, the ASP_i confirms whether parameter set $\{A_3^{**}, FID_j^*, R_1^*\}$ is in its database. The aim is to abort the session if this verification fails. In addition, it derives value $C_2^* = h_1(B_2 || B_3^* || B_4 || C_1 || R_3^* || A_4^{**})$ and checks if $C_2^* \stackrel{?}{=} C_2$. Here, the authentication session is terminated if this verification flops. On its part, the SD_j computes parameters $R_4^* = C_3 \oplus A_4$, session key $\phi_S = h_1(R_3 || R_4^* || B_3 || A_4)$, and $C_4^* = h_1(FID_j || \phi_S)$ upon receiving message $Auth_2$. This is followed by the verification of whether $C_4^* \stackrel{?}{=} C_4$. Essentially, the authentication session is aborted if these two parameters are not the same. As such, the legitimacy of all the communicating entities is verified to thwart impersonations. □

Hypothesis 8: *Robust authentication is executed.*

Proof: At the SD_j side, noncesR_2 and $R3$ are generated and parameters $B_2 = A_1.\ R_2$, $B_3 = A_2.\ R_2$, $B_4 = A_3 \oplus h_3\ (B_3)$, $C_1 = A_4 \oplus R_3$, and $C_2 = h_1(B_2 || B_3 || B_4 || C_1 || R_3 || A_4)$ are computed. These parameters are deployed to construct authentication message $Auth_1 = \{B_2, B_4, C_1, C_2\}$ forwarded to the ASP_i. Similarly, the ASP_i generates random nonce R_4 utilized to derive values $C_3 = R_4 \oplus A_4^{**}$, session key $\phi_A = h_1(R_3^* || R_4 || B_3^* || A_4^{**})$, and $C_4 = h_1(FID_j^* || \phi_A)$. Lastly, authentication message $Auth_2 = \{C_3, C_4\}$ is composed and forwarded to SD_j. During this process of authentication procedures, the legitimacy of SD_j is verified at the ASP_i using parameters $\{A_3^{**}, FID_j^*, R_1^*\}$, C_2^*, and C_2, as demonstrated in Hypothesis 7. Similarly, the authenticity of ASP_i is verified at the SD_j using parameters $C_4^*$and C_4, as illustrated in Hypothesis 7. □

Hypothesis 9: *This protocol prevents de-synchronization and DoS attacks.*

Proof: Most of the authentication protocols incorporate timestamps in the exchanged messages, which renders them susceptible to de-synchronization and DoS attacks. The aim of these timestamps is to uphold the freshness of the transmitted messages. In the proposed scheme, random nonces are utilized to preserve the freshness of the exchanged messages. For instance, the SD_j generates random nonces R_2 and R_3 that are used to derive parameters $B_2 = A_1.\ R_2$, $B_3 = A_2.\ R_2$, $B_4 = A_3 \oplus h_3\ (B_3)$, $C_1 = A_4 \oplus R_3$, and $C_2 = h_1(B_2 || B_3 || B_4 || C_1 || R_3 || A_4)$ of authentication message $Auth_1 = \{B_2, B_4, C_1, C_2\}$ forwarded to the ASP_i. On its part, the ASP_i chooses random nonce R_4, which is incorporated in values $C_3 = R_4 \oplus A_4^{**}$, session key $\phi_A = h_1(R_3^* || R_4 || B_3^* || A_4^{**})$, and $C_4 = h_1(FID_j^* || \phi_A)$ of message $Auth_2 = \{C_3, C_4\}$ forwarded to SD_j. □

Hypothesis 10: *This scheme eliminates the need for verifier tables.*

Proof: Some authentication schemes require that the communicating parties maintain verifier tables, which are queried during the authentication process. If the attackers gain access to these verifier tables, the entire network can be compromised and brought down. In the proposed scheme, the ASP_i authenticates the SD_j using parameter set $\{A_3^{**}, FID_j^*, R_1^*\}$, and $C_2^*$and C_2. Whereas values A_3^{**}, FID_j^* and R_1^* are re-computed and compared to the ones in its database, parameter C_2^* is re-computed and compared to the one received in authentication message $Auth_1 = \{B_2, B_4, C_1, C_2\}$ received from the SD_j. On the other hand, the SD_j authenticates ASP_i using value $C_4^* = h_1(FID_j || \phi_S)$, which is re-calculated and compared with its equivalent C_4 received from ASP_i in authentication message $Auth_2 = \{C_3, C_4\}$. This eliminates the need for the ASP_i and SD_j to maintain verifier tables. □

Hypothesis 11: *Man-in-the-middle and replay attacks are thwarted.*

Proof: Suppose that the adversary is interested in computing and replaying bogus authentication parameters A_3^{**}, FID_j^*, R_1^*, C_2^*, C_3, and C_4 needed to successfully authenticate ASP_i and SD_j. Here, $A_3^{**} = B_4 \oplus h_3 (B_3^*)$, $B_3 = A_2$. R_2, $B_3^* = MK_A$. B_2, $B_4 = A_3 \oplus h_3 (B_3)$, $A_3 = A_3^* \oplus h_1(FPW_j || FID_j)$, $C_2^* = h_1(B_2 || B_3^* || B_4 || C_1 || R_3^* || A_4^{**})$, $C_3 = R_4 \oplus A_4^{**}$, and $C_4 = h_1(FID_j^* || \phi_A)$. Based on *Hypothesis 1*, Ψ has no access to FID_j, while according to *Hypothesis 3*, Ψ has no access to FPW_j. Similarly, it has been shown in *Hypothesis 4* that Ψ does not have access to random nonces incorporated in these parameters. Based on *Hypothesis 5*, master key MK_A is never available to Ψ. Therefore, our scheme can withstand MitM attacks. □

Hypothesis 12: *The session key is set up for message encryption.*

Proof: Upon receiving message $Auth_1$ from SD_j, the ASP_i generates nonce R_4 and computes values $B_3^* = MK_A$. B_2, $A_3^{**} = B_4 \oplus h_3 (B_3^*)$, $A_4^{**} = h_1(h_1(MK_A \oplus R_1^*) || FID_j^*)$, and $R_3^* = C_1 \oplus A_4^{**}$. These parameters are utilized to derive session key $\phi_A = h_1(R_3^* || R_4 || B_3^* || A_4^{**})$. Similarly, after receiving message $Auth_2 = \{C_3, C_4\}$ from ASP_i, the SD_j computes value $R_4^* = C_3 \oplus A_4$ and session key $\phi_S = h_1(R_3 || R_4^* || B_3 || A_4)$. These keys are employed to encipher the exchanged messages. □

Hypothesis 13: *Known session-specific temporary information attacks are prevented.*

Proof: During the authentication phase, the ASP_i computes session key $\phi_A = h_1(R_3^* || R_4 || B_3^* || A_4^{**}$, while the SD_j derives session key $\phi_S = h_1(R_3 || R_4^* || B_3 || A_4)$. Here, $R_3^* = C_1 \oplus A_4^{**}$, $C_1 = A_4 \oplus R_3$, $A_4^{**} = h_1(h_1(MK_A \oplus R_1^*) || FID_j^*)$, $B_3^* = MK_A$. B_2, $R_4^* = C_3 \oplus A_4$, $C_3 = R_4 \oplus A_4^{**}$, $A_4 = h_1(h_1(MK_A \oplus R_1) || FID_j)$, $B_3 = A_2$. R_2, and $A_2 = A_2^* - h_2(FID_j || FPW_j)$. It was demonstrated in *Hypothesis 11* that attacker Ψ has no access to FID_j, MK_A, FPW_j, and random nonces used in these session keys. In addition, the computation of parameters, such as $B_3^* = MK_A$. $B_2 = MK_A$. A_1. $R_2 = MK_A$. $h_2 (FID_j || R_1)$. $R_2 = A_2$. R_2, even when B_2 and A_2 are known, is difficult due to the intractability of the computational Diffie–Hellman (CDH) problem. □

Hypothesis 14: *Key secrecy is upheld.*

Proof: Suppose that attacker Ψ has access to private values such as random nonces R_2, R_3, and R_4. Let us also assume that authentication messages $Auth_1 = \{B_2, B_4, C_1, C_2\}$ and $Auth_2 = \{C_3, C_4\}$ have been captured by the adversary. Using these parameters, an attempt is made to derive messages $Auth_1^{Sub}$ and $Auth_1^{Sub}$ for the subsequent communication session. Here, $B_2 = A_1$. R_2, $A_1 = h_2 (FID_j || R_1)$, $B_3 = A_2$. R_2, $A_2 = h_2 (FID_j || R_1).MK_A$, $B_4 = A_3 \oplus h_3 (B_3)$, $A_3 = h_1(FID_j || R_1 || MK_A)$, $C_1 = A_4 \oplus R_3$, $A_4 = h_1(h_1(MK_A \oplus R_1) || FID_j)$, $C_2 = h_1(B_2 || B_3 || B_4 || C_1 || R_3 || A_4)$,$\phi_A = h_1(R_3^* || R_4 || B_3^* || A_4^{**})$, and $C_4 = h_1(FID_j^* || \phi_A)$. It is clear that even with the captured random nonces, the computation of these authentication messages will still fail. This is because Ψ still needs other parameters, such as FID_j and MK_A. According to *Hypothesis 1*, FID_j is unavailable to Ψ. Similarly, *Hypothesis 5* has shown the difficulty of obtaining master key MK_A. Moreover, *Hypothesis 13* has demonstrated the difficulty of deriving B_3 since it requires solving the CDH problem. □

Hypothesis 15: *Privileged insider and stolen smart device attacks are prevented.*

Proof: Let us assume that Ψ has stolen the farmer's smart device SD_j. Thereafter, the security tokens $\{A_1, A_2^*, A_3^*, A_4^*, B_1\}$ stored in its memory are extracted. This can also happen when Ψ has some privileged access to these parameters. Here, $A_1 = h_2 (FID_j || R_1)$, $A_2^* = A_2 + h_2(FID_j || FPW_j)$, $A_3^* = A_3 \oplus h_1(FPW_j || FID_j)$, $A_4^* = A_4 \oplus h_1(FID_j || FPW_j || P_B)$, and $B_1 = h_1(A_2 || A_3 || A_4)$. The aim of the attacker is to access the secret value set $\{A_2, A_3, A_4\}$, where $A_2 = h_2 (FID_j || R_1)$. MK_A, $A_3 = h_1(FID_j || R_1 || MK_A)$, and $A_4 = h_1(h_1(MK_A$

$\oplus R_1) || FID_j)$. However, all these parameters are encapsulated in other values such FID_j, FPW_j, and P_B; hence, their recovery is challenging. □

Hypothesis 16: *The proposed scheme is highly scalable and adaptable.*

Proof: In the proposed scheme, farmer F_j communicates directly to the service provider ASP_i devoid of any centralized entity. In addition, *Hypothesis 10* describes how the proposed scheme eliminates the need for verifier tables. As such, any farmer smart device SD_K can seamlessly join and leave the network without affecting the performance of the already existing devices. □

5. Performance Evaluation

In this section, three common metrics deployed in the performance evaluation of authentication protocols are used to gauge the proposed scheme. These metrics include computation and communication costs, as well as the supported security characteristics. The specific details about the evaluation procedures are described in the following sub sections.

5.1. Computation Costs

To determine the execution time for the various cryptographic operations, ASP_i is emulated in a multi-precision integer and rational arithmetic cryptographic library (MIRACL) in a server with the specifications in Table 2.

Table 2. Server specifications.

Feature	Description
Operating system	Ubuntu 22.04 LTS
RAM	8 GB
Processor	Intel Core i7-8565U
Operating system type	64-bit
Clock frequency	3.2 GHz

On the other hand, the farmer's SD_j is emulated using Raspberry Pi 3 Model B Rev 1.2, whose specifications are presented in Table 3.

Table 3. Smart device specifications.

Feature	Description
Operating system	Ubuntu 20.04 LTS
RAM	1 GB
Processor	Quad-core
Operating system type	64 bit
Clock frequency	1.4 GHz

Under these conditions, the average execution times for various cryptographic primitives are presented in Table 4.

During the authentication and key negotiation phase, the SD_j executes a single T_{MTP}, six T_H, a single T_{PS}, and two T_{SM} operations. On the other hand, the ASP_i carries out a single T_{SM} and six T_H operations. Table 5 presents the comparisons of the computation cost of the proposed scheme with other related protocols.

Based on the values in Table 5, the protocol in [18] has a computation cost of 31.847 ms, while the scheme in [6] has a computation overhead of 14.838 ms. Similarly, the computation costs for the protocols in [5,20,40] and the proposed scheme are 33.692 ms, 14.97 ms, 77.102 ms, and 12.662 ms, respectively. As shown in Figure 4, the protocol in [40] incurs the highest computation costs.

Table 4. Execution time for various cryptographic operations.

Cryptographic Operation	Time (ms)	
	SD_j	ASP_i
Hashing operation (T_H)	0.314	0.056
Bilinear pairing (T_{BP})	33.051	4.715
Elliptic curve scalar multiplications (T_{SM})	2.256	0.654
Symmetric encryption/Decryption (T_{ED})	0.019	0.002
Elliptic curve point subtraction (T_{PS})	0.0115	0.003
Modular exponentiation (T_{ME})	0.325	0.083
Modular multiplication (T_{MM})	0.015	0.002
Modular addition (T_{MA})	0.012	0.001
Fuzzy extraction (T_{FE})	2.253	0.674
t-degree univariate polynomial evaluation (T_{PL})	13.3	0.3
Map-to-point hashing (T_{MTP})	5.264	2.853
Elliptic curve point addition (T_{PA})	0.017	0.004

Table 5. Computation costs comparisons.

Scheme	Derivations		Total (ms)
	User/Smart Device/Sensor	Server/Gateway Node	
Vangala et al. [18]	$22\,T_H + 8\,T_{SM} + 2\,T_{PA} + T_{FE} = 27.243$	$12\,T_H + 6\,T_{SM} + 2\,T_{PA} = 4.604$	31.847
Rangwani et al. [6]	$8\,T_H + 5\,T_{SM} = 13.792$	$7\,T_H + T_{SM} = 1.046$	14.838
Bera et al. [5]	$7\,T_H + 6\,T_{SM} + 2\,T_{PA} + T_{PL} = 29.068$	$7\,T_H + 6\,T_{SM} + 2\,T_{PA} + T_{PL} = 4.624$	33.692
Vangala et al. [20]	$9\,T_H + 4\,T_{SM} = 11.85$	$9\,T_H + 4\,T_{SM} = 3.12$	14.970
Wu et al. [40]	$2\,T_{BP} + 2\,T_{ME} + 2\,T_{ED} + T_H = 67.446$	$2\,T_{BP} + 2\,T_{ME} + 2\,T_{ED} + T_H = 9.656$	77.102
Proposed	$T_{MTP} + 6\,T_H + 2\,T_{SM} + T_{PS} = 11.672$	$T_{SM} + 6\,T_H = 0.99$	12.662

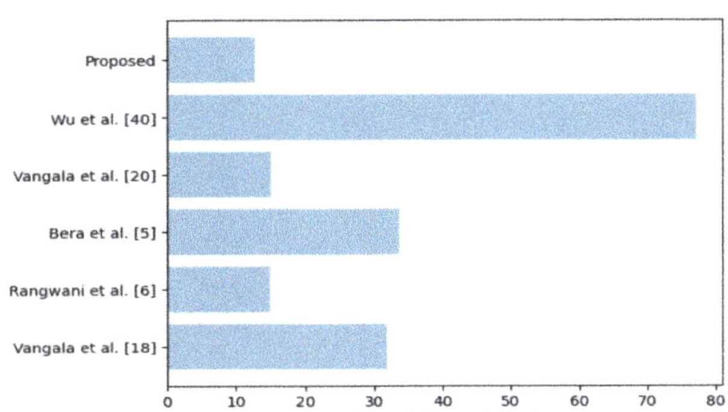

Figure 4. Computation costs comparisons [5,6,18,20,40].

This is attributed to the time-consuming bilinear pairing operations executed in this scheme. This is followed by the schemes in [5,6,18,20] and the proposed protocols in that order. The high computation overhead in [40] is attributed to the time-consuming bilinear pairing operations executed in this scheme. Since the farmer's smart device is battery-powered, our scheme is the most efficient and ensures that the battery for SD_j lasts longer. On the other hand, deploying the protocol in [40] in SD_j will drain its battery within a short time.

5.2. Communication Costs

To derive the number of bits used in the proposed protocol, the sizes of the messages exchanged between the SD_j and ASP_i during the authentication and key agreement phase are taken into consideration. For fair comparison, the values in [5] are used, in which the output sizes of the various cryptographic operations are presented in Table 6 below.

Table 6. Parametric sizes.

Operation	Size (bits)
Real identity	160
Random nonce	160
Hashing output	256
Points in finite group	512
Timestamp	32
Password	160

In our scheme, two messages are exchanged during the authentication and key negotiation phase. Whereas message $Auth_1 = \{B_2, B_4, C_1, C_2\}$ is sent from the SD_j towards the ASP_i, message $Auth_2 = \{C_3, C_4\}$ is transmitted from ASP_i towards SD_j. Here, $B_2 = A_1$. R_2, $B_4 = A_3 \oplus h_3\ (B_3)$, $C_1 = A_4 \oplus R_3$, $C_2 = h_1(B_2 || B_3 || B_4 || C_1 || R_3 || A_4)$, $C_3 = R_4 \oplus A_4^{**}$, and $C_4 = h_1(FID_j^{*} || \phi_A)$. Table 7 illustrates the derivation of the communication cost of this scheme.

Table 7. Message sizes.

Message	Size (bits)
$SD_j \rightarrow ASP_i$ $Auth_1$:$\{B_2, B_4, C_1, C_2\}$ $B_4 = C_1 = C_2 = 160$; $B_2 = 512$	992
$ASP_i \rightarrow SD_j$ $Auth_2$: $\{C_3, C_4\}$ $C_3 = C_4 = 160$	320
Total	1312

On the other hand, the protocol in [18] exchanges four messages, while the scheme in [6] requires five messages during the authentication process, as shown in Table 8. On their part, the schemes in [5,20,40] exchange 2 messages, 3 messages, and 10 messages, respectively. In terms of the total message sizes, the schemes in [5,6,18,20,40] require 5792 bits, 4128 bits, 2016 bits, 2305 bits, and 1600 bits, respectively.

Table 8. Communication costs comparisons.

Scheme	Number of Exchanged Messages	Size (bits)
Vangala et al. [18]	4	5792
Rangwani et al. [6]	5	4128
Bera et al. [5]	2	2016
Vangala et al. [20]	3	2305
Wu et al. [40]	10	1600
Proposed	2	1312

As shown in Figure 5, the scheme in [18] has the highest communication cost of 5792 bits, followed by the protocols in [5,6,20,40] and the proposed scheme, respectively.

Since the farmer's smart device is battery-powered, it has limited communication capability and hence the proposed protocol is the most efficient.

5.3. Security Characteristics

The goal of this section is to compare the security characteristics of the proposed scheme with other related protocols. Table 9 presents the results of this comparative evaluation.

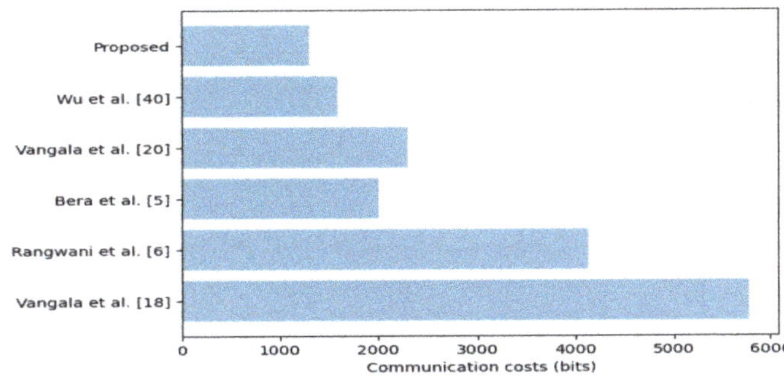

Figure 5. Communication costs comparisons [5,6,18,20,40].

Table 9. Security characteristics comparisons.

	[18]	[6]	[5]	[20]	[40]	Proposed
Security features						
User privacy	√	√	-	√	√	√
Anonymity	√	√	-	√	√	√
Unlinkability	-	-	-	-	-	√
Untraceability	√	√	-	√	√	√
Robust authentication	√	√	√	√	√	√
No verifier tables	×	√	√	×	-	√
Session key agreement	√	√	√	√	√	√
Key secrecy	√	√	√	-	-	√
Robust against:						
Side-channeling	√	√	√	√	×	√
Physical capture	√	√	√	√	×	√
Eavesdropping	×	×	√	×	×	√
Password guessing	√	√	×	×	√	√
Spoofing	×	×	×	×	×	√
Forgery	×	×	√	×	×	√
Replay	√	√	√	√	√	√
Session hijacking	×	×	×	×	×	√
Impersonation	√	√	√	√	×	√
De-synchronization	×	×	×	×	×	√
MitM	√	√	√	√	√	√
Privileged insider	√	√	√	√	√	√
KSSTI	×	√	√	√	×	√
DoS	√	√	-	√	√	√
Stolen smart device	√	√	√	√	×	√

√: supported; ×: not supported; -: not considered.

As shown in Table 9, the schemes in [5,20] each support 14 security characteristics, while the protocol in [18] offers support for 15 security features. On the other hand, the scheme in [6] supports 17 features, while the proposed protocol supports all 23 security features. Therefore, our scheme is the most secure and privacy-preserving.

Based on the results above, it is evident that the proposed scheme results in significant improvements in computation costs, communication costs, and supported security characteristics. Regarding computation overheads, the protocol in [6] with a cost of 14.838 m is used as the baseline. On the hand, the scheme in [40] with a communication cost of 1600 bits is used as the baseline. Similarly, the protocol in [18], which offers support for 15 security features, is deployed as the baseline. Using these baseline values, the proposed protocol results in 14.67% and 18% reductions in computation and communication costs, respectively, and a 35.29% improvement in supported security features.

6. Conclusions

In precision agriculture, numerous sensors such as radiation, air humidity, optimal, soil moisture, and ground sensors are deployed. In addition, intelligent precision agriculture utilizes numerous IoT devices and drones to monitor agricultural surroundings. Although these technologies help boost productivity in the face of limited resources, they are exposed to threats such as eavesdropping, message falsification, DoS, replay, MitM, and impersonations. Therefore, past researchers have seen the development of many security solutions for this environment. However, the attainment of perfect privacy and security at low computation and communication overheads still remains a mirage. The developed scheme has been shown to solve some of these challenges. For example, it has been shown to be resilient against side-channeling, physical capture, eavesdropping, password guessing, spoofing, forgery, replay, session hijacking, impersonation, de-synchronization, man-in-the-middle, privileged insider, denial of service, stolen smart device, and known session-specific temporary information attacks. Using the values in [6,18,40] as baselines, the proposed scheme leads to 14.67% and 18% reductions in computation and communication costs, respectively, and a further 35.29% improvement in supported security features. Future research will revolve around further enhancements of its performance as well as evaluation using metrics that were out of the scope of the current work.

Author Contributions: Z.A.A. and V.O.N.; methodology, and writing—original draft preparation, H.M.J.; software, data curation, validation, and writing—review and editing, J.M. and Z.A.A.; formal analysis, investigation, supervision, project administration, and funding acquisition, M.A.H., Z.A.H. and A.J.Y.A.; resources, visualization, and formal analysis. All authors have read and agreed to the published version of the manuscript.

Funding: This work is supported by the Natural Science Foundation of Top Talent of SZTU (Grant No. 20211061010016).

Institutional Review Board Statement: Not applicable.

Informed Consent Statement: Not applicable.

Data Availability Statement: Not applicable.

Conflicts of Interest: The authors declare no conflict of interest.

References

1. Vangala, A.; Das, A.K.; Mitra, A.; Das, S.K.; Park, Y. Blockchain-Enabled Authenticated Key Agreement Scheme for Mobile Vehicles-Assisted Precision Agricultural IoT Networks. *IEEE Trans. Inf. Forensics Secur.* **2022**, *18*, 904–919. [CrossRef]
2. Shafi, U.; Mumtaz, R.; García-Nieto, J.; Hassan, S.A.; Zaidi, S.A.R.; Iqbal, N. Precision Agriculture Techniques and Practices: From Considerations to Applications. *Sensors* **2019**, *19*, 3796. [CrossRef] [PubMed]
3. Shi, X.; An, X.; Zhao, Q.; Liu, H.; Xia, L.; Sun, X.; Guo, Y. State-of-the-Art Internet of Things in Protected Agriculture. *Sensors* **2019**, *19*, 1833. [CrossRef]
4. Vangala, A.; Das, A.K.; Chamola, V.; Korotaev, V.; Rodrigues, J.J. Security in IoT-enabled smart agriculture: Architecture, security solutions and challenges. *Cluster Comput.* **2022**, *26*, 879–902. [CrossRef]
5. Bera, B.; Vangala, A.; Das, A.K.; Lorenz, P.; Khan, M.K. Private blockchain-envisioned drones-assisted authentication scheme in IoT-enabled agricultural environment. *Comput. Stand. Interfaces* **2022**, *80*, 103567. [CrossRef]
6. Rangwani, D.; Sadhukhan, D.; Ray, S.; Khan, M.K.; Dasgupta, M. An improved privacy preserving remote user authentication scheme for agricultural wireless sensor network. *Trans. Emerg. Telecommun. Technol.* **2021**, *32*, e4218. [CrossRef]
7. Lan, G.; Brewster, C.; Spek, J.; Smeenk, A.; Top, J. *Blockchain for Agriculture and Food*; Findings from the Pilot Study, Report; Wageningen Economic Research: Wageningen, The Netherlands, 2017; p. 34.
8. Nyangaresi, V.O.; Ibrahim, A.; Abduljabbar, Z.A.; Hussain, M.A.; Al Sibahee, M.A.; Hussien, Z.A.; Ghrabat, M.J.J. Provably Secure Session Key Agreement Protocol for Unmanned Aerial Vehicles Packet Exchanges. In Proceedings of the 2021 International Conference on Electrical, Computer and Energy Technologies (ICECET), Cape Town, South Africa, 9–10 December 2021; pp. 1–6.
9. Sontowski, S.; Gupta, M.; Chukkapalli, S.S.L.; Abdelsalam, M.; Mittal, S.; Joshi, A.; Sandhu, R. Cyber attacks on smart farming infrastructure. In Proceedings of the 2020 IEEE 6th International Conference on Collaboration and Internet Computing (CIC), Atlanta, GA, USA, 1–3 December 2020; pp. 135–143.
10. Khanna, A.; Kaur, S. Evolution of Internet of Things (IoT) and its significant impact in the field of Precision Agriculture. *Comput. Electron. Agric.* **2019**, *157*, 218–231. [CrossRef]

11. Van der Merwe, D.; Burchfield, D.R.; Witt, T.D.; Price, K.P.; Sharda, A. Drones in agriculture. *Adv. Agron.* **2020**, *162*, 1–30.
12. Dagar, R.; Som, S.; Khatri, S.K. Smart farming–IoT in agriculture. In Proceedings of the 2018 International Conference on Inventive Research in Computing Applications (ICIRCA), Coimbatore, India, 11–12 July 2018; pp. 1052–1056.
13. Sanjeevi, P.; Prasanna, S.; Kumar, B.S.; Gunasekaran, G.; Alagiri, I.; Anand, R.V. Precision agriculture and farming using Internet of Things based on wireless sensor network. *Trans. Emerg. Telecommun. Technol.* **2020**, *31*, e3978. [CrossRef]
14. Nyangaresi, V.O.; Abduljabbar, Z.A.; Refish, S.H.A.; Al Sibahee, M.A.; Abood, E.W.; Lu, S. Anonymous Key Agreement and Mutual Authentication Protocol for Smart Grids. In *Cognitive Radio Oriented Wireless Networks and Wireless Internet, Proceedings of the 16th EAI International Conference, CROWNCOM 2021, Virtual Event, 11 December 2021, and 14th EAI International Conference, WiCON 2021, Virtual Event, 9 November 2021*; Springer International Publishing: Cham, Switzerland, 2022; pp. 325–340.
15. Wazid, M.; Das, A.K.; Bhat, V.; Vasilakos, A.V. LAM-CIoT: Lightweight authentication mechanism in cloud-based IoT environment. *J. Netw. Comput. Appl.* **2020**, *150*, 102496. [CrossRef]
16. Wang, D.; Li, W.; Wang, P. Measuring Two-Factor Authentication Schemes for Real-Time Data Access in Industrial Wireless Sensor Networks. *IEEE Trans. Ind. Inform.* **2018**, *14*, 4081–4092. [CrossRef]
17. Challa, S.; Das, A.K.; Gope, P.; Kumar, N.; Wu, F.; Vasilakos, A.V. Design and analysis of authenticated key agreement scheme in cloud-assisted cyber–physical systems. *Futur. Gener. Comput. Syst.* **2018**, *108*, 1267–1286. [CrossRef]
18. Vangala, A.; Das, A.K.; Lee, J. Provably secure signature-based anonymous user authentication protocol in an Internet of Things-enabled intelligent precision agricultural environment. *Concurr. Comput. Prac. Exp.* **2021**, *35*, e6187. [CrossRef]
19. Alsamhi, S.H.; Shvetsov, A.V.; Kumar, S.; Shvetsova, S.V.; Alhartomi, M.A.; Hawbani, A.; Rajput, N.S.; Srivastava, S.; Saif, A.; Nyangaresi, V.O. UAV Computing-Assisted Search and Rescue Mission Framework for Disaster and Harsh Environment Mitigation. *Drones* **2022**, *6*, 154. [CrossRef]
20. Vangala, A.; Sutrala, A.K.; Das, A.K.; Jo, M. Smart Contract-Based Blockchain-Envisioned Authentication Scheme for Smart Farming. *IEEE Internet Things J.* **2021**, *8*, 10792–10806. [CrossRef]
21. Akram, S.V.; Malik, P.K.; Singh, R.; Anita, G.; Tanwar, S. Adoption of blockchain technology in various realms: Opportunities and challenges. *Secur. Priv.* **2020**, *3*, e109. [CrossRef]
22. Lin, Y.-P.; Petway, J.R.; Anthony, J.; Mukhtar, H.; Liao, S.-W.; Chou, C.-F.; Ho, Y.-F. Blockchain: The Evolutionary Next Step for ICT E-Agriculture. *Environments* **2017**, *4*, 50. [CrossRef]
23. Almadhoun, R.; Kadadha, M.; Alhemeiri, M.; Alshehhi, M.; Salah, K. A user authentication scheme of IoT devices using blockchain-enabled fog nodes. In Proceedings of the 2018 IEEE/ACS 15th international conference on computer systems and applications (AICCSA), Aqaba, Jordan, 28 October–1 November 2018; pp. 1–8.
24. Wang, L.; Xu, L.; Zheng, Z.; Liu, S.; Li, X.; Cao, L.; Li, J.; Sun, C. Smart Contract-Based Agricultural Food Supply Chain Traceability. *IEEE Access* **2021**, *9*, 9296–9307. [CrossRef]
25. Al Sibahee, M.A.; Nyangaresi, V.O.; Ma, J.; Abduljabbar, Z.A. Stochastic Security Ephemeral Generation Protocol for 5G Enabled Internet of Things. In *IoT as a Service, Proceedings of the 7th EAI International Conference, IoTaaS 2021, Sydney, Australia, 13–14 December 2021*; Springer International Publishing: Cham, Switzerland, 2022; pp. 3–18.
26. Turkanović, M.; Brumen, B.; Hölbl, M. A novel user authentication and key agreement scheme for heterogeneous ad hoc wireless sensor networks, based on the Internet of Things notion. *Ad Hoc Netw.* **2014**, *20*, 96–112. [CrossRef]
27. Chang, C.-C.; Le, H.-D. A Provably Secure, Efficient, and Flexible Authentication Scheme for Ad hoc Wireless Sensor Networks. *IEEE Trans. Wirel. Commun.* **2015**, *15*, 357–366. [CrossRef]
28. Das, A.K.; Kumari, S.; Odelu, V.; Li, X.; Wu, F.; Huang, X. Provably secure user authentication and key agreement scheme for wireless sensor networks. *Secur. Commun. Netw.* **2016**, *9*, 3670–3687. [CrossRef]
29. Shuai, M.; Xiong, L.; Wang, C.; Yu, N. A secure authentication scheme with forward secrecy for industrial internet of things using Rabin cryptosystem. *Comput. Commun.* **2020**, *160*, 215–227. [CrossRef]
30. Tian, Y.; Yuan, J.; Song, H. Efficient privacy-preserving authentication framework for edge-assisted Internet of Drones. *J. Inf. Secur. Appl.* **2019**, *48*, 102354. [CrossRef]
31. Chae, C.-J.; Cho, H.-J. Enhanced secure device authentication algorithm in P2P-based smart farm system. *Peer-to-Peer Netw. Appl.* **2018**, *11*, 1230–1239. [CrossRef]
32. Nyangaresi, V.O.; Abduljabbar, Z.A.; Mutlaq, K.A.-A.; Ma, J.; Honi, D.G.; Aldarwish, A.J.Y.; Abduljaleel, I.Q. Energy Efficient Dynamic Symmetric Key Based Protocol for Secure Traffic Exchanges in Smart Homes. *Appl. Sci.* **2022**, *12*, 12688. [CrossRef]
33. Wu, F.; Xu, L.; Kumari, S.; Li, X. A new and secure authentication scheme for wireless sensor networks with formal proof. *Peer-to-Peer Netw. Appl.* **2015**, *10*, 16–30. [CrossRef]
34. Srinivas, J.; Mukhopadhyay, S.; Mishra, D. Secure and efficient user authentication scheme for multi-gateway wireless sensor networks. *Ad Hoc Netw.* **2017**, *54*, 147–169. [CrossRef]
35. Zeng, X.; Xu, G.; Zheng, X.; Xiang, Y.; Zhou, W. E-AUA: An Efficient Anonymous User Authentication Protocol for Mobile IoT. *IEEE Internet Things J.* **2018**, *6*, 1506–1519. [CrossRef]
36. Liu, C.-H.; Chung, Y.-F. Secure user authentication scheme for wireless healthcare sensor networks. *Comput. Electr. Eng.* **2017**, *59*, 250–261. [CrossRef]
37. Nyangaresi, V.O.; Abduljabbar, Z.A.; Ma, J.; Al Sibahee, M.A. Verifiable Security and Privacy Provisioning Protocol for High Reliability in Smart Healthcare Communication Environment. In Proceedings of the 2022 4th Global Power, Energy and Communication Conference (GPECOM), Cappadocia, Turkey, 14–17 June 2022; pp. 569–574.

38. Challa, S.; Das, A.K.; Odelu, V.; Kumar, N.; Kumari, S.; Khan, M.K.; Vasilakos, A.V. An efficient ECC-based provably secure three-factor user authentication and key agreement protocol for wireless healthcare sensor networks. *Comput. Electr. Eng.* **2018**, *69*, 534–554. [CrossRef]
39. Ali, Z.; Ghani, A.; Khan, I.; Chaudhry, S.A.; Islam, S.H.; Giri, D. A robust authentication and access control protocol for securing wireless healthcare sensor networks. *J. Inf. Secur. Appl.* **2020**, *52*, 102502. [CrossRef]
40. Wu, H.-T.; Tsai, C.-W. An intelligent agriculture network security system based on private blockchains. *J. Commun. Netw.* **2019**, *21*, 503–508. [CrossRef]
41. Abduljaleel, I.Q.; Abduljabbar, Z.A.; Al Sibahee, M.A.; Ghrabat, M.J.J.; Ma, J.; Nyangaresi, V.O. A Lightweight Hybrid Scheme for Hiding Text Messages in Colour Images Using LSB, Lah Transform and Chaotic Techniques. *J. Sens. Actuator Netw.* **2022**, *11*, 66. [CrossRef]
42. Tai, W.-L.; Chang, Y.-F.; Li, W.-H. An IoT notion–based authentication and key agreement scheme ensuring user anonymity for heterogeneous ad hoc wireless sensor networks. *J. Inf. Secur. Appl.* **2017**, *34*, 133–141. [CrossRef]
43. Ali, R.; Pal, A.K.; Kumari, S.; Karuppiah, M.; Conti, M. A secure user authentication and key-agreement scheme using wireless sensor networks for agriculture monitoring. *Futur. Gener. Comput. Syst.* **2018**, *84*, 200–215. [CrossRef]
44. He, D.; Zhang, Y.; Wang, D.; Choo, K.-K.R. Secure and Efficient Two-Party Signing Protocol for the Identity-Based Signature Scheme in the IEEE P1363 Standard for Public Key Cryptography. *IEEE Trans. Dependable Secur. Comput.* **2018**, *17*, 1124–1132. [CrossRef]
45. Feng, Q.; He, D.; Liu, Z.; Wang, D.; Choo, K.K.R. Multi-party signing protocol for the identity-based signature scheme in IEEE P1363 standard. *IET Inf. Secur.* **2020**, *1*, 1–10.
46. Nyangaresi, V.O. Terminal independent security token derivation scheme for ultra-dense IoT networks. *Array* **2022**, *15*, 100210. [CrossRef]
47. Sadhukhan, D.; Ray, S.; Biswas, G.P.; Khan, M.K.; Dasgupta, M. A lightweight remote user authentication scheme for IoT communication using elliptic curve cryptography. *J. Supercomput.* **2020**, *77*, 1114–1151. [CrossRef]
48. Dhillon, P.K.; Kalra, S. A lightweight biometrics based remote user authentication scheme for IoT services. *J. Inf. Secur. Appl.* **2017**, *34*, 255–270. [CrossRef]
49. Chang, C.-C.; Nguyen, N.-T. An Untraceable Biometric-Based Multi-server Authenticated Key Agreement Protocol with Revocation. *Wirel. Pers. Commun.* **2016**, *90*, 1695–1715. [CrossRef]
50. Amin, R.; Islam, S.H.; Kumar, N.; Choo, K.-K.R. An untraceable and anonymous password authentication protocol for heterogeneous wireless sensor networks. *J. Netw. Comput. Appl.* **2018**, *104*, 133–144. [CrossRef]
51. Li, X.; Niu, J.; Alam Bhuiyan, Z.; Wu, F.; Karuppiah, M.; Kumari, S. A Robust ECC-Based Provable Secure Authentication Protocol With Privacy Preserving for Industrial Internet of Things. *IEEE Trans. Ind. Inform.* **2017**, *14*, 3599–3609. [CrossRef]
52. Alotaibi, M. An Enhanced Symmetric Cryptosystem and Biometric-Based Anonymous User Authentication and Session Key Establishment Scheme for WSN. *IEEE Access* **2018**, *6*, 70072–70087. [CrossRef]
53. Moghadam, M.F.; Nikooghadam, M.; Al Jabban, M.A.B.; Alishahi, M.; Mortazavi, L.; Mohajerzadeh, A. An Efficient Authentication and Key Agreement Scheme Based on ECDH for Wireless Sensor Network. *IEEE Access* **2020**, *8*, 73182–73192. [CrossRef]
54. Fadi, A.T.; Deebak, B.D. Seamless authentication: For IoT-big data technologies in smart industrial application systems. *IEEE Trans. Ind. Inform.* **2020**, *17*, 2919–2927.

sustainability

MDPI

Review

A Study of CNN and Transfer Learning in Medical Imaging: Advantages, Challenges, Future Scope

Ahmad Waleed Salehi [1], Shakir Khan [2,3,*], Gaurav Gupta [1,*], Bayan Ibrahimm Alabduallah [4], Abrar Almjally [2], Hadeel Alsolai [4], Tamanna Siddiqui [5] and Adel Mellit [6]

1 Yogananda School of AI, Computers and Data Sciences, Shoolini University, Solan 173212, India
2 College of Computer and Information Sciences, Imam Mohammad Ibn Saud Islamic University (IMSIU), Riyadh 11432, Saudi Arabia
3 Department of Computer Science and Engineering, University Centre for Research and Development, Chandigarh University, Mohali 140413, India
4 Department of Information System, College of Computer and Information Sciences, Princess Nourah bint Abdulrahman University, Riyadh 11564, Saudi Arabia; bialabdullah@pnu.edu.sa (B.I.A.)
5 Department of Computer Science, Aligarh Muslim University, Aligarh 202002, India
6 Faculty of Sciences and Technology, University of Jijel, Jijel 18000, Algeria
* Correspondence: sgkhan@imamu.edu.sa (S.K.); solan.gaurav@gmail.com (G.G.)

Abstract: This paper presents a comprehensive study of Convolutional Neural Networks (CNN) and transfer learning in the context of medical imaging. Medical imaging plays a critical role in the diagnosis and treatment of diseases, and CNN-based models have demonstrated significant improvements in image analysis and classification tasks. Transfer learning, which involves reusing pre-trained CNN models, has also shown promise in addressing challenges related to small datasets and limited computational resources. This paper reviews the advantages of CNN and transfer learning in medical imaging, including improved accuracy, reduced time and resource requirements, and the ability to address class imbalances. It also discusses challenges, such as the need for large and diverse datasets, and the limited interpretability of deep learning models. What factors contribute to the success of these networks? How are they fashioned, exactly? What motivated them to build the structures that they did? Finally, the paper presents current and future research directions and opportunities, including the development of specialized architectures and the exploration of new modalities and applications for medical imaging using CNN and transfer learning techniques. Overall, the paper highlights the significant potential of CNN and transfer learning in the field of medical imaging, while also acknowledging the need for continued research and development to overcome existing challenges and limitations.

Keywords: deep learning; transfer learning; medical imaging; CNN; machine learning

Citation: Salehi, A.W.; Khan, S.; Gupta, G.; Alabduallah, B.I.; Almjally, A.; Alsolai, H.; Siddiqui, T.; Mellit, A. A Study of CNN and Transfer Learning in Medical Imaging: Advantages, Challenges, Future Scope. *Sustainability* **2023**, *15*, 5930. https://doi.org/10.3390/su15075930

Academic Editors: Mourade Azrour, Azidine Guezzaz, Imad Zeroual, Azeem Irshad, Jamal Mabrouki, Said Benkirane and Shehzad Ashraf Chaudhry

Received: 15 January 2023
Revised: 21 February 2023
Accepted: 28 February 2023
Published: 29 March 2023

1. Introduction

People's health is at the center of medical care. The amount of medical data available today is enormous, but to benefit the medical industry [1], it is essential to use this data wisely. Medical images are frequently requested in accordance with a patient's follow-up to ensure that therapy was successful, and it is a critical step in the process of medical diagnosis and treatment [2]. In general, a radiologist examines the obtained medical images and compiles their results in a report [3]. Based on the images and the reports from radiologists, the referring doctor determines a diagnosis and a course of action. The majority of medical professionals, particularly radiologists, interpret medical images. However, human subjectivity, the wide variances among interpreters, and weariness limit human image interpretation. Due to the limited time radiologists have to analyze an ever-growing number of images, missed findings, lengthy turnaround times, and a lack of quantitative data or quantification are common when reviewing cases [4,5]. In turn, this severely restricts the medical profession's potential to expand the use of evidence-based, individualized healthcare [6]. Artificial Intelligence (AI) is a broad field with a wide variety of subfields such as natural language processing (NPL) [7], speech processing [8], machine learning, deep learning, robotics, etc. [9]. AI is applied in various kinds of fields, such as healthcare, agriculture, manufacturing, and the education sector [10]. Machine learning is a branch of AI that can learn from the data itself, automatically identify the patterns in data, and make decisions with minimum human intervention [11,12]. Over recent years, deep learning techniques have gained a lot of attention to solve various problems, especially in medical imaging fields [13]. Deep learning is an advanced field in computer vision. The purpose of computer vision is to carry out multiple tasks such as image detection, image recognition, NPL, image analysis [14], etc. A CNN is a type of artificial neural network specially designed to handle video and image data. It takes input images, extracts them, and classifies the output images after learning the features from the input images based on the learning features [11]. The deep learning CNN technique is used in the majority of AI medical image analyses, especially for the diagnosis of different types of diseases [15] such as breast cancer, Alzheimer's, brain tumors [16], etc. [4,17]. Deep CNN-based algorithms have achieved promising outcomes in the analysis of medical images. Several types of TL have been proposed for medical imaging data and have been very effective, such as Alex Net, SPP-Net, VGGNet, ResNet, GoogLeNet, etc. [4]. The major aim of this review is to highlight the most crucial CNN components so that researchers and students may easily gain a comprehensive understanding of CNN and transfer learning. This article will assist individuals to learn more about recent advancements in the discipline, which will promote DL research. The following list outlines our contributions:

- This review aids in the comprehension of CNN and transfer learning techniques among researchers and students.
- We simply describe the major issues of traditional ML and how DL-based techniques like CNN can come to the rescue and play an important role in diagnostic analysis.
- We describe in-depth the ideas, theories, and cutting-edge architectures of CNN, the most well-known deep learning technique.
- A literature review is provided in this paper to give an overview of related research work done on the use of CNN and TL techniques.
- We discuss the difficulties that deep learning-based techniques currently face, such as the scarcity of training data, overfitting, and vanishing gradient problems.
- The strategies for choosing the right TL-based technique for a problem are discussed.
- We also present a list of medical imaging modalities used in training the model, and we describe computational resources such as GPU, CPU, and TPU by contrasting how each tool affects deep learning algorithms.

CNNs' application to medical image processing is first discussed in this paper. The difference between traditional ML and DL-based algorithms and the analysis of medical images are presented in Section 2. Then we give a detailed overview of the architecture of CNN in Section 3. The corresponding work in the field of diagnosing disease using medical images, including CT, MRI, fMRI, PET, X-ray, and ultrasound, is discussed in Section 4. We also described the significance and importance of transfer learning techniques and explain each with their potential benefits in Section 5. Finally, we discuss the possible problems and predict the development prospects of CNN-based techniques in medical imaging analysis.

2. Imaging Modalities for Analytics and Diagnostics

To create an image, several methods are used. Examples of these measurements are radiofrequency signal capacity in an MRI, sound pressure for ultrasounds, and radiation absorption in X-ray imaging. In a digital image, one measurement is used to determine each image point, while in multi-channel images, several measurements are gathered [18]. To create diagnostic images, a wide variety of imaging modalities are employed, such as computed tomography (CT), X-ray, magnetic and functional resonance imaging (MRI and fMRI) [19], and positron emission tomography (PET) scans [4,5]. Common DL applications using medical imaging include image classification [11], segmentation, synthesis, and regression [12,20]. Figure 1 depicts various imaging modalities used [21]. In the initial phase of using more precise imaging methods to halt the spread of disease, medical imaging techniques are crucial as an aid to early diagnosis in the treatment or eradication of many medical disorders.

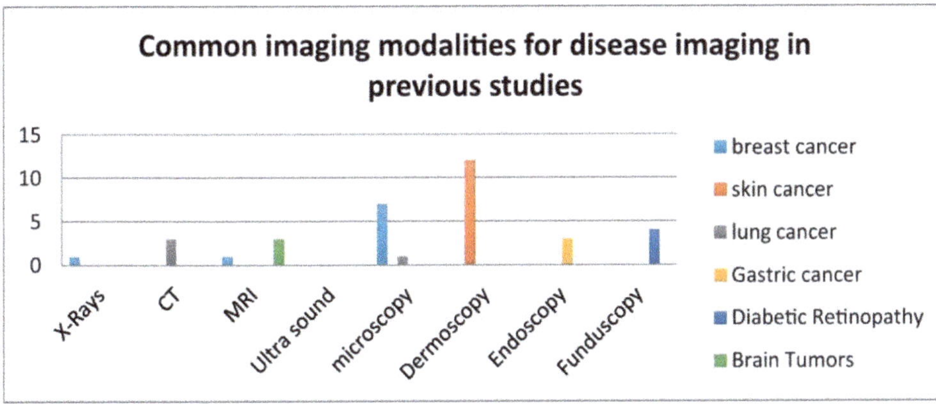

Figure 1. Common imaging modalities for disease imaging [21].

3. Convolutional Neural Network and its Background

David Hubel and Torsten Wiesel, two neurophysiologists, did experimentation in 1959 and eventually published their findings in a work titled "Receptive-Fields of Single-Neurons in cat's straits cortex" [22]. They defined how the neurons in a cat's brain are organized in a tiered pattern or layered form. These are the layers that can learn to detect visual patterns with the help of local features, which are extracted first, and for a higher-level representation, the extracted features are then combined [23]. Consequently, this concept is effectively becoming one of deep learning's core principles. In 1980, another researcher by the name of Kunihiko, who was motivated by the work of T. Wiesel [22], proposed a "Neocognitron". This work proposed a multi-layered neural network for the hierarchical detection of visual patterns learned from data (learning-without-teacher), which is known as a self-organizing neural network. [24]. This design then became the first CNN theoretical model. The Neocognitron develops the ability to classify and accurately detect patterns based on their shape distinctions. Any patterns that we humans consider to be similar are also classified as such by this proposed model. CNN, commonly known as ConvNet, is one of the common types of Artificial Neural Network (ANN) [25] that comes under the supervised method category. This method is known for its ability to discover and interpret patterns. This pattern detection brings up the usefulness of CNN for image analysis [26]. A ConvNet is a series of layers in which each layer performs some unique functions. Furthermore, these layers are usually classified into different categories [27]. The raw data is stored in the first layer, called the input layer. A convolutional layer is the second layer, which is responsible for calculating the output volume by performing a dot product between the image patch and all of the filters, followed by another important function known as activation. The mathematical function is then applied to every element of the convolution layer's output. The next layer comes in to help in reducing the computation costs by making the previous layer's output memory efficient. It is known as the pooling layer. Finally, once the pooling layer computation is done, it will pass its output to the last layer and output the computed 1-D array class score [26]. Two primary tasks must be accomplished when training a deep learning model:

- Forward propagation: To train a neural network, one must first provide it with an input, and then, in light of the outcomes of that processing, an output is produced.

- Backward propagation: Next, the model uses the backpropagation technique, such that the weights of the neural network are modified in response to the error that was obtained in the forward propagation.

3.1. Important Elements of CNN

In this part of the article, we discuss the fundamental components of a CNN in detail with their role in the whole architecture:

3.1.1. Convolutional Layer

The convolution layer, as its name suggests, is crucial to CNN's operation. Where the majority of the calculation is concerned, it is the core unit of a CNN. Since digital image processing is concerned, convolution operations are the most widely used [19]. Convolutional layers are where filters (also known as the set of kernels) are applied or get convolved with the original input images, which can be n-dimensional metrics to generate a feature map as an output [20]. Here, the number of kernels and the size of the kernels are the most critical parameters, which refer to the size of the filter, as shown in below Figure 2. The following mathematical formula is used to determine subsequent feature map values [20], where the kernel is denoted by h and the image input is indicated by f. The result matrix's row and column indexes are denoted by m and n.

$$G[m, n] = (fx)[m, n] = \sum_{j}\sum_{k} h[j,k]f[m-j, n-k] \tag{1}$$

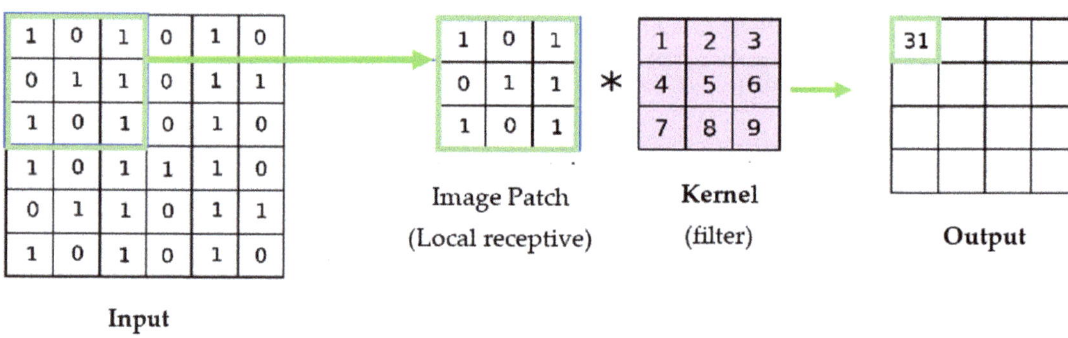

Figure 2. Convolutional process.

3.1.2. Pooling Layer

In CNN, the convolutional operation is applied to learned filters to the input image to summarize and show the presence of those features in the given. This is done in a systematic way to build its feature maps. The feature map is generated by the convolutional layer's output. It has one limitation due to recording the exact location of features in the input. Therefore, in the input image, any small movement that happens to the position of a feature, such as re-cropping, rotation, etc., will cause changes in the feature map. A common solution to this problem can be achieved in the convolution layer using downsampling by altering the convolution stride over the image [28]. This is where the usage of the pooling layer begins. It is nothing but a common and robust approach to the same problem. In a short pooling layer downsample, the previous layers' feature map and pooling operations aid in the creation of an invariant representation for small input translations [29]. Additionally, there are several functions used for specifying the pooling procedure; the most common functions are the following [30]:

(a) Average pooling: This is used when the average value is desired for each patch on the feature map.

(b) Maximum pooling: This is commonly known as Max-pooling, and is used when the maximum value is desired for each patch on the feature map [30]. Below Figure 3, illustrate the working of average and maximum pooling.

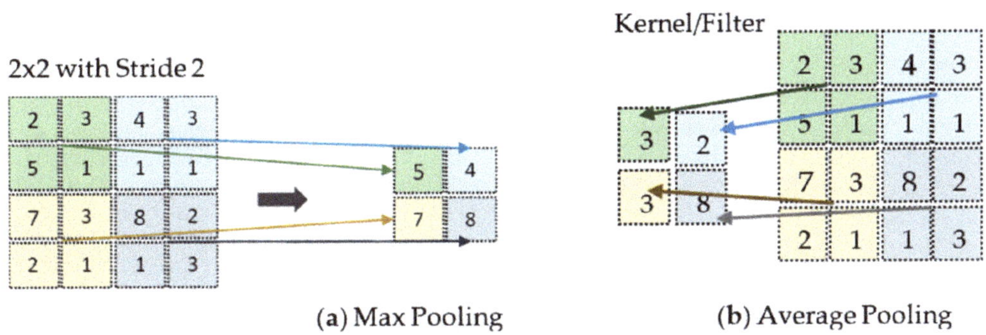

Figure 3. Two different pooling techniques were applied.

3.1.3. Fully Connected Layers

Immediately following the completion of feature extraction and consolidation by the convolutional and pooling layers, another layer comes in, which is known as the fully connected layer [31]. This component is connected to the final node of each network to flatten out the output of the previous layer. Finally, this layer returns the probability of class predictions by building non-linear feature combinations. There are various non-linear functions, such as activation functions, ReLU, and Softmax.

3.2. Important Parameters and Hyperparameters for Building CNN

The following are the important parameters with a high level of description.

- Kernels: The kernel is nothing but a matrix that is used to traverse over the input images to perform a dot product to extract features [32]. By using the stride value, the kernel can move by columns of pixels based on the number assigned to the stride.
- Biases: Before passing the output values through an activation function, the bias is used to adjust the scaled values. For example, in a neural network, the activation function receives an input 'x' which is multiplied by the 'w' weight. Therefore, adding a constant bias to the input will enable you to shift the activation function [33].
- Padding: When a kernel is used with image processing, the image is altered each time a convolution is carried out on the input data. The image shrinks and thus this can be done only a certain number of times before the input image completely disappears [34]. As a result, some of the information contained in the image can be lost. The problem is that when the kernel moves across the image there is a significant impact on the pixels in the outskirts of the image, which are much smaller when compared to the center pixels of the image [35]. Therefore, a more accurate analysis of the image can be achieved by the use of padding, which is added to the image's outer frame to provide more room for the filter to cover the image.
- Stride: Stride is another so-called hyperparameter in the convolutional layer that specifies the pixel count the kernel shifts over the input image matrix. For instance, when two is set as the stride, then the filter or kernel moves two pixels at a time. When three is set as stride, then the filter moves three pixels at a time, and so on [36].
- Dropout for regularization: This is a powerful yet simple regularization technique for deep learning models [37], and CNNs usually have the habit of overfitting. When there are a large number of nodes or neurons in a full-connected layer, it is more likely that co-adaptation occurs. Co-adaption simply means when many neurons in a single layer extract very similar or the same hidden features from the given input data. This

usually happens when two different neurons' connection weights are identical [38]. This technique works based on selecting neurons randomly and ignoring them during training; they will lose their contribution for further processes.

- Learning Rate: The learning rate is a very important parameter in CNN which defines how swiftly a network updates its parameters during backpropagation [39]. Keeping the learning rate low makes the convergence smooth, but the learning process slows down. However, keeping the learning rate larger may speed up the process of learning, but may prevent convergence.

Activation Functions: Nonlinearity is introduced to models via activation functions, allowing deep-learning models to learn nonlinear prediction bounds. In artificial neural networks (ANNs), activation functions are used to transform an input signal into an output signal. This output signal is then used as input by the subsequent layer in the stack. The most common activations used in CNN are described below:

Sigmoid activation function: Because it is a non-linear function, it is the most often utilized activation function. The sigmoid function changes data in the 0 to 1 range and it is widely used for binary classification. It can be summed up as follows [40]:

$$f(x) = \frac{1}{e^{-x}}$$

Tanh activation function: It is a function known as the hyperbolic tangent. The Tanh function is comparable to the sigmoid function; however, it is symmetric concerning the origin [40]. This activation function is smoother, and it is a zero-centered function with a scale that goes from -1 to 1, therefore, the function's output is given as [41]:

$$f(x) = \left(\frac{e^{x} - e^{-x}}{e^{x} - e^{-x}}\right)$$

In contrast to the sigmoid function, the Tanh function became the favored function because it provides higher training performance for a model with multiple layers [42,43].

ReLU function: ReLU stands for the rectified linear unit; it is a non-linear function and very popular in ConvNets. Since all the neurons are not going to be activated at the same time, but rather a small number of neurons are activated at a time, the ReLU function is more efficient than others [40]. According to equation 1, the output of ReLU is the value that is greater than either zero or the value that was fed into the model. When the value of the input is negative, the value of the output is equal to 0. When the value of the input is positive, the output value will be equal to the value of the input [44].

$$f(x) = \max(0, x)$$

An improved version of the ReLU activation function came up after ReLU, where instead of specifying the ReLU function's value as zero for x (negative values), rather it is defined as an x having an extremely insignificant linear component. It can be mathematically stated as [40]:

$$f(x) = 0.01x,\ x < 0 f(x) = x,\ x \geq 0$$

Softmax activation function: For binary (0, 1) classification, the sigmoid function is used, but to deal with multiclass classification Softmax is used. The Softmax function returns a probability for each data point of all individual classes [40]. Therefore, in a deep neural network, when we want to work with a multiclass classification problem, the output layer of the neural network will have an identical amount of network neurons that correspond to the number of target classes. The formula is stated as follows [13]:

$$\sigma(z)j = \frac{e^{z_j}}{\sum_{k=1}^{K} e^{z_j}}\ for\ j = 1, \ldots..K$$

Figure 4 represents the process for these connected layers.

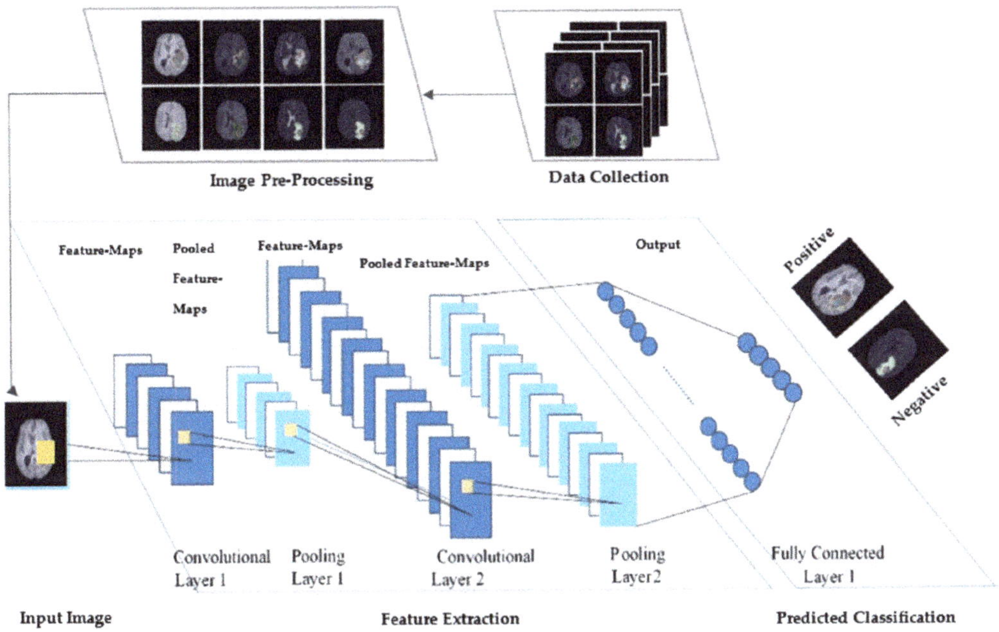

Figure 4. The diagram represents the medical image data collection. After collection, the images are preprocessed then given as input to the CNN model. There are a total of five layers: two conv-layers, two max-pooling layers.and an output layer called fully connected layer. The conv-weights in the first conv-layer are used in extracting feature maps from the input. Each pooled layer reduces the image size by half. Following the completion of each layer of pooling, the number of feature mappings and conv-weights are both increased by one. With the activation function, the last layer of the feature maps is fully connected to data nodes. Using a function, these nodes are then linked together to form a single value. This value was fitted to be the label defined in the training set and finally returned a value range of 0 and 1 [45].

4. ConvNets over Traditional Machine Learning

The process of machine learning involves the use of algorithms to analyze data, draw conclusions from that analysis, and make decisions based on those conclusions. In the case of DL, it uses multiple layers to create an ANN [7]. Each layer provides different information about the data which is fed to them. To perform classification work using machine learning techniques, several preprocessing steps, such as feature selection, [46], feature extraction [47], and classification are required [48]. Even the selection of features can have a significant impact on the efficiency gains achieved through various machine-learning strategies. DL techniques can perform automated feature sets for various tasks. Deep learning has simplified the improvement of object detection, image super-resolution, image classification, and image recognition fields [49].

Typical healthcare applications of classification tasks of images include Alzheimer's disease (AD) classification using MRI [50], dermatological identification of skin conditions [51], breast cancer diagnosis using histopathological images [17], and diagnosis of eye diseases in the field of ophthalmology (such as diabetic retinopathy [52], corneal diseases [53], and glaucoma [54]). With advances in 2021, DL has become a key popular tool for the automatic detection of COVID-19 and classifying healthy and not-healthy individuals using X-rays and CT scan images [50].

4.1. The Problem with Traditional Neural Networks

The main significant distinction between the traditional ANNs and CNNs is the primary usage of ConvNets in the field of pattern recognition, in particular of medical images. This usage enables the developers to encode features of input images into the architecture and makes the convolutional neural network more beneficial for image-specific tasks, while also lowering the number of parameters needed to set up and build the model. Traditional neural networks are known as multilayer perceptrons (MLPs). MLPs have several limitations, particularly when it comes to the processing of images [55]. For each input, MLPs are going to use a single perceptron, which means if we input an RGB image, each pixel is going to be multiplied by three since there are three channels in RGB. Therefore, here is where the problem arises; the number of weights to be used in each perceptron rapidly increases for large images, so it becomes unmanageable for the model. There are approximately 187,000 weights to train for a 250 $\times$ 250-pixel image with three channels. Hence, overfitting can happen, and training becomes difficult [56].

4.2. Feature Extraction

Feature extraction entails the process of obtaining a high level of patterns from raw pixel values to seize the uniqueness of the distinction between the various categories that are being used. The extraction of these features is carried out without the presence of any supervision (unsupervised manner). This indicates that the information that is extracted from the pixels of the image has nothing to do whatsoever with the classes of the image, and, in CNN, the convolution layer is the backbone of feature extraction [57]. This allows for the sharing of parameters. Following the extraction of the features [58], a classifier is then trained using the images and the labels that are associated with them, for example, logistic regression, random forests, decision trees, support vector machines, etc. This pipeline has a problem due to the fact that the feature extraction cannot be changed based on the classes and images. So, no matter what type of classification technique is used, the accuracy of the model is severely compromised as a result if the chosen feature does not give enough information to tell the categories apart [59]. Picking various feature extractors and clubbing them ingeniously to achieve better feature extraction has been a recurrent subject among state of the art studies. However, this necessitates an excessive number of heuristics and tedious manual work to adjust settings depending on the domain. The main philosophy behind deep learning is that there is no predetermined way to extract features (no hard-coding) from data [60]. The CNN learns to extract data by differentiating representations from the input images and to categorize them based on supervised data, all inside a single integrated system.

4.3. Parameter Sharing

With ConvNets, a large dataset like ImageNet can be used to train the whole network from scratch [61]. ImageNet is an ongoing project that has so far collected 14,197,122 images in 21,841 different categories. Sharing parameters cuts down on the total parameters in the network and shortens the training time required for the network [59].

5. Literature Review

For the last few years, researchers have been using CNN-based models to extract unique and useful features for the diagnosis of various diseases, including but not limited to brain cancer, heart disease, Alzheimer's disease (AD), COVID-19, Parkinson's disease, breast cancer, etc. [62], by using medical images. According to previous studies, using convolutional neural network-based models achieved a good level of accuracy when compared with traditional machine learning and volumetric techniques that are manually performed by physicians. Therefore, this section summarized the CNN-based methods using medical images.

In the Alexander et al. [63] study, a CNN model using MRI and diffusion-tensor imaging (DIO) was used for the classification of AD patients. According to their study, the classification performance demonstrates that the size of the hippocampal region of interest (ROI) does not matter when bigger ROIs are combined with using CNN architecture for the classification. Using a six-layered convolutional neural network with 48 × 48 × 48 ROI with a data fusion model achieved a good accuracy of 96.7% for their case (AD-Normal Control).

The Liu et al. [64] study first segmented the MRI image into two segments, grey and white matter. Then they used a Multiscale ConvNet (MSCNet) for the diagnosis of AD. According to their study, white matter is more effective for the detection of AD than gray matter. In terms of accuracy, the MSCNet has a higher performance level than ResNet-50 in the NC and MCI classes of grey and white matter, respectively; however, the standard deviation is lower in ResNet-50. The accuracy of the MSCNet model with grey and white matter is 98.85% and 98.11%, and the ResNet-50 model accuracy is 96.01% and 95.88%, respectively. Therefore, this study shows that with lower computational power and fewer parameters, the CNN-based MSCNet model performs well for the medical image dataset.

Ajagbe et al. [65] wrote an article on their use of Deep CNN and transfer learning models (VGG-16 and VGG-19) for the diagnosis of AD with the help of MRI images. However, in terms of six performance metrics such as Area Under the Curve (AUC), accuracy, F-1 score, precision, computational time, and recall, VGG-16 performed best in one, VGG-19 in three, and CNN best in two metrics. The limitations of this study are the high computational power and the lack of a self-created dataset.

In the study by Villa-Pulgarin et al. [66], the focus was to classify skin lesion cancers by using CNN-based models DensNet-201, Inception-ResNet-V2, and Inception-V3. In their work, they tested the models with different workflows, fine-tuning the optimization and using data augmentation. The best results of their model were obtained by using the HAM10000 dataset, with an accuracy of 98% using the data augmentation stage, and 93% by using the ISIC 2019 dataset using the optimized DenseNet-201 model. El-Din Hemdan et al. [67] presented the COVIDX-Net model for the earlier diagnosis of COVID-19 patients based on seven different CNN-based architectures: MobileNetV2, DenseNet-201, ResNetV2, InceptionV3, Xception, Visual Geometry Group (VGG-19), and InceptionResNetV2. According to their results, the DenseNet-19 and VGG-19 models performed well in determining which cases were COVID-19 negative and which were positive, and the Inception model performed the worst, with an accuracy of 50% and an F1 score of 67 for normal and zero for COVID-19 cases.

The Horry et al. [68] study aimed to focus on important features by removing noise from medical images for the detection of COVID-19 disease. The Horry et al. study selected the VGG-19 with transfer learning in order to classify NC and pneumonia cases accurately. However, the authors reported that the VGG-19 model performed best, with a precision of 100%, using ultrasound images compared with X-ray and CT images. It is a very interesting finding that the pre-trained method tuned very well for the ultrasound data samples, which are very noisy and difficult to interpret by the human eye. Neal Joshua et al. [69] proposed a 3D-CNN architecture for the detection of nonlinear 3D information of the lung nodule using CT scan images. Moreover, they used gradient class activation for visualizing the internal structure of the CT images to get more information. From their lightweight proposed model, they achieved a very good classification accuracy of 97.17% using gradient-weighted class activation when compared with existing AlexNet 2D-CNN and AlexNet 3D-CNN models.

Li et al. [70] used the CNN model for the classification of lung image patches with interstitial lung disease (ILD) patterns. Their proposed architecture can correctly identify the features of the image from the lung patches of ILD. The authors have compared their classification results with three different methods of feature extraction: the rotation-invariant local binary pattens (LBP) feature with three resolutions; the Scale-Invariant Feature Transform (SIFT) feature with a key point located at the patch center; as well as feature learning without supervision through the use of the unsupervised restricted Boltzmann machine (RBM). These three techniques are classified by using SVM. However, the proposed automatic CNN model did not use any extra classifier such as SVM, as their classifier model is trained by the three fully connected layers. Therefore, using the ANN model for the classification training has the advantage that the potential to use the backpropagation method to fine-tune the parameters in each of the layers may achieve a more accurate final classification approach. Out of all three techniques, their customized CNN model performed the best. A multiclass CNN architecture using MRI images was used for the detection of brain tumors. The presented model achieved an accuracy of 99% for classifying the four different classes (glioma, tumor, meningioma, and pituitary tumors). The primary objective of this research was to get a faster learning rate with higher classification results while comparing with traditional deep learning models [71]. Yildirim et al. [72] used the CNN-based MA_ColonNET model for identifying colon cancer using colon histopathological images. The proposed model used 45 layers for classifying two classes of colon cancer with an accuracy rate of 99.75%. Additionally, this CNN-based model is applicable for pre-diagnosis purposes in non-specialist locations and reduces the workload pressure of experts, which can help them to avoid mistakes. Ravi et al. [73] used the penultimate layer (global-avg-pooling) of CNN-based efficient net pre-trained models for the extraction of features, and principal component analysis (PCA) was used to reduce the dimensionality of extracted features. After that, the feature fusion technique was used to combine the features of different data and pass them into the stacked-meta classifier. In the first stage, a stacked meta classifier employed the SVM and random forest (RF) algorithm for prediction. The results of this stage were then passed on to the second stage, where they were classified using logistic regression according to whether or not they contained COVID-19. The proposed model produced an overall result that achieved an accuracy of 0.9946 while maintaining a misclassification rate of 0.0054 for the CT data, and an accuracy of 0.9948 with a 0.0052 misclassification rate for the CRX data. This indicates that the proposed efficient net models are capable of classifying new COVID-19 patients using CT and X-ray images.

A VGG-19 model was trained on 3,797 chest X-ray images [74] for classification of Covid19, pneumonia and healthy cases. An accuracy of 97.11% on the test dataset was obtained. In addition, for further study, the original images and their matching categories were then stored in a Mango DB database.

Another study [75] assessed how well the transfer learning-based CNN models VGG-16, ResNet-50, and Inception-v3 predict the presence of brain tumor cells. The models were trained and tested using a dataset of 233 MRIs. Accuracy was used to measure performance, and the findings revealed that the VGG-16 model gave results that were extremely accurate as compared to the other models. The trainable data for the VGG-16 model, which employs 3 × 3 convolution kernels and 2 × 2 max-pool kernels and includes 138 million hyperparameters, was decreased by 44.9 percent. As a result, learning rates increased and overfitting was decreased. The ResNet50 model is a pretrained CNN model that permits training with more convolution layers without increasing training error rates. The Inception-v3 model uses parallel Inception modules to reduce depth in convolution layers.

EfficientNet, GoogLeNet, and XceptionNet were integrated [76] in a study to classify patients as positive for COVID-19, pneumonia, or tuberculosis, or healthy. For a binary classifier the accuracy was 98%, while for multiclass, the accuracy was 99%. The dataset used for training and testing was taken from two sources. The authors also tried to keep the possible false predictions to a minimum, and hence obtained a better accuracy and generalized model. Another parameter was for no false positives, to have the model maintain a high specificity rate, which keeps the model much more reliable.

For diagnosing monkeypox, the author uses generalization- and regularization-based transfer learning techniques. While ResNet-101 had the best result for multiclass classification, with an accuracy ranging from 84 percent to 99 percent, the proposed strategy paired it with Extreme Inception, which produced an accuracy ranging from 77 percent to 88 percent in binary classification trials.

Transfer learning has been shown to be an effective technique for leveraging pre-trained CNN models to improve the performance of medical image analysis tasks. CNNs have demonstrated high accuracy and robustness in identifying and classifying various medical conditions from medical images. Overall, these findings underscore the crucial role that transfer learning and CNNs play in advancing medical imaging diagnosis, and further research in this area has the potential to significantly improve patient outcomes.

As shown in Table 1, CNN-based models are successful in applications that handle multiple modalities for various tasks involving medical image analysis, such as detection and classification tasks and computer-aided diagnosis. The CNN-based model will be an essential component in the design of upcoming medical image analysis systems, regardless of the number of data, classes, and the deep CNN model used. When compared to other techniques used in comparable application domains, deep ConvNets have demonstrated outstanding performance in the domain of medical image analysis. On the other hand, transfer learning involves leveraging pre-trained CNN models that have been trained on large datasets, such as ImageNet, and fine-tuning them for the specific medical imaging task at hand. Transfer learning has been shown to be an effective technique for reducing the amount of data needed to train a CNN. This is because pre-trained CNN models have already learned general features that are useful for a wide range of computer vision tasks, including medical imaging diagnosis. From a computational perspective, using transfer learning with CNNs can significantly reduce the time and resources needed to train a CNN from scratch, as well as improve the performance of the network on the target task. This is because transfer learning allows for the efficient transfer of knowledge from pre-trained models to new tasks, thereby reducing the amount of data and computation needed to achieve high accuracy. While these techniques show promising results in the medical field, there are still some challenges and limitations like a lack of diversity in the training data. CNNs and transfer learning techniques rely heavily on large and diverse datasets to learn relevant features and patterns. The interpretability of learned features where CNNs and transfer learning techniques are involved is often considered as a "black box," since the features learned by the network are difficult to interpret by medical experts. Another factor can be the limited availability of annotated medical imaging data.

Table 1. Some of the studies that used CNN-based methods for medical images.

Authors	Modalities	Methods	Number of Images	Content	Accuracy
Alexander et al. [63]	sMRI, DTI	CNN	ADNI (Normal data—214, Augmented data—3240)	Hippocampal ROI	AD-NC—96.7%, AD-MCI—80%, MCI-NC—65.8%
Liu et al. [64]	3D-MRI	MSCNet	GM-AD—160, MCI—200, NC—160 WM-AD—160, MCI—200, NC—160	Grey matter and white matter	AD-NC—98.96%, AD-MCI—95.37%, MCI-NC—92.59% (GM—98.85% and WM—98. 11%)
Ajagbe et al. [65]	MRIs	CNN, VGG16, VGG-19	Kaggle (6400)	NA	4 classes-CNN—71%, VGG16—77%, VGG-19—77%
Villa—Pulgarin et al. [66]	Dermatoscopic	DenseNet versin 201, Inception-ResNet version 2, Inception version 3	Human Against Machine (HAM10000) Normal data—10015 Augmented data—42925	8 classes—Akiec, bkl, bcc, mel, df, nv, vasc, and scc	DenseNet—98%, Inception ResNet—97%, Inception—96%

Table 1. *Cont.*

Authors	Modalities	Methods	Number of Images	Content	Accuracy
El—Din Hemdan et al. n.d. [67]	X-ray	COVIDX-Net	Total—50 (Normal—25, Positive—25)	NA	MobileNetV2—60%, DenseNet-201—90%, ResNetV2—70%, InceptionV3—50%, Xception—80%, VGG-19—90%, InceptionResNetV2—80%
Horry et al. [68]	X-ray, ultrasound, and CT scan	VGG-19	Curated dataset—729 (X-ray), 746 (CT), 911 (ultrasound) Augmented dataset—11,680 (X-Ray), 12,000 (CT), 10,880 (ultrasound)	Lung	VGG-19—Precision—100% (Ultrasound), 83% (X—Ray), 84% (CT scan)
Neal Joshua et al. [69]	CT	3D-CNN with Grad-CAM images	LUNA 16 database—888	Lung nodule	3D-CNN—97.17%
Li et al. [70]	HRCT	CNN	Total samples—16,220 (92 HRCT image dataset, 4348 N patches, 1953 G patches, 1047 E patches, 2591 F patches, 6281 M patches)	Lung images	CNN—Precision—76% Recall—77.4%
Tiwari et al. [71]	MRIs	CNN	Total—3264 (MRIs) Four classes (training & testing)—glioma—826 and 100, meningioma—822 and 115, no tumor—395 and 105, pituitary tumor—827 and 74.	Brain images	CNN—99%
Yildirim et al. [72]	Histopathological images	MA_ColonNET	Total—10,000 Two classes (Colon adenocarcinoma—500, Colon benign tissue—9500)	Colon images	MA_ColonNET—99.75%
Ravi et al. [73]	CT scan and Chest X-ray	EfficientNet	Total–CT—8055 (train—5638, test—2417); CXR—9544 (train—6680, test—2864)	Chest	EfficientNet—99%

Table 1. *Cont.*

Authors	Modalities	Methods	Number of Images	Content	Accuracy
Soarov et al. [74]	X-ray Chest	VGG-19	COVID-19 (1,184 images) Pneumonia (1294 images) Healthy (1319 images)	Chest	97.11% accuracy, 97% average precision, 97% average recall
Srinivas, C et al. [75]	MRI scans	Classfiers: VGG-16, Inception-v3, ReseNet50	tumor: 158 malignant tumor: 98	Brain	VGG—16 accuracy 0.96 Inception-v3 0.78 ResNet50x—0.95
N. Kumar et al. [76]	X-ray	Binary and Multiclass Classification	Two datasets: Source 1: null Source 2: 9300 divided for each four class	Chest	Multiclass accuracy 99.21% Binary accuracy 98.95%

6. What is Transfer Learning?

Transfer learning is a method of learning where a model learns about one problem before this serves as the starting point for another task. This is a suitable approach for problems when a procedure near the primary issue already exists and the related task requires a lot of data [77].

Transfer learning uses the technique of feature extraction from a pre-trained model; this eliminates the need for developers to start over when training a model. A TL model is typically trained on a large dataset (for example, ImageNet) [78] and the related parameters obtained from the trained model can then be used with a custom neural network for any other related application. These types of models can be used directly for predictions on new tasks or in any other related application training processes of the model. For instance, in the process of image classification, the model, such as an ANN, which is used for prediction will be trained and learned with a large number of images or datasets of the specific domain [79], like dogs and cats. Model weights are one option for the first step in the process. The traits which the machine has previously mastered for a more extensive mission, such as retrieving shapes, patterns, and lines, are also useful for different objectives.

One more problem to consider with traditional neural networks is that when we apply these kinds of models in clinical practice, the model is likely to fail due to unseen data, which is nothing but data that is not used while training the model. Therefore, the capacity to generalize to previously encountered clinical data is still a major shortcoming of these algorithms. Another shortcoming is when the data is limited; we know that the running performance of a deep network is impacted by the amount of data. One way to overcome this shortcoming is to collect more data, specifically looking for data that is exactly supervised data. Hence, there are transfer learning techniques that may be considered as choices. Rather than starting from scratch, we can use an existing network to train a new one; LeNet-5, AlexNet, VGG-16 Net, Inception Net ResNet, and DenseNet have been widely used as pre-trained networks for the classification of images in medical domains. All these architectures were trained on the well-known dataset (ImageNet) [80] consisting of 1000 object category classifications [81]. There are more than a million images in ImageNet's training set, around fifty thousand in its validation set, and one hundred thousand in its test set. [82]. These models not only reduce training time but also reduce generalization errors. Table 2 shows the main differences between traditional ML and transfer learning.

Table 2. A brief description of traditional ML vs transfer learning.

Traditional ML	Transfer Learning
1. Isolated, single task learning 2. Knowledge is not retained or accumulated. Learning is performed without consideration for knowledge learned from other tasks.	1. The learning of new activities is dependent on previously learnt ones 2. The learning process could be more efficient, more accurate, or need fewer training data sets.

6.1. LeNet5

In 1889, Yann LeCun et al. published a paper that proposed a technique for document recognition which is called gradient-based learning [83]; their work described LeNet-5, which was probably the first widely recognized and effective implementation of CNN. The author trained the model for the recognition of handwritten characters based on a standard famous dataset called MNIST (Modified National Institute of Standards and Technology dataset). As a result, a significant classification result of 99.2% accuracy and a low error rate was achieved. The LeNet-5 architecture receives the input image as a grayscale 32 × 32 image size, and the model is a composite of seven layers, including layers of convolution and average pooling followed by a layer that is fully connected. Figure 5 shows the comparison transfer learning and transfer machine learning. Figure 6b depicts the LeNet-5 architecture. Interestingly, in LeNet-5, the filters used for capturing feature maps are increased as the network progresses in depth [81].

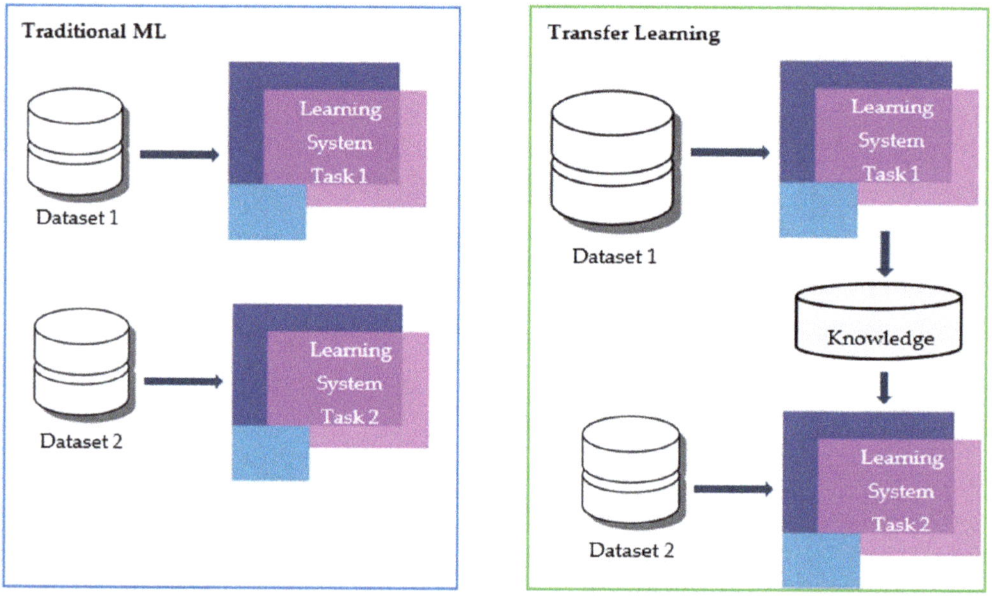

Figure 5. Transfer learning vs traditional ML.

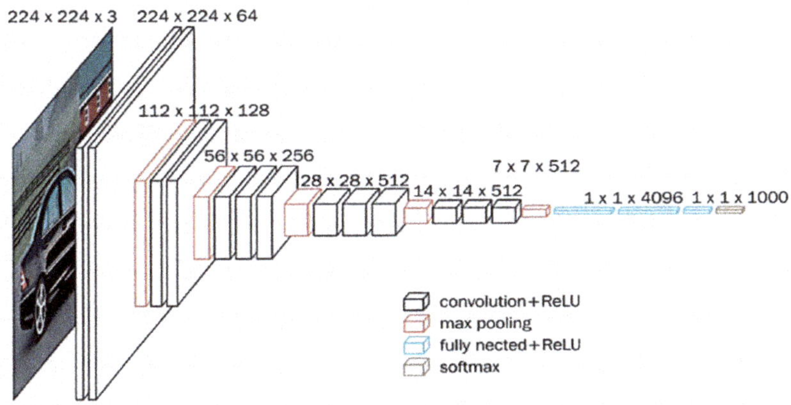

(**a**)

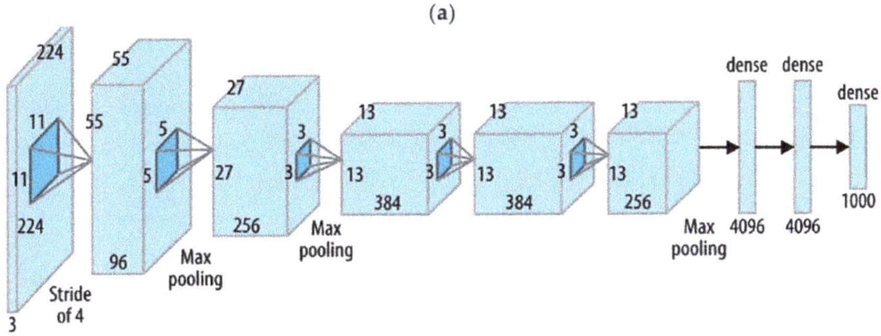

(**b**)

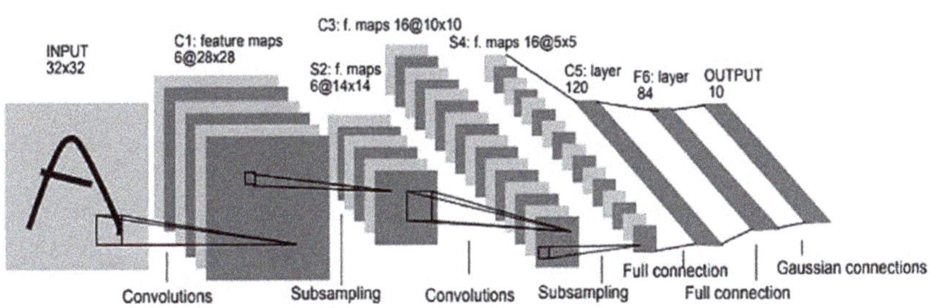

(**c**)

Figure 6. (**a**) A VGG-16 network's structural details are displayed in the figure; (**b**) LeNet-5 architecture; (**c**) AlexNet architecture.

Key facts:

- This network is very easy to understand and served as a good introduction to the field of neural networks. Character recognition works well.
- Due to the shallowness (not deep enough) of the model, it has a difficult time searching for all features, leading to models with poor performance.
- This model does not work with color images.

6.2. AlexNet

In 2012, another researcher named Alex Krizhevsky, and his co-workers developed a model known as AlexNet [84]. The paper proposed and discussed the deep ConvNets for the classification of ImageNet. This was done due to a competition in 2010 called the

ILSVRC (ImageNet Large-Scale Visual Recognition Challenge) [85], whose purpose was to detect and classify objects. This was where, later on, the importance of image classification using CNNs became the buzzword. AlexNet is like LeNet but much larger and with a greater number of filters for each layer. Another major change in AlexNet was to replace the traditional S-shaped functions, like Tanh or logistic, with new nonlinear functions called ReLU (rectified linear), which are placed after every convolutional layer. Additionally, in the output layer of AlexNet, another activation function called Softmax is placed, Figure 6c shows AlexNet architecture. Moreover, this model uses the max-pooling technique instead of average pooling, and a new method called dropout has been utilized between the fully connected layers to address overfitting and enhance generalization error. The AlexNet architecture takes a fixed input of 224x224x3 size and is built upon eight layers. In total, five layers go to convolutional operations and three layers go to fully connected operations.

Key facts:

- The first significant CNN model to use GPU training, which leads to faster training, was AlexNet.
- In comparison to another model like LeNet, the AlexNet model has eight layers and a deeper structural design, making it better able to extract important features. It also performed admirably for color images.
- As compared to future models, it takes longer to obtain results with high accuracy with AlexNet.

6.3. VGG Net

In 2014 [85], two Oxford researchers at the Visual Geometry Group lab came up with an idea of a much deeper CNN with better performance named VGG; this again happened through the ILSVRC 2014 [57] competition. There are different variants of the VGG net architecture, such as the VGG-19 and the VGG-16. Their names refer to the number of learned layers in the architecture. In VGG, before max pooling is performed, several convolutional layers are stacked together, such as two, three, and even four. The reason for stacking the conv-layers together is to define a block. The employment of many tiny filters is the first significant change that has a de facto standard; this CNN utilizes filters of size 1×1 and 3×3, and a stride of one, as opposed to LeNet-5's large filters. The number of filters rises with the mode's depth, starting at 64 and increasing to 128, 256, and 512 filters after extracting features from the model. Figure 6a represent the architecture of VGG Net

Key facts:

- VGG is simple to comprehend and explain.
- A baseline of about 80 percent is recommended for classic problems like classifying cats and dogs.
- A longer inference time is caused by the greater number of weight parameters.

6.4. Inception Net

Christian Szegedy et al. published a paper titled "Going Deeper with Convolutions" [85], which described another complex and heavily engineered architecture named the Inception network. The key goal of the author was to use a lot of techniques to increase performance in terms of precision, accuracy, and speed. The network's ongoing evolution resulted in the production of multiple versions, such as Inception v1, v2, and v3 [86]. Each new version is a step forward from the preceding one [81]. The Inception module is the major innovative element in this network, the model architecture is given in Figure 7a. It is nothing but a parallel block of convolutional layers consisting of 3 different kernel sizes such as 5×5, 1×1, and 3×3, with a max-pooling layer of 3×3. Further, all the results are concatenated. Version 3 of Inception, which is an optimized and upgraded version of Inception, is made of 42 layers and, compared to other versions, it has a lower error rate [81].

Key facts:

- As a result of applying multiple convolution filters to the same input in the case of multi-level feature extraction, computational costs are reduced.
- Increased performance can be achieved on this CNN.
- Inception model can be train more quickly than the VGG model and VGG model is relatively bigger in size as compared to LeNet-5.

6.5. ResNet

Image recognition is further considered using deep residual learning by Kaiming He et al. in 2016 [81]. The ResNet152V2 is built with a total number of 152 layers, and the concept of residual blocks in the network that utilize shortcut connections is crucial to the mode's construction. A residual block is a combination of two conv-layers with an activation function, such as ReLU. The problem of vanishing gradient which exists in deep networks is solved by ResNet skip-or-shortcut connections by letting the gradient flow through an additional path (shortcut path) [81]. The main difference between ResNet v1 and v2 is that the batch-normalization technique is applied before each weight layer. Architecture of Resnet is depicted in Figure 7a.

Key facts:

- Skip-or-shortcut connections aid in addressing the issue of vanishing gradients.
- The structure increases the training pace.
- ResNet provides greater accuracy, particularly in classification.
- It makes an effort to distinguish between learned features, and if a learned feature is not relevant to the decision at hand, its weight is reduced to zero.
- Since it is incorporating skip connections between layers that may take dimensionality into account, it also increases architectural complexity [78].

6.6. DenseNet

Gao Huang and colleagues developed the DenseNet in 2017, which consists of layers that are densely connected to one another and are all associated with one another. This method helps to reuse features, because each layer obtains its input from the levels that came before it and produces its feature mappings to be used by the layers that come after it. Additionally, each layer provides its input to all subsequent layers. The structure of DenseNet includes two dense blocks with two transition blocks in between each pair of dense blocks. Figure 8 shows the DenseNet architecture. [87]. The following are important concepts in DenseNet:

1. Growth rate: This determines the number of feature maps that are output into individual layers within dense blocks;
2. Dense connectivity: Dense connectivity refers to the fact that within a dense block, each successive layer is able to obtain input feature maps from the layer below it [88];
3. Composite functions: The following is an explanation of the order in which operations take place within a layer. First, we begin with batch normalization, then move on to applying activation functions (e.g., ReLU), and finally, arrive at the convolution layer [87];
4. Transition layers: The transition layers reduce the dimensions of the dense block by aggregating the feature maps that are contained within it. Therefore, maximum pooling has been enabled.

Key facts:

- Each subsequent layer adds only a small number of parameters; for example, only about 12 kernels are learned in each subsequent layer. Therefore, parameter efficiency is achieved.
- Better distribution of the gradient throughout the network for all of the feature maps can enable the CNN to directly access the loss function and its gradient, which gives implicit deep supervision [89].

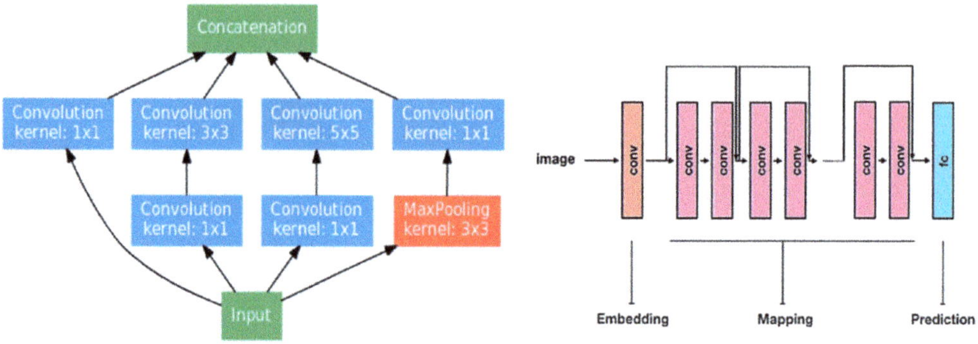

Figure 7. (**a**) Building blocks of the Inception architecture; (**b**) Three building elements make up the ResNet architecture in a schematic: embedding, mapping, and prediction. Convolution procedures and nonlinear activations are described by the terms "conv" and "fc," respectively [90,91].

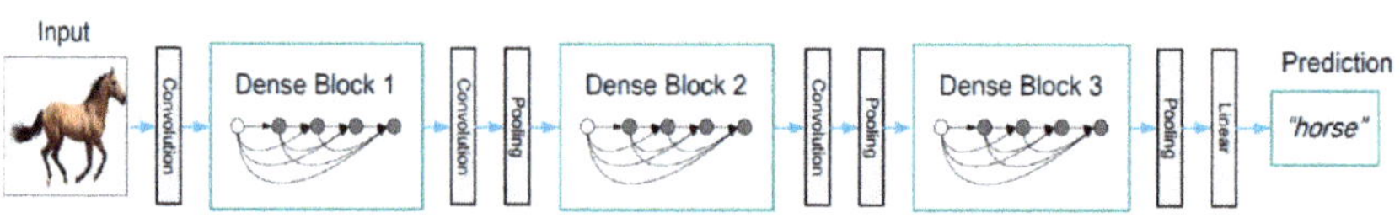

Figure 8. Architecture of DenseNet.

7. Practical Perspective and Fine-Tuning of Transfer Learning Techniques

- **Training the entire model**: In this approach, the entire pre-trained model is used as a starting point, and all the parameters of the model are fine-tuned for the new task. This is suitable when the new task is similar to the original task for which the pre-trained model was trained [92].
- **Freezing some layers**: In this approach, some of the layers in the pre-trained model are frozen and the remaining layers are fine-tuned for the new task. Typically, the lower-level layers of the pre-trained model, which capture low-level features such as edges and corners, are frozen, while the higher-level layers, which capture more complex features, are fine-tuned. This is suitable when the new task is related to the original task but requires some modification of the model [93].
- **Fine-tuning some layers**: In this approach, some of the layers in the pre-trained model are fine-tuned while the remaining layers are frozen. Typically, the higher-level layers of the pre-trained model, which capture more complex features, are fine-tuned, while the lower-level layers are frozen. This is suitable when the new task is significantly different from the original task, but the higher-level features of the pre-trained model can still be useful [93].
- **Freezing the convolutional base**: In this approach, the convolutional base of the pre-trained model is frozen, and a new classifier is added on top of it. The new classifier is then trained on the new task [94]. This is suitable when the new task requires a different classification scheme than the original task, but the pre-trained convolutional base can still be used to extract features from the input data [95].

In general, transfer learning can be a powerful tool for machine learning tasks, as it allows for the reuse of pre-trained models that have already learned useful representations from large amounts of data. The appropriate transfer learning method will depend on the specifics of the new task and the pre-trained model. To choose a pre-trained model for your problem, you can select from a variety of options such as VGG [96], InceptionV3 [97], ResNet5, DenseNet, and so on.

8. Important Things for Consideration

8.1. Generalization: Problems and Key Concepts to Mitigate

We say that a model generalizes well if the model is tested on unlabeled data after training on known labeled data and the machine performs well on testing data. However, when a network performs well on the training set, but poorly overall, it is said to overfit [98]. This problem is common in deep learning models because there are so many parameters for the model to learn; therefore, these types of models are prone to overfitting.

One of the hardest problems is enhancing the generalization capability to prevent the overfitting of machine or deep learning models. Plotting the training and validation accuracy rate at each iteration during training is one method of identifying overfitting [99].

- **Data augmentation**: One approach to avoid overfitting is to simply expand the quantity of data, However, gathering large amounts of data in real-world situations is a laborious and time-consuming task, so collecting new data is not a practical option. Increasing the total size of the dataset [37] used for training is one of the best methods for reducing overfitting. Since we are talking about CNNs for image-based data, the easiest way to add variety to our data and expand it is to add more images to the dataset. This process is referred to as data augmentation [100]. This has potential for narrowing the gap between the training and validation set, as well as between those two sets and any future test sets [98], because the augmented data will represent a comprehensive collection of possible data points. Therefore, augmentation is a highly effective strategy.

Several other common techniques that have been used to tackle the overfitting problem are listed below:

- **Batch normalization**: This approach is the one that is utilized most frequently in deep learning, as it increases the speed at which neural networks learn new information and provides regularization, thereby preventing the problem of overfitting. In CNN convolutions, shared filters follow input feature maps and are the same on every feature map [101]. When this occurs, it is reasonable to normalize the output in the same manner, and then share it across the feature maps. Therefore, each map will have a single standard deviation and mean for all its features [102];
- **Dropout**: This is a training method in which some neurons are selectively ignored. A model with applied dropout cannot rely on any single feature and must instead learn robust features. This method has been shown to effectively decrease overfitting in numerous issues [103]. Tompson [104] expanded on this concept by applying it to a convolutional neural network using a technique called spatial dropout. This technique eliminates entire feature maps as opposed to individual neurons;
- **Weight decay**: In model training, large weights mean that the prediction relies heavily on one pixel; therefore, a more interesting method comes to the picture, which is weight decay, which says that large weights are penalized [105]. Intuitively, the classification of an image based on one or a few pixels seems to not make sense;
- **Transfer learning**: This involves training a machine model on a large amount of data like ImageNet and using those weights in a new classification task [106].

8.2. Computational Accelerators within the Scope of DL

The majority of ConvNets have extremely high memory and computation requirements, particularly while they are being trained. As a result, this should be one of your primary concerns. For example, deploying a model to run locally on mobile, you should give careful consideration to the size of the trained model after it has been completed. Increasing the amount of computational work done by a network is necessary to achieve higher levels of accuracy [59]. Therefore, there is always a compromise that needs to be made between accuracy and computational speed. In addition to these factors, there are a great number of other considerations, such as the simplicity of training, the capacity of a network to generalize data effectively, etc.

It is stated [107] that the increased ratio of the total amount of layers accumulated over time appears to be significantly quicker than the growth ratio predicted by Moore's Law. The use of graphics processing units (GPUs) and tensor processing units (TPUs) acts as a means of enhancing the performance of central processing units (CPUs) when running deep nets; therefore, it is necessary to have an understanding of the technologies that underpin the CPU, GPU, and TPU in order to maintain a competitive edge in terms of both performance and efficiency.

The main difference between CPU, GPU, and TPU is thatwhile the CPU is used for general-purpose processing, the GPU and TPU, on the other hand, are more like the computer's muscles. GPUis a performance accelerator that helps computer graphics and artificial intelligence work [108], while TPUs are Google's processors designed to speed up machine learning tasks using frameworks such as TensorFlow.

9. Discussion

Medical imaging is an essential tool that unites societal and scientific requirements and can create a significant synergy that could promote research in both fields. Machine learning, particularly deep learning (DL), is a fast-moving research subject with promising imaging and therapeutic applications. DL has already permeated medical image analysis. Although recent advances in DL approaches have been astounding, there are still obstacles to their implementation in healthcare. Because it does not leave an audit trail to explain the decisions it makes, DL is often referred to as a black box. Image analysis was not meant to replace radiologists, but to serve as a second opinion. There is no denying that improvements in the digital imaging industry have had and will continue to have a favorable impact on medical imaging. CNNs have positively contributed to many fields, including medical research and radiology, and they are becoming more and more popular. As a result, with the capability to learn high-level features from medical images without requiring a step of feature engineering, they become a viable alternative to machine learning algorithms. In this manuscript, a comprehensive review of the strengths, performance, and limitations of the latest DL-based approaches is presented [109] for applications dealing with medical imaging domains.

As per the data from MetaAI [110], we have found that the majority of work has been done on classification problems, specifically in image classification as shown in below Figure 9 [111].

We also determined that, after reviewing many research articles, the most common and efficient activation function which has been adopted in the building of CNN is ReLU. In the field of DL, ReLU is a non-linear function, meaning that it converts all negative values to 0, and it has become an increasingly prominent activation function. The primary benefit that the ReLU function has over other activations is that it does not fire all of the neurons at the same time, as other functions do [112]. This is mostly used on every conv-layer, as well as each and every dense layer [113]. The following are the most common reasons behind using this activation function:

- **Vanishing gradient:** Since the derivative of this activation can only be either on the value 0 or 1, it cannot fall within the range [0, 1] [114]. As a consequence of this result, the product of various derivatives would also be either 0 or 1. Therefore, the problem of vanishing gradients does not arise when backpropagation is being performed;
- **Sparsity:** An ReLU will always produce an output value of 0 in response to negative input. This indicates that a smaller percentage of the network's neurons are actively firing. As a result, the neural network possesses activations that are both sparse and efficient;
- **Speedier training:** Better convergence performance is typically demonstrated by networks that have the ReLU function and offer faster training. As a result, our total running time is significantly shorter [111].

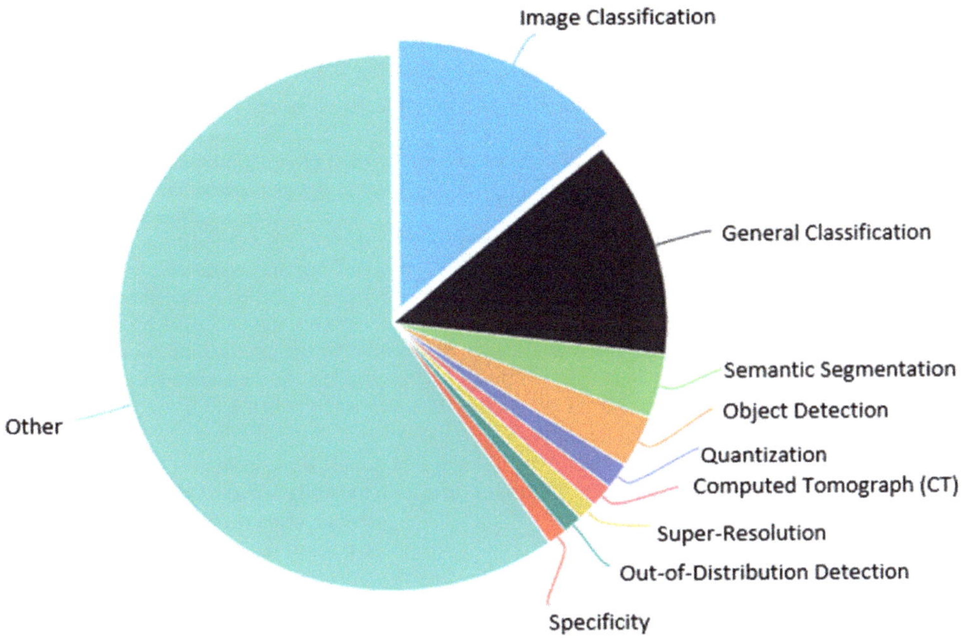

Task	Papers	Share
• Image Classification	61	13.83%
• General Classification	58	13.15%
• Semantic Segmentation	17	3.85%
• Object Detection	14	3.17%
• Quantization	7	1.59%
• Computed Tomography	6	1.36%
• Super-Resolution	5	1.13%
• Out-of-Distribution Detection	5	1.13%
• Specificity	5	1.13%

Figure 9. The proportion of papers on various tasks [111].

Even though there are a few methods, like the ones listed above, that make it easier to learn from smaller datasets, it is still important to have large, well-annotated medical datasets because most of the big achievements of deep learning are usually based on huge datasets. Building these kinds of medical datasets is expensive, takes a lot of work from experts, and may have ethical and privacy problems. However, once such datasets are made accessible, specialized medical pre-trained networks would likely be presented, which might encourage deep-learning research on medical imaging.

- Furthermore, due to the complex structures of data, training a deep learning model is extremely expensive. They often require expensive GPUs and a large number of computers, which raises the cost for users.
- Training performance worsens as a result of the large computational load required by the growing complexity of multiple layers. To tackle the vanishing gradient issue, overfitting concerns, improved activation, and cost function design, dropout techniques have been employed [113].
- Utilizing hardware with a high degree of parallelism, such as GPUs and normalization techniques, allowed for the large computational-weight issue to be resolved [61].

From a practical point of view, transfer learning techniques for image classification problems are mainly based on the size and similarity of the dataset. The strategies are summarized into four categories as follows [92]:

- **Category 1:** Large dataset, but different from the pre-trained model's dataset. Strategy 1 is recommended, which involves training a model from scratch but using the architecture and weights of a pre-trained model to initialize the model.
- **Category 2**: Large dataset and similar to the pre-trained model's dataset. Any option can work, but Strategy 2 is the most efficient. This involves training the classifier and top layers of the convolutional base, leveraging previous knowledge.
- **Category 3**: Small dataset and different from the pre-trained model's dataset. Strategy 2 is recommended, but it can be challenging to find the right balance between the number of layers to train and freeze. Data augmentation techniques may be necessary.
- **Category 4**: Small dataset, but similar to the pre-trained model's dataset. Strategy 3 is the best option, which involves using the pre-trained model as a fixed feature extractor, removing the last fully connected layer, and training a new classifier using the resulting features.

A learning model that is subject to supervision, either for classification or regression problems, must learn from training data to produce accurate predictions. Unfortunately, the problem arises that whenever we make an effort to train a complicated model with an insufficient amount of data for training purposes, overfitting occurs [114]. Overfitting is the most critical issue in deep learning, so understanding, finding, and avoiding it is important. Researchers have described many methods to tackle this problem such as data augmentation, weight decay, transfer learning, batch normalization, dropout, etc.

As for the computational approach, is a concern. It has been demonstrated by researchers from Harvard that different platforms give advantages to different models based on their individual qualities. These advantages might be advantageous for the model's overall performance. Below are the key takeaways:

- **CPU:** It is responsible for achieving the highest FLOPS utilization for Recurrent Neural Networks and is capable of supporting the largest models due to its vast memory capacity;
- **GPU**: For irregular computations, such as tiny batches and nonMatMul computations, the GPU demonstrates more flexibility and programmability than other processing units;
- **TPU:** It is highly optimized for large batches and boasts the best possible training throughput.

10. Conclusions

This article gives people in the field of DL a place to start. It could also help them to choose the best way to go about their work in order to come up with more accurate models. The study further provides an analysis of the various architectures of CNNs used in the classification of medical images, and also demonstrates how developments in deep learning algorithms produce promising findings that can assist and act as a second eye to many radiologists. CNN-based architectures have been utilized in the medical domain in various disease detection and prediction cases. The following points are given to wrap up our review and show where things are going in the current and the future.

- To make accurate predictions and train deep learning models, these models need access to large datasets, preferably with labels. When processing data in real-time is necessary, to be specific in the case of healthcare data, this problem becomes more difficult. Over the past few years, researchers have investigated potential solutions to this problem, such as data augmentation and pre-trained CNN models.
- Changes to the hyperparameter settings will have a significant impact on the deep learning-based models' overall performance. Therefore, developing an optimization

technique requires careful consideration of parameter choices; for example, there are various techniques to mitigate this problem such as Keras Tuner, Ray Tuner, etc.

- In order to train a CNN model effectively, powerful computational approaches are required like GPUs or TPUs. Therefore, there is a significant amount of ongoing work being conducted to think of ways to speed up these resources.
- Generalizability of the CNN in the case of medical imaging is very important; therefore, concepts like dropout, batch-normalization, weight decay, transfer learning, and data augmentation are presented.
- To find the solution to not having enough data for training, we discussed data augmentation, which is one way to help in the creation of more data from the existing data, and it is likely that different pre-trained CNN models will utilize this solution. For example, a CNN could be trained on a huge amount of unlabeled data, and then that knowledge could be used to train that CNN on a smaller amount of labeled data for the same job.
- It is anticipated that a variety of approaches to learning through transfer will be taken into consideration and choosing the right strategy for utilizing such models in image classification depends on the similarity and the amount of the dataset.
- While utilizing a CNN alone can be computationally costly, using transfer learning with pre-trained CNN models can greatly lower the cost of training a CNN for medical imaging diagnosis while simultaneously enhancing its performance.

Author Contributions: Conceptualization, A.W.S., G.G. and S.K.; Methodology, A.W.S., G.G. and B.I.A.; Software, A.W.S., G.G., A.A. and H.A.; Validation, A.W.S., G.G. and T.S.; Formal analysis, A.W.S., G.G., A.M. and T.S.; Investigation, S.K., A.W.S. and H.A.; Resources, S.K.; Writing—original draft, A.W.S. and G.G.; Writing—review & editing, T.S. and A.M.; Supervision, S.K. and B.I.A.; Funding acquisition, S.K. All authors have read and agreed to the published version of the manuscript.

Funding: This research was funded by the Deanship of Scientific Research at Imam Mohammad Ibn Saud Islamic University (IMSIU) for funding and supporting this work through Research Partnership Program no. RP-21-07-06.

Institutional Review Board Statement: Not applicable.

Informed Consent Statement: Not applicable.

Data Availability Statement: Not applicable.

Acknowledgments: The authors extend their appreciation to the Deanship of Scientific Research at Imam Mohammad Ibn Saud Islamic University (IMSIU) for funding and supporting this work through Research Partnership Program no. RP-21-07-06. The authors would also like to thank Aleem Ali from Department of Computer Science and Engineering, Chandigarh University for serving as a consultant to critically reviewed the study proposal and participated in technical editing of the manuscript.

Conflicts of Interest: The authors declare that they have no conflict of interest.

References

1. Cai, L.; Gao, J.; Zhao, D. A review of the application of deep learning in medical image classification and segmentation. *Ann. Transl. Med.* **2020**, *8*, 713. [CrossRef]
2. Chopra, P.; Junath, N.; Singh, S.K.; Khan, S.; Sugumar, R.; Bhowmick, M. Cyclic GAN Model to Classify Breast Cancer Data for Pathological Healthcare Task. *BioMed Res. Int.* **2022**, *2022*, 6336700. [CrossRef] [PubMed]
3. Zhou, S.K.; Greenspan, H.; Davatzikos, C.; Duncan, J.S.; Van Ginneken, B.; Madabhushi, A.; Prince, J.L.; Rueckert, D.; Summers, R.M. A Review of Deep Learning in Medical Imaging: Imaging Traits, Technology Trends, Case Studies with Progress Highlights, and Future Promises. *Proc. IEEE* **2021**, *109*, 820–838. [CrossRef]
4. Dutta, P.; Upadhyay, P.; De, M.; Khalkar, R. Medical Image Analysis using Deep Convolutional Neural Networks: CNN Architectures and Transfer Learning. In Proceedings of the 5th International Conference on Inventive Computation Technologies, ICICT 2020, Coimbatore, India, 26–28 February 2020; pp. 175–180. [CrossRef]
5. Singh, S.P.; Wang, L.; Gupta, S.; Goli, H.; Padmanabhan, P.; Gulyás, B. 3D Deep Learning on Medical Images: A Review. *Sensors* **2020**, *20*, 5097. [CrossRef] [PubMed]

6. Guezzaz, A.; Azrour, M.; Benkirane, S.; Mohy-Eddine, M.; Attou, H.; Douiba, M. A Lightweight Hybrid Intrusion Detection Framework Using Machine Learning for Edge-Based IIoT Security. *Int. Arab. J. Inf. Technol.* **2022**, *19*, 822–830. [CrossRef]
7. Khan, S. Business Intelligence Aspect for Emotions and Sentiments Analysis. In Proceedings of the 2022 First International Conference on Electrical, Electronics, Information and Communication Technologies (ICEEICT), Trichy, India, 16–18 February 2022; pp. 1–5. [CrossRef]
8. Khan, S.; Fazil, M.; Sejwal, V.K.; Alshara, M.A.; Alotaibi, R.M.; Kamal, A.; Baig, A.R. BiCHAT: BiLSTM with deep CNN and hierarchical attention for hate speech detection. *J. King Saud Univ. Comput. Inf. Sci.* **2022**, *34*, 4335–4344. [CrossRef]
9. Azrour, M.; Mabrouki, J.; Fattah, G.; Guezzaz, A.; Aziz, F. Machine learning algorithms for efficient water quality prediction. *Model. Earth Syst. Environ.* **2021**, *8*, 2793–2801. [CrossRef]
10. Mutasa, S.; Sun, S.; Ha, R. Understanding artificial intelligence based radiology studies: CNN architecture. *Clin. Imaging* **2021**, *80*, 72–76. [CrossRef]
11. Lee, J.-G.; Jun, S.; Cho, Y.-W.; Lee, H.; Kim, G.B.; Seo, J.B.; Kim, N. Deep Learning in Medical Imaging: General Overview. *Korean J. Radiol.* **2017**, *18*, 570–584. [CrossRef]
12. Bhatt, D.; Patel, C.; Talsania, H.; Patel, J.; Vaghela, R.; Pandya, S.; Modi, K.; Ghayvat, H. CNN Variants for Computer Vision: History, Architecture, Application, Challenges and Future Scope. *Electronics* **2021**, *10*, 2470. [CrossRef]
13. Kim, M.; Yun, J.; Cho, Y.; Shin, K.; Jang, R.; Bae, H.-J.; Kim, N. Deep Learning in Medical Imaging. *Neurospine* **2019**, *16*, 657–668. [CrossRef]
14. Haq, A.U.; Li, J.P.; Khan, I.; Agbley, B.L.Y.; Ahmad, S.; Uddin, M.I.; Zhou, W.; Khan, S.; Alam, I. DEBCM: Deep Learning-Based Enhanced Breast Invasive Ductal Carcinoma Classification Model in IoMT Healthcare Systems. *IEEE J. Biomed. Health Inform.* **2022**, 1–12. [CrossRef]
15. Abdelhafiz, D.; Yang, C.; Ammar, R.; Nabavi, S. Deep convolutional neural networks for mammography: Advances, challenges and applications. *BMC Bioinform.* **2019**, *20*, 281. [CrossRef] [PubMed]
16. Salehi, W.; Gupta, G.; Bhatia, S.; Koundal, D.; Mashat, A.; Belay, A. IoT-Based Wearable Devices for Patients Suffering from Alzheimer Disease. *Contrast Media Mol. Imaging* **2022**, *2022*, 3224939. [CrossRef] [PubMed]
17. IEEE. *2017 IEEE 19th International Conference on e-Health Networking, Applications and Services (Healthcom)*; IEEE: Piscataway, NJ, USA, 2017.
18. Kushnure, D.T.; Tyagi, S.; Talbar, S.N. LiM-Net: Lightweight multi-level multiscale network with deep residual learning for automatic liver segmentation in CT images. *Biomed. Signal Process. Control.* **2023**, *80*, 104305. [CrossRef]
19. Haq, A.U.; Li, J.P.; Khan, S.; Alshara, M.A.; Alotaibi, R.M.; Mawuli, C. DACBT: Deep learning approach for classification of brain tumors using MRI data in IoT healthcare environment. *Sci. Rep.* **2022**, *12*, 15331. [CrossRef] [PubMed]
20. Torres-Velazquez, M.; Chen, W.-J.; Li, X.; McMillan, A.B. Application and Construction of Deep Learning Networks in Medical Imaging. *IEEE Trans. Radiat. Plasma Med. Sci.* **2020**, *5*, 137–159. [CrossRef]
21. Mukhlif, A.A.; Al-Khateeb, B.; Mohammed, M.A. An extensive review of state-of-the-art transfer learning techniques used in medical imaging: Open issues and challenges. *J. Intell. Syst.* **2022**, *31*, 1085–1111. [CrossRef]
22. Wiesel, T.N. Receptive fields and functional architecture of monkey striate cortex. *J. Physiol.* **1968**, *195*, 215–243.
23. Ghosh, A.; Sufian, A.; Sultana, F.; Chakrabarti, A.; De, D. Fundamental Concepts of Convolutional Neural Network. In *Recent Trends and Advances in Artificial Intelligence and Internet of Things*; Springer: Berlin/Heidelberg, Germany, 2019; pp. 519–567. [CrossRef]
24. Fukushima, K.; Miyake, S. Neocognitron learning by backpropagation. *Syst. Comput. Jpn.* **1995**, *26*, 19–28. [CrossRef]
25. Khan, S.; Kamal, A.; Fazil, M.; Alshara, M.A.; Sejwal, V.K.; Alotaibi, R.M.; Baig, A.R.; Alqahtani, S. HCovBi-Caps: Hate Speech Detection Using Convolutional and Bi-Directional Gated Recurrent Unit with Capsule Network. *IEEE Access* **2022**, *10*, 7881–7894. [CrossRef]
26. Jogin, M.; Mohana; Madhulika, M.S.; Divya, G.D.; Meghana, R.K.; Apoorva, S. Feature Extraction using Convolution Neural Networks (CNN) and Deep Learning. In Proceedings of the 2018 3rd IEEE International Conference on Recent Trends in Electronics, Information and Communication Technology, RTEICT 2018, Bangalore, India, 18–19 May 2018; pp. 2319–2323. [CrossRef]
27. Zhang, S.; Zhang, M.; Ma, S.; Wang, Q.; Qu, Y.; Sun, Z.; Yang, T. Research Progress of Deep Learning in the Diagnosis and Prevention of Stroke. *BioMed Res. Int.* **2021**, *2021*, 5213550. [CrossRef] [PubMed]
28. Brownlee, J. A Gentle Introduction to Pooling Layers for Convolutional Neural Networks. *Mach. Learn. Mastery* **2019**, *22*. Available online: https://machinelearningmastery.com/pooling-layers-for-convolutional-neural-networks/ (accessed on 5 December 2022).
29. Naranjo-Torres, J.; Mora, M.; Hernández-García, R.; Barrientos, R.J.; Fredes, C.; Valenzuela, A. A Review of Convolutional Neural Network Applied to Fruit Image Processing. *Appl. Sci.* **2020**, *10*, 3443. [CrossRef]
30. Sun, M.; Song, Z.; Jiang, X.; Pan, J.; Pang, Y. Learning Pooling for Convolutional Neural Network. *Neurocomputing* **2017**, *224*, 96–104. [CrossRef]
31. Liu, T.; Fang, S.; Zhao, Y.; Wang, P.; Zhang, J. Implementation of Training Convolutional Neural Networks. *arXiv* **2015**, arXiv:1506.01195, preprint.
32. Mac, S.; Products, S.; Also, C. Convolutional Kernel Networks Julien. *arXiv* **2014**, arXiv:1406.3332, preprint.

33. Corvil. The Role of Bias in Neural Networks. 2018. Available online: https://www.pico.net/kb/the-role-of-bias-in-neural-networks/ (accessed on 16 March 2022).
34. Skalski, P. Gentle Dive into Math Behind Convolutional Neural Networks. *Data Sci.* 2019. Available online: https://towardsdatascience.com/https-medium-com-piotr-skalski92-deep-dive-into-deep-networks-math-17660bc376ba (accessed on 20 February 2023).
35. Hashemi, M. Enlarging smaller images before inputting into convolutional neural network: Zero-padding vs. interpolation. *J. Big Data* **2019**, *6*, 1–13. [CrossRef]
36. Prabhu. Understanding of Convolutional Neural Network (CNN)—Deep Learning. *Medium.* 2022. Available online: https://medium.com/@RaghavPrabhu/understanding-of-convolutional-neural-network-cnn-deep-learning-99760835f148 (accessed on 15 February 2023).
37. Srivastava, N.; Hinton, G.; Krizhevsky, A.; Sutskever, I.; Salakhutdinov, R. Dropout: A Simple Way to Prevent Neural Networks from Overfitting. *J. Mach. Learn. Res.* **2014**, *15*, 1929–1958.
38. Adrian, G. Dropout in Recurrent Neural Networks. 2018. Available online: https://adriangcoder.medium.com/a-review-of-dropout-as-applied-to-rnns-72e79ecd5b7b (accessed on 12 October 2022).
39. Radhakrishnan, P. What are Hyperparameters? And How to tune the Hyperparameters in a Deep Neural Network? *Data Sci.* 2017. Available online: https://towardsdatascience.com/what-are-hyperparameters-and-how-to-tune-the-hyperparameters-in-a-deep-neural-network-d0604917584a (accessed on 20 February 2023).
40. Sharma, S.; Sharma, S.; Anidhya, A. Understanding Activation Functions in Neural Networks. *Int. J. Eng. Appl. Sci. Technol.* **2020**, *4*, 310–316.
41. Nwankpa, C.; Ijomah, W.; Gachagan, A.; Marshall, S. Activation Functions: Comparison of trends in Practice and Research for Deep Learning. *arXiv* **2018**, arXiv:1811.03378. preprint.
42. Neal, R.M. Connectionist learning of belief networks. *Artif. Intell.* **1992**, *56*, 71–113. [CrossRef]
43. DeepAI. ReLu Definition. In *Deep AI Machine Learning Glossary*; DeepAI; Available online: https://deepai.org/machine-learning-glossary-and-terms/relu (accessed on 22 February 2023).
44. Agostinelli, F.; Hoffman, M.; Sadowski, P.; Baldi, P. Learning activation functions to improve deep neural networks. In Proceedings of the 3rd International Conference on Learning Representations, ICLR 2015—Workshop Track Proceedings, San Diego, CA, USA, 7–9 May 2015; pp. 1–9.
45. IEEE. Engineering in Medicine and Biology Society. In Proceedings of the IECBES, IEEE-EMBS Conference on Biomedical Engineering and Science, Kuching, Malaysia, 3–6 December 2018.
46. Khan, S.; AlSuwaidan, L. Agricultural monitoring system in video surveillance object detection using feature extraction and classification by deep learning techniques. *Comput. Electr. Eng.* **2022**, *102*, 108201. [CrossRef]
47. Boutahir, M.K.; Farhaoui, Y.; Azrour, M. Machine Learning and Deep Learning Applications for Solar Radiation Predictions Review: Morocco as a Case of Study. In *Digital Economy, Business Analytics, and Big Data Analytics Applications*; Springer: Berlin/Heidelberg, Germany, 2022; pp. 55–67. [CrossRef]
48. Alzubaidi, L.; Zhang, J.; Humaidi, A.J.; Al-Dujaili, A.; Duan, Y.; Al-Shamma, O.; Santamaría, J.; Fadhel, M.A.; Al-Amidie, M.; Farhan, L. Review of deep learning: Concepts, CNN architectures, challenges, applications, future directions. *J. Big Data* **2021**, *8*, 53. [CrossRef]
49. Jain, R.; Jain, N.; Aggarwal, A.; Hemanth, D.J. Convolutional neural network based Alzheimer's disease classification from magnetic resonance brain images. *Cogn. Syst. Res.* **2019**, *57*, 147–159. [CrossRef]
50. Wu, H.; Yin, H.; Chen, H.; Sun, M.; Liu, X.; Yu, Y.; Tang, Y.; Long, H.; Zhang, B.; Zhang, J.; et al. A Deep Learning, Image Based Approach for Automated Diagnosis for Inflammatory Skin Diseases. Available online: https://www.ncbi.nlm.nih.gov/pmc/articles/PMC7290553 (accessed on 5 December 2022).
51. Ting, D.S.W.; Cheung, C.Y.-L.; Lim, G.; Tan, G.S.W.; Quang, N.D.; Gan, A.; Hamzah, H.; Garcia-Franco, R.; Yeo, I.Y.S.; Lee, S.Y.; et al. Development and Validation of a Deep Learning System for Diabetic Retinopathy and Related Eye Diseases Using Retinal Images from Multiethnic Populations with Diabetes. *JAMA* **2017**, *318*, 2211–2223. [CrossRef]
52. Gu, H.; Guo, Y.; Gu, L.; Wei, A.; Xie, S.; Ye, Z.; Xu, J.; Zhou, X.; Lu, Y.; Liu, X.; et al. Deep Learning for Identifying Corneal Diseases from Ocular Surface Slit-Lamp Photographs. Available online: https://www.nature.com/articles/s41598-020-75027-3 (accessed on 5 December 2022).
53. Bai, X.; Niwas, S.I.; Lin, W.; Ju, B.-F.; Kwoh, C.K.; Wang, L.; Sng, C.C.; Aquino, M.C.; Chew, P.T.K. Learning ECOC Code Matrix for Multiclass Classification with Application to Glaucoma Diagnosis. *J. Med. Syst.* **2016**, *40*, 78. [CrossRef] [PubMed]
54. Xin, M.; Wang, Y. Research on image classification model based on deep convolution neural network. *EURASIP J. Image Video Process.* **2019**, *2019*, 40. [CrossRef]
55. Brown, M.; An, P.E.; Harris, C.J.; Wang, H. How Biased is Your Multi-Layered Perceptron? *World Congr. Neural Netw.* **1993**, 507–511. Available online: https://eprints.soton.ac.uk/250244/ (accessed on 22 February 2023).
56. Haq, A.U.H.; Li, J.P.L.; Agbley, B.L.Y.; Khan, A.; Khan, I.; Uddin, M.I.; Khan, S. IIMFCBM: Intelligent Integrated Model for Feature Extraction and Classification of Brain Tumors Using MRI Clinical Imaging Data in IoT-Healthcare. *IEEE J. Biomed. Health Inform.* **2022**, *26*, 5004–5012. [CrossRef]
57. Guezzaz, A.; Benkirane, S.; Azrour, M.; Khurram, S. A Reliable Network Intrusion Detection Approach Using Decision Tree with Enhanced Data Quality. *Secur. Commun. Networks* **2021**, *2021*, 1230593. [CrossRef]

58. ResNet; AlexNet; VGGNet. *Inception: Understanding Various Architectures of Convolutional Networks.* 2022. Available online: https://cv-tricks.com/cnn/understand-resnet-alexnet-vgg-inception/ (accessed on 23 February 2023).
59. Wang, P.; Fan, E.; Wang, P. Comparative analysis of image classification algorithms based on traditional machine learning and deep learning. *Pattern Recognit. Lett.* **2020**, *141*, 61–67. [CrossRef]
60. Suganyadevi, S.; Seethalakshmi, V.; Balasamy, K. A review on deep learning in medical image analysis. *Int. J. Multimed. Inf. Retr.* **2021**, *11*, 19–38. [CrossRef]
61. Shamshirband, S.; Fathi, M.; Dehzangi, A.; Chronopoulos, A.T.; Alinejad-Rokny, H. A review on deep learning approaches in healthcare systems: Taxonomies, challenges, and open issues. *J. Biomed. Inform.* **2020**, *113*, 103627. [CrossRef] [PubMed]
62. Khvostikov, A.; Aderghal, K.; Benois-Pineau, J.; Krylov, A.; Catheline, G. 3D CNN-based classification using sMRI and MD-DTI images for Alzheimer disease studies. *arXiv* **2018**, arXiv:1801.05968, preprint.
63. Liu, Z.; Lu, H.; Pan, X.; Xu, M.; Lan, R.; Luo, X. Diagnosis of Alzheimer's disease via an attention-based multi-scale convolutional neural network. *Knowl. Based Syst.* **2022**, *238*, 107942. [CrossRef]
64. Ajagbe, S.A.; Amuda, K.A.; Oladipupo, M.A.; Afe, O.F.; Okesola, K.I. Multi-classification of alzheimer disease on magnetic resonance images (MRI) using deep convolutional neural network (DCNN) approaches. *Int. J. Adv. Comput. Res.* **2021**, *11*, 51–60. [CrossRef]
65. Villa-Pulgarin, J.P.; Ruales-Torres, A.A.; Arias-Garz, D.; Bravo-Ortiz, M.A.; Arteaga-Arteaga, H.B.; Mora-Rubio, A.; Alzate-Grisales, J.A.; Mercado-Ruiz, E.; Hassaballah, M.; Orozco-Arias, S.; et al. Optimized Convolutional Neural Network Models for Skin Lesion Classification. *Comput. Mater. Contin.* **2022**, *70*, 2131–2148. [CrossRef]
66. Hemdan, E.E.-D.; Shouman, M.A.; Karar, M.E. COVIDX-Net: A Framework of Deep Learning Classifiers to Diagnose COVID-19 in X-ray Images. *arXiv* **2020**, arXiv:2003.11055. preprint.
67. Horry, M.J.; Chakraborty, S.; Paul, M.; Ulhaq, A.; Pradhan, B.; Saha, M.; Shukla, N. COVID-19 Detection through Transfer Learning Using Multimodal Imaging Data. *IEEE Access* **2020**, *8*, 149808–149824. [CrossRef]
68. Joshua, E.S.N.; Bhattacharyya, D.; Chakkravarthy, M.; Byun, Y.-C. 3D CNN with Visual Insights for Early Detection of Lung Cancer Using Gradient-Weighted Class Activation. *J. Healthc. Eng.* **2021**, *2021*, 6695518. [CrossRef]
69. Li, Q.; Cai, W.; Wang, X.; Zhou, Y.; Feng, D.D.; Chen, M. Medical image classification with convolutional neural network. In Proceedings of the 2014 13th International Conference on Control Automation Robotics & Vision (ICARCV), Singapore, 10–12 December 2014; pp. 844–848.
70. Tiwari, P.; Pant, B.; Elarabawy, M.M.; Abd-Elnaby, M.; Mohd, N.; Dhiman, G.; Sharma, S. CNN Based Multiclass Brain Tumor Detection Using Medical Imaging. *Comput. Intell. Neurosci.* **2022**, *2022*, 1830010. [CrossRef] [PubMed]
71. Yildirim, M.; Cinar, A. Classification with respect to colon adenocarcinoma and colon benign tissue of colon histopathological images with a new CNN model: MA_ColonNET. *Int. J. Imaging Syst. Technol.* **2021**, *32*, 155–162. [CrossRef]
72. Ravi, V.; Narasimhan, H.; Chakraborty, C.; Pham, T.D. Deep learning-based meta-classifier approach for COVID-19 classification using CT scan and chest X-ray images. *Multimed. Syst.* **2021**, *28*, 1401–1415. [CrossRef]
73. Chakraborty, S.; Paul, S.; Hasan, K.M.A. A Transfer Learning-Based Approach with Deep CNN for COVID-19- and Pneumonia-Affected Chest X-ray Image Classification. *SN Comput. Sci.* **2021**, *3*, 17. [CrossRef] [PubMed]
74. Srinivas, C.; Prasad, N.K.S.; Zakariah, M.; Alothaibi, Y.A.; Shaukat, K.; Partibane, B.; Awal, H. Deep Transfer Learning Approaches in Performance Analysis of Brain Tumor Classification Using MRI Images. *J. Healthc. Eng.* **2022**, *2022*, 3264367. [CrossRef] [PubMed]
75. Kumar, N.; Gupta, M.; Gupta, D.; Tiwari, S. Novel deep transfer learning model for COVID-19 patient detection using X-ray chest images. *J. Ambient. Intell. Humaniz. Comput.* **2021**, *14*, 469–478. [CrossRef]
76. Vijaysinh, L. A Comparison of 4 Popular Transfer Learning Models. *AIM* **2021**. Available online: https://arxiv.org/abs/1603.08631 (accessed on 23 February 2023).
77. Sarraf, S.; Tofighi, G. Classification of Alzheimer's Disease Using fMRI Data and Deep Learning Convolutional Neural Networks. *arXiv* **2016**, arXiv:1603.08631. preprint.
78. He, K.; Zhang, X.; Ren, S.; Sun, J. Deep residual learning for image recognition. In Proceedings of the IEEE Computer Society Conference on Computer Vision and Pattern Recognition (CVPR), Las Vegas, NV, USA, 27–30 June 2016; pp. 770–778. [CrossRef]
79. ImageNet Large Scale Visual Recognition Challenge (ILSVRC). Available online: https://www.image-net.org/ (accessed on 24 February 2023).
80. Russakovsky, O.; Deng, J.; Su, H.; Krause, J.; Satheesh, S.; Ma, S.; Huang, Z.; Karpathy, A.; Khosla, A.; Bernstein, M.; et al. ImageNet Large Scale Visual Recognition Challenge. *Int. J. Comput. Vis.* **2015**, *115*, 211–252. [CrossRef]
81. Lecun, Y.; Bottou, L.; Bengio, Y.; Haffner, P. Gradient-Based Learning Applied to Document Recognition. *Proc. IEEE* **1998**, *86*, 2278–2324. [CrossRef]
82. Krizhevsky, A.; Sutskever, I.; Hinton, G.E. Imagenet classification with deep convolutional neural networks. *Commun. ACM* **2017**, *60*, 84–90. [CrossRef]
83. Szegedy, C.; Vanhoucke, V.; Ioffe, S.; Shlens, J.; Wojna, Z. Rethinking the Inception Architecture for Computer Vision. In Proceedings of the IEEE Computer Society Conference on Computer Vision and Pattern Recognition (CVPR), Las Vegas, NV, USA, 27–30 June 2016; pp. 2818–2826. [CrossRef]
84. Huang, G.; Liu, Z.; van der Maaten, L.; Weinberger, K.Q. Densely Connected Convolutional Networks. 2017. Available online: https://github.com/liuzhuang13/DenseNet (accessed on 12 December 2022).

85. Liu, W.; Zeng, K. SparseNet: A Sparse DenseNet for Image Classification. *arXiv* **2018**, arXiv:1804.05340. preprint.
86. Introduction to DenseNets (Dense CNN)—Analytics Vidhya. Available online: https://www.analyticsvidhya.com/blog/2022/03/introduction-to-densenets-dense-cnn/ (accessed on 12 December 2022).
87. Tra, V.; Kim, J.; Khan, S.A.; Kim, J.-M. Bearing Fault Diagnosis under Variable Speed Using Convolutional Neural Networks and the Stochastic Diagonal Levenberg-Marquardt Algorithm. *Sensors* **2017**, *17*, 2834. [CrossRef]
88. Rousseau, F.; Drumetz, L.; Fablet, R. Residual Networks as Flows of Diffeomorphisms. *J. Math. Imaging Vis.* **2019**, *62*, 365–375. [CrossRef]
89. Marcelino, P. Transfer learning from pre-trained models. *Medium* **2018**. Available online: https://towardsdatascience.com/transfer-learning-from-pre-trained-models-f2393f124751 (accessed on 23 February 2023).
90. Taresh, M.M.; Zhu, N.; Ali, T.A.A.; Hameed, A.S.; Mutar, M.L. Transfer Learning to Detect COVID-19 Automatically from X-ray Images Using Convolutional Neural Networks. *Int. J. Biomed. Imaging* **2021**, *2021*, 8828404. [CrossRef]
91. Ezzat, D.; Hassanien, A.E.; Ella, H.A. An optimized deep learning architecture for the diagnosis of COVID-19 disease based on gravitational search optimization. *Appl. Soft Comput.* **2020**, *98*, 106742. [CrossRef]
92. Keykhaie, S. POLYTECHNIQUE MONTRÉAL Affiliée à l'Université de Montréal Secure Authentication for Mobile Users SEPEHR KEYKHAIE Département de Génie Informatique Et Génie Logiciel". Available online: https://publications.polymtl.ca/5604/ (accessed on 12 December 2022).
93. Simonyan, K.; Zisserman, A. Very deep convolutional networks for large-scale image recognition. *arXiv* **2014**, arXiv:1409.1556. preprint.
94. Salman, S.; Liu, X. Overfitting Mechanism and Avoidance in Deep Neural Networks. *arXiv* **2019**, arXiv:1901.06566. preprint.
95. Guide to Prevent Overfitting in Neural Networks—Analytics Vidhya. Available online: https://www.analyticsvidhya.com/blog/2021/06/complete-guide-to-prevent-overfitting-in-neural-networks-part-1/ (accessed on 13 December 2022).
96. Khan, S.; Alqahtani, S. Big Data Application and its Impact on Education. *Int. J. Emerg. Technol. Learn. (IJET)* **2020**, *15*, 36–46. [CrossRef]
97. Shorten, C.; Khoshgoftaar, T.M. A survey on Image Data Augmentation for Deep Learning. *J. Big Data* **2019**, *6*, 60. [CrossRef]
98. Batch Normalization in Convolutional Neural Networks | Baeldung on Computer Science. Available online: https://www.baeldung.com/cs/batch-normalization-cnn (accessed on 13 December 2022).
99. Ioffe, S.; Szegedy, C. Batch Normalization: Accelerating Deep Network Training by Reducing Internal Covariate Shift. *arXiv* **2015**, arXiv:1502.03167.
100. Tompson, J.; Goroshin, R.; Jain, A.; LeCun, Y.; Bregler, C. Efficient object localization using Convolutional Networks. *arXiv* **2015**, arXiv:1411.4280. preprint.
101. Toronto. Preventing Overfitting. 2022. Available online: https://www.cs.toronto.edu/~lczhang/360/lec/w05/overfit.html (accessed on 13 December 2022).
102. Deng, J.; Dong, W.; Socher, R.; Li, L.J.; Li, K.; Fei-Fei, L. ImageNet: A large-scale hierarchical image database. In Proceedings of the 2009 IEEE Conference on Computer Vision and Pattern Recognition, Miami, FL, USA, 20–25 June 2009. [CrossRef]
103. Premio Inc. What Are Accelerators in the Context of Computing Hardware? Available online: https://premioinc.com/blogs/blog/performance-accelerators-in-the-context-of-computing-hardware (accessed on 12 December 2022).
104. Ways AI is Changing our World for the Better. Available online: https://www.salzburgglobal.org/news/latest-news/article/5-ways-ai-is-changing-our-world-for-the-better?gclid=CjwKCAjw-rOaBhA9EiwAUkLV4n43oOgEhJktVevO_iWDnbu-GSMYPvDosnD4Is8nmt76SS_NeHDpFxoCh8wQAvD_BwE (accessed on 18 October 2022).
105. Meta AI. Available online: https://ai.facebook.com/#notable-papers (accessed on 12 December 2022).
106. DenseNet Explained | Papers with Code. Available online: https://paperswithcode.com/method/densenet (accessed on 12 December 2022).
107. Feng, J.; Lu, S. Performance Analysis of Various Activation Functions in Artificial Neural Networks. *J. Phys. Conf. Ser.* **2019**, *1237*, 022030. [CrossRef]
108. Chung, H.; Lee, S.J.; Park, J.G. Deep neural network using trainable activation functions. In Proceedings of the 2016 International Joint Conference on Neural Networks (IJCNN), Vancouver, BC, Canada, 24–29 July 2016; pp. 348–352. [CrossRef]
109. Tan, H.H.; Lim, K.H. Vanishing Gradient Mitigation with Deep Learning Neural Network Optimization. In Proceedings of the 7th International Conference on Smart Computing & Communications (ICSCC), Sarawak, Malaysia, 28–30 June 2019; pp. 1–4. [CrossRef]
110. Hu, Y.; Huber, A.; Anumula, J.; Liu, S.-C. Overcoming the vanishing gradient problem in plain recurrent networks. *arXiv* **2018**, arXiv:1801.06105. preprint.
111. Training, Testing & Deploy of Classification Model Using CNN & ML. Available online: https://www.turing.com/kb/training-testing-deployment-of-classification-model-using-convolutional-neural-networks-and-machine-learning-classifiers (accessed on 12 December 2022).
112. Akselrod-Ballin, A.; Karlinsky, L.; Alpert, S.; Hasoul, S.; Ben-Ari, R.; Barkan, E. A region based convolutional network for tumor detection and classification in breast mammography. In *Deep Learning and Data Labeling for Medical Applications*; Springer: Cham, Switzerland, 2016; pp. 197–205.

113. Anavi, Y.; Kogan, I.; Gelbart, E.; Geva, O.; Greenspan, H. Visualizing and enhancing a deep learning framework using patients age and gender for chest X-ray image retrieval. *SPIE* **2016**, *9785*, 249–254. [CrossRef]
114. Zakharchuk, I. Generalization, Overfitting, and Under-fitting in Supervised Learning | MLearning.ai | Medium. Available online: https://medium.com/mlearning-ai/generalization-overfitting-and-underfitting-in-supervised-learning-a21f02ebf3df (accessed on 13 December 2022).

sustainability

Article

Provably Secure Dynamic Anonymous Authentication Protocol for Wireless Sensor Networks in Internet of Things

Zixuan Ding and Qi Xie *

Key Laboratory of Cryptography of Zhejiang Province, Hangzhou Normal University, Hangzhou 311121, China
* Correspondence: qixie68@126.com

Abstract: Wireless sensor networks are a promising application of the Internet of Things in the sustainable development of smart cities, and have been afforded significant attention since first being proposed. Authentication protocols aim to protect the security and confidentiality of legitimate users when accessing and transmitting data. However, existing protocols may suffer from one or more security flaws. Recently, Butt et al. proposed an energy-efficient three-factor authentication protocol for wireless sensor networks. However, their protocol is vulnerable to several attacks, and lacks certain security properties. In this paper, the causes of these design flaws are analyzed. Furthermore, we propose a novel three-factor authentication protocol (password, smart card, and biometric information) for wireless sensor networks in Internet of Things contexts. A dynamic anonymous strategy is designed to prevent privacy disclosure and to resist sensor node capture attacks, tracking attacks, and desynchronization attacks. The Find–Guess model and random oracle model are combined to prove the security of the proposed protocol. A comparative analysis with related schemes shows that the proposed protocol has higher security and is able to maintain a low computational overhead.

Keywords: authentication protocol; wireless sensor network; Internet of Things; privacy-preserving; provably secure

Citation: Ding, Z.; Xie, Q. Provably Secure Dynamic Anonymous Authentication Protocol for Wireless Sensor Networks in Internet of Things. *Sustainability* **2023**, *15*, 5734. https://doi.org/10.3390/su15075734

Academic Editor: Juan Miguel Navarro

Received: 7 February 2023
Revised: 21 March 2023
Accepted: 22 March 2023
Published: 24 March 2023

1. Introduction

A wireless sensor network is a network formed by multi-sensor cooperative detection of the complex physical environment through wireless communication technology. The lightweight nature of the Internet of Things makes wireless sensor network more practical. Nodes in traditional sensor networks realize point-to-point transmission through a wired channel. The function and application scenario of traditional sensor networks are relatively simple and limited. Until the end of the last century, circuit bus technology and wireless technology were applied to sensor networks, and the current Internet of Things-based wireless sensor network architecture was gradually formed. Compared with traditional sensor networks, the cost of sensors in Internet of Things contexts is greatly reduced, while the reliability and scalability of the system are significantly improved. In addition, as the most recent sensor networks, Internet of Things-based wireless sensor networks have the advantages of self-organizing dynamic topology [1], the development and application of which have a far-reaching impact on all fields. However, communication based on wireless channels makes wireless sensor networks in Internet of Things contexts vulnerable to passive eavesdropping, active intrusion, message replay, and other attacks [2]. Therefore, the security issues involving Internet of Things-based wireless sensor networks are urgent and need to be solved.

In the past few decades, authentication protocols for wireless sensor networks for the Internet of Things have made great progress. In 2006, Wong et al. [3] proposed a dynamic password-based authentication scheme for wireless sensor network, in which the users request sensors via the gateway node. However, their scheme is vulnerable to multiple attacks, including replay attacks, impersonation attacks, and stolen-verifier attacks. Das [4] pointed out these defects in Wong et al.'s scheme and proposed a security-enhanced scheme based on a smart card and password, allowing it to resist the attacks above. Although the scheme proposed by Das is more secure than that of Wong et al., it is not satisfactory in that the secret parameters are stored as plaintext in the sensor node and smart card, which makes it vulnerable to node-capture attacks or smart card loss attacks. In 2010, the research of Khan and Alghathbar [5] indicated that the scheme proposed by Das fails to realize mutual authentication and suffers from vulnerability to privileged-insider attacks. They proposed an authentication scheme that resists privileged-insider attacks. In addition, multiple password-based authentication schemes have been proposed [6–9]. Nevertheless, most password-based schemes are generally vulnerable to off-line password guessing attacks.

To overcome the shortcomings of password-based schemes, biometrics can be adopted in identity verification. In 2010, Yuan et al. [10] proposed an authentication protocol based on biometrics, passwords, and smart cards. However, their protocol lacks message confidentiality and integrity verification. Moreover, their scheme suffers from sensor node capture attacks and impersonation attacks. In the following year, Yoon et al. [11] improved on Yuan et al.'s scheme without using passwords. Legitimacy verification in the scheme proposed by Yoon et al. is based on secret parameters. However, there are security flaws in this scheme, such as confidentiality issues and vulnerability to denial of service attacks. He et al. [12] proposed a protocol in 2012 that fixes these flaws. Chen et al. [13] proposed an authentication scheme suitable for wireless sensor networks in Internet of Things environments; however, Hu [14] pointed out that their scheme is vulnerable to off-line password guessing attacks and impersonation attacks, and fails to achieve perfect forward secrecy, user anonymity, and unlinkability. They instead proposed a novel security-enhanced scheme for wireless sensor networks in Internet of Things contexts.

Compared with password-based two-factor authentication schemes, three-factor schemes with the participation of biometrics have higher security. Generally speaking, three-factor schemes can resist password-guessing attacks and user impersonation attacks. In 2021, Shuai et al. [15] proposed a three-factor authentication scheme for wireless sensor networks in Internet of Things environments. However, Xie et al. pointed out that this scheme is vulnerable to stolen-verifier attacks and desynchronization attacks, and has no perfect forward secrecy. Instead, they proposed a security-enhanced anonymous three-factor authentication scheme [16] based on elliptic curve cryptography [17]. Unfortunately, if an adversary captures a sensor node, it can be used to recover the user's identity in the schemes of both Xie et al. [16] and Shuai et al. [15]. Generally speaking, if the designed protocols does not use Diffie-Hellman key exchange algorithm [18] to generate the session key, such schemes cannot achieve perfect forward secrecy [19–23]. Recently, Butt et al. [24] proposed a three-factor authentication scheme based on elliptic curve cryptography for wireless sensor network. However, their scheme is vulnerable to replay attacks, sensor node capture attacks, and off-line password guessing attacks, and fails to preserve session key secrecy, perfect forward secrecy, anonymity, and unlinkability. In 2021, Xie et al. [25] proposed a wireless sensor network authentication protocol for a smart city that addressed a number of open issues, such as the inability to resist offline password guessing attacks and impersonation attacks as well as the lack of session key secrecy, identity unlinkability, and perfect forward secrecy. In 2022, Ouni and Saleem [26] proposed a framework for wireless sensor networks for the Internet of Things that detects environmental data through sensors connected to the cloud and feeds these data back to users.

Wireless sensor networks for the Internet of Things are considered to be one of the most crucial technologies affecting the future development of mankind. They can provide a new way to obtain and process information. However, wireless sensor networks face two major challenges that traditional sensor networks do not have. First, in terms of security, information transmitted via open channels is subject to eavesdropping, tampering, and other attacks [27]. Second, in terms of computing power, the limited resources of sensors demand the design of lightweight authentication protocols for wireless sensor networks. Previous researchers have often focused on only one of these issues, not both, resulting in proposed solutions with either insufficient security or high computational cost.

Most protocols are likely to suffer from one or more attacks. In particular, sensor node captured attacks can easily lead to user privacy disclosure and vulnerability to forgery attacks. Therefore, we take Butt et al.'s scheme as an example. We analyze their scheme for security flaws and investigate their causes. In addition, we propose a protocol using the Diffie-Hellman algorithm and symmetric encryption algorithm. The contributions of this paper are listed as follows:

- The security flaws of Butt et al.'s protocol are typical, which implies that the design strategies used in our proposed protocol can be applied to similar schemes in general.
- The proposed strategies can avoid existing design deficiencies. Symmetric encryption and elliptic curve cryptography are combined to design a secure protocol for wireless sensor networks. A dynamic pseudonym strategy is developed to resist desynchronization attacks and sensor node capture attacks, while symmetric encryption is used to protect the privacy of transmitted messages.
- The proposed protocol is formally proved by combining a Find–Guess model and a random oracle model.

The remainder of this paper is organized as follows. In the next Section, the scheme of Butt et al. is reviewed and analyzed. Section 3 introduces the system model, the adversary model, and the proposed protocol. A security analysis and proof of the proposed protocol are presented in Section 4. Section 5 provides comparisons of performance and security properties with related works. Finally, the paper is concluded in Section 6.

2. Security Analysis of Butt et al. Scheme

In this section, the scheme of Butt et al. is reviewed and the flaws in their scheme are pointed out. In addition, the causes of these design flaws are analyzed, implying design strategies that can be used to overcome them.

2.1. Review of Butt et al. Scheme

2.1.1. Registration Phase

In this phase, the user registers with the gateway and the sensor nodes. The gateway generates a secret number x_0, which is shared with users and sensor nodes. The steps are as follows.

Step RP1: The user first enters his/her identity ID_i, password PW_i, and biometric B_i into the device with a biometric reader. The device calculates $ID_i^* = ID_i \oplus x_0$, $(\sigma_i, \tau_i) = Gen(B_i)$, $A_i = h(\sigma_i)$, and $HPW = h(PW_i)$, where the secret parameter x_0 is known to all nodes. Then, $M_1 = \{ID_i^*, A_i, HPW\}$ is transmitted to the gateway.

Step RP2: Upon receiving $M_1 = \{ID_i^*, A_i, HPW\}$, the gateway computes $ID_i = ID_i^* \oplus x_0$, $N_i = HPW \oplus x_0$, $S_i = ID_i \oplus x_0$, and $M_i = S_i \oplus h(A_i \oplus HPW)$. Then, the gateway sends $M_2 = \{N_i, S_i, M_i\}$ to the user and sensor nodes.

2.1.2. Login and Authentication Phase

The user follows the steps below to log in and authenticate.

Step LA1: The user first inputs the identity ID_i, the password PW_i^*, and the biometric B_i^* into the device. Then, the device calculates $\sigma_i^* = Rep(B_i^*, \tau_i)$, $A_i^* = h(\sigma_i^*)$, $HPW^* = h(PW_i^*)$, $X_i = (r_u \cdot P_{ec}) \oplus x_0$, $Y_i = ID_i \oplus N_i$, and $ID_i^* = Y_i \oplus HPW^*$, where P_{ec} is the generator point on an elliptic curve, r_u is a random number generated by the device. After that, the user sends $M_3 = \{ID_i^*, A_i^*, X_i, HPW^*, RI, N_a\}$ to the gateway, where RI is the request information and N_a is a nonce generated by the user.

Step LA2: On receiving the message $M_3 = \{ID_i^*, A_i^*, X_i, HPW^*, RI, N_a\}$, the gateway calculates and verifies $ID_i = ID_i^* \oplus x_0$. If the verification is passed, the gateway compares A_i^* and HPW^* with A_i and HPW to check their equality, then generates a random number r_s and computes $S_i^* = ID_i^* \oplus x_0$, $M_i^* = S_i^* \oplus h(A_i^* \oplus HPW^*)$,

$X_i^* = X_i \oplus x_0$, $D_i = r_s \cdot P_{ec}$, and $C_i = r_s \cdot X_i^*$. Then, the gateway transmits the message $M_4 = \{ID_i^*, X_i, A_i^*, C_i, D_i, HPW^*, RI, M_i^*, N_a\}$ to the sensor node.

Step LA3: After receiving the message M_4, the sensor node first verifies whether $M_i^* = M_i$. If the verification is passed, the sensor node calculates $X_i^* = X_i \oplus x_0$, $SK_s = h(C_i \parallel ID_i^* \parallel N_a \parallel x_0)$, $E_{SK_s}(RI) = e$, and $E_{SK_s}(N_a) = J_i$. Then, the message $M_5 = \{D_i, e, N_s, J_i\}$ is sent to the user, where N_s is a nonce generated by the sensor node.

Step LA4: Upon receiving the message M_5, the user calculates $C_i = r_u \cdot D_i$, $SK_u = h(C_i \parallel ID_i^* \parallel N_a \parallel x_0)$, and $RI = D_{SK_u}(e)$.

Step LA5: The user sends $(N_s \oplus x_0)SK_u$ to the sensor node as an acknowledgment.

2.2. Security Analysis of Butt et al. Scheme

The scheme proposed by Butt et al. cannot resist replay attacks, sensor node capture attacks, or off-line password guessing attacks, and fails to preserve session key secrecy, perfect forward secrecy, anonymity, or unlinkability. These weaknesses are described in the following subsections, and the causes of these issues are presented.

2.2.1. Replay Attack/User Impersonation Attack

By replaying $\{ID_i^*, A_i^*, HPW^*, RI\}$ and generating $\{X_i, N_a\}$ to forge the message M_3, an adversary can impersonate a user to complete the authentication process and negotiate the session key, where $X_i = (r_u \cdot P_{ec}) \oplus x_0$, P_{ec} is the generator point, r_u is a random number, N_a is a nonce, and x_0 is shared among all nodes and users, and can be obtained by registering or capturing nodes.

The cause of vulnerability to replay attacks is that the proposed scheme does not use timestamps and makes improper use of random numbers. The cause of vulnerability to user impersonation attacks is that the secret key x_0 is shared among the users, sensor nodes, and gateway.

2.2.2. Sensor Node Capture Attacks

In Butt et al.'s scheme, the secret parameter x_0 and user's information $\{N_i, S_i, M_i\}$ are known to all nodes, where $N_i = HPW \oplus x_0$, $S_i = ID_i \oplus x_0$, and $M_i = S_i \oplus h(A_i \oplus HPW)$. An adversary can obtain them by capturing a sensor node. Therefore, the adversary can obtain the session key by calculating $SK_s = h(C_i \parallel ID_i^* \parallel N_a \parallel x_0)$, where C_i, ID_i^*, and N_a are transmitted in public. In addition, the adversary can corrupt a user's privacy and impersonate the user.

The reason for the above defect is that sensors store a generic secret value x_0 and users' private information $\{N_i, S_i, M_i\}$.

2.2.3. No Perfect Forward Secrecy

If the long-term key x_0 is known by an adversary, he/she can acquire the previous session keys and the later session keys by calculating $SK_s = h(C_i \parallel ID_i^* \parallel N_a \parallel x_0)$, where C_i, ID_i^*, and N_a are transmitted in public. Therefore, the scheme lacks perfect forward secrecy.

The reason for the lack of perfect forward security is the absence of a Diffie–Hellman key agreement algorithm.

2.2.4. No Anonymity and Unlinkability

In Step LA2 of Butt et al.'s scheme, the gateway transmits $M_4 = \{ID_i^*, X_i, A_i^*, C_i, D_i, HPW^*, RI, M_i^*, N_a\}$ to the sensor node in public. The adversary can recover S_i^* by calculating $M_i^* \oplus h(A_i^* \oplus HPW^*) = S_i^*$, where A_i^* and HPW^* are available from $M_3 = \{ID_i^*, A_i^*, X_i, HPW^*, RI, N_a\}$. By computing $S_i^* \oplus ID_i^* = x_0$ and $x_0 \oplus ID_i^* = ID_i$, the identity ID_i of the user is known to the adversary. Therefore, the scheme fails to provide anonymity.

ID_i^*, A_i^*, and HPW^* are fixed in each session, and the adversary can trace the user by eavesdropping. Consequently, the scheme fails to achieve unlinkability.

The reason for the lack of anonymity and unlinkability is that the identity ID_i can be easily recovered and the parameter $S_i^* = ID_i^* \oplus x_0$ is fixed.

2.2.5. Off-Line Password Guessing Attacks

In Step LA1 of the login and authentication phase, the user sends $M_3 = \{ID_i^*, A_i^*, X_i, HPW^*, RI, N_a\}$ to the gateway via the open channel, where $HPW^* = h(PW_i^*)$, PW_i^* is the user's password. An adversary can verify the correctness of the guessed password PW_A by comparing HPW^* with $h(PW_A)$.

The reason for this flaw is that the adversary can use the public parameter HPW^* to verify the correctness of the guessed password. Making the verification of guessed passwords infeasible can ensure password security.

2.2.6. No Session Key Secrecy

As per the steps shown in Section 2.2.4, the secret parameter x_0 can be obtained by calculating $M_i^* \oplus h(A_i^* \oplus HPW^*) \oplus ID_i^* = x_0$. Therefore, an adversary can compute the session key $SK_s = h(C_i \parallel ID_i^* \parallel N_a \parallel x_0)$, where C_i, ID_i^*, and N_a are transmitted via the open channel, implying that the scheme has no session key secrecy.

This deficiency exists because all negotiation parameters of the session key are accessible to third parties over the public channel. In order to fulfill session key secrecy, it must be ensured that the key composition parameters cannot be easily calculated and obtained by attackers.

3. The Proposed Scheme

In this section, a lightweight dynamic anonymous authentication scheme for wireless sensor networks in Internet of Things environments is proposed. It consists of a system setup phase, registration phase, and login and authentication phase. The system model and the adversary model are shown as follows. The notation mentioned in the scheme is listed in Table 1.

Table 1. Notation.

Notation	Description
U_i	i^{th} User
ID_i	The identity of U_i
SN_j	j^{th} Sensor node
SID_j	The identity of SN_j
GWN	The gateway node
PW_i	The password of U_i
SK	The session key
B_i	The biometric of U_i
N_a, N_s	Nonces
T_1, T_2	Timestamps
x_0	Secret parameter shared with trusted nodes
τ_i, σ_i	Reproduction parameter and biometric key of fuzzy extractor
$Rep(), Gen()$	Reproduction and generation function of fuzzy extractor
$\parallel$	Concatenation
$\oplus$	XOR operation
$h(.)$	Hash function
ΔT	Transmission delay time
K_{GWN}	The secret key of the gateway
r_i, u_i, c_j	Random numbers
DID_i	The temporary identity of the user

3.1. System Model and Adversary Model

3.1.1. System Model

There are three types of participants involved in wireless sensor network systems: the user, the gateway, and the sensor. Users and sensors are registered in the gateway via the secure channel. After registration, mutual authentication and communication between users and sensors are established through the public channel. The system model is shown in Figure 1.

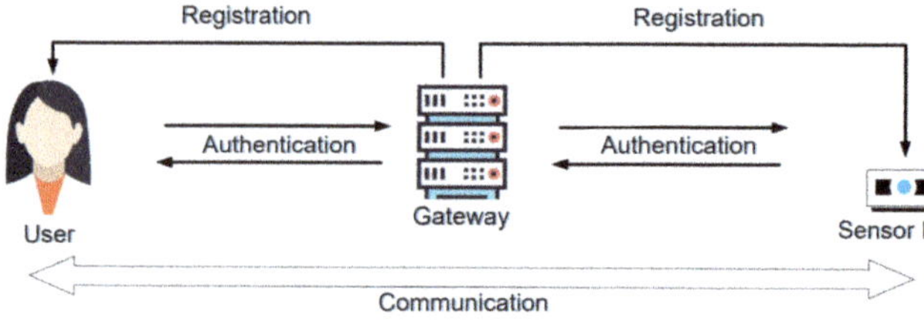

Figure 1. The system model.

User: The registered user has a smart card to store the registration information. The smart card can be obtained and analyzed by the adversary. While the adversary may guess the user's password, the user's biometric information cannot be obtained by the adversary.

Gateway: The gateway is assumed to be trustful, and its key cannot be obtained by an adversary. The gateway never conspires with other participants.

Sensor node: The storage and computing capabilities of sensor nodes are limited, and adversaries can capture sensors and analyze their stored information.

3.1.2. Adversary Model

According to Dolev and Yao's attack model [28] as well as to a previous survey [29], the adversary model of the proposed protocol is presented as follows:

- All messages transmitted publicly can be captured by the adversary, and he/she can replay, modify, and reroute the messages.
- Off-line password guessing attacks can be launched, and the identity of users can be obtained.
- Sensor nodes and smart cards can be captured, and all information stored in captured nodes and stolen cards can be obtained.
- Adversaries may be privileged insider users.
- The only trusted entity is the gateway.
- The master key of the gateway and the biological keys of the users cannot be obtained.

3.2. *The Proposed Protocol*

3.2.1. System Initialization Phase

The gateway chooses an elliptic curve $E(GF_q)$ based on a finite field $GF(q)$, where q is a large prime number and P is a generator point. The gateway then publishes P, the generation function $Gen(.)$ and reproduction function $Rep(.)$ of the fuzzy extractor, and the hash function $h(.)$.

3.2.2. Registration Phase

User registration phase:

Step UR1: The user first inputs their identity ID_i, biometrics B_i, and password PW_i into the device. The device computes $(\sigma_i,\ \tau_i) = Gen(B_i)$ and $HPW_i = h(ID_i \parallel PW_i \parallel \sigma_i)$, and sends $\{ID_i\}$ to the gateway via secure channel.

Step UR2: The gateway verifies the uniqueness of ID_i; if it cannot, it requests a new identity from the user. Otherwise, the gateway generates a random number r_i and calculates $S_i = h(ID_i \parallel h(K_{GWN}))$ and $DID_i = E_{K_{GWN}}(ID_i, r_i)$, where K_{GWN} is the gateway's secret key. Then, $\{S_i, DID_i\}$ is sent to the user via secure channel and $h(ID_i)$ is stored for uniqueness verification.

Step UR3: The user's device computes $A_i = S_i \oplus h(ID_i \parallel \sigma_i \parallel PW_i)$ and $HID_i = DID_i \oplus h(S_i \parallel PW_i \parallel \sigma_i)$, and stores $\{HPW_i, \tau_i, A_i, HID_i\}$ on the smart card.

The process of the user registration phase is shown in Figure 2.

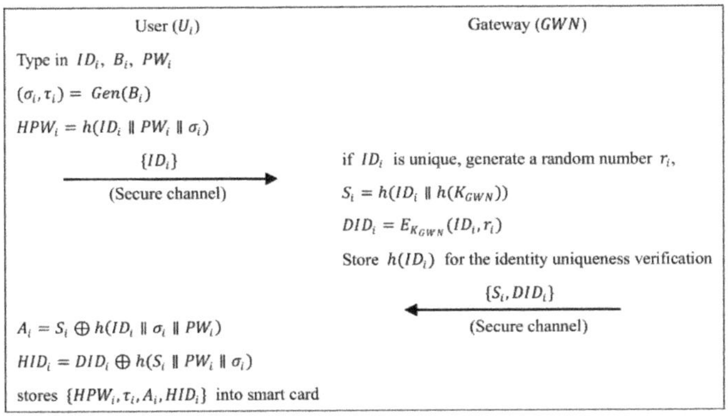

Figure 2. User registration phase.

Sensor node registration phase:

Step SR1: The gateway first chooses a unique identity SID_j of the sensor node and computes $b_j = h(SID_j \parallel h(K_{GWN}))$. Then, $\{b_j, SID_j\}$ is securely transmitted to the sensor node.

Step SR2: The sensor node stores $\{b_j, SID_j\}$.

Figure 3 shows the registration phase of the sensor node.

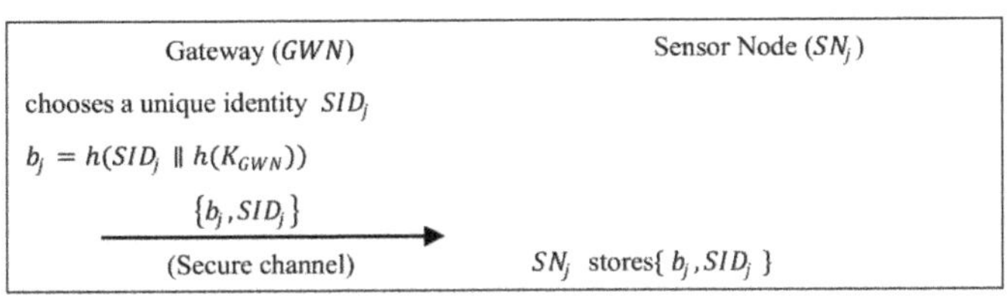

Figure 3. Registration phase of the sensor node.

3.2.3. Authentication and Session Key Agreement Phase

Step AP1: The user first inserts the smart card and inputs their identity ID_i^*, password PW_i^*, and biometrics B_i^* into the device. The device calculates $\sigma_i^* = Rep(B_i^*, \tau_i)$ and $HPW_i^* = h(ID_i^* \parallel PW_i^* \parallel \sigma_i^*)$. If $HPW_i^* \neq HPW_i$, the user login fails; otherwise, the device computes $S_i^* = A_i \oplus h(ID_i^* \parallel \sigma_i^* \parallel PW_i^*)$, $DID_i^* = HID_i \oplus h(S_i^* \parallel PW_i^* \parallel \sigma_i^*)$, $M_1 = u_i \cdot P$, and $M_2 = E_{S_i^*}(ID_i^*, SID_j, M_1, T_1)$, where u_i is a random number and T_1 is a timestamp. After that, the device sends the message $MES_1 = \{DID_i^*, M_2\}$ to the gateway via the public channel.

Step AP2: On receiving $\{DID_i^*, M_2\}$, the gateway recovers the identity of the user by calculating $(ID_i', r_i') = D_{K_{GWN}}(DID_i^*)$, then computes $S_i' = h(ID_i' \parallel h(K_{GWN}))$ and $(ID_i^*, SID_j, M_1, T_1) = D_{S_i'}(M_2)$. The gateway generates the current timestamp T_1^* and verifies whether $|T_1^* - T_1| \leq \Delta T$ and $ID_i^* = ID_i'$. If not, the gateway aborts this session. Otherwise, the gateway generates the timestamp T_2 and computes $b_j = h(SID_j \parallel h(K_{GWN}))$ and $M_3 = E_{b_j}(DID_i^*, SID_j, M_1, T_2)$. Then, the message $MES_2 = \{M_3\}$ is transmitted to the sensor node.

Step AP3: After MES_2 is received, the sensor node calculates $\left(DID_i^*, SID_j^*, M_1, T_2\right) = D_{b_j}(M_3)$ and verifies whether $|T_2^* - T_2| \leq \Delta T$ and $SID_j^* = SID_j$. If not, the sensor node terminates this session. Otherwise, the sensor node generates a random number c_j and timestamp T_3, then computes $M_4 = c_j \cdot M_1$, $M_5 = c_j \cdot P$, $SK = h(DID_i^* \parallel SID_j \parallel M_4 \parallel T_3)$, $V_1 = h(SK \parallel SID_j \parallel T_3)$, and $M_6 = E_{b_j}(M_1, V_1, M_5, T_3)$. Then, the sensor node transmits the message $MES_3 = \{M_6\}$ to the gateway.

Step AP4: On receiving the message MES_3, the gateway calculates $(M_1', V_1, M_5, T_3) = D_{b_j}(M_6)$ and checks whether $|T_3^* - T_3| \leq \Delta T$ and $M_1' = M_1$. If not, the gateway aborts the session. Otherwise, the gateway generates a random number r_i'' and a timestamp T_4. Then, the gateway updates the temporary identity of the user by computing $DID_i^{new} = E_{K_{GWN}}(ID_i^*, r_i'')$. Finally, M_7 is transmitted to the user as the message MES_4, where $M_7 = E_{S_i'}(DID_i^{new}, M_1', M_5, V_1, T_3, T_4)$ is computed by the gateway.

Step AP5: After receiving M_7, the user's device calculates $(DID_i^{new}, M_1', M_5, V_1, T_3, T_4) = D_{S_i^*}(M_7)$ and verifies whether $|T_4^* - T_4| \leq \Delta T$ and $M_1' = M_1$. If not, the session is terminated. Otherwise, the device calculates $M_8 = u_i \cdot M_5$, $SK' = h(DID_i^* \parallel SID_j \parallel M_8 \parallel T_3)$, and $V_1' = h(SK' \parallel SID_j \parallel T_3)$. If $V_1' \neq V_1$, the device aborts the session. Otherwise, the smart card updates the HID_i with HID_i^{new}, where $HID_i^{new} = h(S_i^* \parallel PW_i^* \parallel \sigma_i^*) \oplus DID_i^{new}$.

Figure 4 shows the authentication and session key agreement phases.

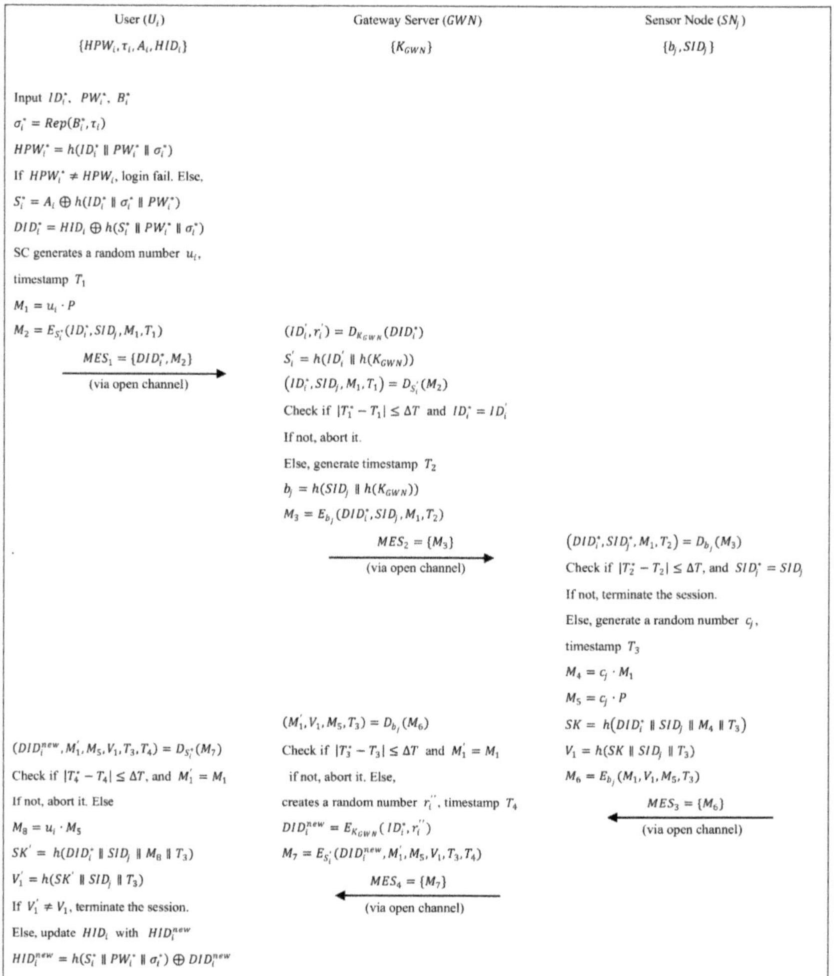

Figure 4. Authentication and session key agreement.

4. Security Analysis

In this section, the security of the proposed protocol is analyzed and proven. Methods used for security proof and validation include the random oracle model, BAN logic, ProVerif, AVISPA, etc. The proposed protocol is based on computational the Diffie–Hellman problem and symmetric encryption; hence, the Find–Guess Model is adopted to prove the security of symmetric encryption, and is combined with the random oracle model to formally prove the security of the proposed protocol.

4.1. Security Model and Proof

4.1.1. Security Model

Definition 1 (Participants): *There are three participants (P) in the proposed scheme (denoted as* Π*): the user, the gateway, and the sensor node, denoted as* U, GWN, *and* SN, *respectively. For the i-th instance, they can be recorded as* I^i_U, I^i_{GWN}, *and* I^i_{SN}, *which are collectively known as* I^i_P.

Definition 2 (Oracle states): *In the proposed scheme, the oracle has only three states:* $accept$, $reject$,*and* $\perp$. $accept$ *represents that an oracle receives a correct request. If the request is illegal, the oracle is in* $reject$. *If the above conditions have not occurred, the state of the oracle is* $\perp$. *We identify that if the state of the oracle* I^i_U *(or* I^i_{SN}*) is* $accept$ *and the session key denoted as* SK^i_U (SK^i_{SN}) *is negotiated already,* I^i_U *(or* I^i_{SN}*) obtains a session identity* Sid^i_U *(or* Sid^i_{SN}*) and corresponding partner identity* $Pid^i_{SN}(Pid^i_U)$.

Definition 3 (Partnering): *If the states of* I^i_U *and* I^i_{SN} *are* $accept$ *and the session key has been negotiated between the user and the sensor node,* I^i_U *and* I^i_{SN} *are considered partners. Meanwhile, they meet the following conditions:*

(1) $I^i_U = Pid^i_{SN}$ and $I^i_{SN} = Pid^i_U$,
(2) $Sid^i_U = Sid^i_{SN} \neq NULL$,
(3) $SK^i_U = SK^i_{SN}$.

Definition 4 (Queries): *Queries are listed as follows to simulate various attacks.*

- *Send* (I^i_P, MES_i): This query simulates an adversary A sending a message to an oracle I^i_P. If the message is correct, the oracle responds with A based on Π. Otherwise, I^i_P ignores the message.
- *Execute* (I^i_P): An execute query represents a passive attack; if the query is executed, A receives all the messages transmitted openly.
- *Reveal* (I^i_P): If I^i_U and I^i_{SN} have negotiated the session key, and all are in the *accept* state while A has not launched a *Test* query, the *Reveal* query reveals the session key SK^i; otherwise, it reveals $null$.
- *Corrupt* (I^i_U): This query offers the credentials of the user. In the proposed scheme, A executes the *Corrupt* query to obtain the information $\{HPW_i, \tau_i, A_i, HID_i\}$ stored in a smart card.
- *Test* (I^i_P): This query can only be executed once by A. If the session key has not been generated, A obtains $null$. Otherwise, the *Test* query creates a random bit r. If $r = 1$, the correct session key is sent to A; otherwise, A obtains a random number.
- *Hash* (*string*): This query outputs the hash value of the input *string*.
- *SymmetricEncryption* (*string1*, *string2*): The output of this query is the symmetric encryption value, where *string1* is the symmetric key and *string2* is the input.

Definition 5 (Freshness): *If an instance* I^i_P *satisfies the following conditions, it can be regarded as freshness.*

(1) The *Corrupt* query can only be executed once at most. (2) The *Reveal* query has not been executed yet. (3) The states of I^i_U and I^i_{SN} are $accept$.

Definition 6 (Semantic Security): *In the random oracle model, A is allowed to execute a Test query once at most, and can use multiple Execute, Send, and Reveal queries. According to the definition of the Test query, the random bit r determines the correctness of the returned session key. Meanwhile, A generates a random bit r′, and if r′ = r, then A knows the correctness of the output. In this case, the semantic security of Π is broken. The possibility of breaking the semantic security is portrayed as* $Adv_{\Pi}^{A} = |2\Pr[r = r'] - 1| = |2\Pr[succ(A)] - 1|$. *Π is not secure unless* $Adv_{\Pi}^{A} < \eta$, *where η is sufficiently small.*

Definition 7 (CDHP&ECDLP): *CDHP refers to the given generator point P of an elliptic curve aP and bP, where a, b* $\in Z_p$. *Computing abP is computationally infeasible for A in probabilistic polynomial time (PPT). The advantage of solving CDHP in PPT can be described as* $Adv_{A}^{CDHP} = Pr[A(P, aP, bP) = abP : P \in E(F_p); a, b \in Z_p]$, $Adv_{A}^{CDHP} < \eta$.

Given P and aP, it is computationally infeasible to calculate a. This is called the Elliptic Curve Discrete Logarithm Problem (ECDLP). The probability of solving ECDLP in PPT is expressed as $Adv_{A}^{\mathrm{ECDLP}} = \Pr[A(P, aP) = a : P \in E(F_p); a \in Z_p]$. $Adv_{A}^{\mathrm{ECDLP}} < \eta$.

Definition 8 (Symmetric Encryption & Find-Guess Model): *Suppose a symmetric encryption scheme denoted as* $\Gamma = (E, D, KSP, MSP)$, *where E is the encryption algorithm, D is the decryption algorithm, KSP(k) and MSP(k) are finite sets, and* $KSP(k),\ MSP(k) \subseteq \{0,1\}^{l}$, $k \in \mathbb{N}$. *For any* $k \in \mathbb{N}$, *given* $a \in KSP(k)$, $x \in MSP(k)$, *the result of encryption* $y = E_a(x)$, *and* $x = D_a(y)$.

The Find–Guess model is a security notion for symmetric encryption [30]. In the Find stage, an adversary A adaptively produces distinct messages x_0, x_1, and information i. In the Guess stage, the encryption oracle selects $a \in {}_R KSP(k)$ as the symmetric key, $b \in {}_R\{0,1\}$, and calculates $y = E_a(x_b)$. A does not know a. The advantage of A guessing b from y is defined as follows:

$$Adv_{A,\Gamma}^{\mathrm{FG}}(k) = 2\Pr[a \leftarrow {}_R KSP(k); (x_0, x_1, i) \leftarrow A(Find); b \leftarrow {}_R\{0,1\}; y = E_a(x_b) : A(Guess, i, y) = b] - 1 < \eta.$$

It should be noted that $x_0 \neq x_1$ and $x_0, x_1 \in MSP(k)$. Here, we defined the adversary $A(PPT, \eta)$ breaking through Γ in the sense of the Find–Guess model if A runs it in Probabilistic Polynomial Time (PPT) and $Adv_{A,\Gamma}^{\mathrm{FG}}(k) \geq \eta$, where η is sufficiently small. Otherwise, Γ is (PPT, η)-secure.

Furthermore, in the proposed scheme, the inputs of the encryption algorithm are mixed with random numbers and timestamps; the adversary A does not know the inputs or the symmetric keys. Therefore, given $a \in {}_R KSP(k)$, $m \in {}_R MSP(k)$, and $x_m = E_a(m)$, the advantage of calculating m from x_m by A is described as follows:

$$Adv_{A}^{\mathrm{SE}}(k) = \Pr[A(x_m) = m : a \leftarrow {}_R KSP(k); m \leftarrow {}_R MSP(k); x_m = E_a(m)]$$

Therefore, $Adv_{A}^{\mathrm{SE}}(k) < Adv_{A,\Gamma}^{\mathrm{FG}}(k) < \eta$.

4.1.2. Security Proof

Theorem 1: *We define an adversary A to attack the proposed scheme Π;* q_{hash}, q_{SE}, q_{send}, *and* q_{exe} *are the executed times of the hash operation, symmetric encryption, and the Send and Execute queries, respectively, while* l_{out}, l_{in}, *and* l_k *are the bit lengths of the output, input, and symmetric key of the symmetric encryption, respectively, the biometric key of the user is* l_{bio} *bits, and C′ and s′ are the regression constants of the distributed password dictionary. The advantage of breaking Π by A in PPT* is

$$Adv_{A}^{\Pi} \leq 2q_{hash} \cdot Adv_{A}^{SE}(k) \cdot Adv_{A}^{CDHP} + \frac{q_{SE}^{2}}{2^{l_k + l_{in}}} + \frac{(q_{send} + q_{exe})^{2}}{2^{l_{out}}} + 2C' \cdot q_{send}^{s'} \cdot \frac{q_{send}}{2^{l_{bio}}}$$

Proof : The goal of the adversary A is to break the security of Π in PPT. To simulate the attacks executed by A, we can define various games, which are donated as $Game_i$ $(0 \leq i \leq 4)$. The events Eve_i $(0 \leq i \leq 4)$ signify that a random bit r of the query *Text* is obtained by A in $Game_i$. The games are described as follows.

Game$_0$: This game simulates a real attack on Π implemented by A. In the beginning, A has to guess the random bit r. Therefore, we have

$$Adv_A^{\Pi} = |2\Pr[Eve_0] - 1| \tag{1}$$

Game$_1$: In this game, an eavesdropping attack is simulated following the *Execute* query. Meanwhile, A is permitted to execute the *Test* query once at most. After obtaining the output of the *Test* query and the transmitted messages, A must determine whether the output is the session key. In the proposed scheme, M_6 and M_7 are related to the session key, where $M_6 = E_{b_j}(ID_i^*, M_1, V_1, M_5, T_3)$, $M_7 = E_{S_i'}(DID_i^{new}, M_1', M_5, V_1, T_3, T_4)$, and $SK = h(ID_i^* \parallel SID_j \parallel c_j \cdot u_i \cdot P_{ec} \parallel T_3)$. First, A cannot decrypt M_6 and M_7 without the symmetric key. Even if A knows the encrypted information in M_6 and M_7, he/she cannot compute $c_j \cdot u_i \cdot P_{ec}$ in PPT because of the CDHP and hash function. Therefore, A cannot determine whether the output is the session key, and we have

$$\Pr[Eve_0] = \Pr[Eve_1] \tag{2}$$

Game$_2$: In the proposed scheme, all the transmitted messages are encrypted using symmetric encryption. This game simulates A executing *Execute* queries in an attempt to break the symmetric encryption and calculate the session key. The advantage of breaking the symmetric encryption by A is $Adv_A^{\mathrm{SE}}(k)$. The session key $SK = h(ID_i^* \parallel SID_j \parallel c_j \cdot u_i \cdot P_{ec} \parallel T_3)$, that is, even if A knows all the encrypted information $\{ID_i, r_i', SID_j, M_1, T_1, T_2, V_1, M_5, T_3, DID_i^{new}, T_4\}$ after breaking the symmetric encryption, he/she cannot calculate the session key using $M_5 = c_j \cdot P_{ec}$ and $M_1 = u_i \cdot P_{ec}$ because of CDHP. Therefore, we have

$$\Pr[Eve_2] - \Pr[Eve_1] \leq q_{hash} \cdot Adv_A^{\mathrm{SE}}(k) \cdot Adv_A^{\mathrm{CDHP}} \tag{3}$$

This shows that symmetric encryption greatly reduces the probability of the adversary winning *Game*$_2$.

Game$_3$: The *Execute*, *Send*, and *SymmetricEncryption* queries are executed to simulate an active game committed to implementing collision attacks in the transmitted messages. The length of the output of the encryption algorithm is based on the input length, and the relationship between them is $l_{out} = 128 \cdot \frac{l_{in}}{64} \cdot$ bits, where l_{in} is the bit-length of the input and l_{out} is the bit-length of the output. According to the birthday paradox, the collision probability of the result of symmetric encryption is at most $\frac{q_{SE}^2}{2^{l_k + l_{in} + 1}}$, where l_k is the bit-length of the symmetric key and q_{SE} is the execute number of the symmetric encryption. The probability of collisions in the transcripts is at most $\frac{(q_{send} + q_{exe})^2}{2^{l_{out}+1}}$. Therefore, we have

$$\Pr[Eve_3] - \Pr[Eve_2] \leq \frac{q_{SE}^2}{2^{l_k + l_{in} + 1}} + \frac{(q_{send} + q_{exe})^2}{2^{l_{out}+1}} \tag{4}$$

where q_{send} and q_{exe} are the execution times of the *Send* and *Execute* queries, respectively.

Game$_4$: The *Corrupt* query is executed in this game to simulate a smart card being obtained and analyzed by A, In this case, A knows the card's stored information $\{HPW_i, \tau_i, A_i, HID_i\}$, where $HPW_i = h(ID_i \parallel PW_i \parallel \sigma_i)$, $A_i = S_i \oplus h(ID_i \parallel \sigma_i \parallel PW_i)$, $HID_i = DID_i \oplus h(S_i \parallel PW_i \parallel \sigma_i)$, and τ_i is the reproduction parameter of the fuzzy extractor. Then, A tries to forge $MES_1 = \{DID_i^*, M_2\}$ and execute the *Send* query to impersonate the user, where $DID_i^* = HID_i \oplus h(S_i^* \parallel PW_i^* \parallel \sigma_i^*)$ and $M_2 = E_{S_i^*}(ID_i^*, SID_j, M_1, T_1)$.

Therefore, A has to guess the user's biometric key σ_i^* and password PW_i^*. According to Zipf's law [31], we have

$$\Pr[Event_4] - \Pr[Event_3] \leq C' \cdot q_{send}^{s'} \cdot \frac{q_{send}}{2^{l_{bio}}} \tag{5}$$

where C' and s' are the constants [31] corresponding to the dictionary space of the passwords and l_{bio} is the bit-length of the biometric key.

The probability that the returned session key from the *Test* query is correct is equal to the probability of guessing the right answer of the random bit r. Consequently, we have

$$\Pr[Eve_4] = \frac{1}{2} \tag{6}$$

Taking the formulas from (2) into (1), we obtain

$$Adv_A^{\Pi} = |2\Pr[Eve_1] - 1| \tag{7}$$

That is,

$$\frac{1}{2} Adv_A^{\Pi} = \left|\Pr[Eve_1] - \frac{1}{2}\right| \tag{8}$$

Taking (6) into (8), we have

$$\frac{1}{2} Adv_A^{\Pi} = |\Pr[Eve_1] - \Pr[Eve_4]| \tag{9}$$

Moreover, because $|\Pr[Eve_1] - \Pr[Eve_4]| \leq |\Pr[Eve_3] - \Pr[Eve_4]| + |\Pr[Eve_2] - \Pr[Eve_3]| + |\Pr[Eve_1] - \Pr[Eve_2]|$, we have

$$\frac{1}{2} Adv_A^{\Pi} \leq q_{hash} \cdot Adv_A^{\mathrm{SE}}(k) \cdot Adv_A^{\mathrm{CDHP}} + \frac{q_{SE}^2}{2^{l_k + l_{in} + 1}} + \frac{(q_{send} + q_{exe})^2}{2^{l_{out}+1}} + C' \cdot q_{send}^{s'} \cdot \frac{q_{send}}{2^{l_{bio}}} \tag{10}$$

Now, we can rewrite (10) as

$$Adv_A^{\Pi} \leq 2 q_{hash} \cdot Adv_A^{\mathrm{SE}}(k) \cdot Adv_A^{\mathrm{CDHP}} + \frac{q_{SE}^2}{2^{l_k + l_{in}}} + \frac{(q_{send} + q_{exe})^2}{2^{l_{out}}} + 2C' \cdot q_{send}^{s'} \cdot \frac{q_{send}}{2^{l_{bio}}} \tag{11}$$

□

4.2. Security Analysis

4.2.1. Stolen Verifier Attacks

In the proposed scheme, the gateway does not store any valuable information or verification tables related to the user. Therefore, the proposed scheme can resist stolen verifier attacks.

4.2.2. Off-Line Password Guessing Attacks

Even assuming that an adversary knows the information $\{HPW_i, \tau_i, A_i, HID_i\}$ stored in a smart card, where $HPW_i = h(ID_i \parallel PW_i \parallel \sigma_i)$, he/she cannot verify the correctness of the guessed password PW_i without knowing the biometric key σ_i.

4.2.3. Replay Attacks

As timestamps are varied between any two different sessions, the proposed scheme can resist replay attacks.

4.2.4. Man-In-The-Middle and Impersonation Attacks

Supposing that an adversary tries to impersonate a user, he/she has to forge or replay M_2, where $M_2 = E_{S_i^*}(ID_i^*, SID_j, M_1, T_1)$ and $S_i^* = A_i \oplus h(ID_i^* \parallel \sigma_i^* \parallel PW_i^*) =$

$h(ID_i^* \parallel h(K_{GWN}))$. However, such forgery is impossible without knowing S_i^*. Similarly, the adversary cannot forge M_7 to impersonate the gateway, where $M_7 = E_{S_i'}(DID_i^{new}, M_1', M_5, V_1, T_3, T_4)$.

If the adversary tries to forge $MES_2 = \{M_3\}$ to impersonate the gateway, where $M_3 = E_{b_j}(DID_i^*, SID_j, M_1, T_2)$ and $b_j = h(SID_j \parallel h(K_{GWN}))$, this is infeasible because he/she does not know the symmetric key b_j. Therefore, the proposed scheme is robust against impersonation and man-in-the-middle attacks.

4.2.5. Smart Card Loss Attacks

Supposing that an adversary obtains a smart card and the information $\{HPW_i, \tau_i, A_i, HID_i\}$ stored in the smart card, where $HPW_i = h(ID_i \parallel PW_i \parallel \sigma_i)$, $A_i = S_i \oplus h(ID_i \parallel \sigma_i \parallel PW_i)$, $HID_i = DID_i \oplus h(S_i \parallel PW_i \parallel \sigma_i)$, and τ_i is the reproduction parameter of the fuzzy extractor, the adversary cannot recover any information from the encrypted data without the biometric key σ_i. Consequently, the proposed scheme resists smart card loss attacks.

4.2.6. Sensor Node Capture Attacks

The sensor node stores $\{b_j, SID_j\}$, where $b_j = h(SID_j \parallel h(K_{GWN}))$, SID_j is the identity of the sensor node, b_j is the secret parameter shared between the sensor node and the gateway, and SID_j is public. Supposing that an adversary captures a sensor node and obtains the information $\{b_j, SID_j\}$, he/she can decrypt M_3 by calculating $(DID_i^*, SID_j^*, M_1, T_2) = D_{b_j}(M_3)$, where DID_i^* is the user's temporary pseudo-identity, $M_1 = u_i \cdot P$, and T_2 is the timestamp. As the adversary cannot obtain any information from M_3, robustness against node capture attacks is guaranteed.

4.2.7. Known Session Key Secrecy

The negotiation of session keys in the proposed scheme is based on ECDLP, which is hard to solve because of the intractability of CDHP. The random numbers used for session key agreement in each session are different and unlinkable, meaning that the session keys are as well. Even if the adversary obtains the current session key, this does not provide any information about previous or future session keys.

4.2.8. Perfect Forward Secrecy

In the proposed scheme, the session key is $SK = h(ID_i^* \parallel SID_j \parallel c_j \cdot u_i \cdot P \parallel T_3)$, where c_j and u_i are random numbers generated by the user and the sensor node, respectively. Even if the adversary knows all the long-term keys, including the password PW_i, the biometric key σ_i, and the shared secret parameters S_i and b_j, he/she cannot calculate the current session key or previous session keys due to the intractability of ECDLP. Therefore, the proposed protocol provides perfect forward secrecy.

4.2.9. Desynchronization Attacks

In the proposed scheme, the new temporary identity DID_i^{new} of the user is updated by the gateway. However, the gateway does not store DID_i^{new}. The real identity ID_i' of the user can only be revealed by the gateway through computing $(ID_i' \parallel r_i') = DID_i^* \oplus h(K_{GWN})$. Even if the adversary intercepts or modifies MES_4 that contains DID_i^{new} to prevent the user from updating her/his temporary identity, the user can use their previous temporary identity to pass the authentication. Hence, the proposed scheme can resist desynchronization attacks.

4.2.10. Anonymity and Unlinkability

In the proposed protocol, the user sends a temporary identity $DID_i = E_{K_{GWN}}(ID_i, r_i)$ to the gateway instead of their real identity. The user updates their temporary identity at the end of each session, which is unlinkable because of the random number r_i. Meanwhile, through combination with fresh timestamps and random numbers generated in each session, the messages transmitted are different as well. Therefore, the proposed protocol preserves anonymity and unlinkability.

5. Performance Comparisons

In this section, the proposed scheme is compared with several related schemes (Shuai et al. [15], Xie et al. [16], Masud et al. [22], Kou et al. [23], Butt et al. [24], and Xie et al. [25]) in terms of computational costs and security properties, which are shown in Tables 2 and 3, respectively.

Table 2. Comparison of the computational costs.

Scheme	User	Gateway (Server)	Sensor	Total	Time (ms)
Shuai [15]	$7T_H + 2T_{SE}$	$4T_H$	$10T_H + 2T_{SE}$	$21T_H + 4T_{SE}$	3.688 ms
Xie [16]	$6T_H + 3T_{ECC}$	$7T_H + T_{ECC}$	$4T_H + 2T_{ECC}$	$17T_H + 6T_{ECC}$	16.162 ms
Masud [22]	$3T_H$	$3T_H$	$2T_H$	$8T_H$	0.544 ms
Kou [23]	$7T_H + 2T_{SE}$	$12T_H + 2T_{SE}$	$6T_H$	$25T_H + 4T_{SE}$	3.94 ms
Butt [24]	$3T_H + T_{SE} + 2T_{ECC}$	$12T_H + 2T_{ECC}$	$T_H + 2T_{SE}$	$5T_H + 3T_{SE} + 4T_{ECC}$	12.024 ms
Xie [25]	$8T_H + T_{SE} + 3T_{ECC}$	$7T_H + 2T_{SE} + T_{ECC}$	$5T_H + T_{SE} + 2T_{ECC}$	$20T_H + 4T_{SE} + 6T_{ECC}$	18.606 ms
Ours	$6T_H + 2T_{SE} + 2T_{ECC}$	$4T_H + 6T_{SE}$	$2T_H + 2T_{SE} + 2T_{ECC}$	$12T_H + 10T_{SE} + 4T_{ECC}$	16.42 ms

Table 3. Comparison of security and properties.

Attacks/Properties	Shuai [15]	Xie [16]	Masud [22]	Kou [23]	Butt [24]	Xie [25]	Ours
Privileged-Insider Attack	✓	✓	✗	✗	✗	✓	✓
Off-line Password Guessing Attack	✓	✓	✗	✓	✗	✓	✓
Impersonation Attack	✓	✓	✗	✓	✗	✓	✓
Replay Attack	✓	✓	✓	✓	✗	✓	✓
Man-in-Middle Attack	✓	✓	✓	✓	✓	✓	✓
Smart Card (Device) Loss Attack	✓	✓	✗	✗	✗	✓	✓
Sensor (Edge) Node Captured Attack	✗	✗	✓	✓	✗	✗	✓
Stolen-Verifier Attack	✗	✓	✗	✓	✓	✓	✓
Update asynchronous Attack	✗	✓	✗	✓	✓	✓	✓
Identity Anonymity	✓	✓	✗	✗	✗	✓	✓
Mutual Authentication	✓	✓	✓	✓	✓	✓	✓
Session key secrecy	✓	✓	✗	✓	✗	✓	✓
Know session key attack	✓	✓	✗	✓	✓	✓	✓
Perfect forward secrecy	✗	✓	✗	✗	✗	✓	✓
Unlinkability	✓	✓	✗	✗	✗	✓	✓

✓: Resist (Attacks)/Possess (Properties). ✗: Suffer (Attacks)/No (Properties)

The time costs of the hash operation, symmetric encryption/decryption, and elliptic curve scalar multiplication are recorded as T_H, T_{SE}, and T_{ECC}, respectively. A Raspberry Pi 4B environment was used to simulate the computational cost of operation on Internet of Things devices in practical applications. According to the test, $T_H = 0.068$ ms (millisecond), $T_{SE} = 0.56$ ms, and $T_{ECC} = 2.501$ ms.

Compared with Butt et al.'s scheme, the computational cost of the proposed protocol is slightly higher, as more hash operations and symmetric encryptions are used to ensure security. Table 3 and Section 3 indicate that the scheme of Butt et al. is vulnerable to multiple attacks and fails to maintain perfect forward secrecy, unlinkability, anonymity, and session key secrecy. The proposed scheme not only resists various attacks, it provides the properties above.

The proposed protocol requires sensors, gateways, and user devices to support point multiplication on elliptic curves and symmetric encryption/decryption. The sensor nodes need to be capable of performing point multiplication on elliptic curves and symmetric encryption and decryption. According to the survey in [29], more and more sensors support symmetric encryption and point multiplication on elliptic curves; thus, this issue around limited computing capacity is expected to be solved in the near future.

6. Conclusions

In this paper, we demonstrate that the protocol of Butt et al. is unable to resist replay attacks, sensor node capture attacks, or off-line password guessing attacks, and that it fails to provide session key secrecy, perfect forward secrecy, anonymity, or unlinkability. The causes of these security issues are presented, and we describe how the same analysis can be adopted more generally as a reference in the design of other related schemes. On this basis, we propose an elliptic curve cryptography-based three-factor authentication protocol for wireless sensor networks in Internet of Things environments. The proposed protocol uses a dynamic anonymous strategy and symmetrical encryption technology, and its security is proven by combining the Find–Guess and random oracle models. Comparisons between the proposed protocol and several related protocols show that the proposed protocol achieves higher security with acceptable computational cost; thus, it is suitable for wireless sensor network applications in Internet of Things environments.

Author Contributions: Conceptualization, Q.X. and Z.D.; methodology, Q.X. and Z.D.; validation, Q.X. and Z.D.; formal analysis, Z.D.; investigation, Q.X.; writing—original draft preparation, Q.X. and Z.D.; writing—review and editing, Q.X.; supervision, Q.X.; project administration, Q.X.; funding acquisition, Q.X. All authors have read and agreed to the published version of the manuscript.

Funding: This research was supported by the National Natural Science Foundation of China (Grant No. U21A20466).

Institutional Review Board Statement: Not applicable.

Informed Consent Statement: Not applicable.

Data Availability Statement: Not applicable.

Conflicts of Interest: The authors declare no conflict of interest.

References

1. Abdollahi, A.; Rejeb, K.; Rejeb, A.; Mostafa, M.M.; Zailani, S. Wireless Sensor Networks in Agriculture: Insights from Bibliometric Analysis. *Sustainability* **2021**, *13*, 12011. [CrossRef]
2. Azrour, M.; Mabrouki, J.; Guezzaz, A.; Kanwal, A. Internet of Things security: Challenges and key issues. *Secur. Commun. Netw.* **2021**, *2021*, 5533843. [CrossRef]
3. Wong, K.H.; Zheng, Y.; Cao, J.; Wang, S. A dynamic user authentication scheme for wireless sensor networks. In Proceedings of the IEEE International Conference on Sensor Networks, Ubiquitous, and Trustworthy Computing (SUTC'06), Taichung, Taiwan, 5–7 June 2006; Volume 1, p. 8.
4. Das, M.L. Two-factor user authentication in wireless sensor networks. *IEEE Trans. Wirel. Commun.* **2009**, *8*, 1086–1090. [CrossRef]
5. Khan, M.K.; Alghathbar, K. Cryptanalysis and security improvements of 'two-factor user authentication in wireless sensor networks'. *Sensors* **2010**, *10*, 2450–2459. [CrossRef] [PubMed]
6. Xie, Q.; Wong, D.S.; Wang, G.; Tan, X.; Chen, K.; Fang, L. Provably Secure Dynamic ID-based Anonymous Two-factor Authenticated Key Exchange Protocol with Extended Security Model. *IEEE Trans. Inf. Secur.* **2017**, *12*, 1382–1392. [CrossRef]
7. Chaudhry, S.A.; Yahya, K.; Garg, S.; Kaddoum, G.; Hassan, M.M.; Zikria, Y.B. LAS-SG: An Elliptic Curve-Based Lightweight Authentication Scheme for Smart Grid Environments. *IEEE Trans. Ind. Inform.* **2023**, *19*, 1504–1511. [CrossRef]
8. Irshad, A.; Chaudhry, S.A.; Sher, M.; Alzahrani, B.A.; Kumari, S.; Li, X.; Wu, F. An Anonymous and Efficient Multiserver Authenticated Key Agreemen t With Offline Registration Centre. *IEEE Syst. J.* **2019**, *13*, 436–446. [CrossRef]
9. Turkanovic, M.; Holbl, M. An improved dynamic password-based user authentication scheme for hierarchical wireless sensor networks. *Elektron. Elektrotechnika* **2013**, *19*, 109–116. [CrossRef]
10. Yuan, J.; Jiang, C.; Jiang, Z. A biometric-based user authentication for wireless sensor networks. *Wuhan Univ. J. Nat. Sci.* **2010**, *15*, 272–276. [CrossRef]

11. Yoon, E.J.; Yoo, K.Y. A new biometric-based user authentication scheme without using password for wireless sensor networks. In Proceedings of the 2011 IEEE 20th International Workshops on Enabling Technologies: Infrastructure for Collaborative Enterprises, Paris, France, 27–29 June 2011; pp. 279–284.
12. He, D. Robust Biometric-Based User Authentication Scheme for Wireless Sensor Networks. Cryptology ePrint Archive, 2012. Available online: https://eprint.iacr.org/2012/203 (accessed on 13 April 2012).
13. Chen, C.T.; Lee, C.C.; Lin, I.C. Efficient and secure three-party mutual authentication key agreement scheme for WSNs in IoT environments. *PLoS ONE* **2020**, *15*, e0232277.
14. Bin, H.; Tang, W.; Xie, Q. A Two-factor Security Authentication Scheme for Wireless Sensor Networks in IoT Environments. *Neurocomputing* **2022**, *500*, 741–749.
15. Shuai, M.; Yu, N.; Wang, H.; Xiong, L.; Li, Y. A lightweight three-factor Anonymous authentication scheme with privacy protection for personalized healthcare applications. *J. Organ. End User Comput. JOEUC* **2021**, *33*, 1–18. [CrossRef]
16. Xie, Q.; Ding, Z.; Hu, B. A secure and privacy-preserving three-factor anonymous authentication scheme for wireless sensor networks in Internet of Things. *Secur. Commun. Netw.* **2021**, *2021*, 4799223. [CrossRef]
17. Hankerson, D.; Menezes, A.J.; Vanstone, S. *Guide to Elliptic Curve Cryptography*; Springer Science & Business Media: Berlin/Heidelberg, Germany, 2006.
18. Boneh, D. The decision diffie-hellman problem. In Proceedings of the Algorithmic Number Theory: Third International Symposiun, ANTS-III, Portland, OR, USA, 21–25 June 1998; pp. 48–63.
19. Irshad, A.; Chaudhry, S.A.; Ghani, A.; Mallah, G.A.; Bilal, M.; Alzahrani, B.A. A low-cost privacy preserving user access in mobile edge computing framework. *Comput. Electr. Eng.* **2022**, *98*, 107692. [CrossRef]
20. Fan, K.; Zhu, S.; Zhang, K.; Li, H.; Yang, Y. A lightweight authentication scheme for cloud-based RFID healthcare systems. *IEEE Netw.* **2019**, *33*, 44–49. [CrossRef]
21. Almulhim, M.; Zaman, N. Proposing secure and lightweight authentication scheme for IoT based E-health applications. In Proceedings of the 2018 20th International Conference on Advanced Communication Technology (ICACT), Chuncheon, Republic of Korea, 11–14 February 2018; pp. 481–487.
22. Masud, M.; Gaba, G.S.; Choudhary, K.; Hossain, M.S.; Alhamid, M.F.; Muhammad, G. Lightweight and anonymity-preserving user authentication scheme for IoT-based healthcare. *IEEE Internet Things J.* **2021**, *9*, 2649–2656. [CrossRef]
23. Kou, L.; Shi, Y.; Zhang, L.; Liu, D.; Yang, Q. A lightweight three-factor user authentication protocol for the information perception of IoT. *CMC-Comput. Mater. Contin.* **2019**, *58*, 545–565. [CrossRef]
24. Butt, T.M.; Riaz, R.; Chakraborty, C.; Rizvi, S.S.; Paul, A. Cogent and energy efficient authentication protocol for wsn in iot. *Comput. Mater. Contin.* **2021**, *68*, 1877–1898.
25. Xie, Q.; Li, K.; Tan, X.; Han, L.; Tang, W.; Hu, B. A secure and privacy-preserving authentication protocol for wireless sensor networks in smart city. *EURASIP J. Wirel. Commun. Netw.* **2021**, *2021*, 119. [CrossRef]
26. Ouni, R.; Saleem, K. Framework for Sustainable Wireless Sensor Network Based Environmental Monitoring. *Sustainability* **2022**, *14*, 8356. [CrossRef]
27. Chaganti, R.; Mourade, A.; Ravi, V.; Vemprala, N.; Dua, A.; Bhushan, B. A Particle Swarm Optimization and Deep Learning Approach for Intrusion Detection System in Internet of Medical Things. *Sustainability* **2022**, *14*, 12828. [CrossRef]
28. Dolev, D.; Yao, A. On the security of public key protocols. *IEEE Trans. Inf. Theory* **1983**, *29*, 198–208. [CrossRef]
29. Wang, C.; Wang, D.; Tu, Y.; Xu, G.; Wang, H. Understanding node capture attacks in user authentication schemes for wireless sensor networks. *IEEE Trans. Dependable Secur. Comput.* **2020**, *19*, 507–523. [CrossRef]
30. Fujisaki, E.; Okamoto, T. Secure integration of asymmetric and symmetric encryption schemes. In Proceedings of the Advances in Cryptology—CRYPTO'99: 19th Annual International Cryptology Conference, Santa Barbara, CA, USA, 15–19 August 1999; Springer: Berlin/Heidelberg, Germany; pp. 537–554.
31. Wang, D.; Cheng, H.; Wang, P.; Huang, X.; Jian, G. Zipf's law in passwords. *IEEE Trans. Inf. Secur.* **2017**, *12*, 2776–2791. [CrossRef]

sustainability

MDPI

Article

A Multi-Criteria Analysis Approach to Identify Flood Risk Asset Damage Hotspots in Western Australia

Pornpit Wongthongtham [1], Bilal Abu-Salih [2,3,*], Jeff Huang [4], Hemixa Patel [4] and Komsun Siripun [5]

1 School of Arts & Sciences, The University of Notre Dame Australia, Fremantle, WA 6160, Australia; ponnie.clark@nd.edu.au
2 King Abdullah II School of Information Technology, The University of Jordan, Amman 11972, Jordan
3 School of Management and Marketing, Faculty of Business and Law, Curtin University, Bentley, WA 6102, Australia
4 School of Physics, Mathematics and Computing, Computer Science and Software Engineering, The University of Western Australia, Crawley, WA 6009, Australia; jeff.w.key@gmail.com (J.H.); hemixa.patel@uwa.edu.au (H.P.)
5 Main Roads Western Australia, East Perth, WA 6004, Australia; kom.siripun@mainroads.wa.gov.au
* Correspondence: b.abusalih@ju.edu.jo

Abstract: Climate change is contributing to extreme weather conditions, which transform the scale and degree of flood events. Therefore, it is important for relevant government agencies to effectively respond to both extreme climate conditions and their impacts by providing more efficient asset management strategies. Although international research projects on water-sensitive urban design and rural drainage design have provided partial solutions to this problem, road networks commonly serve unique combinations of urban-rural residential and undeveloped areas; these areas often have diverse hydrology, geology, and climates. Resultantly, applying a one-size-fits-all solution to asset management is ineffective. This paper focuses on data-driven flood modelling that can be used to mitigate or prevent floodwater-related damage in Western Australia. In particular, a holistic and coherent view of data-driven asset management is presented and multi-criteria analysis (MCA) is used to define the high-risk hotspots for asset damage in Western Australia. These state-wide hotspots are validated using road closure data obtained from the relevant government agency. The proposed approach offers important insights with regard to factors influencing the risk of damage in the stormwater management system.

Keywords: multi-criteria analysis; flood modelling; stormwater management system; Western Australia

check for updates

Citation: Wongthongtham, P.; Abu-Salih, B.; Huang, J.; Patel, H.; Siripun, K. A Multi-Criteria Analysis Approach to Identify Flood Risk Asset Damage Hotspots in Western Australia. *Sustainability* **2023**, *15*, 5669. https://doi.org/10.3390/su15075669

Academic Editors: Mourade Azrour, Azidine Guezzaz, Imad Zeroual, Azeem Irshad, Jamal Mabrouki, Said Benkirane and Shehzad Ashraf Chaudhry

Received: 9 February 2023
Revised: 17 March 2023
Accepted: 20 March 2023
Published: 23 March 2023

1. Introduction

Severe flooding has become increasingly widespread in Australia in recent years. Climate change has contributed to the recent prevalence of extreme weather conditions. The magnitude and extent of flood risks has also greatly increased, with unexpected heavy rain that brings fast-moving and rapidly rising stormwater the primary causal factor [1]. Flood disasters have caused significant damage to road and infrastructure assets, resulting in social disruption. Stormwater management is hence an important challenge. Stormwater management is the practice of managing the flow of rainwater or snowmelt runoff in urban and suburban areas, to mitigate the negative impacts of stormwater on the environment and public health [2,3]. Stormwater management is necessary because urbanisation and development have significantly increased the extent of impervious surfaces, such as roads, parking lots, and buildings, which prevent rainwater from infiltrating the ground [4].

Flood risk hotspot analysis is a method used to identify areas that are most vulnerable to flooding. The analysis is based on a combination of data sources, such as historical flood records, topographical information, land use patterns, and climate data. By analysing these factors, flood risk hotspots can be identified and prioritised for flood risk management and disaster preparedness efforts [5,6]. Flood risk hotspot analysis aims to locate the areas experiencing the greatest risk of flood occurrence [7]. Thorough flood risk analysis, by identifying the worst-affected areas, is a vital part of risk control [8]. By implementing a maintenance plan, it is possible to control and reduce the potential harm that flooding can cause to drainage infrastructure. Specifically, at locations with a high risk of flooding, effective measures can be implemented to lower the flood risk and avoid costly asset damage. Flood risk hotspots are identified by considering factors such as climate, terrain, geology, infrastructure layout, and stormwater capacity. To support informed decision making, data-driven and analytics-based methods are utilised in the development of a comprehensive maintenance strategy [9].

Flood risk hotspots are determined by combining flood hazard and drainage asset pressure data [10–12]. In this paper, we focus on data-driven flood modelling that can be used to support informed decision making for professionals who design strategic maintenance plans. Having a strategic management plan is essential for organisations to navigate the complex and dynamic business environment, make informed decisions, and achieve long-term success in vulnerable areas [13–15]. A strategic management plan for stormwater is a critical component of a sustainable and effective stormwater management programme. The plan provides a roadmap for achieving stormwater management goals and objectives, while ensuring that resources are allocated efficiently and effectively [16]. This would result in mitigation or avoidance of floodwater impact and damage. In particular, this study aims to determine flood risk hotspots for Western Australia, using case studies compiled in the Great Southern Region area. As the main factors of risk assessment, flood risk and drainage asset pressure indices (together producing stormwater index) were developed in this research using a GIS-based multi-criteria analysis (MCA) and analytic hierarchy process (AHP). These research methods provide an empirical analysis of stormwater catchment information and stormwater parameters. The results of this research define flood risk hotspot areas in Western Australia. Information systems played an important role in many aspects, including data integration and visualisation in a holistic coherent view, flood risk assessment through data analysis, and data-driven asset management through defect analysis. The outcomes will accommodate governmental agencies' asset management plans with flood-related data integration and data-driven decision support. This study aims to address the research gaps in the literature by introducing the following contributions:

i. This is the first paper that aims to design a heterogeneous and integrated data island for asset management that provides a holistic and coherent view of the current scattered data sources;
ii. This study aims to implement an effective MCA model, which indicates hotspots that embody a high probability of asset damage due to runoff;
iii. The practical impact of this study is to provide relevant governmental agencies with a data-driven asset management system, so that agencies can launch timely and effective responses to extreme climate conditions and their impacts.

The remainder of this paper is organised into five sections. Section 2 provides an overview background of the preliminaries of this research with a focus on relevant state-of-the-art approaches, while Section 3 presents the methodology followed in this paper. The experimental results are discussed in Section 4 followed by a discussion on the key contributions of this paper, while Section 5 discusses the study's limitations. We conclude the paper in Section 6.

2. Literature Review

2.1. Water Resources Models

The Australian Water Resources Assessment Landscape (AWRA-L) model provides credible estimates of landscape water runoff, evapotranspiration, soil moisture, and aquifer recharge across Australia. The AWRA-L model originally aimed to monitor water availability and use due to drought in Australia during the millennium drought, which occurred between 1997 and 2009 [17]. It has now extended the range of applications [18] to include flood prediction. The Australian Bureau of Meteorology (BoM) was given charge for collating water data, and analysing and reporting on water status, through the AWRA-L model [19]. The model runs on a daily timestamp over an approximately 5 km grid simulating the Australian landscape water balance from 1911 until the present day [20]. The WaterDyn model is a national water balance model providing a daily national 5 km grid-based biophysical model of the water stability between the soil and atmosphere [21]. For the WaterDyn model, monthly simulation values were available for between 1900 and 2014 [21]. The CSIRO Atmosphere Biosphere Land Exchange (CABLE) model is a global biogeochemical land-surface model [22].

Computing fluxes in energy, water and momentum fluxes can be carried out using the CABLE model. This model can be also used to calculate generated carbon that occurs between land surfaces and the atmosphere [23]. For the CABLE model, monthly simulation values were available between 1900 and 2013 [20]. Table 1 summarises the model's features.

Table 1. Feature summary of WaterDyn, CABLE, and AWRA-L models.

Water Resources Models	WaterDyn	CABLE	AWRA-L Version 4.5	AWRA-L Version 5	AWRA-L Version 6
Reference(s)	[21]	[22,23]	[17]	[18]	[20]
Organisation(s)	CSIRO, BoM, ABARES	CSIRO, BoM, universities	CSIRO, BoM		
Purposes	Monitoring water balance	Land surface scheme	Water resources reporting, assessment, and monitoring		
Key outputs	Parameter calibration and sensitivity analysis to 6 catchments in Murrumbidgee	Calibration to derived evapo-transpiration (50 catchments within Australia) and flux tower data	Streamflow ~300 catchment	Streamflow ~300 catchment, satellite soil moisture, and evapotranspiration	Streamflow ~300 catchment, satellite soil moisture, evapotranspiration, and deep drainage
Periods for available simulation values	January 1900 to February 2014 (114 years)	January 1900 to December 2013 (114 years)	1911–now (109+ years)		

Frost et al. [20] reported the hydrologic performance of the AWRA-L model against the WaterDyn and CABLE models. The AWRA-L model performs best for streamflow due to better nationwide calibration (approximately 300 catchments across Australia) and conceptual hydrological structure. The AWRA-L model performs similarly to CABLE for profile soil moisture. Overall, streamflow or runoff is the dominant hydrological variable used in surface water resource assessment and is critical to stormwater management. Another key variable is soil moisture, against which AWRA-L also performs well. The AWRA-L model is considered most fit for purpose with its comprehensive calibration. In addition, the AWRA-L model has been continuously developed, with version 6 being the current version at the time of writing. Hence, in this paper, AWRA-L version 6 model data (including raw, input, and output data) are incorporated.

2.2. Flood Prediction Models

Due to the dynamic nature of climate conditions, flood prediction models can be complex, with physical processes and basin behaviour determining the dynamics of water throughout the hydrologic systems. Physically based models have conventionally been used to forecast a diverse range of flooding scenarios through a storm, rainfall/runoff, hydraulic models of flow, etc. Physically based models have a large degree of uncertainty because of the physiographic and geomorphic characteristics of complicated hydrologic systems [24]. To predict hydrological events, various types of hydro-geomorphological monitoring are required and it can be highly challenging to collect comprehensive hydrological parameters. Moreover, the cost of data acquisition and the lack of necessary data have caused inadequacy in applying physically based models. On several occasions, the physically based models failed to generate proper predictions [25]. Likewise, numerical prediction models reporting deterministic calculation have not been wholly reliable due to systematic errors [26]. Model spin-up time problem is also a significant limitation of numerical models [27]. A compromise between computational time regarding the level of precision and vigour [28] overcomes the numerical model limitation by offering more precise predictions with extended lead times [29].

Advanced data-driven models are used as a alternative approach for flood modelling and have gained more popularity than physical and numerical models. Data-driven methods integrate the climate data and hydro-meteorological parameters to deliver insight into flood prediction [24]. The models numerically formulate the flood prediction based on historical data. Machine learning (ML) is one of the data-driven models applied in water resources systems and hydrology. ML, a subset of artificial intelligence (AI), refers to techniques identifying regularities and patterns that enable program systems to learn from experience.

In hydrology, the preferred type of prediction depends on the lead-time requirements. Short-term predictions for floods are often used as warning systems and are not a focus of this project. Rather, long-term predictions are the focus for flood analysis and management purposes. In addition, this project is concerned with lead times, regardless of the type of flood. Table 2 shows a review of relevant papers reporting water resources variables, type of flood model (physical, numerical, and/or data-driven flood models), and case study region, against the proposed approach for this project.

Table 2. A review of related papers against this project's proposed approach, which intended to bridge the gap in the literature.

Reference	Water Resources Variable(s)	Region
[25]	River gauge height & weather forecasting	Brisbane, Australia
[30]	Rainfall & river flow	Europe (Italy, Austria)
[31]	Flood	Kelantan, Malaysia
[32]	Flood & asset management	NSW, Australia
[33]	Water storage, infiltration, & runoff	Korea
[34]	Flood	Queensland, Australia
[35]	Flood	Athens, Greece
[26]	Rainfall	Southeast Australia
[36]	Rainfall	Oklahoma, USA
[28]	Flood	Italy, Spain
[37]	Rainfall & runoff	China
[29]	Rainfall & flood forecasting	Korea
[38]	Streamflow & water level	Utah, USA
[39]	River inflow	Iran
[40]	Water levels	Sudan
[41]	Precipitation	Eastern Australia
[42]	Rainfall	India
[43]	Reservoir levels	Turkey
[44]	Streamflow	Canada
[14]	Streamflow	China
[45]	Floodplain forests	Melbourne, Australia
[46]	Streamflow	South-eastern Australia
[47]	Rainfall–runoff	Italy
[48]	Rainfall	Thailand
[49]	Rainfall	South-eastern Australia

Table 2. *Cont.*

Reference	Water Resources Variable(s)	Region
[50]	Streamflow	Turkey
[51]	Flood forecasting	China
[52]	Streamflow	China
[53]	Rainfall–runoff	USA
[54]	Runoff forecast	China
[55]	Streamflow	Iran
Our Study	Rainfall, runoff, Drainage assets	Western Australia

Van et al. [25] reported the failure of the physically based model in predicting floods in Queensland, Australia in 2011. The authors then suggested developing advanced data-driven models in a spatial and temporal pattern. Flood risk management was also discussed to measure flood risk by way of flood maps and identify high-risk areas. Shakirah et al. [31] documented the intensity/duration of precipitation that contributed to extreme hydrological events. The intense rainfall event, as well as the flood events' patterns, re demonstrate an escalating trend. Hydrological flood events can be assessed, and relevant parameters can be extracted and used, to build flood forecasting models. This is particularly important as it leads to better stormwater management and planning. In this direction, the implications of rain and flood events in Queensland are analysed using sensitivity analyses; hence, the cost is measured in major flooding occurrences [34]. Kim et al. [33] focused on the necessity to improve flood detention practices and storage mechanisms, thereby mitigating flood impacts in densely developed areas and developing a hydraulic model for stormwater management.

Hydraulic simulation systems allow water operators to update or change gate opening and reservoir/canal discharge strategies based on the real-time status of the irrigation system. Nevertheless, the use of these types of technologies is not suitable in harsh environments, e.g., agricultural areas, because the automation system is prone to carry along distortions or errors in the measurements, along with the gates, canals, and other structures that can affect the accuracy of the simulation results. Torres et al. [29] developed coupled ML and hydraulic simulation systems to reduce the impact of the above-mentioned errors. Kenley et al. [32] raised road network damage from the extreme flood events. Effective infrastructure management practices are crucial to enhance road network conditions, and will allow agencies to better confront unpredictable climate conditions. A series of case studies were conducted to minimise the life-cycle cost of major flooding and rain events. Kenley et al. [32] proposed a location-based framework to provide an essential concept of collective efforts, ensuring that reactive maintenance activities can be well linked to the outcomes of predictive models.

2.3. *Digital Elevation Model*

The elevation of land plays an important role in determining flood hotspots because the water flow is gravity driven. This causes maximum flood events in low-lying regions [56]. There are various terms used to describe elevation including DEM, Topography, and Elevation contours. Various researchers have used DEM for the delineation of flood-prone areas [57–59].

DEM derivatives are parameters, such as slope, flow accumulation, curvature, topographic Wetness Index (TWI), Stream Power Index (SPI), and Sediment Transport Index (STI). These parameters are obtained from the DEM through respective formulae. These parameters affect important flood characteristics, such as the speed of water, the direction of the flow, and the intensity of the flow. Therefore, these parameters play important role in determining the flood-risk zone [60–62].

2.4. Multi-Criteria Analysis

Multiple factors influence the flow accumulation of water and the risk of damage to assets. Analysis of zones that have a higher probability of such events requires methods that can include a variety of factors. Multi-criteria analysis (MCA) is one decision making tool that can be used for delineating zones of interest in the fields of environment, waste management, and hazard risk zones [63–66]. MCA was created in 1977 to enable the inclusion of multiple criteria based on human understanding [67]. Analytic hierarchical process (AHP), analytic network process (ANP), technique for order of preference by similarity to ideal solution (TOPSIS), outranking, multi-attribute utility theory (MAUT), and multi-attribute value theory (MAVT) are all MCA methods [8]. There is an increasing trend of applying MCA to zoning areas prone to flooding or analysing the high flood risk zone [58,63,66,68]. Ellis et al. [69] applied MCA for supporting the treatment of highways and runoff. Various parameters, including slope, soil permeability, and other asset parameters, were ranked. Despite high interest among researchers in the application of multi-criteria analysis, the validation of results is currently an untouched aspect. This research attempts to validate the high-risk spots with the road closure data.

MCA is a robust and promising technique that can help decision making through the application of GIS technology [70]. This can provide a grid-based map of the high-risk zones for decision making regarding the management of assets in the context of runoff and flooding.

2.5. Current and Future Sustainable Environment

Detecting natural disasters is essential for sustainability because it mitigates the impact of such events on human lives, the environment, and the economy [71]. Timely detection of natural disasters can save lives, reduce property damage, and ensure a more efficient and effective disaster response [72]. For instance, early warning systems for hurricanes, tornadoes, floods, and other natural disasters allow people to evacuate and seek shelter in advance, which can reduce the number of fatalities and injuries [73]. Similarly, detecting earthquakes and tsunamis allows for early deployment of emergency response teams, medical personnel, and rescue teams, allowing them to reach the affected areas quickly [74].

Moreover, detecting natural disasters also allows us to understand the root causes and patterns of such events, which can help prevent or mitigate future occurrences. This understanding can help professionals and organisations develop better risk management strategies, improve infrastructure design, and promote sustainable land-use practices [75]. Detecting natural disasters is critical for ensuring a sustainable future. It allows us to improve preparations for and responses to natural disasters, reduce the impact of such events, and develop strategies to prevent or mitigate future occurrences.

Merging strategic stormwater management with an effective maintenance plan can be very complex, particularly in areas where drainage maintenance is commonly reactive, with limited targeted early intervention. In short, the regional drainage maintenance approach is rather ad hoc, being based on the occurring event without involving climate change risks, asset management resilience, and a sustainability framework. Moreover, current efforts to develop flood risk assessments using GIS-based road networks that provide a long-term solution for vulnerability, asset criticality assessment, and adaptation responses are inadequate [69].

Although international research projects on water-sensitive urban design and rural drainage design provide a partial solution to this problem [76,77], road networks commonly serve unique combinations of urban-rural residential and undeveloped areas; these areas often have diverse hydrology, geology, and climates. Therefore, applying a one-size-fits-all solution is practically ineffective. Although local knowledge and historical data are available in certain regions, there has been limited efforts to consolidate and analyse the data for stormwater management [78,79]. Moreover, from a theoretical perspective, flood risk modelling using computational intelligent techniques is an ongoing and active area of research [80].

Practically, although knowledge sharing and historical data are available across regions in Western Australia, limited efforts have been made to consolidate and analyse the data for stormwater management. Hence the objective of this paper is to identify regions that are likely vulnerable to extensive damage to their assets. In the provided case study, we provide valuable insights into the flood-related parameters in Western Australia that can heighten the risk of water damage to road and stormwater management assets in 2022.

3. Methodology

Figure 1 outlined the conceptual framework for this study, which comprises four stages. In particular, data acquisition, assimilation, and integration were carried out in Stage 1. In Stage 2, the collected data was analysed to contrast runoff data against asset capacity and attributes. Stage 3 aimed to identify risk hotspots through a probabilistic approach. Finally, in Stage 4, the outcome results were validated with defect and road closure data. The Methodology elaborates on these designated stages.

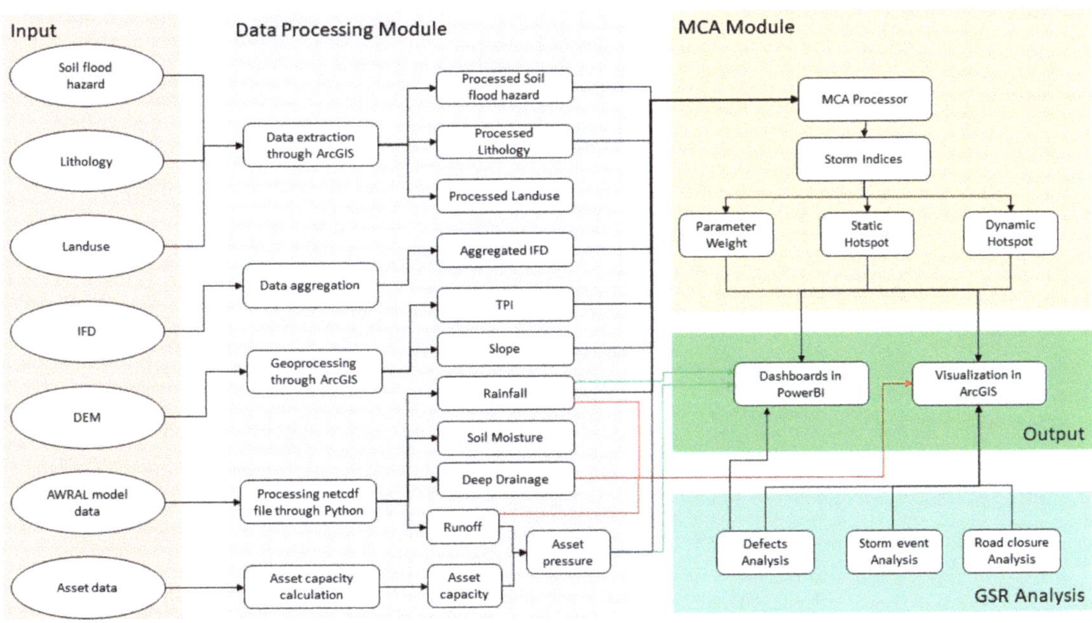

Figure 1. A Conceptual Framework.

3.1. Data Acquisition, Assimilation, and Integration

Data sources. Various data sources were explored, acquired and assimilated. In particular, stormwater management and asset management data were obtained from Main Roads Western Australia (MRWA) (https://www.mainroads.wa.gov.au/, accessed on 20 December 2022). These included asset inventory data of stormwater drainage and capacity (i.e., culverts, state road network, floodway). We also collected publicly available data in relation to hydro-informatics from several sources, including the Bureau of Meteorology (BoM) (http://www.bom.gov.au/, accessed on 20 December 2022), Shuttle Radar Topography Mission (SRTM) (https://dwtkns.com/srtm30m/, accessed on 20 December 2022), and Landgate (https://www0.landgate.wa.gov.au/, accessed on 20 December 2022). Data from BoM included precipitation, soil moisture, surface runoff, slope of land surface, streamflow, topographic, vegetation, deep drainage, and intensity frequency duration. The data collected from Landgate included state-wide flood records. Lastly, data collected from SRTM included the elevation.

The Australian Landscape Water Balance model (AWRA-L) is a high-resolution (5 km × 5 km) hydrological model developed by the Commonwealth Scientific and Industrial Research Organisation (CSIRO), in collaboration with the BoM and other partners. It estimates the water balance components, such as precipitation, evapotranspiration, soil moisture, and runoff, at a daily time step for the entire Australian continent. The model incorporates a range of datasets, including satellite and radar data, ground-based measurements, and hydrological modelling. The collection of spatial data was processed and handled in Geographic Information Systems (GIS) through ArcGIS (https://www.arcgis.com/index.html, accessed on 20 December 2022) to illustrate the geographical data of the case study area.

Data Processing and Preparation: Slope, topographic wetness index (TWI) and topographic position index (TPI) were derived from DEM. These DEM Derivatives were calculated using ArcGIS Pro and the calculation was automated in Model Builder. Figure 2 shows the final workflow of the model builder. Slope was calculated using an inbuilt tool called "slope" in the geoprocessing toolbox.

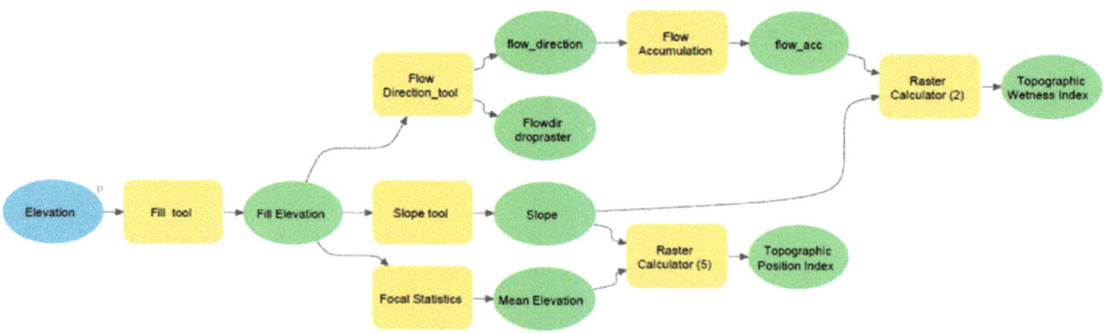

Figure 2. Workflow from model builder in ArcGIS Pro v3.0.1.

TPI was calculated from the focal statistic and raster calculator [64]. TPI is the difference in elevation at the central point (X_0) and the mean elevation across the central point within the local window ($\overline{X}$) (Equation (1)). Furthermore, raster statistics were used to calculate the TPI value using Equation (2) [64]. Mean elevation across the central point within the local window ($\overline{X}$) was calculated using focal statistics. X_i is the elevation value of ith cell and n is the total number of surrounding points employed in the evaluation. The final result produces a tiff file for each pixel showing the topographic position index of that location.

$$\overline{X} = \frac{\sum_i^n X_i}{n} \quad (1)$$

$$TPI = X_0 - \overline{X} \quad (2)$$

TWI was calculated using the raster calculator in ArcGIS Pro. by applying the following formula in Equation (3) [81]. The final result produces a tiff file for each pixel showing the topographic wetness index of that location. A_s is a specific area and slope angle measured in degree (β) from elevation.

$$TWI = \ln\left(\frac{A_s}{\tan\beta}\right) \quad (3)$$

3.2. Pressure on Assets

Pressure on assets was analysed by comparing the capacity of the asset with the runoff received by the asset. Asset capacity was calculated using the dimensions of assets, such as length, radius, height, and slope of culvert or floodway. The flow of water was calculated using the runoff parameters from the AWRAL model. This was used to compare with the capacity of assets. Finally, the pressure was calculated in percentage.

Culvert Capacity: To calculate the capacity (Q) of the culvert, manning's formula was used (Manning 1891). Equation (4) was used to calculate culvert capacity. Manning's coefficient value (n) was obtained using Table 3 for every corresponding type of material in the floodway. Values for flow area (A), wetted perimeter (P), and channel slope (S) were obtained from the dimension data of the culvert.

$$Q = \frac{1}{n} A \left(\frac{A}{P} \right)^{\frac{2}{3}} \sqrt{S} \tag{4}$$

Table 3. Manning's coefficient value for different types of material used in construction.

Material	Mannings Value
Precast reinforced concrete	0.013
Prestressed concrete	0.013
Masonry	0.013
Timber	0.017
Steel	0.01
Plastic	0.009
Aluminium	0.009
In situ reinforced concrete	0.013
Mass concrete	0.015

In Equation (5), a simplified method was used to calculate the floodway capacity. Here, H is the height of water and L is the length of the floodway. Floodway capacity was calculated using the length of the floodway and assuming the height of 0.3 m.

$$Q = 1.69 H^{3/2} L \tag{5}$$

Furthermore, the catchment data and runoff from AWRA-L model were used to obtain the flow of water at that particular location on the culvert and floodway. The flow of water at the asset's location and the capacity of the asset were compared to calculate the pressure on the asset in percentage terms.

3.3. Develop Risk Hotspots through a Probabilistic Approach

The factors that could cause the floods were also the factors that could lead to a higher flow of water in asset locations. These parameters were considered for the assessment of hotspots that have a high risk of damage to these assets. The risk is calculated from the hazard and vulnerability following the methodology from [63] using Equation (6). In this study, hazard was the probability of water accumulation conditions calculated through elevation, land use, soil, and factors that affect water flow and accumulation. Similarly, vulnerability was the probability of damage to the assets calculated from the assets receiving high pressure.

$$Risk = Hazard * Vulnerability \tag{6}$$

The calculation methodology for hazard and vulnerability used in this study is as follows:

Hazard: Hazard referred to the physical parameters that contribute to the potential damage of an asset. In this study, the hazard was calculated based on morphometric and hydrometeorological parameters, which are sourced from various data portals. These parameters included slope, elevation, catchment area, average rainfall intensity, and rainfall duration. To calculate the hazard, the following steps were followed: (i) derived the morphometric parameters, such as slope, elevation, and catchment area, using SRTM data; (ii) obtained the hydrometeorological parameters, such as average rainfall intensity and rainfall duration, from the BoM data, and; (iii) calculated the hazard by combining the morphometric and hydrometeorological parameters using a weighting factor. The weighting factor was derived from AHP analysis, which involved stakeholder input to assess the relative importance of each parameter.

Vulnerability: Vulnerability referred to the susceptibility of an asset to damage from flood events. In this study, the vulnerability was calculated based on the asset pressure, which was the ratio of the water flow (inundation status) in the location of the asset to the capacity of the asset. To calculate the vulnerability, the following steps were followed; (i) obtained the water flow data (inundation status) from the Australian Landscape Water Balance model (AWRA-L) dataset; (ii) determined the capacity of the asset based on asset inventory data regarding stormwater drainage and capacity obtained from MRWA; (iii) calculated the asset pressure by dividing the water flow in the location of the asset by the asset capacity, and; (iv) calculated the vulnerability by normalising the asset pressure value using a sigmoid function, which maps the pressure values across a range between 0 and 1. The sigmoid function was used to model the nonlinear relationship between the asset pressure and the vulnerability. By combining the hazard and vulnerability calculations, a risk map was generated that identified the high-risk hotspots for asset damage from flooding.

Level 1 multi-criteria analysis: The MCA approach was used to create thematic layers of various contributing parameters to generate the risk hotspot map. These parameters for hazard calculation included elevation, slope, aspect, curvature of land, land use, runoff, rainfall, aquifer, and streamflow. Similarly, vulnerability was calculated using asset pressure. This method was the Level 1 option for choosing the parameters based on criteria. In Level 2, the analytical hierarchy process was used for assigning relative importance weight to every parameter. This weight was obtained by creating a pairwise comparison matrix [63]. This matrix was flexible because of its amenability and consideration of the interest of stakeholders by allowing them to input their views and experiences [63].

Level 2 analytical hierarchy process: As mentioned above, this process generated a pairwise comparison matrix [82]. Every parameter was listed in the first column and first row. The value of importance was assigned using the Saaty scale [82]. The format for calculating the pairwise comparison matrix is given in Table 4. Higher weight was given to the most important parameters.

Table 4. Pairwise comparison matrix calculation.

	Lithology	Land Use	TWI	Soil	TPI	Elevation	Slope	IFD	EE	RWI	RIW* Colum Totals	
Lithology	a_a	a_b	a_c	a_d	a_e	a_f	a_g	a_h	EE_1	RIW_1	RIW_1^*	$\sum n_a$
Land use	b_a	b_b	b_c	b_d	b_e	b_f	b_g	b_h	EE_2	RIW_2	RIW_2^*	$\sum n_a$
TWI	c_a	c_b	c_c	c_d	c_e	c_f	c_g	c_h	EE_3	RIW_3	RIW_3^*	$\sum n_a$
Soil	d_a	d_b	d_c	d_d	d_e	d_f	d_g	d_h	EE_4	RIW_4	RIW_4^*	$\sum n_a$
TPI	e_a	e_b	e_c	e_d	e_e	e_f	e_g	e_h	EE_5	RIW_5	RIW_5^*	$\sum n_a$
Elevation	f_a	f_b	f_c	f_d	f_e	f_f	f_g	f_h	EE_6	RIW_6	RIW_6^*	$\sum n_a$
Slope	g_a	g_b	g_c	g_d	g_e	g_f	g_g	g_h	EE_7	RIW_7	RIW_7^*	$\sum n_a$
IFD	h_a	h_b	h_c	h_d	h_e	h_f	h_g	h_h	EE_8	RIW_8	RIW_8^*	$\sum n_a$
	$\sum n_a$	$\sum n_b$	$\sum n_c$	$\sum n_d$	$\sum n_e$	$\sum n_f$	$\sum n_g$	$\sum n_h$	EE_n		λmax	

In MCA, stakeholders could provide their inputs to create criteria for parameter selection via a pairwise comparison matrix; known as the analytic hierarchy process (AHP), it is a widely used method for capturing stakeholders' preferences and priorities. In our study, stakeholders, including government personnel and academics, were asked to evaluate the importance of each criterion or parameter relative to all the other criteria. Stakeholders were presented with a pairwise comparison matrix, which listed all the criteria in rows and columns; they were then asked to indicate the relative importance of each criterion by assigning numerical values that reflected the strength of the relationship between the criteria.

The numerical values typically ranged from 1 (indicating equal importance) to 9 (indicating very strong importance). The values could also be fractional, such as 1/3, 1/5, etc.; fractional values indicated intermediate levels of importance. Once stakeholders had completed the pairwise comparison matrix, the data were analysed to derive weights for each criterion. These weights reflected the relative importance of each criterion in the decision-making process and were used to weigh the scores of each option in the MCA.

Further estimated eigenvalue (EE) was calculated using Equation (7) [63]. EE_a can be the value for row 1 of the pairwise comparison matrix. Similarly, EE value can be calculated for every row in the pairwise comparison matrix termed as EE_1, EE_2 and EE_n.

$$EE_1 = \sqrt[n]{a_a * a_b * .. * a_n} \tag{7}$$

Here, each element of the row is denoted by the values of a_a, a_b and a_n. Furthermore, relative importance weight (RIW) was calculated using all eigenvalues in pairwise comparison matrix. Equation (8) is used to calculate relative importance weight of each row in the pairwise comparison matrix [63]

$$RIW_1 = \frac{EE_1}{EE_1 + EE_2 + \ldots + EE_n} \tag{8}$$

This matrix was checked for consistency using consistency ratio (C_R). Consistency ratio is calculated from relative consistency index (R_I) and consistency index (C_I), as shown in Equation (9). The relative consistency index depends on the number of parameters (as shown in Table 5) for n = 1, 2 . . . 8 (adapted from [82]).

$$CR = \frac{CI}{RI} \tag{9}$$

Table 5. Random consistency index (RI).

n	1	2	3	4	5	6	7	8
RI	0	0	0.58	0.9	1.12	1.24	1.32	1.41

The consistency index was calculated using Equation (10) [63]. Here λmax is calculated by summing up the product of RIW and the sum of every column representing the criteria. This is shown in the right most column of Table 4.

$$CI = \frac{\lambda max - 1}{n - 1} \tag{10}$$

A consistency ratio of less than 0.1 shows consistent importance is assigned to the parameters in the respective pairwise comparison matrix.

Level 3 data classification and weighting: Data for every parameter consisted of unique values or numerical ranges. For example, the lithology parameter's data was in .shp format (Shapefile). It had a unique value depending on whether rock's classification as sedimentary, metasedimentary, metamorphic, metai, or igneous; this classification was represented by a polygon. Each different type of rock had a different impact on the water flow. Contrastingly, the data of elevation was in .tif format (image file). The values were all numbers ranging from 58 to 864 m. Different elevations all had a different impact on flow of water.

To incorporate this variation into the final hotspot, a pairwise comparison matrix was generated for every parameter. RIW is calculated for every range, while the consistency of that pairwise comparison matrix was confirmed using CI. Steps in level 2 and level 3 were repeated to calculate the vulnerability, using the asset pressure data to generate accurate results. Hazard and vulnerability were calculated using the following equation:

$$Hazard = Reclassified\ Raster\ of\ Hazard\ Parameter * \ RIW\ of\ the\ Parameter \quad (11)$$

$$Vulnerability = Reclassified\ Raster\ of\ Vulnerability\ Parameter * RIW\ of\ the\ Parameter \quad (12)$$

Risk hotspot map: A raster image was generated for each parameter using the original data and relative importance weight. Data was imported into ArcGIS software v.10.8.2. Table 6 shows the source, type of data, and process for each parameter.

Table 6. Multi-criteria analysis parameters processing.

Parameter	Source Data	Source Data Type	Process
Land use	Land use	Vector—Polygon feature	• Road Network line was split using the land use polygons. • New split road network lines were joined with land use data. • Lines were converted to raster using "feature to raster" to generate land use raster
Soil flood hazard	Soil	Vector—Polygon Shapefile	• Road Network line was split using the land use polygons. • New split road network lines were joined with land use data. • Lines were converted to raster using "feature to raster"
Lithology	Lithology	Vector—Polygon feature	• Road Network line was split using the land use polygons. • New split road network lines were joined with land use data. • Lines were converted to raster using "feature to raster"
Elevation	DEM	Raster—Tiff	• Not required
Slope	DEM	Raster—Tiff	• Calculated from DEM using Slope tool
Topographic wetness index	DEM	Raster—Tiff	• Calculated from DEM using raster calculator.
Topographic position index	DEM	Raster—Tiff	• Calculated from DEM using "focal statistics" and raster calculator
Rainfall depth	Australia rainfall and runoff—Intensity frequency duration	Pont file—CSV	• Point file was clipped for road buffer • Converted to raster using "feature to raster"
Asset pressure	Calculation in this project	Pont file—CSV	• Point file was clipped for road buffer • Converted to raster using "feature to raster"

A shapefile of a 50 m road buffer was generated using a buffer tool. We clipped the data using the road buffer shapefile, reducing the processing time by only focusing on the required area. The clipped data were assigned relative importance weight using the results calculated from the pairwise comparison matrix. Using summary statistics, new hazard layer and vulnerability layer data were generated from all parameter raster layers. The final raster data were converted to vector data to make a hotspot map.

3.4. Validation of Hotspot

The risk map was generated using the vulnerability data from 2020; hence, the road closure that occurred in 2020 was screened from the road closure map in Figure 3. The road closure and risk for 2020 can be observed in Figure 3. To understand the temporal behaviour of risk hotspots, the time series analysis of risk and road closure was performed. Figure 3 shows the maps of road closure and risk from January to June 2020. The large size map for all size months can be observed from Figures 4–9. High risk and road closure was identified for Kimberley region, and southern areas of the Mid West-Gascoyne regions. High density of risk spots could be clearly observed around the road closure. However, the road closure in north Kimberley was observed to be persistent for long time duration. Road closure in Mid West-Gascoyne, and high-risk spots, were identified in January 2020 and February 2020.

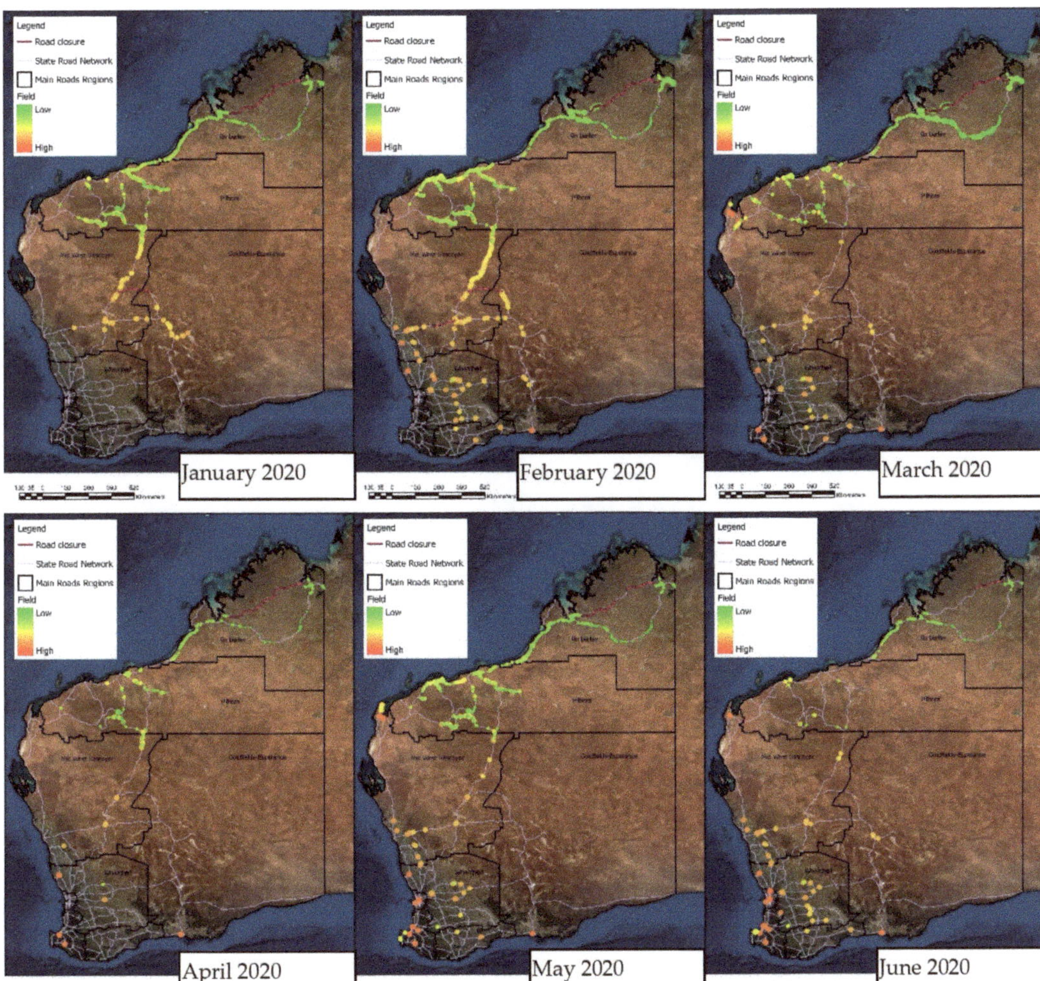

Figure 3. Time series map of risk and road closure from January to June 2020.

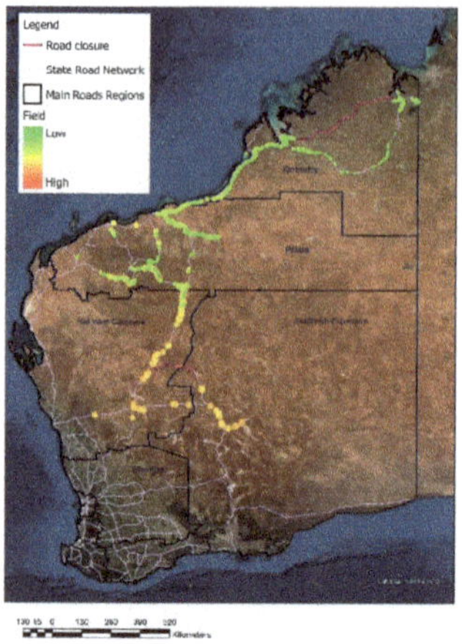

Figure 4. Risk and road closure in January 2020.

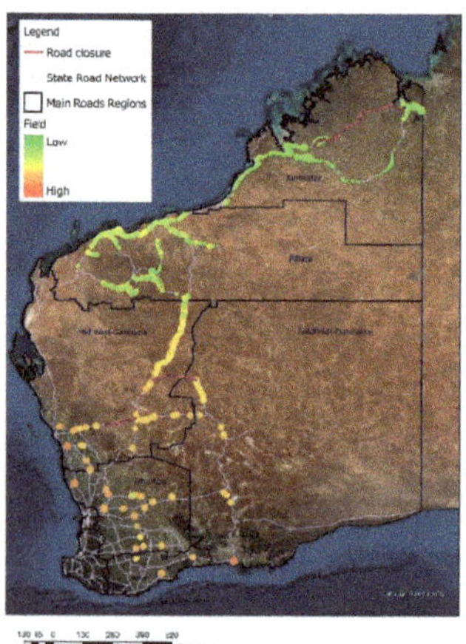

Figure 5. Risk and road closure in February 2020.

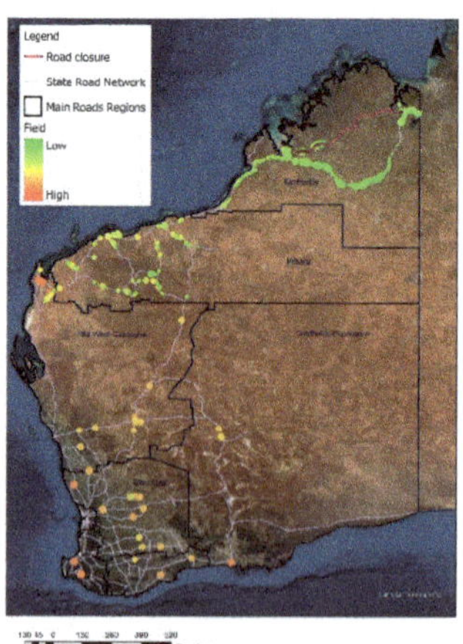

Figure 6. Risk and road closure in March 2020.

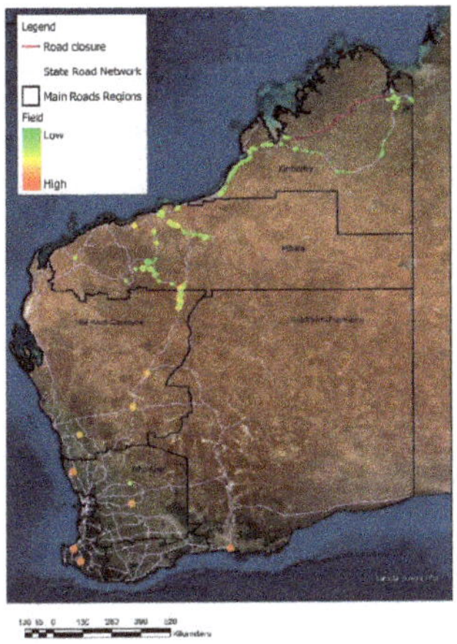

Figure 7. Risk and road closure in April 2020.

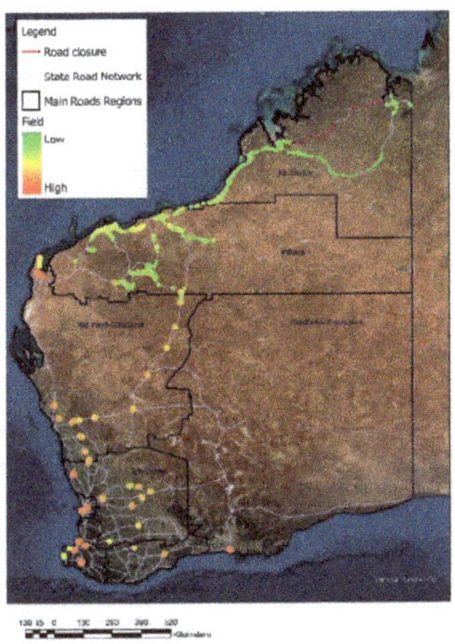

Figure 8. Risk and road closure in May 2020.

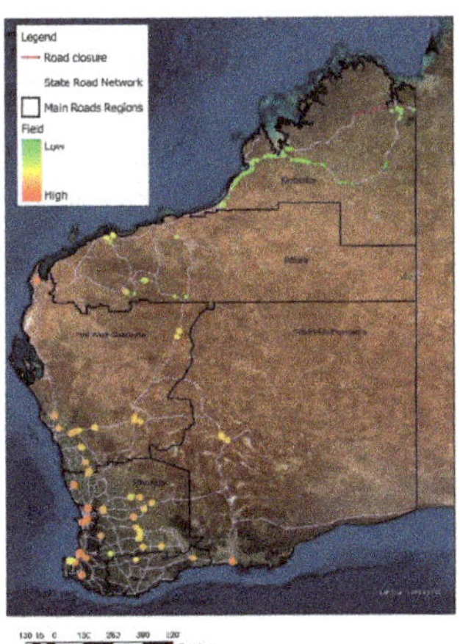

Figure 9. Risk and road closure in June 2020.

Road closures were a useful metric for verifying the identified hotspots of high risk for asset damage in the study because they are a direct consequence of flooding and can be easily tracked and verified. When a road was closed due to flooding, it was an indication that the floodwater had exceeded the capacity of the stormwater drainage system, leading to the flooding of the road and potential damage to the surrounding assets. Thus, road closures could serve as a reliable proxy for identifying areas of potential asset damage due to flooding. By comparing the identified hotspots with road closures, this study helped to verify the accuracy of the risk maps and provided evidence of the effectiveness of the data-driven approach. If the identified hotspots aligned with areas that experience road closures due to flooding, it provided strong evidence that the risk maps are reliable and could be used to prioritise asset management interventions. On the other hand, if there was a mismatch between the identified hotspots and areas of road closure, it may suggest that there were other factors at play for which the model failed to account, or that the model itself needs to be refined.

4. Results and Discussion

4.1. Data Acquisition

To validate the utility of the proposed approach, a case study was been carried out in Western Australia (WA) state. WA is the largest state in Australia, spanning for around 2.5 million square kilometres. Figure 10 includes the regional map of WA.

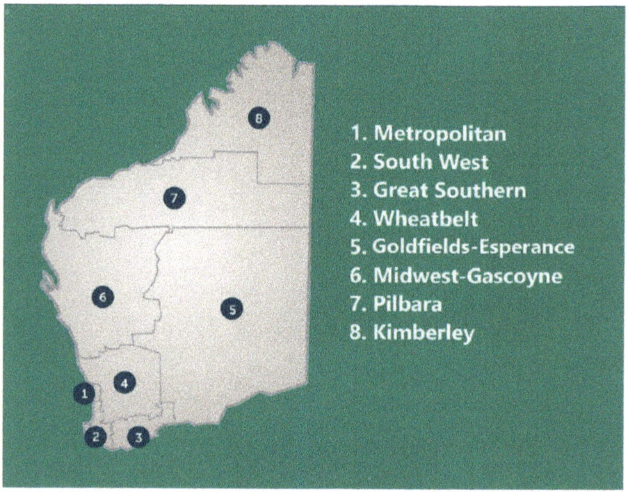

Figure 10. Regional map of Western Australia [83].

Maps for required parameters were generated from data collection, assimilation, and integration. Figure 11 shows the elevation map of Western Australia. Elevation in Western Australia ranges from 66.8 to 1247 m. Higher elevation is observed inland, decreasing gradually towards the coastal area. Moreover, in northern Western Australia, a high peak can be observed. Figure 12 shows the map of the slope of terrain across Western Australia. Slope varies between 0 to 75.3. Slope can be observed as high in north, north-west and southwestern parts of Western Australia, with the elevation map showing stiff parches of land in these areas. The topographic position index shows the presence of a valley or ridge in the terrain. From Figure 13, it can be observed that regions with varying slope values also have a higher variation in TPI values. TWI is found to be constantly varying across study regions, with higher TWI values recorded in inland regions (as depicted in Figure 14). An Inverse relationship between TWI and slope is found. The flat regions are found to have higher TWI values than can be observed from Figures 12 and 13.

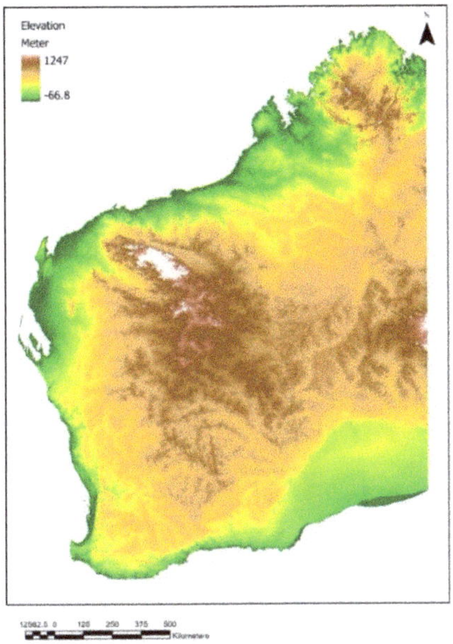

Figure 11. Elevation map of Western Australia.

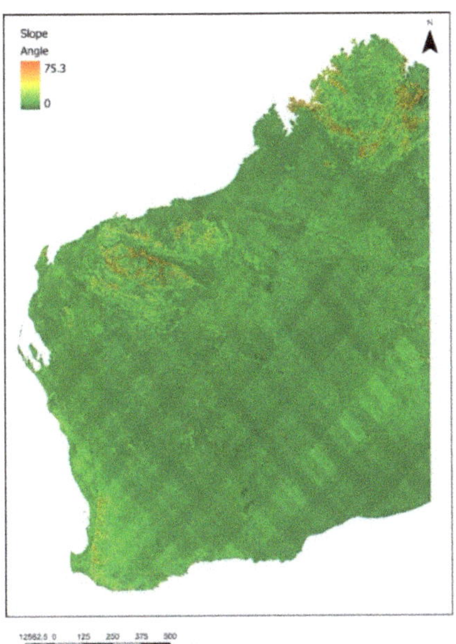

Figure 12. Map of terrain slope of Western Australia.

Figure 13. Topographic position index map of Western Australia.

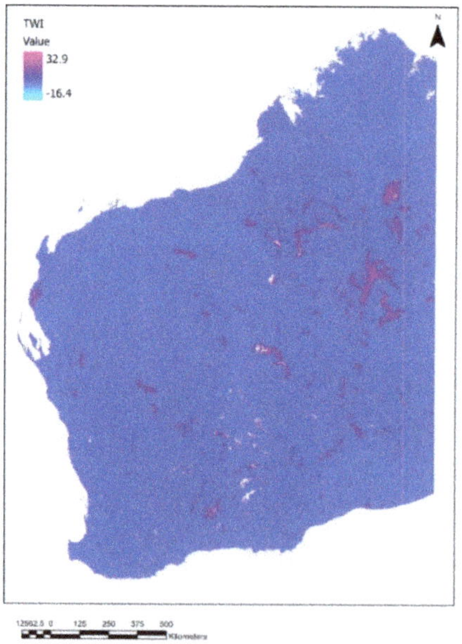

Figure 14. Topographic wetness index map of Western Australia.

The road closure data was obtained from the Main Roads data portal. The data available dates from after 1996. Figure 15 shows the road closure stretch on the state road network. Frequent road closures can be observed in Kimberley, Pilbara, Mid West-Gascoyne, and Goldfields-Esperance. No road closure was observed in the South West region, while road closures were rare in the Wheatbelt and Great Southern region.

Figure 15. Road closure due to rain events.

Intensity Frequency duration: BoM provides the design rainfall data in the form of intensity frequency duration; results ranged from a very frequent probability of 12 exceedances per yar (EY) to 0.2 EY, frequent and infrequent probability of 1 EY to 1% Annual Exceedance Probability (AEP), and rare probability of 1-in-100 AEP to 1-in-2000 AEP. Figure 16 illustrates both intensity frequency duration and asset capacity. The graph shows frequency duration with respect to intensity in unit of mm per hours. It can be observed that lower intensity and same intensity both cause frequent exceedance when such rainfall happens in smaller time duration (1 h), compared to the same intensity in a longer time period (24 h).

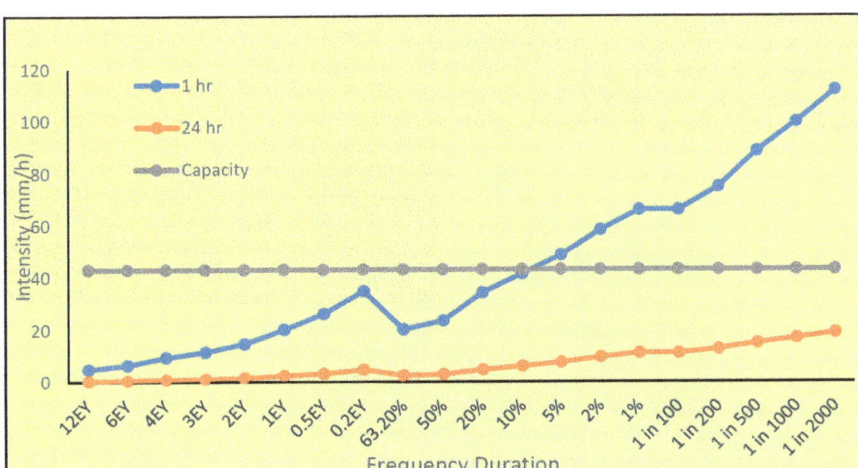

Figure 16. Illustration of intensity frequency duration and asset capacity.

4.2. Pressure on Assets

Percentage pressure on culverts and floodways in 2020 can be seen in Figures 17 and 18, respectively. It can be observed that culverts received higher pressure compared to the floodways. Culverts received higher pressure in Kimberley and the southwestern region, while floodway pressure was highest in the northern region.

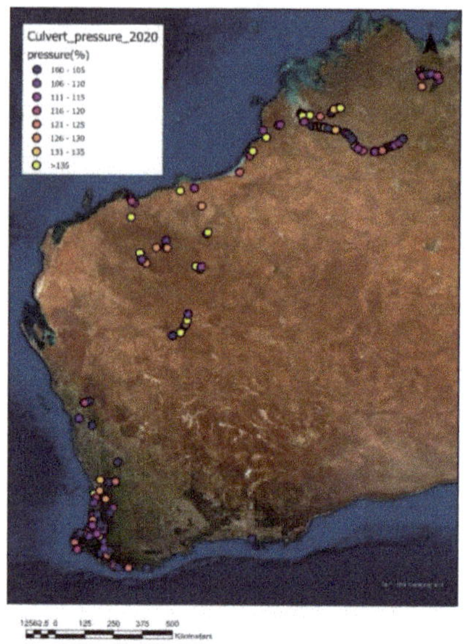

Figure 17. Pressure on culverts.

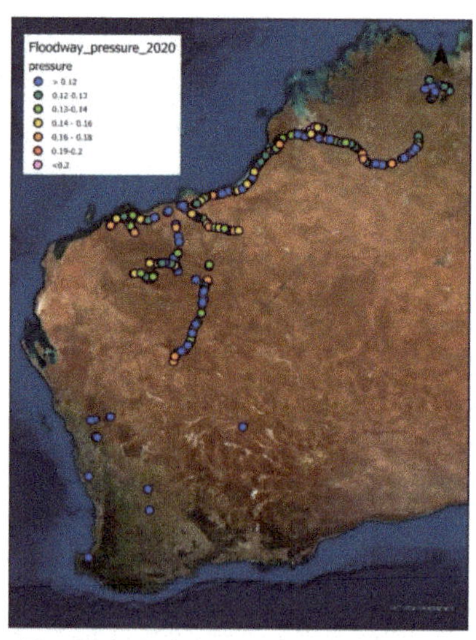

Figure 18. Pressure on Floodways.

4.3. Multi-Criteria Analysis

The pairwise comparison matrix for all parameters can be seen in Table 7. It can be observed that IFD received the highest weight, followed by slope, elevation, and TPI. In contrast, lithology, land use, and TWI received lower relative importance based on their impact on the behaviour of water, such as flow, infiltration, evapotranspiration, and accumulation.

Table 7. Pairwise comparison matrix results.

	Lithology	Land Use	TWI	Soil	TPI	Elevation	Slope	IFD	Estimated Eigen-value	Relative Importance Weight	RIW* Column Totals
Lithology	1.00	0.33	0.20	0.20	0.17	0.14	0.13	0.11	0.2141	0.02	0.73
Land use	3.00	1.00	0.33	0.20	0.20	0.17	0.14	0.13	0.3232	0.02	0.89
TWI	5.00	3.00	1.00	0.33	0.20	0.20	0.17	0.14	0.5125	0.04	1.10
Soil	5.00	5.00	3.00	1.00	0.33	0.20	0.20	0.17	0.7993	0.06	1.29
TPI	6.00	5.00	5.00	3.00	1.00	0.33	0.20	0.20	1.2510	0.10	1.45
Elevation	7.00	6.00	5.00	5.00	3.00	1.00	0.33	0.20	1.9511	0.15	1.52
Slope	8.00	7.00	6.00	5.00	5.00	3.00	1.00	0.33	3.0941	0.24	1.24
IFD	9.00	8.00	7.00	6.00	5.00	5.00	3.00	1.00	4.6715	0.36	0.83

λmax = 8.35; count = 8.00; consistency index = 0.05; relative consistency index = 1.41; consistency ratio = 0.03

Furthermore, this pairwise consistency matrix was checked for consistency using a consistency ratio. Here, a consistency ratio of 0.03 was observed; as that ratio is less than 0.1, this pairwise comparison matrix can be said to have consistent weighting.

Relative importance weights for all parameters were produced using a pairwise comparison matrix. The resultant RIW for various classes or types of data is shown in Table 8. The consistency ratio of the pairwise comparison matrix was found to be less than 0.1 for consistency. Similarly, for the vulnerability map, the pressure on culvert and floodway was used. Table 9 shows the relative importance weight of different pressure ranges calculated from the pairwise comparison matrix. It is also important to note that the factors do not cause multicollinearity due to the independency between factors and their low correlation.

4.4. Hazard Vulnerability

Relative importance weights from Tables 8 and 9 were used to reclassify every parameter. New values were assigned by converting the original value to the allocated weight. The reclassified raster images were then multiplied by the relative importance weight of respective parameters given in Table 9. Figure 19 shows the hazard map generated from parameters in Table 9 including Land use, lithology, soil hazard, IFD, elevation, slope, TPI, and TWI. A higher level of hazard is observed along the coastline of Western Australia. From the map of elevation in Figure 11, it can be observed that lower elevation is found along the coastline. It can be also observed that there is an inverse relation of elevation with the hazard level.

The map in Figure 20 shows the vulnerability which was generated from the reclassified value of pressure on culverts and floodways in 2020. A high vulnerability level can be observed to have a direct relation to the topographic position index (Figure 13).

Table 8. Relative importance weight and consistency ratio for unique values and range of data.

Parameter	Consistency Ratio	Classes	RWI
Land use	0.06	Conservation and natural environments	0.03
		Production from relatively natural environments	0.05
		Production from dryland agriculture and plantations	0.09
		Production from irrigated agriculture and plantations	0.16
		Intensive uses	0.26
		water	0.40
Soil	0.05	0–20	0.03
		20–40	0.06
		40–60	0.13
		60–80	0.26
		80–100	0.51
Lithology	0.02	Sedimentary	0.06
		Meta sedimentary	0.10
		Metamorphic	0.16
		Meta-Igneous	0.26
		Igneous	0.42
TWI	0.05	$2.7 \leq TWI \leq 5.7$	0.06
		$5.8 \leq TWI \leq 7.7$	0.10
		$7.8 \leq TWI \leq 14.9$	0.16
		$15 \leq TWI \leq 22.8$	0.26
		$22.9 \leq TWI \leq 28.1$	0.42
Slope	0.03	88–90	0.05
		77–88	0.11
		34–77	0.26
		0–34	0.59
		699–1440	0.03
Depth IFD	0.02	12–6 EY	0.02
		4–3 EY	0.02
		2–1 EY	0.04
		0.5–0.2 EY	0.05
		63–50%	0.07
		20–10%	0.11
		5–2%	0.16
		1 in 100–1 in 200	0.22
		1 in 500–1 in 2000	0.31
TPI	0.04	≤ -1	0.40
		$-1 < TPI \leq -0.5$	0.25
		$-0.5 < TPI < 0.5$	0.16
		$-0.5 < TPI < 0.5$	0.09
		$0.5 < TPI \leq 1$	0.06
		>1	0.03
Elevation	0.04	476–699	0.06
		288–476	0.13
		124–288	0.26
		0–124	0.51

Table 9. Relative importance weight and consistency ratio for range pressure of assets.

Parameter	Consistency Ratio	Classes	RWI
Pressure on culvert	0.05	100–200	0.03
		200–400	0.06
		400–600	0.13
		600–800	0.26
		800–1000	0.51
Pressure on floodway	0.05	0–10	0.03
		10–20	0.06
		20–30	0.13
		30–40	0.26
		40–50	0.51

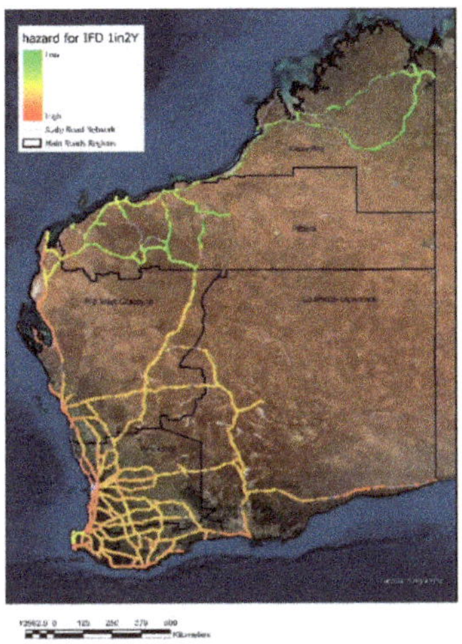

Figure 19. Map of flood hazard.

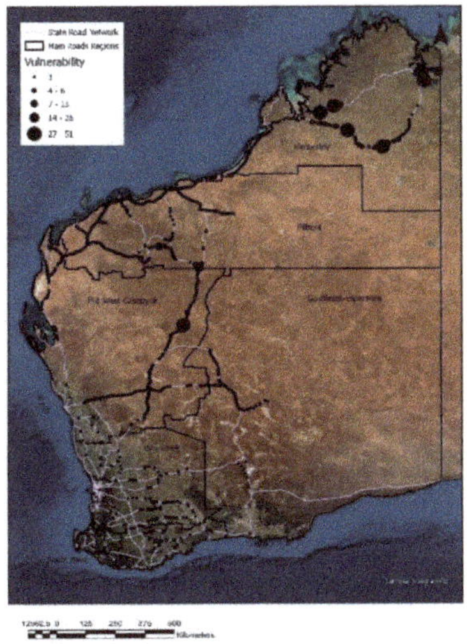

Figure 20. Map of the vulnerability of assets to damage.

Overall, a higher level of hazard can be observed in the south of Western Australia compared to north regions, such as including Pilbara and Kimberley. On the other hand, high vulnerability spots can be seen in the Kimberley and northern Midwest-Gascoyne regions. Despite major differences, the presence of vulnerability and high levels of hazard can be observed along the western cost and the south west region. It can be noted that regions with a high level of hazard may not have a higher level of vulnerability [63].

Risk map: In this study, the variation in risk across the Western Australia in 2020 was measured. Hazard map and vulnerability in 2020 were merged to visualise the risk map (Figure 21). The resultant map shows the asset damage risk level in 2020. High-risk areas can be observed in Kimberley, the north-west road stretch of Midwest-Gascoyne region, and the South west region.

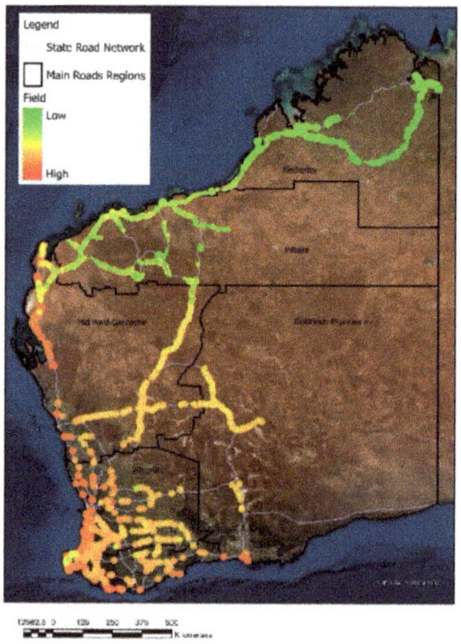

Figure 21. Map of risk hotspots for asset damage.

Note that the differences in data points and lines, known as mapping scale, are shown in the figures above.

Sensitivity analysis in MCA has been performed using different techniques, such as varying the weights of the criteria within a specified range, or examining the impact of varying weights on the ranking of the alternatives. This has helped us to identify the range of values for the weights of each factor, producing the most stable and reliable rankings of the alternatives. Furthermore, the authors found the consistency ratio for the pairwise comparison matrix to be less than 0.1, which is considered an acceptable level of consistency. This indicates that the judgments made by the decision makers in the pairwise comparison process were reasonably consistent. Moreover, we incorporated a well-established method for deriving the pairwise comparison matrix, and followed standard procedures to ensure consistency and validity. This can increase the reliability and robustness of the results.

5. Discussion and Limitations

The multi-criteria analysis (MCA) technique is a powerful decision making tool that can be used to assess and compare multiple criteria or factors that may influence a particular decision. When MCA is combined with Geographic Information System (GIS) technology, it can enhance decision making processes by incorporating spatial data, analysing spatial relationships and patterns, and visualising the results on maps. Incorporating MCA techniques provide several benefits to regional governments in Australia in terms of preparations for and mitigations during future flood events. MCA can help regional governments make informed decisions by evaluating the different criteria that are relevant to flood preparedness and mitigation. Benefits of using MCA for flood preparedness and mitigation include: (i) identifying and prioritising critical criteria: MCA can help regional governments identify and prioritise the criteria that are most important for flood preparedness and mitigation. This can help to ensure that resources are allocated to the most critical areas, and that the most effective strategies are implemented to address flood risks; (ii) evaluating trade-offs: MCA can help regional governments evaluate the trade-offs between different strategies and criteria. For example, governments can use MCA to evaluate the trade-offs between the costs of flood mitigation strategies and the potential benefits in terms of reducing the impact of floods; (iii) incorporating stakeholder input: MCA can help regional governments incorporate stakeholder input into the decision making process. By involving stakeholders in the development of criteria and the evaluation of strategies, regional governments can ensure that the decision making process is transparent and reflects the interests and concerns of all stakeholders; (iv) supporting data-driven decision making: MCA can help regional governments make data-driven decisions by providing a structured approach to evaluating criteria and strategies. This can help to ensure that decisions are based on objective criteria and data, rather than subjective opinions or biases, and; (v) enhancing communication and transparency: MCA can help to enhance communication and transparency by providing a clear and structured approach to evaluating flood preparedness and mitigation strategies. This can help to build trust and credibility with stakeholders, and improve public understanding of and support for flood preparedness and mitigation efforts.

This paper aims to identify regions that are likely vulnerable to a higher risk of damage to their assets. In the provided case study, we indicate valuable insights into the flood-related parameters in Western Australia that can increase the risk of damage to road and stormwater management assets in 2020. The findings of this study also provide the relationship between parameters and the risk. To sum up, the following are the outcomes of the study:

The topographic index has a relationship to the vulnerability of assets to damage. The presence of ridges and valleys might cause water accumulation.

Intensity frequency duration of design rainfall studies revealed higher pressure on the assets when there is rain in a small duration of time compared to rain in a longer duration. Culverts receive significantly higher pressure compared to the floodways.

The hazard distribution was found to be related to elevation. According to this study lower elevations regions were found to have a higher level of hazard.

Regions with high vulnerability might not have a higher level of hazard.

A high number of road closures and risk spots were found in January 2020 and February 2020. However, parameters were observed at a low level in March–June 2020. The road closure stretch may not align with the hotspot map because of the nature of the risk map.

The focus on damage to roads in the context of flood risk and asset management is due to the critical role that roads play in maintaining access and mobility in urban and rural areas. When roads are damaged or closed due to flooding, it can severely impact the ability of people and goods to move, resulting in economic and social disruptions. Moreover, damage to roads can also have safety implications, such as increased risk of accidents due to damaged road surfaces, or reduced visibility in flooded areas. Roads are also often the most visible and easily quantifiable asset in the floodplain; as a result, they can serve as an important indicator of the potential impacts of flooding on other assets and infrastructure in the area. By focusing on damage to roads, asset managers can prioritise interventions that help reduce the risk of road damage and closure, which can in turn mitigate the impact of flooding on other assets, such as buildings and farmland. In addition, damage to roads can be expensive to repair and have a significant impact on local and regional economies. By focusing on reducing the risk of road damage, asset managers can help minimise the economic impact of flooding events and promote faster recovery in affected areas.

The limitations of this study include the computation capacity; if the hotspot map is generated without limiting it to asset location, this can also provide a stretch of road where there is higher risk. However, because of the limitation of intensity frequency duration data, the assessment calculation is limited to the asset location. Furthermore, studying the state-wise risk distribution can give more insights into the asset management system. Multi-criteria analysis can incorporate the rainfall and a greater number of DEM derivatives that are linked to the flow and accumulation of water. This study could also be improved by adding parameters, such as the road traffic and vegetation in the region, from satellite sources, such as NDVI from Landgate. Road traffic also places stress on the road and alters the strength of the road. Similarly, the presence of vegetation prevents the build-up of fast-moving water that can create the conditions required for flash flooding and increase the risks posed to the stormwater management system.

6. Conclusions

Floodwater is a destructive phenomenon that causes substantial damage to road assets and infrastructure. Sustainable flood management focusing on prevention, protection, and preparedness is the focus of this paper. Sustainable flood management would help government agencies to effectively respond to extreme climate conditions and their impacts by enabling more efficient asset management strategies. The proposed approach serves as a useful tool for government agencies in Western Australia to effectively manage and mitigate the risks associated with flood events. By combining hazard and vulnerability analysis using MCA, the approach offers a comprehensive view of the factors affecting the risk of asset damage and can help prioritise efforts and resources in the most vulnerable areas. With the increasing influence of climate change on extreme weather events, this approach can provide important insights for data-driven asset management, and the development of more accurate flood models and early warning systems.

The data used in this study were collected from various data sources. In particular, rainfall and runoff data were obtained from Australian Water Resources Assessment Landscape (AWRA-L) model. Stormwater management and asset management data were obtained from Main Roads WA. These include asset inventory data of stormwater drainage and capacity. Publicly available data in relation to hydro-informatics were collected from BoM, SRTM, and Landgate. MCA was then used to define the high-risk hotspots for asset damage. These state-wide hotspots were validated using road closure data from Main Roads WA. Road damage risk maps were generated using hazard and vulnerability raster maps. Road damage hazard was calculated from morphometric and hydrometeorological parameters sourced from various data portals. Vulnerability was calculated based on pressure on assets. The hotspot analysis shows high risk can be observed in Kimberley, the north-west road stretch in the Midwest-Gascoyne region, and the South West region. Moreover, after validation with the road closure data. High flooding risk and road closures can be observed in Kimberley region and south of Mid WestGascoyne region. The time series analysis is undertaken in order to understand the temporal variation in risk map. Alignment in the occurrence of risk spots and road closure is observed. However, further work on multiple years could help understand risk distribution more accurately.

Author Contributions: Conceptualization, P.W. and H.P.; Methodology, P.W., J.H. and H.P.; Software, P.W., B.A.-S. and H.P.; Validation, P.W., B.A.-S. and H.P.; Formal analysis, B.A.-S.; Investigation, B.A.-S. and J.H.; Resources, P.W. and K.S.; Writing—original draft, H.P.; Writing—review & editing, B.A.-S. and K.S.; Visualization, B.A.-S. and J.H.; Supervision, P.W. and K.S.; Project administration, K.S. All authors have read and agreed to the published version of the manuscript.

Funding: This research received no external funding.

Institutional Review Board Statement: Not applicable.

Informed Consent Statement: Not applicable.

Data Availability Statement: The data presented in this study are available on request from the corresponding author.

Conflicts of Interest: The authors declare no conflict of interest.

References

1. Liu, J.; Zhang, Y.; Yang, Y.; Gu, X.; Xiao, M. Investigating relationships between Australian flooding and large-scale climate indices and possible mechanism. *J. Geophys. Res. Atmos.* **2018**, *123*, 8708–8723. [CrossRef]
2. Berland, A.; Shiflett, S.A.; Shuster, W.D.; Garmestani, A.S.; Goddard, H.C.; Herrmann, D.L.; Hopton, M.E. The role of trees in urban stormwater management. *Landsc. Urban Plan.* **2017**, *162*, 167–177. [CrossRef] [PubMed]
3. Khayan, K.; Husodo, A.H.; Astuti, I.; Sudarmadji, S.; Djohan, T.S. Rainwater as a source of drinking water: Health impacts and rainwater treatment. *J. Environ. Public Health* **2019**, *2019*, 1760950. [CrossRef] [PubMed]
4. Newhart, K.B.; Holloway, R.W.; Hering, A.S.; Cath, T.Y. Data-driven performance analyses of wastewater treatment plants: A review. *Water Res.* **2019**, *157*, 498–513. [CrossRef] [PubMed]
5. Matheswaran, K.; Alahacoon, N.; Pandey, R.; Amarnath, G. Flood risk assessment in South Asia to prioritize flood index insurance applications in Bihar, India. *Geomat. Nat. Hazards Risk* **2019**, *10*, 26–48. [CrossRef]
6. Pallathadka, A.; Sauer, J.; Chang, H.; Grimm, N.B. Urban flood risk and green infrastructure: Who is exposed to risk and who benefits from investment? A case study of three US Cities. *Landsc. Urban Plan.* **2022**, *223*, 104417. [CrossRef]
7. Molinari, D.; De Bruijn, K.M.; Castillo-Rodríguez, J.T.; Aronica, G.T.; Bouwer, L.M. Validation of flood risk models: Current practice and possible improvements. *Int. J. Disaster Risk Reduct.* **2019**, *33*, 441–448. [CrossRef]
8. Scaini, A.; Stritih, A.; Brouillet, C.; Scaini, C. Flood risk and river conservation: Mapping citizen perception to support sustainable river management. *Front. Earth Sci.* **2021**, *9*, 510. [CrossRef]
9. Abu-Salih, B.; Wongthongtham, P.; Zhu, D.; Chan, K.Y.; Rudra, A.; Abu-Salih, B.; Wongthongtham, P.; Zhu, D.; Chan, K.Y.; Rudra, A. Predictive analytics using Social Big Data and machine learning. In *Social Big Data Analytics: Practices, Techniques, and Applications*; Springer: Singapore, 2021; pp. 113–143.
10. Hawchar, L.; Naughton, O.; Nolan, P.; Stewart, M.G.; Ryan, P.C. A GIS-based framework for high-level climate change risk assessment of critical infrastructure. *Clim. Risk Manag.* **2020**, *29*, 100235. [CrossRef]
11. Kumar, N.; Poonia, V.; Gupta, B.B.; Goyal, M.K. A novel framework for risk assessment and resilience of critical infrastructure towards climate change. *Technol. Forecast. Soc. Chang.* **2021**, *165*, 120532. [CrossRef]
12. Pathak, S.; Liu, M.; Jato-Espino, D.; Zevenbergen, C. Social, economic and environmental assessment of urban sub-catchment flood risks using a multi-criteria approach: A case study in Mumbai City, India. *J. Hydrol.* **2020**, *591*, 125216. [CrossRef]
13. Brimfield, B.E.; Myers, S.D. An integrated approach to benefits realisation of railway condition monitoring innovations. In Proceedings of the 5th IET Conference on Railway Condition Monitoring and Non-Destructive Testing (RCM 2011), Derby, UK, 29–30 November 2011.
14. Li, Q.; Kumar, A. *National & International Practices in Decision Support Tools in Road Asset Management*; CRC for Construction Innovation: Brisbane, QLD, Australia, 2003.
15. Christodoulou, S.; Deligianni, A. A neurofuzzy decision framework for the management of water distribution networks. *Water Resour. Manag.* **2010**, *24*, 139–156. [CrossRef]
16. Sobieraj, J.; Bryx, M.; Metelski, D. Stormwater Management in the City of Warsaw: A Review and Evaluation of Technical Solutions and Strategies to Improve the Capacity of the Combined Sewer System. *Water* **2022**, *14*, 2109. [CrossRef]
17. Hafeez, F.; Frost, A.; Vaze, J.; Dutta, D.; Smith, A.; Elmahdi, A. A new integrated continental hydrological simulation system. *Water J. Aust. Water Assoc.* **2015**, *42*, 75–82.
18. Viney, N.; Vaze, J.; Crosbie, R.; Wang, B.; Dawes, W.; Frost, A. *AWRA-L v5.0: Technical Description of Model Algorithms and Inputs*; CSIRO: Perth, SA, Australia, 2015. [CrossRef]
19. Elmahdi, A.; Hafeez, M.; Smith, A.; Frost, A.; Vaze, J.; Dutta, D. Australian Water Resources Assessment Modelling System (AWRAMS)-informing water resources assessment and national water accounting. In Proceedings of the 36th Hydrology and Water Resources Symposium: The Art and Science of Water, Hobart, TAS, Australia, 7–10 December 2015; Engineers Australia: Barton, CBR, Australia, 2015; p. 979.
20. Frost, A.J.; Ramchurn, A.; Hafeez, M. *Evaluation of the Bureau's Operational AWRA-L Model*; Bureau of Meteorology: Melbourne, VIC, Australia, 2016; p. 80.
21. Raupach, M.R.; Briggs, P.R.; Haverd, V.; King, E.A.; Paget, M.; Trudinger, C.M. *Australian Water Availability Project (AWAP): CSIRO Marine and Atmospheric Research Component: Final Report for Phase 3*; Centre for Australian Weather and Climate Research (Bureau of Meteorology and CSIRO): Melbourne, VIC, Australia, 2009; p. 67.
22. Wang, Y.P.; Kowalczyk, E.; Leuning, R.; Abramowitz, G.; Raupach, M.R.; Pak, B.; van Gorsel, E.; Luhar, A. Diagnosing errors in a land surface model (CABLE) in the time and frequency domains. *J. Geophys. Res. Biogeosci.* **2011**, *116*, G01034. [CrossRef]
23. Kowalczyk, E.A.; Wang, Y.P.; Law, R.M.; Davies, H.L.; McGregor, J.L.; Abramowitz, G. The CSIRO Atmosphere Biosphere Land Exchange (CABLE) model for use in climate models and as an offline model. *CSIRO Mar. Atmos. Res. Pap.* **2006**, *13*, 42.

24. Mosavi, A.; Ozturk, P.; Chau, K.-W. Flood prediction using machine learning models: Literature review. *Water* **2018**, *10*, 1536. [CrossRef]
25. Van den Honert, R.C.; McAneney, J. The 2011 Brisbane floods: Causes, impacts and implications. *Water* **2011**, *3*, 1149–1173. [CrossRef]
26. Shrestha, D.L.; Robertson, D.E.; Wang, Q.J.; Pagano, T.C.; Hapuarachchi, H.A. Evaluation of numerical weather prediction model precipitation forecasts for short-term streamflow forecasting purpose. *Hydrol. Earth Syst. Sci.* **2013**, *17*, 1913–1931. [CrossRef]
27. Daley, R. *Atmospheric Data Analysis*; Cambridge University Press: Cambridge, UK, 1993.
28. Echeverribar, I.; Morales-Hernández, M.; Brufau, P.; García-Navarro, P. 2D numerical simulation of unsteady flows for large scale floods prediction in real time. *Adv. Water Resour.* **2019**, *134*, 103444. [CrossRef]
29. Yoon, S.-S. Adaptive blending method of radar-based and numerical weather prediction QPFs for urban flood forecasting. *Remote Sens.* **2019**, *11*, 642. [CrossRef]
30. Kundzewicz, Z.W.; Pińskwar, I.; Brakenridge, G.R. Changes in river flood hazard in Europe: A review. *Hydrol. Res.* **2018**, *49*, 294–302. [CrossRef]
31. Shakirah, J.A.; Sidek, L.M.; Hidayah, B.; Nazirul, M.Z.; Jajarmizadeh, M.; Ros, F.C.; Roseli, Z.A. A review on flood events for kelantan river watershed in malaysia for last decade (2001–2010). *IOP Conf. Ser. Earth Environ. Sci.* **2016**, *32*, 012070. [CrossRef]
32. Kenley, R.; Harfield, T.; Bedggood, J. Road asset management: The role of location in mitigating extreme flood maintenance. *Procedia Econ. Financ.* **2014**, *18*, 198–205. [CrossRef]
33. Kim, B.; Sanders, B.F.; Han, K.; Kim, Y.; Famiglietti, J.S. Calibration of stormwater management model using flood extent data. In Proceedings of the Institution of Civil Engineers-Water Management; Thomas Telford Ltd.: London, UK, 2014; pp. 17–29.
34. Beecroft, A.; Peters, E.; Toole, T. Life-cycle costing of rain and flood events in Queensland—Case studies and network-wide implications. *Road Transp. Res. J. Aust. New Zealand Res. Pract.* **2017**, *26*, 22–35.
35. Costabile, P.; Costanzo, C.; Macchione, F. Comparative analysis of overland flow models using finite volume schemes. *J. Hydroinform.* **2012**, *14*, 122–135. [CrossRef]
36. Lee, T.H.; Georgakakos, K.P. Operational Rainfall Prediction on Meso-γ Scales for Hydrologic Applications. *Water Resour. Res.* **1996**, *32*, 987–1003. [CrossRef]
37. Hu, Z.-J.; Wang, L.-L.; Tang, H.-W.; Qi, X.-M. Prediction of the future flood severity in plain river network region based on numerical model: A case study. *J. Hydrodyn. Ser. B* **2017**, *29*, 586–595. [CrossRef]
38. Costabile, P.; Costanzo, C.; Macchione, F. A storm event watershed model for surface runoff based on 2D fully dynamic wave equations. *Hydrol. Process.* **2013**, *27*, 554–569. [CrossRef]
39. Valipour, M.; Banihabib, M.E.; Behbahani, S.M.R. Comparison of the ARMA, ARIMA, and the autoregressive artificial neural network models in forecasting the monthly inflow of Dez dam reservoir. *J. Hydrol.* **2013**, *476*, 433–441. [CrossRef]
40. Elsafi, S.H. Artificial neural networks (ANNs) for flood forecasting at Dongola Station in the River Nile, Sudan. *Alex. Eng. J.* **2014**, *53*, 655–662. [CrossRef]
41. Deo, R.C.; Şahin, M. Application of the artificial neural network model for prediction of monthly standardized precipitation and evapotranspiration index using hydrometeorological parameters and climate indices in eastern Australia. *Atmos. Res.* **2015**, *161*, 65–81. [CrossRef]
42. Ramana, R.V.; Krishna, B.; Kumar, S.R.; Pandey, N.G. Monthly rainfall prediction using wavelet neural network analysis. *Water Resour. Manag.* **2013**, *27*, 3697–3711. [CrossRef]
43. Shamim, M.A.; Hassan, M.; Ahmad, S.; Zeeshan, M. A comparison of artificial neural networks (ANN) and local linear regression (LLR) techniques for predicting monthly reservoir levels. *KSCE J. Civ. Eng.* **2016**, *20*, 971–977. [CrossRef]
44. Araghinejad, S.; Azmi, M.; Kholghi, M. Application of artificial neural network ensembles in probabilistic hydrological forecasting. *J. Hydrol.* **2011**, *407*, 94–104. [CrossRef]
45. Cunningham, S.C.; Griffioen, P.; White, M.D.; Nally, R.M. Assessment of ecosystems: A system for rigorous and rapid mapping of floodplain forest condition for Australia's most important river. *Land Degrad. Dev.* **2018**, *29*, 127–137. [CrossRef]
46. Prasad, R.; Deo, R.C.; Li, Y.; Maraseni, T. Input selection and performance optimization of ANN-based streamflow forecasts in the drought-prone Murray Darling Basin region using IIS and MODWT algorithm. *Atmos. Res.* **2017**, *197*, 42–63. [CrossRef]
47. Cannas, B.; Fanni, A.; Sias, G.; Tronci, S.; Zedda, M.K. River flow forecasting using neural networks and wavelet analysis. *Geophys. Res. Abstr* **2005**, *7*, 08651.
48. Tantanee, S.; Patamatammakul, S.; Oki, T.; Sriboonlue, V.; Prempree, T. Coupled wavelet-autoregressive model for annual rainfall prediction. *J. Environ. Hydrol.* **2005**, *13*, 1–8.
49. Mekanik, F.; Imteaz, M.A.; Talei, A. Seasonal rainfall forecasting by adaptive network-based fuzzy inference system (ANFIS) using large scale climate signals. *Clim. Dyn.* **2016**, *46*, 3097–3111. [CrossRef]
50. Kisi, O.; Nia, A.M.; Gosheh, M.G.; Tajabadi, M.R.J.; Ahmadi, A. Intermittent streamflow forecasting by using several data driven techniques. *Water Resour. Manag.* **2012**, *26*, 457–474. [CrossRef]
51. Li, C.; Guo, S.; Zhang, J.; Guo, J. A modified NLPM-ANN model and its application to flood forecasting. *Eng. J. Wuhan Univ.* **2009**, *42*, 1–5.
52. Huang, S.; Chang, J.; Huang, Q.; Chen, Y. Monthly streamflow prediction using modified EMD-based support vector machine. *J. Hydrol.* **2014**, *511*, 764–775. [CrossRef]
53. Bass, B.; Bedient, P. Surrogate modeling of joint flood risk across coastal watersheds. *J. Hydrol.* **2018**, *558*, 159–173. [CrossRef]

54. Tan, Q.-F.; Lei, X.-H.; Wang, X.; Wang, H.; Wen, X.; Ji, Y.; Kang, A.-Q. An adaptive middle and long-term runoff forecast model using EEMD-ANN hybrid approach. *J. Hydrol.* **2018**, *567*, 767–780. [CrossRef]
55. Ravansalar, M.; Rajaee, T.; Kisi, O. Wavelet-linear genetic programming: A new approach for modeling monthly streamflow. *J. Hydrol.* **2017**, *549*, 461–475. [CrossRef]
56. Shafapour Tehrany, M.; Shabani, F.; Jebur, M.N.; Hong, H.; Chen, W.; Xie, X. GIS-based spatial prediction of flood prone areas using standalone frequency ratio, logistic regression, weight of evidence and their ensemble techniques. *Geomat. Nat. Hazards Risk* **2017**, *8*, 1538–1561. [CrossRef]
57. Manfreda, S.; Nardi, F.; Samela, C.; Grimaldi, S.; Taramasso, A.C.; Roth, G.; Sole, A. Investigation on the use of geomorphic approaches for the delineation of flood prone areas. *J. Hydrol.* **2014**, *517*, 863–876. [CrossRef]
58. Papaioannou, G.; Vasiliades, L.; Loukas, A. Multi-criteria analysis framework for potential flood prone areas mapping. *Water Resour. Manag.* **2015**, *29*, 399–418. [CrossRef]
59. Majumder, R.; Bhunia, G.S.; Patra, P.; Mandal, A.C.; Ghosh, D.; Shit, P.K. Assessment of flood hotspot at a village level using GIS-based spatial statistical techniques. *Arab. J. Geosci.* **2019**, *12*, 409. [CrossRef]
60. Aksoy, H.; Kirca, V.S.O.; Burgan, H.I.; Kellecioglu, D. Hydrological and hydraulic models for determination of flood-prone and flood inundation areas. *Proc. Int. Assoc. Hydrol. Sci.* **2016**, *373*, 137–141. [CrossRef]
61. Falguni, M.; Singh, D. Detecting flood prone areas in Harris County: A GIS based analysis. *GeoJournal* **2020**, *85*, 647–663.
62. Pratidina, G.; Santoso, P.B. Detection of satellite data-based flood-prone areas using logistic regression in the central part of Java Island. *J. Phys. Conf. Ser.* **2019**, *1367*, 012086. [CrossRef]
63. Mishra, K.; Sinha, R. Flood risk assessment in the Kosi megafan using multi-criteria decision analysis: A hydro-geomorphic approach. *Geomorphology* **2020**, *350*, 106861. [CrossRef]
64. Rana, V.K.; Suryanarayana, T.M.V. GIS-based multi criteria decision making method to identify potential runoff storage zones within watershed. *Ann. GIS* **2020**, *26*, 149–168. [CrossRef]
65. Rozos, D.; Bathrellos, G.D.; Skillodimou, H.D. Comparison of the implementation of rock engineering system and analytic hierarchy process methods, upon landslide susceptibility mapping, using GIS: A case study from the Eastern Achaia County of Peloponnesus, Greece. *Environ. Earth Sci.* **2011**, *63*, 49–63. [CrossRef]
66. Stefanidis, S.; Stathis, D. Assessment of flood hazard based on natural and anthropogenic factors using analytic hierarchy process (AHP). *Nat. Hazards* **2013**, *68*, 569–585. [CrossRef]
67. Saaty, T.L. A scaling method for priorities in hierarchical structures. *J. Math. Psychol.* **1977**, *15*, 234–281. [CrossRef]
68. Cegan, J.C.; Filion, A.M.; Keisler, J.M.; Linkov, I. Trends and applications of multi-criteria decision analysis in environmental sciences: Literature review. *Environ. Syst. Decis.* **2017**, *37*, 123–133. [CrossRef]
69. Ellis, J.B.; Deutsch, J.-C.; Mouchel, J.-M.; Scholes, L.; Revitt, M.D. Multicriteria decision approaches to support sustainable drainage options for the treatment of highway and urban runoff. *Sci. Total Environ.* **2004**, *334*, 251–260. [CrossRef]
70. Ouma, Y.O.; Yabann, C.; Kirichu, M.; Tateishi, R. Optimization of urban highway bypass horizontal alignment: A methodological overview of intelligent spatial MCDA approach using fuzzy AHP and GIS. *Adv. Civ. Eng.* **2014**, *2014*, 182568. [CrossRef]
71. Zhao, X.-X.; Zheng, M.; Fu, Q. How natural disasters affect energy innovation? The perspective of environmental sustainability. *Energy Econ.* **2022**, *109*, 105992. [CrossRef]
72. Lu, L.-C.; Chiu, S.-Y.; Chiu, Y.-H.; Chang, T.-H. Sustainability efficiency of climate change and global disasters based on greenhouse gas emissions from the parallel production sectors—A modified dynamic parallel three-stage network DEA model. *J. Environ. Manag.* **2022**, *317*, 115401. [CrossRef] [PubMed]
73. Zhang, K.; Chen, G.; Xia, Y.; Wang, S. An Ensemble-Based, Remote-Sensing-Driven, Flood-Landslide Early Warning System. In *Remote Sensing of Water-Related Hazards*; John Wiley & Sons: Hoboken, NJ, USA, 2022; pp. 123–134.
74. Rais, A.M.; Nur, A.N.P.; Tricahyaningati, D. Android Real Time Earthquake & Tsunami Warning Alert System Based on Open Data of Indonesia Government Agency of Geophysics. *IOP Conf. Ser. Earth Environ. Sci.* **2022**, *1095*, 012008.
75. Sufi, F.K.; Khalil, I. Automated disaster monitoring from social media posts using AI-based location intelligence and sentiment analysis. *IEEE Trans. Comput. Soc. Syst.* 2022; *early access*. [CrossRef]
76. Kuller, M.; Bach, P.M.; Ramirez-Lovering, D.; Deletic, A. Framing water sensitive urban design as part of the urban form: A critical review of tools for best planning practice. *Environ. Model. Softw.* **2017**, *96*, 265–282. [CrossRef]
77. Radcliffe, J.C. History of water sensitive urban design/low impact development adoption in Australia and internationally. In *Approaches to Water Sensitive Urban Design*; Elsevier: Amsterdam, The Netherlands, 2019.
78. Yuan, Z.; Liang, C.; Li, D. Urban stormwater management based on an analysis of climate change: A case study of the Hebei and Guangdong provinces. *Landsc. Urban Plan.* **2018**, *177*, 217–226. [CrossRef]
79. Adugna, D.; Lemma, B.; Jensen, M.B.; Gebrie, G.S. Evaluating the hydraulic capacity of existing drain systems and the management challenges of stormwater in Addis Ababa, Ethiopia. *J. Hydrol. Reg. Stud.* **2019**, *25*, 100626. [CrossRef]
80. Pham, B.T.; Luu, C.; Van Phong, T.; Nguyen, H.D.; Van Le, H.; Tran, T.Q.; Ta, H.T.; Prakash, I. Flood risk assessment using hybrid artificial intelligence models integrated with multi-criteria decision analysis in Quang Nam Province, Vietnam. *J. Hydrol.* **2021**, *592*, 125815. [CrossRef]
81. Pourali, S.H.; Arrowsmith, C.; Chrisman, N.; Matkan, A.A.; Mitchell, D. Topography wetness index application in flood-risk-based land use planning. *Appl. Spat. Anal. Policy* **2016**, *9*, 39–54. [CrossRef]

82. Saaty, T.L. The Analytic Hierarchy Process. In *Agricultural Economics Review*; Mcgraw Hill: New York, NY, USA, 1980; p. 70.
83. MRWA. Regional Map. 2021. Available online: https://portal-mainroads.opendata.arcgis.com/datasets/main-roads-regions/explore (accessed on 24 September 2021).

sustainability

MDPI

Article

A Cost-Effective Fall-Detection Framework for the Elderly Using Sensor-Based Technologies

Ch. Anwar Ul Hassan [1], Faten Khalid Karim [2,*], Assad Abbas [3], Jawaid Iqbal [4], Hela Elmannai [5], Saddam Hussain [6], Syed Sajid Ullah [7,*] and Muhammad Sufyan Khan [8]

1 Department of Creative Technologies, Air University, Islamabad 44000, Pakistan
2 Department of Computer Sciences, College of Computer and Information Sciences, Princess Nourah bint Abdulrahman University, P.O. Box 84428, Riyadh 11671, Saudi Arabia
3 Department of Computer Science, COMSATS University, Islamabad 44000, Pakistan
4 Faculty of Computing, Riphah International University, Islamabad 45210, Pakistan
5 Department of Information Technology, College of Computer and Information Sciences, Princess Nourah bint Abdulrahman University, P.O. Box 84428, Riyadh 11671, Saudi Arabia
6 School of Digital Science, Universiti Brunei Darussalam, Jalan Tungku Link, Gadong BE1410, Brunei
7 Department of Information and Communication Technology, University of Agder (UiA), N-4898 Grimstad, Norway
8 Department of Software Engineering, Capital University of Science and Technology, Islamabad 44000, Pakistan
* Correspondence: fkdiaaldin@pnu.edu.sa (F.K.K.); syed.s.ullah@uia.no (S.S.U.)

Abstract: Falls are critical events among the elderly living alone in their rooms and can have intense consequences, such as the elderly person being left to lie for a long time after the fall. Elderly falling is one of the serious healthcare issues that have been investigated by researchers for over a decade, and several techniques and methods have been proposed to detect fall events. To overcome and mitigate elderly fall issues, such as being left to lie for a long time after a fall, this project presents a low-cost, motion-based technique for detecting all events. In this study, we used IRA-E700ST0 pyroelectric infrared sensors (PIR) that are mounted on walls around or near the patient bed in a horizontal field of view to detect regular motions and patient fall events; we used PIR sensors along with Arduino Uno to detect patient falls and save the collected data in Arduino SD for classification. For data collection, 20 persons contributed as patients performing fall events. When a patient or elderly person falls, a signal of different intensity (high) is produced, which certainly differs from the signals generated due to normal motion. A set of parameters was extracted from the signals generated by the PIR sensors during falling and regular motions to build the dataset. When the system detects a fall event and turns on the green signal, an alarm is generated, and a message is sent to inform the family members or caregivers of the individual. Furthermore, we classified the elderly fall event dataset using five machine learning (ML) classifiers, namely: random forest (RF), decision tree (DT), support vector machine (SVM), naïve Bayes (NB), and AdaBoost (AB). Our result reveals that the RF and AB algorithms achieved almost 99% accuracy in elderly fall-d\detection.

Keywords: fall detections; fall-detection; cost efficiency; machine learning

Citation: Hassan, C.A.U.; Karim, F.K.; Abbas, A.; Iqbal, J.; Elmannai, H.; Hussain, S.; Ullah, S.S.; Khan, M.S. A Cost-Effective Fall-Detection Framework for the Elderly Using Sensor-Based Technologies. *Sustainability* **2023**, *15*, 3982. https://doi.org/10.3390/su15053982

Academic Editors: Mourade Azrour, Azidine Guezzaz, Imad Zeroual, Azeem Irshad, Jamal Mabrouki, Said Benkirane and Shehzad Ashraf Chaudhry

Received: 26 January 2023
Revised: 19 February 2023
Accepted: 19 February 2023
Published: 22 February 2023

1. Introduction

Elderly falling is a serious issue and has been investigated by researchers for over a decade. In a Centers for Disease Control report, it was reported that, in the United States, falls among the elderly each year cause multiple deaths and injuries. Approximately 61% of falls occurred indoors, which caused around 10,000 deaths [1]. Providing quick assistance to the victims of such events significantly reduces the hospitalization risk (by around 26%) and decreases the death rate to around 80% [1]. In favor of this, several mechanisms that detect fall incidents and provide quick assistance have been introduced to control and overcome this problem instead of leaving the person lying for a long time after falling. In [1], the author proposed smart tiles for elderly fall detection. However, it is an expensive solution.

The World Health Organization stated that accidental falls of the elderly are the main reason for or the cause of injuries and even death in some cases [2]. When a fall happens, instant help is a need. For this, researchers have put in a lot of effort and appreciable work to detect and prevent elderly falls [3–5]; however, some limitations exist in their work. In [6–8], the authors used a pressure- and vibration-based scheme in which sensors were placed on the wall or floor in a specific direction to analyze patient movement. However, these sensors, RFID systems, and smart tiles are not cost-effective solutions for detecting elderly falls. As in the case of RFID technologies, they detect elderly falls but do not protect them from fall injury. RFID fall solutions are not cost-effective solutions [9]. We need to provide the elderly with better shelter to provide quality of life and minimize the risks of living alone. Furthermore, an accelerometer system has been introduced to accelerate wearable airbags to prevent the patient from fall-related injuries [10–12]. In a wearable scenario, wearing the airbag all the time is frustrating and annoying to human nature. The elderly also need to remember to wear the device, which, with growing age, is difficult to remember. Over the last few decades, sensors and technologies have gained importance as they do not affect the privacy of either the elderly or patients who have never worn wearable devices. The PIR sensor is the most useable technique for detecting elderly falls; the authors used PIR sensors to detect human motion in [13,14].

1.1. Motivation

Elderly falling is a serious issue. In relation to this, several mechanisms have been introduced to control and overcome this problem by detecting fall incidents and providing quick assistance instead of leaving the person lying for a long time after falling. In this regard, several techniques and methods have been proposed by researchers to detect fall events, such as smart tiles [1], RFID tags [9], accelerometers [12], and home monitoring-based techniques [13]. However, always keeping an eye on the elderly affects the privacy of the elderly, and they do not feel comfortable in life; additionally, these are expensive solutions. To overcome and mitigate elderly fall issues, such as being left to lie for a long time after a fall, this study presents low-cost, sensor-based approaches for fall event detection.

1.2. Paper Organization

The remainder of the paper is prepared as follows: Related work is described in Section 2. Section 3 describes the proposed methodology. Section 4 gives implementation details. ML classifiers are described in Section 5. The results of the experiment are presented in Section 6. Finally, in Section 7, conclusions and future work are defined.

2. Related Work

In this section, we focus especially on research efforts that are related to elderly fall detection. Several papers have been written on the topic of detecting elderly falls. Researchers are working on elderly fall detection. Researchers have designed and implemented several motion detection systems to detect elderly fall events using radio-frequency identification tags (RFID). In the tags, the frequency values of static, normal movements and sudden falls are varied, and impact is shown in received signal strength (RSS) and Doppler frequency value (DFV), and, in this way, the authors could detect elderly fall events. Battery-less technologies, such as RFID and readers, were incorporated into smart floor carpets to detect elderly falls and, subsequently, to inform caregivers [15]. However, the authors used distinct equipment and devices, and the accuracy was not adequately high. In [16], the author also used the floor-based RFID technique, with RFID arranged in a two-dimensional grid on a smart carpet to form an unobtrusive monitoring zone. Heuristic algorithms and

ML are used to identify falls and prevent a situation where the elderly person is left to lie for a long time.

The vision-based system is frequently used to detect patient falls; researchers have used a digital camera to monitor the patient's activities and detect fall events. In [17], cameras were installed in the ceiling which could see 77% of fall cases. In [18–21], the authors also used the vision-based system and analyzed the data using the classifiers PCA and SVM, and they achieved 89.2% and 90.3% sensitivity and specificity, respectively, in the detection of patient fall events. In [22], through the SVM classifier, the author achieved 83.2% accuracy in the detection of fall events. Some other practitioners used neural network techniques [23–26] to analyze patient fall data obtained through a vision-based system [27–30]. Three-dimensional depth image investigation, i.e., elderly fall detection based on 3D image shape analysis, of images captured by the kinetic sensor in the room environment was proposed in [31]. Subtraction methods analyze depth images. The human body centroids measure the angle between the human body and the floor level. When the angle is smaller than the defined thresholds, the fall events are detected, and several other techniques have been proposed [32,33]. However, in all the abovementioned cases, particular attention has not been devoted to addressing fall-relevant injuries.

In [34–37], the authors proposed wearable-based solutions, for example, sensors and wearable devices, to detect fall events and prevent those individuals from falling injuries. The abovementioned methods introduced wearable devices to protect the wearers from damage. The key feature of fall sensors is the fall-sensing algorithm, which uses both an accelerometer and angular velocity [38]. The fall-detection algorithm detects the fall signal. These airbags protect the head and thighs of a falling person [39]. However, it seems impractical to always wear an airbag and device. A micro-inertial measurement unit (mIMU) that consisted of three-dimensional MEMS Bluetooth, gyroscopes, accelerometers, a microcontroller unit (MCU), and high-speed cameras was used to record and analyze human motion [37–40]. The proposed technique was used to protect the thighs of a falling person; it may never protect the elder from a head injury as in [41]. In [42], the Elman neural network algorithm was used for fall detection. This algorithm was implemented in a wearable device combined with low-energy Bluetooth, which detected fall signals based on the second-order train method and sent a response to a remote PC.

In [43], the authors established an intellectual fall detector system based on infrared array detectors. This system monitors the elders in the house and generates an alarm in case a person is out of range or falls. Infrared sensors are extensively used to detect human motion due to their sensitivity. This novel sensing scheme also uses PIR sensors to detect elderly falls. Stereo infrared is intended to expand the sensing pitch to capture IRC signals entirely [44]. Fall-based image classification achieved by observing the human activities and classifying them using ML techniques to send the notification to the caretakers was performed in [45–48] to detect elderly falls. As discussed above, multiple methods have been proposed to detect falls, but very few services are used in practical life.

A vision-based system to detect falls was highlighted in [49,50]. It is a type of nonobtrusive fall-detection structure. This system uses image-processing methods for fall detection based on an image series or images captured from video clips recorded through a camera. The image analysis identifies the image of the elder so efficiently identifies an elderly fall. However, a visual system cannot provide good results in dark and shady environments [51–53].

Furthermore, connecting a camera affects elder privacy. A researcher used RFID to design numerous motion detection systems to detect elderly fall events [54–58]. Instead of utilizing on-body sensors, the proposed project uses PIR sensors and mounts them on the room's walls. In [59,60], the author used a wearable accelerometer to detect elderly falls using three datasets to detect the elderly fall accurately, while, in another, a Bluetooth-based, low-energy-enabled accelerometer sensor for the detection of patient bed falls was used. However, the authors worked on detecting falls and did not inform caregivers that these solutions were not cost efficient. In [61,62], the authors used a video camera to take human skeleton images and ultrasonic sensors and hybridized video techniques to detect elderly falls. The author used the same video-based fall-detection method as used in [63]. First, a person's silhouette is extracted through a feature extraction technique and then the characteristics set is measured to determine whether or not a fall occurred. A finite-state machine (FSM) was introduced to estimate the head position to compute the head vertical velocity. This algorithm was evaluated using the L2ei dataset, containing over 2700 labelled frames, to train three distinct classifiers. In [64,65], the authors also used video-based fall-detection techniques using the neural network approach. The authors of [66] proposed intellectual healthcare frameworks to examine the contribution and effectiveness of the axes of a tri-axial accelerometer sensor for accurate activity recognition [67]. Machine-learning-assisted cognitive strength assessment in smart houses is a multi-agent for healthcare systems based on contextual logic. Situational awareness in BDI, perceptive to detecting early COVID-19 symptoms through smartwatches, for improving the healthcare sector was explored in [68]. However, these techniques are not cost effective. We use a sensor-based, cost-effective technique for elderly fall detection and generate the alarm to inform caregivers. In this way, we can save the patient's life and enhance the patient's fall accuracy. Another solution is a vision-based solution [69,70], used to detect the elderly fall, and it is ideal as it covers an extensive area. However, always keeping an eye on the elderly affects the privacy of the elderly, and they do not feel comfortable in life. ML and deep learning models have been successfully used in several areas for classifying fall datasets [71–75].

In our proposed system, we use a low-cost, sensor-based system. We mounted the sensor on the wall to detect patient falls and analyzed the obtained data using highly mature, state-of-the-art, and representative algorithms.

In this regard, we used a PIR sensor and Arduino Uno to detect elderly falls. We collected datasets of normal motions and fall events for classification. So, to form the dataset, 20 persons contributed to performing normal motion and fall events, and the intensity of the signal generated by the PIR sensors in normal motions and the fall events was varied. This data were saved in Arduino SD. Additionally, a set of parameters was extracted to build datasets. The dataset which was obtained was classified using DT, SVM, RF, naïve Bayes (NB), and AdaBoost (AB) to increase the elderly fall detection event accuracy. Our proposed solution is cost effective. We can detect elderly falls at a very cheap cost, such as approximately PKR 1945/- per feet square. The cost of the equipment is shown in Table 1.

Table 1. Equipment cost.

Equipment	Price
Infrared Sensor	500/- × 3 =1500/-
Arduino Uno Microcontroller	200/- × 1 = 200/-
Transistor	=20/- × 2 = 40/-
Photodiode	=20/- × 3 = 60/-
Transformer	=80/- × 1 = 80/-
Capacitors	=5/- × 5 = 25/-
Resistors	=5/- × 5 = 25/-
Diodes	=3/- × 5 = 15/-
LCD Display	=400/- × 1 =400/-
Buzzer	=20/- × 1 =20/-

3. Proposed Methodology

In the proposed solution, we intend to reduce health issues to support the autonomous living of elders. The proposed project aims to use pyroelectric infrared (PIR) sensors to identify and detect human motions. The analog of PIR sensors is contingent on numerous aspects, such as the distance from the body, the direction of the sensor, the movement speed, and the body's shape, etc. By using the output of infrared sensors, we can detect near-to-fall events of a body in a unit of time in a unit area.

In our proposed system, the MC programing language is used to program Arduino Uno using the Arduino tool and is integrated with the circuit. The human-motion-detecting device consists of PIR sensors, Arduino Uno, Arduino SD, and Arduino GSM. The stages of our working methodology are revealed in Figure 1.

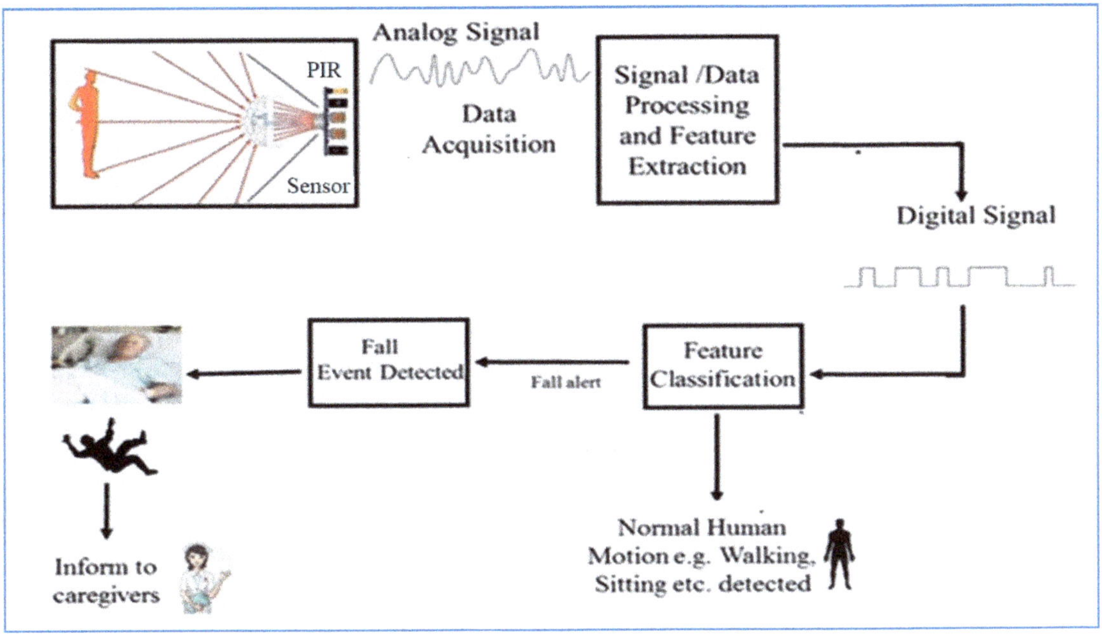

Figure 1. A low-cost fall-detection framework for the elderly.

The PIR sensors are mounted on the wall near the patient's or elderly person's bed. First, the data of normal and falling motions are collected through the PIR sensors. Typically, when a human walks around the room usually, the sensors identify the movements as usual motion based on speed and displacement in unit time. When an individual falls, the sign of the body increases significantly, and the sensors detect the greater speed than usual that the human body has increased to; consequently, the distance between the body and sensors decreases very fast. To differentiate normal walking from fall actions, machine-learning-based classifiers, such as SVM, DT, NV, RF, and AB for data reduction and classification, are used to classify the events as walking or falling. The distance and speed of the body are compared with the already stored datasets. In case the classifier detects an event such as a fall, a green signal is issued, and an alarm is generated to inform the family members or caregivers of the elderly person to prompt further assistance.

The workflow of our projected architecture is shown in Figure 2. Figure 2a shows the view of the PIR sensors; the experimental setup for statistics collection of normal motion is revealed in Figure 2b. PIR sensors generate an analog signal by detecting the speed of normal motion. We developed an operational amplifier (op-amp) circuit to amplify the PIR signal. In the proposed architecture, we use the following equipment to detect elderly fall, as shown in Figure 3: PIR sensors, Arduino Uno, Arduino SD, and Arduino GSM, transistor, photodiode, transformer, capacitors, resistors, diodes, buzzer, and jumper wires male–female.

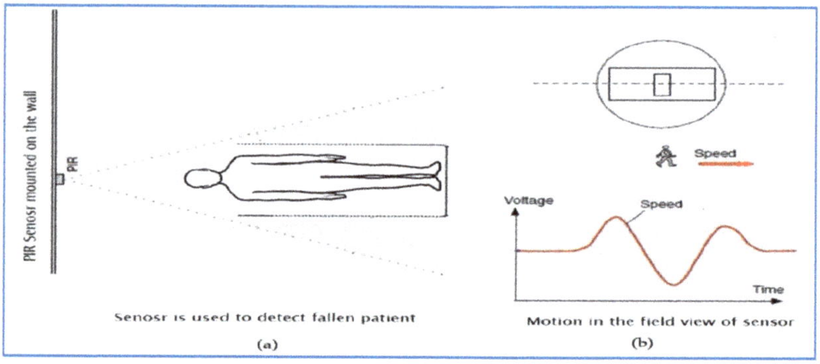

Figure 2. Data collection based on normal motion.

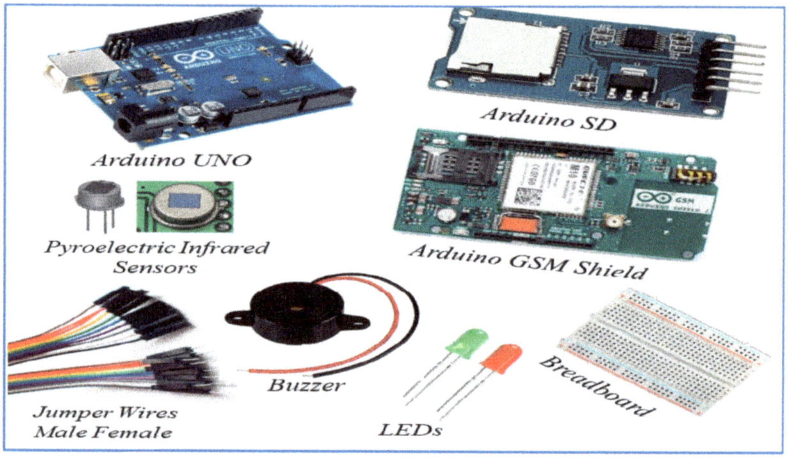

Figure 3. Equipment used to detect elderly falls.

Data Extraction Framework

We collected the datasets of regular motions and fall events for classification. For data collection, 20 persons contributed by performing normal movement and falling. From the signals generated by the PIR sensors, a set of parameters was extracted in different scenarios to build datasets. Experiments were performed in a living room. The PIR sensor module was mounted to the wall in the horizontal field of view to cover the maximum area, especially around the elderly person's bed. When an elderly person makes a regular motion, the PIR sensor detects the thermal radiation emitted from the human body and shows the different intensity of the analog output signals. These signals are compared with existing datasets and analyzed if certain conditions, such as current output signal range or intensity and near-to-fall dataset range intensity, are the same. This means that, if the intensity of the current signal is equal to the intensity of falling signals, then an alarm is generated and informs the caregivers. In addition, ML classifiers, i.e., NB, SVM, DT, RF, and AB, are used to increase elderly fall detection accuracy.

We collected data on normal motion and falling events using the PIR sensors and extracting the data generated by the PIR sensors to classify the data of falling events and the data of normal motion and detect elderly fall events. Feature extraction identified the signal intensity of falling and normal motion data range from the collected data. Unlike normal motion, fall motion activity consists of abrupt, fast motion velocity v. If the elder is near to the sensor, the field of view decreases, and, if the elder is far from the sensor, the field of view expands, as the angular size appears smaller to the sensor. PIR sensor p identifies and detects the normal motion, generating a different output signal. Further, we classified the normal motion and fall event from the obtained dataset using ML classifiers. In vector form, the productivity of the sensor p is as follows in Equation (1):

$$v_{p,t} = \varphi_p.x_t \sum_{i=1}^{D} \varphi_{p,\,i} \cdot x_{i,\,t,}\; p \in \{1,2,3\,,\ldots,\,O\} \quad (1)$$

where the "." symbol stands for the dot product, the D—dimensional row vector φ, p, and O represent the number of PIR sensors, respectively, and denote the visibility of all D cells to the path PIR sensor. The visibility function $\varphi\ p$, and i specify whether the ith cell is visible to the PIR sensor (p). To be more precise, the ith cell is invisible to sensor $\varphi\ p$ if i is valued at 0 and visible if p, i is valued at 1.

4. Implementation Details

We created a circuit board as shown in Figure 4 using jumper male–female wires to obtain amplified PIR sensor signals with op-amp circuits and PIR sensors, the outputs of which were associated to the analog inputs of an Arduino Uno, Arduino SD. Arduino GSM was used to inform caregivers. The analog signals were shown on an oscillator or a laptop by installing an Arduino PIR plotter, as shown in Figure 5. Arduino Uno was constituted to maintain each analog input as a time sequence on an Arduino SD card or memory card.

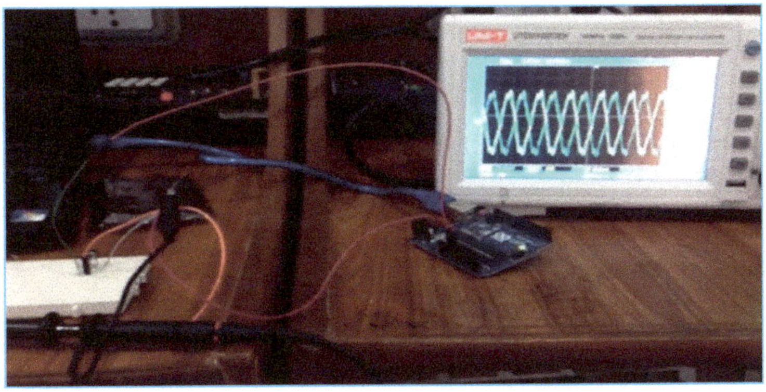

Figure 4. Circuit created to detect elderly falls.

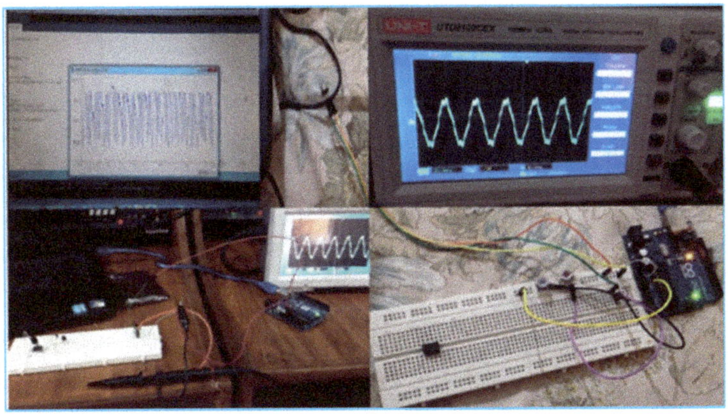

Figure 5. Circuit and analog signals are shown on the laptop and oscillator.

5. Analysis and Discussion

In this section, a state-of-the-art analysis is shown in Table 2, and experimental results are discussed.

Table 2. State of the art.

Author	Methods/Classifiers	Hardware/Evaluation Parameters	Limitations
[15]	Radio-frequency identification tags (RFID)	RFID tags, received signal strength (RSS), and Doppler frequency value (DFV)	The authors used distinct equipment and devices; however, the accuracy was not adequately high
[16]	Metaheuristic algorithms are used	Floor-based RFID technique, RFID tags arranged in a two-dimensional grid on a smart carpet	Inadequate accuracy
[17–31]	Digital camera, 3D image shape analysis, analysis using the PCA, SVM, NN algorithms	Vision-based system	Cameras were installed in the ceiling, detecting 77% of fall cases with 90% accuracy. Additionally, this affected the privacy of the elderly by monitoring the patient activities
[34–37]	Wearable-based solutions to protect the head and thighs	Sensors and wearable devices which use both an accelerometer and angular velocity to detect fall events	It seems impractical to wear an airbag and device all the time
[37–41]	Three-dimensional MEMS Bluetooth, accelerometers, microcontroller unit (MCU), gyroscopes, and high-speed cameras	High-speed cameras are used to record and analyze human motion	It seems impractical to use the device all the time
[42]	Neural network algorithm is used for fall detection	Implemented in a wearable device combined with low-energy Bluetooth	It seems impractical to wear the expedient all the time
[43]	Array-based detectors of smart inactivity	Intelligent fall indicator system based on infrared array detectors	Infrared radiation changes impact on elderly fall detection
Proposed Methodology	IRA-E700ST0 pyroelectric infrared sensors (PIR), Arduino Uno, SVM, DT, RF, NB, AB	Accuracy, precision (specificity), recall (sensitivity). RF achieves 99% accuracy in the detection of elderly fall events	Low-cost, sensor-based system with highly mature, state-of-the-art, and representative algorithms

5.1. PIR Analog Signals

The PIR analog signal simulation results for normal motion and high-intensity motion (falling/moving fast) are revealed in Figures 6 and 7. As you can see, the difference between Figures 6 and 7 is that, when the movement speed increased, the length of the signal increased compared to the normal motion. The analog data were extracted using one PIR sensor, then two and three PIR sensors, respectively, to make it more precise and accurate and then were converted into numeric data for analysis and classification. The PIR sensors generated the data in analog signals, as shown in Figures 6 and 7, and then Arduino converted the generated data from an analog signal to digital form as Arduino has a built-in analog-to-digital converter. Then, we used hand-crafted feature techniques. The main reason for adopting these hand-crafted features is their efficient, state-of-the-art performance. The data were used to train the approaches to detect and classify elderly falls. When we gained the data through Arduino, we applied ML approaches and split the data into training and testing at 80:20 ratios using the Python Jupyter Notebook platform to spread the ML approaches to gain accuracy according to different approaches, e.g., NB, AB, SVM, DT, and RF.

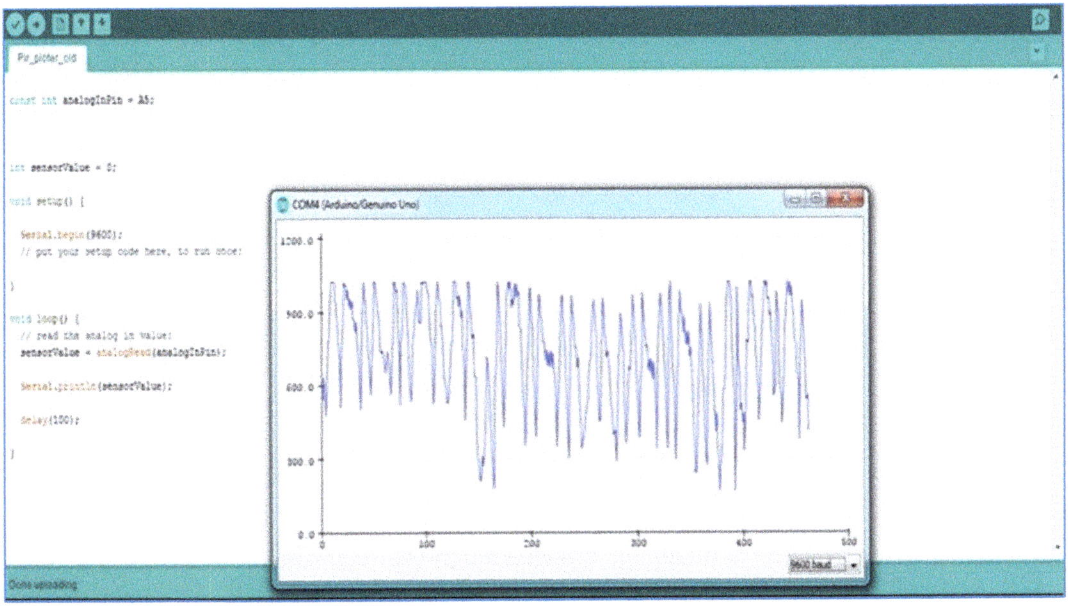

Figure 6. Simulation results of normal motion analog signals to detect elderly falls.

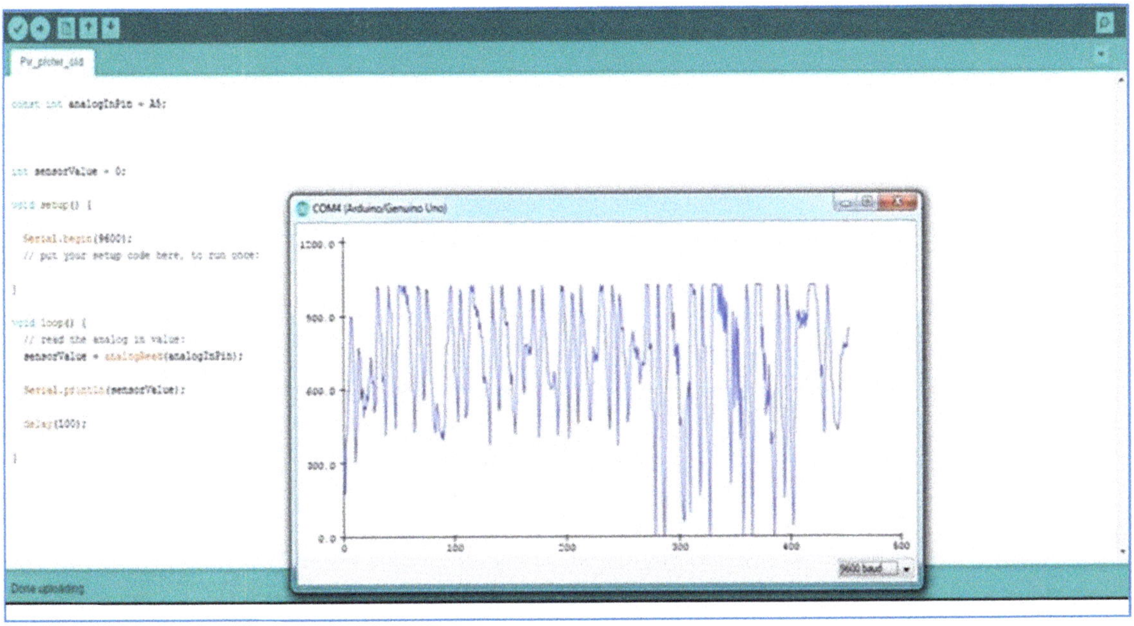

Figure 7. Simulation results of high-intensity motion analog signals to detect elderly falls.

5.2. Machine Learning Classifiers

ML is the use of computerized perception to give context and has the capacity to certainly take in and improve a fact deprived of being expressly modified. ML centers around the advancement of PC curricula to obtain and utilize information for their learning. By applying ML, classifiers compared the current data value to the predefined threshold. Based on this threshold, it could distinguish between different postures, falling, sitting, and standing. Feature or parameter extraction relevant to attributes or characteristics was identified from existing data and collected data [76]. Feature extraction efficiently identified and classified the normal motion and falling events. The feature had to be carefully picked to obtain smaller and more descriptive output datasets. Next, we extracted the practical value.

5.3. Support Vector Machine (SVM)

This section briefly introduces the SVM technique of ML for binary classification. The binary classifier can be articulated as a function f: Rn $\rightarrow \pm 1$, which maps patterns y onto their accurate classification x as $x = f(y)$. In the case of the SVM, the function f is formed as in [77] Equation (2):

$$f(x) = \sum_{i=1}^{N} x_i a_i \; k(y, \; y_i \;) + b \tag{2}$$

where N represents the training patterns, i is its classification, (x_i, y_i) are training patterns, learned α_i and b represent weights, and $k()$ shows the kernel function. We used the linear function $k(y, \; y_i \;) \; = \; y - y_i$ and the radial basis function $(y, \; y_i \;) \; = \; e^{-||y-y_i \; ||/2\sigma 2}$. The patterns $\alpha_i > 0$ are symbolized support vectors.

The kernel $k()$ communicates a hyperplane through into the component space through the surface $f(y) = 0$. The ranges from the support vectors and the hyperplane are maximized, and the weights I and b are chosen to reduce the number of incorrect classifications inside this training set. Solving the optimization model allows for this [78] to be maximized using Equation (3):

$$L_D = \sum_{i=1}^{N} a_i - \frac{1}{2} \sum_{i=1}^{N} \sum_{i=1}^{N} y_i \; y_j \; a_i a_j k\left(x_i, \; x_j \;\right) \tag{3}$$

Subject to Equation (4):

$$0 \leq a_i \leq C, \sum_{i=1}^{N} y_i a_i \tag{4}$$

The tolerance to incorrect classifications is influenced by the constant C. Equation (2), with either support vector (x_i, y_i), as in the data, can be used to find b using the ideal parameters i for a comprehensive explanation of the SVM. The SVM recognizes the event that has occurred and detects the signal change [79].

5.4. Decision Tree

A DT classifier is used for regression as well as classification. In a decision analysis system, a DT can be utilized to explicitly and visually represent decisions and decision making. In a DT, the continuous values or real numbers taken in the target variable are called regression trees. A DT is a tree-like or flowchart-like structure, where internal nodes represent an examination of a feature, the class label is represented by each leaf node (computing all features for decision), and conjunctions represent the branches of the structures that govern the class labels. The paths or the way from root towards leaf show the classification principals [80].

5.5. Random Forest

RF is a supervised ML classifier. RF classifiers are built up of multiple trees and merged to make a decision. So, the decision becomes more accurate and precise. It makes a forest and makes it somehow random. The "forest" it hypothesizes is a group of DT, more often than not equipped with the "bagging" approach. The universal thought of the bagging technique is that a combination of learning models increases the overall result [81].

5.6. Naïve Bayes

An NB approach is a probabilistic classifier; it is used for the classification of relevant tasks. The NB classifier is based on the Bayes theorem. By using the Bayes theorem, we determine the probability of c occurrence given that x has happened. Here, x is the proof, and c is the speculation. The supposition is made that the predictors or the features are self-determining. It means that the presence of one precise feature does not influence or affect the other feature. However, it may help to discover more. Subsequently, it is called naïve [82]. As shown in Equation (5), we use $P(c \mid x)$, showing posterior probability. $P(c)$ is class prior probability, $P(x \mid c)$ is likelihood, and $P(x)$ is predictor prior probability.

$$P(c|x) = \frac{P(x|c)P(c)}{P(x)} \tag{5}$$

5.7. Adaboost

AB is commonly used for classification problems and intends to change a weak classifier into a set of strong classifiers. The final classification can be represented in Equation (6) as:

$$F(x) = \mathrm{sign}(\textstyle\sum_{i=1}^{N} 0_m\, f_m(\mathrm{x})) \tag{6}$$

where $f(n)$ donates for the nth weak approach, and theta(n) is the correspondence weight. It is precisely the weighted blend of N weak classifiers given a dataset containing "n" points, where:

$$xi \in \mathrm{R}\, d,\ \in y_i \in \{-1,\ 1\} \tag{7}$$

In Equation (7), the negative class is denoted by -1, and the positive class is represented by 1. The weight of each data point is initialized as shown in Equation (8):

$$w(xi, yi) = \frac{1}{\mathrm{n}},\ i = 1\,,\ \ldots,\ n \tag{8}$$

Most detection approaches are based on hypothesis testing and statistical detection [83]. In our proposed approach, we use SVM, DT, RF, NB, and AB to detect the change that occurs in the signal characteristics and notify that the elder falling event has happened. As, by programming, we obtain the value "1" on the falling event, this value generates the alarm and informs the caregivers (using Arduino GSM) by sending a message.

5.8. Classifier Performance Assessment

Five ML algorithms (SVM, DT, NB, RF, and AB) were used to analyze the PIR sensor data we collected; their performance was then compared using the precision, recall, accuracy, and F-measure metrics shown in Table 3.

In this study, we primarily use accuracy, specificity, sensitivity, and the F-measure [84] to assess the ML classifier's efficiency. As a result, we calculate the focused class's specificity (precision) and sensitivity (recall) to evaluate the algorithm's predicted accuracy. For ML, the "TP—true positive", "FP—false positive", "TN—true negative", and "FN—false negative" rate is used to determine accuracy, recall, precision, and the F-measure. Each set of accurate predictions is further divided into all positive and all negative forecasts. TP, TN, FN, and FP were all predicted by all models. TP represents the vertical distance that an object actually falls. A predicted non-fall, or FN, is an expected non-fall. Falling prey to FP is like planning for a fall that never happens. In the real world, and in the foreseeable future, TN does not constitute a fall.

Accuracy is calculated as the number of accurately classified instances separated by the total number of cases inr the dataset, as shown in Equation (9):

$$\mathrm{Accuracy} = \frac{\mathrm{TP} + \mathrm{TN}}{\mathrm{TP} + \mathrm{FP} + \mathrm{TN} + \mathrm{FN}} \tag{9}$$

Specificity: the average possibility of relevant retrieval, as described in Equation (10):

$$\text{Specificity} = \frac{\text{TP}}{\text{TP} + \text{FP}} \quad (10)$$

Sensitivity is the average prospect of complete retrieval, defined in Equation (11):

$$\text{Sensitivity} = \frac{\text{TP}}{\text{TP} + \text{FN}} \quad (11)$$

F-measure is calculated through precision, as well as recall, as shown in Equation (12):

$$\text{F} - \text{Measure} = \frac{(2 * (\text{Precision} * \text{Recall}))}{\text{Precision} * \text{Recall}} \quad (12)$$

Table 3. Outcomes of ML models.

ML Algorithm	Accuracy	Precision (Specificity)	Recall (Sensitivity)	F-Measure
		Three PIR Sensors Dataset		
SVM	0.9711 (97%)	0.97	0.97	0.97
DT	0.9708 (97%)	0.97	0.97	0.97
NB	0.8942 (89%)	0.90	0.89	0.89
RF	0.9809 (98%)	0.98	0.98	0.98
AB	0.9904 (99%)	0.99	0.99	0.99
		Two PIR Sensors Dataset		
SVM	93%	0.93	0.93	0.93
DT	92%	0.91	0.92	0.92
NB	86%	0.86	0.86	0.86
RF	96%	0.96	0.96	0.96
AB	98%	0.98	0.98	0.98
		One PIR Sensor Dataset Accuracy		
SVM	89%	0.89	0.89	0.89
DT	89%	0.89	0.89	0.89
NB	82%	0.82	0.81	0.81
RF	87%	0.87	0.87	0.87
AB	86%	0.86	0.86	0.86

6. Experimental Results and Discussion

We experimented by applying one, two, and three PIR sensors. The results demonstrate that, by growing the number of sensors, we obtained positive effects by increasing the accuracy, having a more significant number of sensors covering the maximum area of the fall detection, and the matching ratio was increased, and we obtained the signal promptly. Moreover, when we considered the classifiers, we observed that the boosting and ensemble algorithms performed better because they were built by combining two or more than two classifiers, e.g., RF classifiers worked like the DT. Still, they had groups of DT, so the chance of accuracy was increased, and the SVM was the vector classification method. In such datasets, vector classification gives better results as it makes the classification better. The advantage of the proposed framework is that it is a low-cost framework, and the elderly do not need to wear a device all the time, and it also maintains the privacy of the elderly.

The background circumstance may alter the results, such as birds' movement, interior walls, and substances causing muddles and ghost boards due to scattered signals. We tested our system in such situations to obtain the effectiveness of our proposed model. In this regard, we obtained the minimum and maximum accuracy of one, two, and three PIR sensors by applying ML classifiers, as revealed in Figures 8–10. Moreover, in such situations, our experimental results indicated that the algorithm AdaBoost (AB) performed best, achieving a minimum accuracy of 87% on the PIR sensor elderly fall dataset classification, as shown in Figure 8.

In normal scenarios, the experimental results revealed that the algorithm AdaBoost (AB) produced better results, respectively, 98% and 99% in the situations where two or three IR sensors were installed or mounted in the room, as shown in Figures 11 and 12. However, in the case of a single sensor, the SVM gave 89% accuracy on the elderly fall dataset classification, as shown in Figure 13. The specificity and sensitivity of these ML classifiers are shown in Figures 14–16, respectively, for three, two, and one planted sensor.

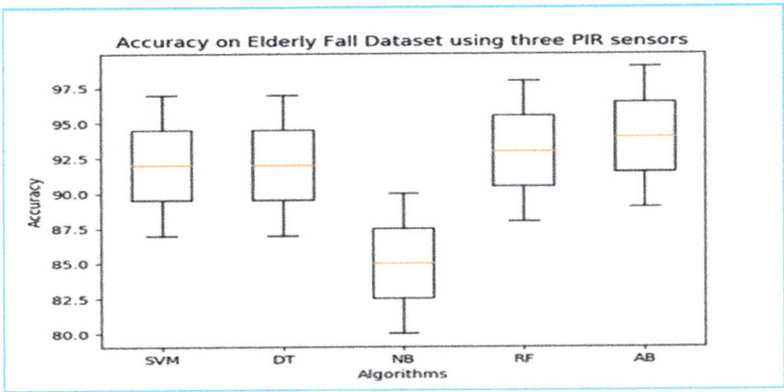

Figure 8. Accuracy on elderly fall dataset using three PIR sensors.

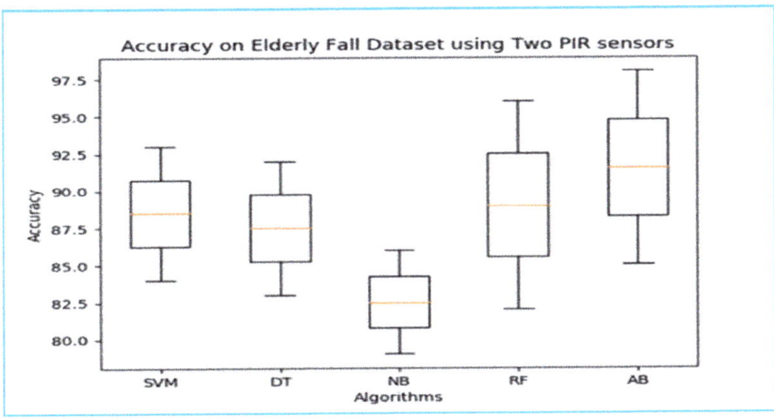

Figure 9. Accuracy on elderly fall dataset using two PIR sensors.

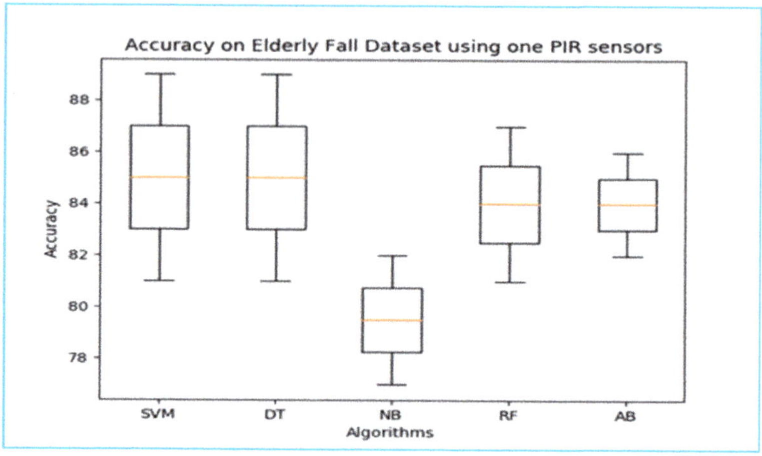

Figure 10. Accuracy on elderly fall dataset using one PIR sensor.

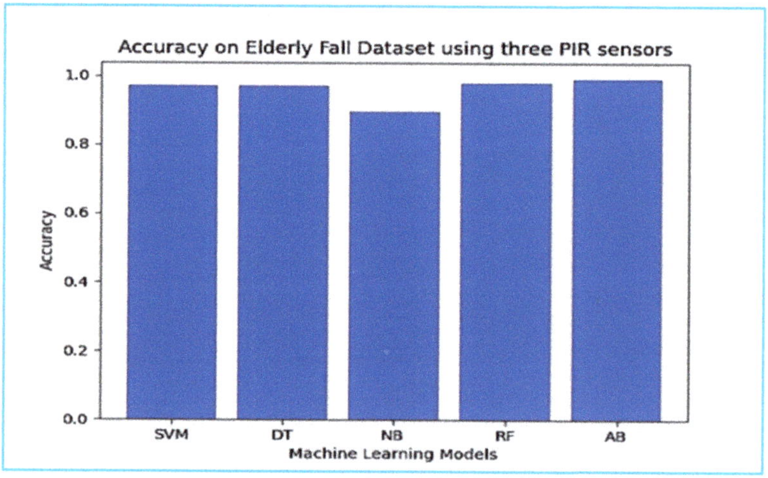

Figure 11. Accuracy using three IR sensors.

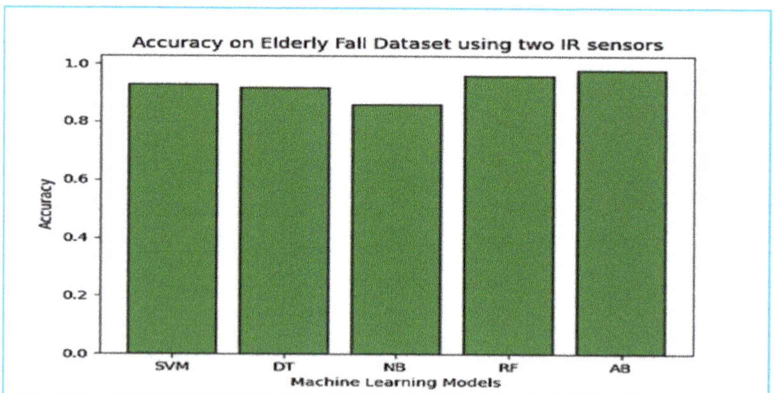

Figure 12. Accuracy using two IR sensors.

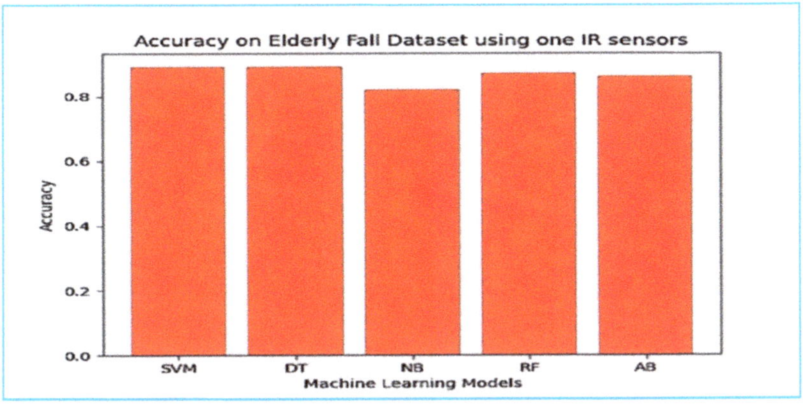

Figure 13. Accuracy using one IR sensor.

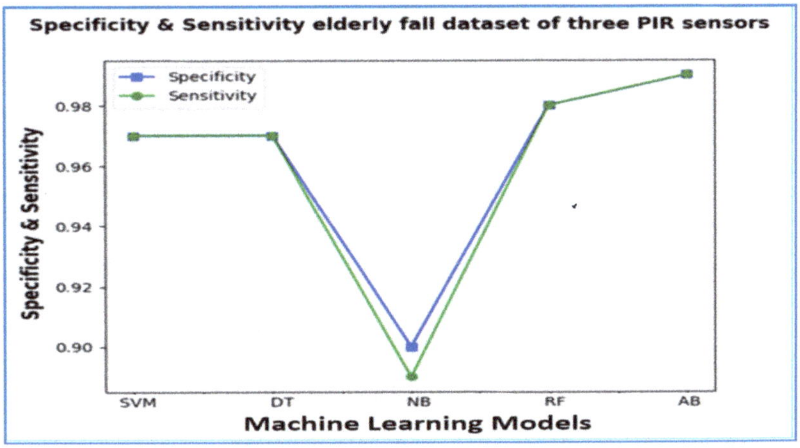

Figure 14. Specificity and sensitivity of elderly fall dataset of three PIR sensors.

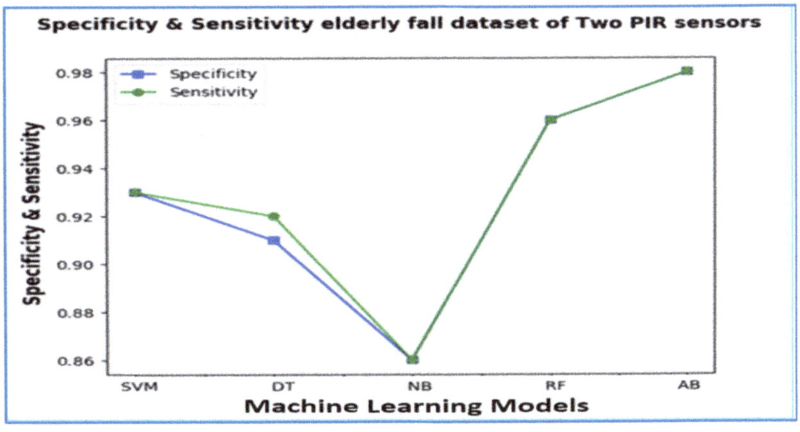

Figure 15. Specificity and sensitivity of elderly fall dataset of two PIR sensors.

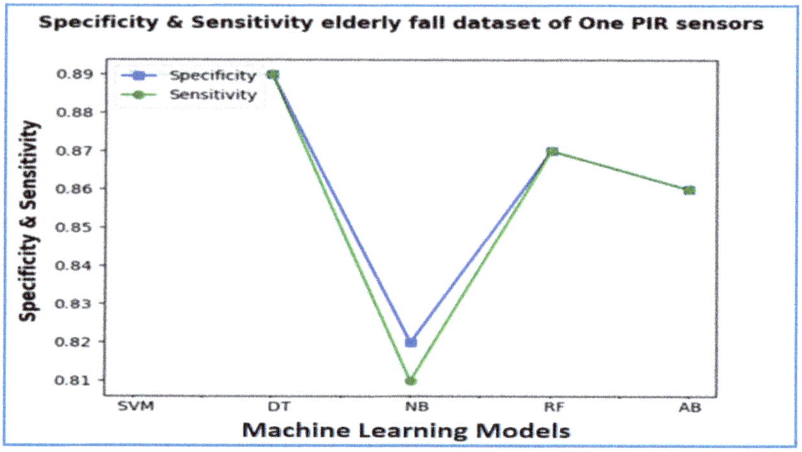

Figure 16. Specificity and sensitivity of elderly fall dataset of one PIR sensor.

7. Conclusions and Future Work

This paper proposed a sensor-based fall-detection scheme. The system detects elderly falls using IRA-E700ST0 pyroelectric infrared sensors which are mounted on the wall in a horizontal field of view. We considered physiological falls, lower-level falls, falls on a single level, and swing falls, enhancing the patient's falling accuracy and saving the patient's life. When the person is falling, the PIR sensor detects the high-intensity signal, turns on the green light, and generates an alarm to inform the caregivers that a fall event is happening. Furthermore, fall event dataset classification was performed using SVM, DT, NB, RF, and AB ML classifiers, and it was observed that ensemble and boosting algorithms more effectively classify fall event data.

In future work, we will integrate an ultrasonic sensor into the system to increase fall detection accuracy and enhance privacy and data security by combining blockchain technologies with the system to save data relating to the elderly person. Moreover, we will also try to incorporate floor airbags with PIR sensors operated using IoT devices for reliable and efficient services so protection from injuries as a result of falls is controlled. Earlier researchers used wearable airbags, and they lacked the ability to protect the entire body. Moreover, having to wear airbags at all times seems bothersome for the elderly who may already have declining health. So, in future, we will try to integrate floor airbags with the PIR sensors. When the elderly person falls, the sensor perceives the elderly person's fall motion and starts the motor to fill the airbag. So, when the elderly person falls, the airbag will blow up, the elderly person will fall on it, and, this way, we will prevent the elderly person from falling injuries.

Author Contributions: Writing—original draft, C.A.U.H., F.K.K., A.A., J.I., H.E., S.H., S.S.U. and M.S.K. All authors have read and agreed to the published version of the manuscript.

Funding: This research project was funded by Princess Nourah bint Abdulrahman University Researchers Supporting Project number (PNURSP2023R300), Princess Nourah bint Abdulrahman University, Riyadh, Saudi Arabia.

Data Availability Statement: Not applicable.

Acknowledgments: Authors acknowledge the Princess Nourah bint Abdulrahman University Researchers Supporting Project number (PNURSP2023R300), Princess Nourah bint Abdulrahman University, Riyadh, Saudi Arabia.

Conflicts of Interest: The authors declare no conflict of interest.

References

1. Daher, M.; Diab, A.; El Najjar, M.E.B.; Khalil, M.A.; Charpillet, F. Elder tracking and fall detection system using smart tiles. *IEEE Sens. J.* **2016**, *17*, 469–479. [CrossRef]
2. Falls. Available online: http://www.who.int/mediacentre/factsheets/fs344/en/ (accessed on 24 January 2019).
3. Igual, R.; Medrano, C.; Plaza, I. Challenges, issues and trends in fall detection systems. *Biomed. Eng. Online* **2013**, *12*, 66. [CrossRef]
4. Zhang, Z.; Conly, C.; Athitsos, V. A survey on vision-based fall detection. In Proceedings of the 8th ACM International Conference on PErvasive Technologies Related to Assistive Environments, Corfu, Greece, 1–3 July 2015; pp. 1–7.
5. Mubashir, M.; Shao, L.; Seed, L. A survey on fall detection: Principles and approaches. *Neurocomputing* **2013**, *100*, 144–152. [CrossRef]
6. Alwan, M.; Rajendran, P.J.; Kell, S.; Mack, D.; Dalal, S.; Wolfe, M.; Felder, R. A smart and passive floor-vibration based fall detector for elderly. In Proceedings of the 2006 2nd International Conference on Information & Communication Technologies, Damascus, Syria, 24–28 April 2006; IEEE: New York, NY, USA, 2006; Volume 1, pp. 1003–1007.
7. Chaccour, K.; Darazi, R.; El Hassans, A.H.; Andres, E. Smart carpet using differential piezoresistive pressure sensors for elderly fall detection. In Proceedings of the 2015 IEEE 11th International Conference on Wireless and Mobile Computing, Networking and Communications (WiMob), Abu Dhabi, United Arab Emirates, 19–21 October 2015; IEEE: New York, NY, USA, 2015; pp. 225–229.
8. Tzeng, H.W.; Chen, M.Y.; Chen, J.Y. Design of fall detection system with floor pressure and infrared image. In Proceedings of the 2010 International Conference on System Science and Engineering, Taipei, Taiwan, 1–3 July 2010; IEEE: New York, NY, USA, 2010; pp. 131–135.
9. Zhu, L.; Wang, R.; Wang, Z.; Yang, H. TagCare: Using RFIDs to monitor the status of the elderly living alone. *IEEE Access* **2017**, *5*, 11364–11373. [CrossRef]
10. Palmerini, L.; Bagalà, F.; Zanetti, A.; Klenk, J.; Becker, C.; Cappello, A. A wavelet-based approach to fall detection. *Sensors* **2015**, *15*, 11575–11586. [CrossRef]
11. Álvarez-García, J.A.; Soria Morillo, L.M.; Concepción, M.Á.Á.D.L.; Fernández-Montes, A.; Ortega Ramírez, J.A. Evaluating wearable activity recognition and fall detection systems. In Proceedings of the 6th European Conference of the International Federation for Medical and Biological Engineering, Dubrovnik, Croatia, 7–11 September 2014; Springer: Cham, Switzerland, 2015; pp. 653–656.
12. Aziz, O.; Musngi, M.; Park, E.J.; Mori, G.; Robinovitch, S.N. A comparison of accuracy of fall detection algorithms (threshold-based vs. machine learning) using waist-mounted tri-axial accelerometer signals from a comprehensive set of falls and non-fall trials. *Med. Biol. Eng. Comput.* **2017**, *55*, 45–55. [CrossRef]
13. Guan, Q.; Li, C.; Guo, X.; Shen, B. Infrared signal based elderly fall detection for in-home monitoring. In Proceedings of the 2017 9th International Conference on Intelligent Human-Machine Systems and Cybernetics (IHMSC), Hangzhou, China, 26–27 August 2017; IEEE: New York, NY, USA, 2017; Volume 1, pp. 373–376.
14. Shinmoto Torres, R.L.; Wickramasinghe, A.; Pham, V.N.; Ranasinghe, D.C. What if your floor could tell someone you fell? A device free fall detection method. In Proceedings of the Conference on Artificial Intelligence in Medicine in Europe, Pavia, Italy, 17–20 June 2015; Springer: Cham, Switzerland, 2015; pp. 86–95.
15. Wickramasinghe, A.; Torres, R.L.S.; Ranasinghe, D.C. Recognition of falls using dense sensing in an ambient assisted living environment. *Pervasive Mob. Comput.* **2017**, *34*, 14–24. [CrossRef]
16. Lee, T.; Mihailidis, A. An intelligent emergency response system: Preliminary development and testing of automated fall detection. *J. Telemed. Telecare* **2005**, *11*, 194–198. [CrossRef]
17. Rougier, C.; Meunier, J.; St-Arnaud, A.; Rousseau, J. Robust video surveillance for fall detection based on human shape deformation. *IEEE Trans. Circuits Syst. Video Technol.* **2011**, *21*, 611–622. [CrossRef]
18. Wang, S.; Chen, L.; Zhou, Z.; Sun, X.; Dong, J. Human fall detection in surveillance video based on PCANet. *Multimed. Tools Appl.* **2016**, *75*, 11603–11613. [CrossRef]
19. Lee, D.W.; Jun, K.; Naheem, K.; Kim, M.S. Deep Neural Network–Based Double-Check Method for Fall Detection Using IMU-L Sensor and RGB Camera Data. *IEEE Access* **2021**, *9*, 48064–48079. [CrossRef]
20. Gutiérrez, J.; Rodríguez, V.; Martin, S. Comprehensive review of vision-based fall detection systems. *Sensors* **2021**, *21*, 947. [CrossRef]
21. Zhou, Z.; Stone, E.E.; Skubic, M.; Keller, J.; He, Z. Nighttime in-home action monitoring for eldercare. In Proceedings of the 2011 Annual International Conference of the IEEE Engineering in Medicine and Biology Society, Boston, MA, USA, 30 August–1 September 2011; IEEE: New York, NY, USA, 2011; pp. 5299–5302.
22. Alhimale, L.; Zedan, H.; Al-Bayatti, A. The implementation of an intelligent and video-based fall detection system using a neural network. *Appl. Soft Comput.* **2014**, *18*, 59–69. [CrossRef]
23. Chen, J.; Romero, R.; Thompson, L.A. Motion Analysis of Balance Pre and Post Sensorimotor Exercises to Enhance Elderly Mobility: A Case Study. *Appl. Sci.* **2023**, *13*, 889. [CrossRef]
24. Wang, X.; Ellul, J.; Azzopardi, G. Elderly fall detection systems: A literature survey. *Front. Robot. AI* **2020**, *7*, 71. [CrossRef]
25. Khraief, C.; Benzarti, F.; Amiri, H. Elderly fall detection based on multi-stream deep convolutional networks. *Multimed. Tools Appl.* **2020**, *79*, 19537–19560. [CrossRef]
26. Yacchirema, D.; de Puga, J.S.; Palau, C.; Esteve, M. Fall detection system for elderly people using IoT and ensemble machine learning algorithm. *Pers. Ubiquitous Comput.* **2019**, *23*, 801–817. [CrossRef]

27. Hussain, F.; Umair, M.B.; Ehatisham-ul-Haq, M.; Pires, I.M.; Valente, T.; Garcia, N.M.; Pombo, N. An efficient machine learning-based elderly fall detection algorithm. *arXiv* **2019**, arXiv:1911.11976.
28. Bridenbaugh, S.A.; Kressig, R.W. Laboratory review: The role of gait analysis in seniors' mobility and fall prevention. *Gerontology* **2011**, *57*, 256–264. [CrossRef]
29. Baldewijns, G.; Claes, V.; Debard, G.; Mertens, M.; Devriendt, E.; Milisen, K.; Tournoy, J.; Croonenborghs, T.; Vanrumste, B. Automated in-home gait transfer time analysis using video cameras. *J. Ambient. Intell. Smart Environ.* **2016**, *8*, 273–286. [CrossRef]
30. Yang, L.; Ren, Y.; Zhang, W. 3D depth image analysis for indoor fall detection of elderly people. *Digit. Commun. Netw.* **2016**, *2*, 24–34. [CrossRef]
31. Leone, A.; Rescio, G.; Caroppo, A.; Siciliano, P.; Manni, A. Human Postures Recognition by Accelerometer Sensor and ML Architecture Integrated in Embedded Platforms: Benchmarking and Performance Evaluation. *Sensors* **2023**, *23*, 1039. [CrossRef]
32. Youssfi Alaoui, A.; Tabii, Y.; Oulad Haj Thami, R.; Daoudi, M.; Berretti, S.; Pala, P. Fall Detection of Elderly People Using the Manifold of Positive Semidefinite Matrices. *J. Imaging* **2021**, *7*, 109. [CrossRef]
33. Tamura, T.; Yoshimura, T.; Sekine, M.; Uchida, M.; Tanaka, O. A wearable airbag to prevent fall injuries. *IEEE Trans. Inf. Technol. Biomed.* **2009**, *13*, 910–914. [CrossRef]
34. Shi, G.; Chan, C.S.; Luo, Y.; Zhang, G.; Li, W.J.; Leong, P.H.; Leung, K.S. Development of a human airbag system for fall protection using mems motion sensing technology. In Proceedings of the 2006 IEEE/RSJ International Conference on Intelligent Robots and Systems, Beijing, China, 9–15 October 2006; IEEE: New York, NY, USA, 2006; pp. 4405–4410.
35. Shi, G.; Chan, C.S.; Li, W.J.; Leung, K.S.; Zou, Y.; Jin, Y. Mobile human airbag system for fall protection using MEMS sensors and embedded SVM classifier. *IEEE Sens. J.* **2009**, *9*, 495–503. [CrossRef]
36. Saleh, M.; Jeannès, R.L.B. Elderly fall detection using wearable sensors: A low cost highly accurate algorithm. *IEEE Sens. J.* **2019**, *19*, 3156–3164. [CrossRef]
37. Wang, Z.; Ramamoorthy, V.; Gal, U.; Guez, A. Possible life saver: A review on human fall detection technology. *Robotics* **2020**, *9*, 55. [CrossRef]
38. Chuma, E.L.; Roger, L.L.B.; De Oliveira, G.G.; Iano, Y.; Pajuelo, D. Internet of things (IoT) privacy–protected, fall-detection system for the elderly using the radar sensors and deep learning. In Proceedings of the 2020 IEEE International Smart Cities Conference (ISC2), Piscataway, NJ, USA, 28 September–1 October 2020; IEEE: New York, NY, USA, 2020; pp. 1–4.
39. Yu, X.; Jang, J.; Xiong, S. Machine learning-based pre-impact fall detection and injury prevention for the elderly with wearable inertial sensors. In Proceedings of the International Conference on Applied Human Factors and Ergonomics, virtually, 25–29 July 2021; Springer: Cham, Switzerland, 2021; pp. 278–285.
40. Sankaran, S.; Thiyagarajan, A.P.; Kannan, A.D.; Karnan, K.; Krishnan, S.R. Design and Development of Smart Airbag Suit for Elderly with Protection and Notification System. In Proceedings of the 2021 6th International Conference on Communication and Electronics Systems (ICCES), Coimbatre, India, 8–10 July 2021; IEEE: New York, NY, USA, 2021; pp. 1273–1278.
41. Chu, C.T.; Chang, C.H.; Chang, T.J.; Liao, J.X. Elman neural network identify elders fall signal base on second-order train method. In Proceedings of the 2017 6th International Symposium on Next Generation Electronics (ISNE), Keelung, Taiwan, 23–25 May 2017; IEEE: New York, NY, USA, 2017; pp. 1–4.
42. Sixsmith, A.; Johnson, N. A smart sensor to detect the falls of the elderly. *IEEE Pervasive Comput.* **2004**, *3*, 42–47. [CrossRef]
43. Hayashida, A.; Moshnyaga, V.; Hashimoto, K. New approach for indoor fall detection by infrared thermal array sensor. In Proceedings of the 2017 IEEE 60th International Midwest Symposium on Circuits and Systems (MWSCAS), Boston, MA, USA, 6–9 August 2017; IEEE: New York, NY, USA, 2017; pp. 1410–1413.
44. Jeffin Gracewell, J.; Pavalarajan, S. Fall detection based on posture classification for smart home environment. *J. Ambient. Intell. Humaniz. Comput.* **2021**, *12*, 3581–3588. [CrossRef]
45. Badgujar, S.; Pillai, A.S. Fall Detection for Elderly People using Machine Learning. In Proceedings of the 2020 11th International Conference on Computing, Communication and Networking Technologies (ICCCNT), Kharagpur, India, 1–3 July 2020; IEEE: New York, NY, USA, 2020; pp. 1–4.
46. Kalinga, T.; Sirithunge, C.; Buddhika, A.G.; Jayasekara, P.; Perera, I. A Fall Detection and Emergency Notification System for Elderly. In Proceedings of the 2020 6th International Conference on Control, Automation and Robotics (ICCAR), Singapore, 20–23 April 2020; IEEE: New York, NY, USA, 2020; pp. 706–712.
47. Nahian, M.; Raju, M.H.; Tasnim, Z.; Mahmud, M.; Ahad, M.A.R.; Kaiser, M.S. Contactless fall detection for the elderly. In *Contactless Human Activity Analysis*; Springer: Cham, Switzerland, 2021; pp. 203–235.
48. Edgcomb, A.; Vahid, F. Automated fall detection on privacy-enhanced video. In Proceedings of the 2012 Annual International Conference of the IEEE Engineering in Medicine and Biology Society, San Diego, CA, USA, 28 August–1 September 2012; IEEE: New York, NY, USA, 2012; pp. 252–255.
49. Ngo, Y.T.; Nguyen, H.V.; Pham, T.V. Study on fall detection based on intelligent video analysis. In Proceedings of the 2012 International Conference on Advanced Technologies for Communications, Ha Noi, Vietnam, 10–12 October 2012; IEEE: New York, NY, USA, 2013; pp. 114–117.
50. Shieh, W.Y.; Huang, J.C. Speedup the multi-camera video-surveillance system for elder falling detection. In Proceedings of the 2009 International Conference on Embedded Software and Systems, Hangzhou, China, 25–27 May 2009; IEEE: New York, NY, USA, 2009; pp. 350–355.

51. Anderson, D.; Luke, R.H.; Keller, J.M.; Skubic, M.; Rantz, M.; Aud, M. Linguistic summarization of video for fall detection using voxel person and fuzzy logic. *Comput. Vis. Image Underst.* **2009**, *113*, 80–89. [CrossRef]
52. Foroughi, H.; Aski, B.S.; Pourreza, H. Intelligent video surveillance for monitoring fall detection of elderly in home environments. In Proceedings of the 2008 11th International Conference on Computer and Information Technology, Khulna, Bangladesh, 24–27 December 2008; IEEE: New York, NY, USA, 2009; pp. 219–224.
53. Ruan, W.; Yao, L.; Sheng, Q.Z.; Falkner, N.J.; Li, X. Tagtrack: Device-free localization and tracking using passive rfid tags. In Proceedings of the 11th International Conference on Mobile and Ubiquitous Systems: Computing, Networking and Services, London, UK, 2–5 December 2014; pp. 80–89.
54. Su, B.Y.; Ho, K.C.; Rantz, M.J.; Skubic, M. Doppler radar fall activity detection using the wavelet transform. *IEEE Trans. Biomed. Eng.* **2014**, *62*, 865–875. [CrossRef]
55. Liu, L.; Popescu, M.; Rantz, M.; Skubic, M. Fall detection using doppler radar and classifier fusion. In Proceedings of the 2012 IEEE-EMBS International Conference on Biomedical and Health Informatics, Hong Kong, China, 5–7 January 2012; IEEE: New York, NY, USA, 2012; pp. 180–183.
56. Liu, L.; Popescu, M.; Skubic, M.; Rantz, M.; Cuddihy, P. An automatic in-home fall detection system using Doppler radar signatures. *J. Ambient. Intell. Smart Environ.* **2016**, *8*, 453–466. [CrossRef]
57. Kim, Y.; Moon, T. Human detection and activity classification based on micro-Doppler signatures using deep convolutional neural networks. *IEEE Geosci. Remote Sens. Lett.* **2015**, *13*, 8–12. [CrossRef]
58. Goodfellow, I.; Bengio, Y.; Courville, A. Machine learning basics. In *Deep Learning*; MIT Press: Cambridge, MA, USA, 2016; pp. 98–164.
59. Lin, W.Y.; Chen, C.H.; Lee, M.Y. Design and Implementation of a Wearable Accelerometer-Based Motion/Tilt Sensing Internet of Things Module and Its Application to Bed Fall Prevention. *Biosensors* **2021**, *11*, 428. [CrossRef]
60. Ramirez, H.; Velastin, S.A.; Meza, I.; Fabregas, E.; Makris, D.; Farias, G. Fall detection and activity recognition using human skeleton features. *IEEE Access* **2021**, *9*, 33532–33542. [CrossRef]
61. Hsu, F.S.; Chang, T.C.; Su, Z.J.; Huang, S.J.; Chen, C.C. Smart Fall Detection Framework Using Hybridized Video and Ultrasonic Sensors. *Micromachines* **2021**, *12*, 508. [CrossRef]
62. Zheng, L.; Zhao, J.; Dong, F.; Huang, Z.; Zhong, D. Fall detection algorithm based on inertial sensor and hierarchical decision. *Sensors* **2023**, *23*, 107. [CrossRef]
63. Sehairi, K.; Chouireb, F.; Meunier, J. Elderly fall detection system based on multiple shape features and motion analysis. In Proceedings of the 2018 International Conference on Intelligent Systems and Computer Vision (ISCV), Fez, Morocco, 2–4 April 2018; IEEE: New York, NY, USA, 2018; pp. 1–8.
64. Amsaprabhaa, M. Multimodal spatiotemporal skeletal kinematic gait feature fusion for vision-based fall detection. *Expert Syst. Appl.* **2023**, *2*, 118–681.
65. Chen, Z.; Wang, Y.; Yang, W. Video Based Fall Detection Using Human Poses. In Proceedings of the Big Data: 9th CCF Conference, BigData 2021, Guangzhou, China, 8–10 January 2022; Springer: Singapore, 2022; pp. 283–296.
66. Javed, A.R.; Sarwar, M.U.; Beg, M.O.; Asim, M.; Baker, T.; Tawfik, H. A collaborative healthcare framework for shared healthcare plan with ambient intelligence. *Hum. Cent. Comput. Inf. Sci.* **2020**, *10*, 40. [CrossRef]
67. Javed, A.R.; Sarwar, M.U.; Khan, S.; Iwendi, C.; Mittal, M.; Kumar, N. Analyzing the effectiveness and contribution of each axis of tri-axial accelerometer sensor for accurate activity recognition. *Sensors* **2020**, *20*, 2216. [CrossRef]
68. Javed, A.R.; Shahzad, F.; ur Rehman, S.; Zikria, Y.B.; Razzak, I.; Jalil, Z.; Xu, G. Future smart cities requirements, emerging technologies, applications, challenges, and future aspects. *Cities* **2022**, *129*, 103794. [CrossRef]
69. Hussain, T.; Hussain, D.; Hussain, I.; AlSalman, H.; Hussain, S.; Ullah, S.S.; Al-Hadhrami, S. Internet of Things with Deep Learning-Based Face Recognition Approach for Authentication in Control Medical Systems. *Comput. Math. Methods Med.* **2022**, *2022*, 5137513. [CrossRef]
70. Hussain, I.; Hussain, D.; Kohli, R.; Ismail, M.; Hussain, S.; Sajid Ullah, S.; Alroobaea, R.; Ali, W.; Umar, F. Evaluation of Deep Learning and Conventional Approaches for Image Recaptured Detection in Multimedia Forensics. *Mob. Inf. Syst.* **2022**, *2022*, 2847580. [CrossRef]
71. Bourouis, S.; Sallay, H.; Bouguila, N. A Competitive Generalized Gamma Mixture Model for Medical Image Diagnosis. *IEEE Access* **2021**, *9*, 13727–13736. [CrossRef]
72. Faruk, O.; Ahmed, E.; Ahmed, S.; Tabassum, A.; Tazin, T.; Bourouis, S.; Khan, M.M. A Novel and Robust Approach to Detect Tuberculosis Using Transfer Learning. *J. Healthc. Eng.* **2021**, *2021*, 1002799. [CrossRef]
73. Bourouis, S.; Laalaoui, Y.; Bouguila, N. Bayesian frameworks for traffic scenes monitoring via view-based 3D cars models recognition. *Multimed. Tools Appl.* **2019**, *78*, 18813–18833. [CrossRef]
74. Bourouis, S.; Pawar, Y.; Bouguila, N. Entropy-Based Variational Scheme with Component Splitting for the Efficient Learning of Gamma Mixtures. *Sensors* **2022**, *22*, 186. [CrossRef]
75. Kumar, S.; Jain, A.; Rani, S.; Alshazly, H.; Ahmed Idris, S.; Bourouis, S. Deep neural network based vehicle detection and classification of aerial images. *Intell. Autom. Soft Comput.* **2022**, *34*, 119–131. [CrossRef]
76. Sidenbladh, H. Detecting human motion with support vector machines. In Proceedings of the 17th International Conference on Pattern Recognition ICPR, Cambridge, UK, 26 August 2004; IEEE: New York, NY, USA, 2004; Volume 2, pp. 188–191.

77. Grahn, J.; Kjellstromg, H. Using SVM for efficient detection of human motion. In Proceedings of the 2005 IEEE International Workshop on Visual Surveillance and Performance Evaluation of Tracking and Surveillance, Beijing, China, 15–16 October 2005; IEEE: New York, NY, USA, 2006; pp. 231–238.
78. Cao, D.; Masoud, O.T.; Boley, D.; Papanikolopoulos, N. Human motion recognition using support vector machines. *Comput. Vis. Image Underst.* **2009**, *113*, 1064–1075. [CrossRef]
79. Charbuty, B.; Abdulazeez, A.; Abdulazeez, A. Classification based on decision tree algorithm for machine learning. *J. Appl. Sci. Technol. Trends* **2021**, *2*, 20–28. [CrossRef]
80. Liu, J.; Ulishney, C.; Dumitrescu, C.E. Random forest machine learning model for predicting combustion feedback information of a natural gas spark ignition engine. *J. Energy Resour. Technol.* **2021**, *143*, 012301. [CrossRef]
81. Vangara, V.; Vangara, S.P.; Thirupathur, K. Opinion Mining Classification using Naive Bayes Algorithm. *Int. J. Innov. Technol. Explor. Eng. (IJITEE)* **2020**, *9*, 495–498. [CrossRef]
82. Shahraki, A.; Abbasi, M.; Haugen, Ø. Boosting algorithms for network intrusion detection: A comparative evaluation of Real AdaBoost, Gentle AdaBoost and Modest AdaBoost. *Eng. Appl. Artif. Intell.* **2020**, *94*, 103770. [CrossRef]
83. Al Nahian, J.; Ghosh, T.; Al Banna, H.; Aseeri, M.A.; Uddin, M.N.; Ahmed, M.R.; Mahmud, M.; Kaiser, M.S. Towards an accelerometer-based elderly fall detection system using cross-disciplinary time series features. *IEEE Access* **2021**, *9*, 39413–39431. [CrossRef]
84. Hashim, H.A.; Mohammed, S.L.; Gharghan, S.K. Accurate fall detection for patients with Parkinson's disease based on a data event algorithm and wireless sensor nodes. *Measurement* **2020**, *156*, 107573. [CrossRef]

sustainability MDPI

Article

Efficient Scheduling of Home Energy Management Controller (HEMC) Using Heuristic Optimization Techniques

Zafar Mahmood [1], Benmao Cheng [2,*], Naveed Anwer Butt [1], Ghani Ur Rehman [3], Muhammad Zubair [3], Afzal Badshah [4] and Muhammad Aslam [5,6]

1 Department of Computer Science, University of Gujrat, Gujrat 50700, Pakistan
2 Jiangsu Key Lab of IoT Application Technology, Wuxi Taihu University, Wuxi 214064, China
3 Department of Computer Science and Bioinformatics, Khushal Khan Khattak University, Karak 27000, Pakistan
4 Department of Computer Science & Software Engineering, International Islamic University, Islamabad 44000, Pakistan
5 School of Computing Engineering and Physical Sciences, University of West of Scotland, Paisley G72 0LH, UK
6 Scotland Academy, Wuxi Taihu University, Wuxi 214064, China
* Correspondence: chengbm@wxu.edu.cn

Abstract: The main problem for both the utility companies and the end-used is to efficiently schedule the home appliances using energy management to optimize energy consumption. The microgrid, macro grid, and Smart Grid (SG) are state-of-the-art technology that is user and environment-friendly, reliable, flexible, and controllable. Both utility companies and end-users are interested in effectively utilizing different heuristic optimization techniques to address demand-supply management efficiently based on consumption patterns. Similarly, the end-user has a greater concern with the electricity bills, how to minimize electricity bills, and how to reduce the Peak to Average Ratio (PAR). The Home Energy Management Controller (HEMC) is integrated into the smart grid, by providing many benefits to the end-user as well to the utility. In this research paper, we design an efficient HEMC system by using different heuristic optimization techniques such as Genetic Algorithm (GA), Binary Particle Swarm Optimization (BPSO), and Wind Driven Optimization (WDO), to address the problem stated above. We consider a typical home, to have a large number of appliances and an on-site renewable energy generation and storage system. As a key contribution, here we focus on incentive-based programs such as Demand Response (DR) and Time of Use (ToU) pricing schemes which restrict the end-user energy consumption during peak demands. From the results figures, it is clear that our HEMC not only schedules all the appliances but also generates optimal patterns for energy consumption based on the ToU pricing scheme. As a secondary contribution, deploying an efficient ToU scheme benefits the end-user by paying minimum electricity bills, while considering user comfort, at the same time benefiting utilities by reducing the peak demand. From the graphs, it is clear that HEMC using GA shows better results than WDO and BPSO, in energy consumption and electricity cost, while BPSO is more prominent than WDO and GA by calculating PAR.

Keywords: optimization techniques; demand-supply system; energy consumption patterns; genetic algorithm; particle swarm optimization; wind driven optimization; home energy management controller

check for updates

Citation: Mahmood, Z.; Cheng, B.; Butt, N.A.; Rehman, G.U.; Zubair, M.; Badshah, A.; Aslam, M. Efficient Scheduling of Home Energy Management Controller (HEMC) Using Heuristic Optimization Techniques. *Sustainability* **2023**, *15*, 1378. https://doi.org/10.3390/su15021378

Academic Editors: Mourade Azrour, Azidine Guezzaz, Imad Zeroual, Azeem Irshad, Jamal Mabrouki, Said Benkirane and Shehzad Ashraf Chaudhry

Received: 2 December 2022
Revised: 27 December 2022
Accepted: 9 January 2023
Published: 11 January 2023

1. Introduction

The energy requirement of commercial as well as residential users is increasing day by day. At the same time, a different generating unit of electricity faces a shortage of energy due to line losses and unpredictable energy demand from the end-users. For this reason, engineers and scientists are looking to adopt and implement a strategy that must be safe with reliable transmission and delivery. The researchers in this field are also looking to establish interactive communication between the end-user and utility by introducing advanced information and control technologies [1]. As a result of different efforts, they proposed and developed different solutions based on smart grid technology to improve the reliability of the grid, thus providing communication between the end-user and the utility by minimizing the environmental issues caused by fossil-fueled generation [2]. Around the globe, the energy requirements of end-users and industry are increasing linearly, while power generation from different sources and the reliable transmission of the power is much slower than the consumption [3]. The traditional grid lacks communication between the users and the utility causing inefficient operation on both the demand and supply sides.

Smart grid technology majorly relays on different types of renewable energy and efficient demand-supply management. With the increasing energy consumption demand, requirements to switch from traditional fossil-fueled generation to smart grid technology will become a prominent research area [4–6]. As a result of the smart grid, the resultant energy will be environment-friendly, cheaper, and easy-to-use on-site energy, by addressing the stability and irregular nature of Renewable Energy systems (RES) [7–10].

A survey was held in the United States to determine household electricity usage. Based on the survey results, different household appliances consumed almost 42% of residential energy [11]. Researchers in this field propose and design new prototypes and standards based on energy consumption patterns for the residential electricity market by coping with energy optimization [12,13]. For this purpose, they introduced and deployed new technologies such as on-site RES and grid power, an advanced metering system, controllable appliances, an energy storage system, and intercommunication between utility companies. For this purpose, they introduced and deployed new technologies such as distributed energy generation (on-site RES and grid power), an advanced metering system, controllable appliances, an energy storage system, intercommunication between utility companies and end-users, and a stand-alone storage system. Using the two-way communication mechanism, end-users will be permitted to access their energy usage and pricing information. On the other hand, introducing the different pricing schemes at the retailer level allows an opportunity for the electricity consumer to minimize his electricity bills by shifting from peak hour to shoulder [14]. As we do not have a large-scale energy storage system, a balanced mechanism for energy generation and consumption must be implemented to avoid complete shutter-down and load-shedding problems [15]. In this article, we propose a mathematical model and implement a controlling mechanism:

- The Home Energy Management Controller (HEMC) enhances energy efficiency and improves the comfort level within a single residential home, without taking into account the retailer side data for load forecasts and different pricing schemes for scheduling.
- Using a two-way communication mechanism a user can shift appliances from high to off-peak time by increasing high demand at a particular time interval.
- To achieve this shift from high to low-peak, another mechanism Demand Response (DR) is introduced which is in response to the changes in the price of electricity over a certain time interval offered by the utility company, end-users also change their electricity consumption patterns from their normal routine to achieve many benefits from the subsidized patterns. In this work, we proposed HEMC based on load shifting technique and user preferences.
- We consider a typical home with 10 different types of electrical appliances, having a variable length of operation time to show the effectiveness of the proposed algorithms.

For energy optimization problems different researchers proposed different techniques such as Linear Programming (LP), Dynamic Programming (DP), Artificial Intelligence (AI) inspired techniques such as Genetic Algorithm (GA), Binary Particle Swarm Optimization (BPSO), Ant Colony Optimization (ACO), Wind Driven Optimization (WDO), etc., for energy consumption patterns, minimizing electricity cost calculation, maximizing user comfort and PAR based on different pricing schemes, different types of appliances having different operation time. We solve the same problems using GA, BPSO, and WDO with a different load-shifting strategy to achieve better results. Our proposed algorithm considers two types of energy, grid energy and RES, with the storage system. For grid energy, we use the ToU pricing scheme which is fixed for a duration or season.

A typical home with an energy management system is depicted in Figure 1. This figure also represents the flow of the paper and the system model deployed in the paper. The proposed system model consists of RES which may be Photo Voltaic Cell (PVC), wind energies, and power utilities that supply electricity from the main grid. Electricity from the power grid is directly transmitted to the smart meter, whereas renewable energy is first transmitted to the proposed HEMC and then stored in the storage system deployed in the smart home. All the schedulable appliances in a typical smart home are connected with the proposed HEMC system, which optimally scheduled their operation to switch them to renewable and power grid to save costs while mainlining the user comfort levels. On the other hand, non-schedulable appliances which are fixed to be operated or based on their demands (while maintaining user comfort levels) also directly communicate with the proposed HEMC system to switch their operation on the RES system or power grid to reduce the cost and PAR values.

In the future, more incentives will be offered by different retailer companies to fascinate the user by reducing the peak-to-average ratio [16–18]. Both customers and the utility companies take advantage by using a demand response program [19,20]; the utility companies introduce different pricing schemes based on energy consumption at a certain time unit, which encourages the customer to shift their requirement from peak demand in response to the subsidized incentive [21,22]. A typical home with an energy management system is depicted in Figure 1.

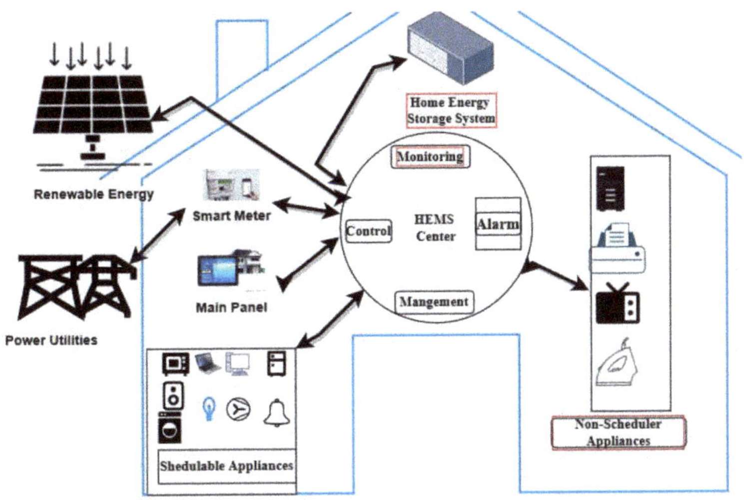

Figure 1. Overall structure of the proposed system.

Figure 1 also represents the flow of the paper and the system model deployed in the article. The proposed system model consists of RES which may be Photo Voltaic Cell (PVC), wind energies, and power utilities that supply electricity from the main grid. Electricity from the power grid is directly transmitted to the smart meter, whereas renewable energy is first transmitted to HEMC and then store in the storage system deployed in the smart home. All the schedulable appliances in a typical smart home are connected with the HEMC system, which optimally scheduled their operation to switch them to renewable and power grid to save costs while mainlining the user comfort level. On the other hand, non-schedulable appliances which are fixed to be operated or based on their demands (while maintaining user comfort levels) also directly communicate with the HEMC system to switch their operation on the RES system or power grid to reduce the cost and PAR values.

The rest of the paper material is structured as follows: Section 2 discusses the related work. Section 3 discusses the proposed schemes and system model architecture, and the appliance energy consumption patterns are discussed, respectively. In Section 4, we describe the different algorithms used in the smart scheduler, and a detailed discussion of the load optimization problem, WDO, and PSO algorithms is presented. The results of the simulations and performance evaluation metrics are given in Section 5, and Section 6 highlights the conclusion of the paper.

2. Literature Review

Energy consumption and optimization are the core issues in designing a smart grid while maintaining user comfort and reducing costs, as pointed out by [23–25]. These papers give an overview of PSO and its applications when used in different fields of life. The authors propose and design an optimization technique based on particle swarm optimization due to its robust nature and easy implementation in complex and non-linear problems. Due to its fast convergence nature, PSO-based scheduling offers an easy way to shift peak and high-peak demand to low-peak time intervals [26], at the same time allowing the end-user to use an incentive-based package to reduce the cost of electricity. This work rationalizes the importance and usefulness of DSM in efficient home energy management systems. To manage more expanded and elaborated ideas related to efficient energy management in the smart grid, such robust PSO optimization techniques can be used.

To cope with increasing energy demand effectively and efficiently, electric power supply and consumption patterns can be simulated for effective grid management. Over time, the generation capacity of electricity is decreasing day by day in traditional grids, facing several challenges related to delivering the increasing power demand. In traditional grids, data/information can only flow in one direction; from the grid to the end-used. To make two-way communication possible between the grid and the customer, smart grid technology is introduced, which intelligently interacts with the grid and user to efficiently carry and distribute viable, maintainable, cost-effective, cheap, and secure electricity supplies [27,28]. Following the smart grid mechanism, a scheduler is proposed and designed using different AI techniques to schedule different types of home appliances. AI-based techniques such as GA, PSO, ACO, WDO, etc., have been in use for the last couple of years for energy optimization problems. In the article [17], the author designed and implement a scheduling scheme based on PSO for different home appliances with a predefined length of operation time. The paper also presents a comprehensive review of PSO with its other counterpart algorithms. The proposed scheme takes both the interruptable and uninterruptible load for the scheduling. The authors in [29] integrate a fuzzy logic-based thermostat with the home energy management system to minimize battery degradation to prevent unnecessary renewable energy arbitrage. The proposed system employs day-ahead load scheduling to save costs, offers the best possible Demand Response (DR) and Photovoltaic (PV) self-consumption, and the fuzzy logic-based controller aims for effective DR of air conditioning while maintaining thermal comfort. Reinforcement learning is used by the authors in [30] to construct a home energy management system. The home electric appliance systems, which contribute to the most important loads in a household, are regulated by HEMC based on machine learning and reinforcement learning, allowing consumers to save power while still improving their comfort. By correctly optimizing and addressing the optimal use of renewable energy sources, the proposed method is examined for monitoring home electric appliances to reduce energy consumption.

Heuristic algorithms play an important role in energy optimization problems [31]. Paper [32] highlights a genetic algorithm-based smart scheduler for optimal power scheduling, and scheduled home usage appliances of different types in the HEM system. The author used time-of-use pricing signals and real-time pricing schemes for scheduling purposes, to minimize total energy cost and energy consumption. A comparison with unscheduled loads is carried out in the paper, which proves that GA shows more prominent results in cost reduction and energy consumption. The DR program based on GA and PSO is proposed in [33], by sightseeing the appliance scheduling schemes to help the end-user ease by minimizing the electricity bill, without compromising his/her comfort. At the same time, the proposed DR program facilitates retailer companies to stabilize the grid by optimally reducing peak demand. The simulation was carried out for unscheduled and scheduled DR programs by implementing GA and PSO as a hybrid approach. The results show that the hybrid approach has better insight into energy consumption patterns as compared to BPSO, whereas in cost analysis, the hybrid approach shows more prominent results than unscheduled ones. A similar concept is also highlighted in [30].

The authors in [34,35] propose and implement a demand-side management scheme to reduce PAR and minimize the electricity cost while considering user preferences. The proposed scheme is based on a heuristic-based evolutionary algorithm by shifting the load during high-peak hours to facilitate both the customer and grid. Both papers deploy on-site energy generation and storage unit, which is integrated with the grid energy. The results demonstrate that during high-peak hours, some of the load is shifted to RES by reducing PAR and minimizing the cost. The technique in the papers scheduled a large number of appliances of different types. The authors in [36,37] use a heuristic-based evolutionary algorithm to reduce the PAR and minimize the cost by shifting the peak hour load to different sources.

The author in [38] studied a typical home load management problem for different classes of appliances using the ToU pricing scheme. Appliances are categorized based on the length of operation time and energy consumption. The author first proposed and developed an efficient mathematical model for different classes, using this model an efficient and optimal algorithm is designed to condense and decrease the overall electricity consumption and bill as well as peak lessening and saving during subsidized hours by maintaining user comfort levels. Time scheduling flexibility is introduced for each class of appliances so that users can adopt any model based on requirement and priority. Simulation results demonstrate that the proposed algorithm optimally scheduled the home appliances based on energy consumption requirements and patterns, the ToU pricing scheme, and operation time.

In [39], the author proposed a demand-side management technique using WDO and BPSO algorithms. The smart meter communicates with both the grid and the end-used, taking price signals directly from the grid and energy demand and requests from the different appliances. The energy management controller takes this information from the smart meter and performs its calculation to schedule all the appliances, by keeping in view the peak hour and price signal. The schedule was transmitted to the end-used's appliance and grid company. Furthermore, in the suggested model, the author tries to balance electricity cost and waiting time for different appliances to provide benefits not only to the utility company but also to the end-user. To further reduce and minimize the cost, a well-known mathematical problem formulation technique "min-max regret knapsack" is used, and this technique is compared to that of simple optimization algorithms. The simulation results show that WDO gives better results than BPSO for the waiting time of appliances and cost reduction.

The authors in [40] proposed an HEM system for residential users to reduce their electricity costs and PAR using two approaches, HEM with microgrid and HEM without microgrid. The proposed HEM not only scheduled the home appliances but also electric vehicle charging and discharging optimally while maintaining user comfort. Each end-user has its microgrid which is connected to their grid, having a solar panel, gas turbine, wind turbine, and energy storage system (ESS). The authors use linear programming techniques to formulate the scheduling problems. The simulated results demonstrate that linear programming techniques can efficiently schedule different smart appliances and electric vehicles according to electricity generated by the microgrid. Using microgrid generation causes a reduction of the PAR and total cost of electricity. The authors in [41] designed a mathematical model to integrate different energy sources having small-scale generation capacity. The model, which is based on an intelligent multi-objective named home energy management (MOHEM), aims to reduce the end-user electricity bill along with the system's peak demand by efficiently scheduling a smart residential home. The authors use the super criterion approach and the Pareto optimal solution ideas to deploy the cooperative game theory approach.

To improve the resiliency of the system, the authors in [42] proposed a new approach that is based on a genetic algorithm to empower the system planners to effectively handle different resiliency matrices in a bi-objective optimization planning model. A novel mixed-integer model was proposed by the authors in [43] to control the performance and efficiency of a LESS when used in conjunction with a DR scheme. The suggested approach includes cutting-edge managerial options, such as different DR activities that are permitted and the quantity of charging and discharging that is permitted. Furthermore, the model is built to be able to filter out the times when the demand side is prohibited from engaging in DR. The authors in [44] designed a multi-objective-based model to efficiently operate the demand and supply of a Smart Microgrid (SMG). The basic aim of the proposed model is to minimize the operation cost of the model, discharging pollution chemicals, and customer desired demand and usage curve in the daytime. A major contribution of the proposed model is to introduce an objective function used by SMG operators to balance the customer demands according to the supply with shiftable loads. The author Finlay uses fuzzy logic and the weighted sum approaches to choose the best solution.

The authors in [45] proposed a model based on multi-objective optimization, using a hydrogen storage system (HSS) while considering responsive consumers (RC). The basis of the proposed objective function is to increase the reliability of the system and minimize the operational cost and the gap between the demand and supply of the electricity. The author further used the Monte Carlo simulation model to effectively deal with uncertainties in the system. The end model was deployed by using the Shuffled Frog Leaping Algorithm (SFLA), through which the non-dominated solution is generated. Fuzzy logic and the weighted sum approach were used for the best solution. The authors in [46] designed a technique called MORL (Multi-Objective Reinforcement Learning algorithm) to effectively deal with the demand response to reduce the energy usage pattern while maintaining user comfort. If the proposed scheme is compared with conventional approaches, the earlier scheme alleviates the result of different end-user preferences and handles the indecision of future prices and renewable energy generation. Table 1 presents the comparison of the existing schemes and the proposed scheme.

Table 1. Comparison of Existing Schemes and Proposed Scheme.

Article with Authors	Limitations of Existing Schemes	Novelties of Proposed Scheme
Zhao et al. [32]	The authors used time-of-use pricing signals and real-time pricing schemes for scheduling purposes, to minimize total energy cost and energy consumption.	The proposed HEMC system is based on GA, BPSO, and WDO to reduce the PAR which will minimize the electricity bills and thus increase the user comfort level.
Aslam et al. [40]	HEM system based on linear programming is used for residential and electric vehicles, to reduce the electricity cost and optimize power consumption.	The proposed HEMC system uses an offline storage system (ESS), and a different heuristic algorithm to optimally schedule the home appliances to reduce the overall electricity bill designed for residential purposes.
Lokeshgupta et al. [41]	Multi-Objective Home Energy Management (MOHEM) is proposed to handle a small home energy demand. The authors use the cooperative game theory approach in their study based on super-criterion and a Pareto optimal solution concept.	The proposed HEMC system is stimulated with a Smart Scheduler (SS) which act intelligently, and all the appliances use two communication mechanism to communicate with the utility companies and SS using a smart meter.

Table 1. *Cont.*

Article with Authors	Limitations of Existing Schemes	Novelties of Proposed Scheme
Chen et al. [46]	Multi-Objective Reinforcement Learning algorithm (MORL) to design a DR program to minimize energy usage while maintaining user comfort. Energy consumption based on a 24-h equal interval pricing scheme is important for both end-user and utility companies, by distributing the load properly in the H-hour horizon to offer maximum offer in terms of electricity cost.	The proposed system uses a different heuristic algorithm based on a linear programming model to optimize the appliance's scheduling.
Bina et al. [38]	A typical home load management problem for different classes of appliances, using the ToU pricing scheme. Appliances are categorized based on the length of operation time and energy consumption. The author first proposed and developed an efficient mathematical model for different classes, using this model an efficient and optimal algorithm is designed to condense and decrease the overall electricity consumption and bill as well as peak lessening and saving during subsidized hours by maintaining user comfort levels.	The proposed HEMC implements a demand-side management technique, using WDO and BPSO algorithms. The smart meter communicates with both the grid and the end-user, taking price signals directly from the grid and energy demand and requests from the different appliances. The energy management controller takes this information from the smart meter and performs its calculation to schedule all the appliances, by keeping in view the peak hour and price signal. The schedule was transmitted to the end-user's appliance and grid company.

3. Proposed Scheme and System Model for HEMC

This section carries the discussion of an ideal and ultimate approach for scheduling and managing the required power and power consumption of an ideal home having a large number of appliances is proposed based on a specific pricing scheme. Since most of the end-users still use traditional electromechanical meters, utility companies use Fixed Retailer Price (FRP) models for the end-user which is a fixed price all the time. Smart meters as a replacement for old and traditional methods will be used to record energy consumption reading in a real environment with high accuracy and minimum effort. These utility companies use different incentives and subsidy-based pricing schemes for the customer to reduce energy demand and thus stabilize the grid. Some of the important and more used pricing schemes are ToU and Real-Time Pricing (RTP): In the earlier pricing schemes, 24 h of the day are equally divided into equal intervals and the price for each interval is known in advance by the users. Thus, a user can schedule his/her appliances based on the price signals, i.e., in peak hours, the user tries to turn on fewer appliances to reduce energy consumption, and hence minimize the cost. While in the shoulder and off-peak interval, most appliances are turned on. The RTP is somehow like ToU, where the price is based on end-user energy demand varying each hour. In this work, we present an inventive prototypical strategy to determine the required energy and electricity usage pattern for typical home electrical appliances in advance. The proposed HEMC system is stimulated with a Smart Scheduler(SS) which acts intelligently, and all the appliances use two communication mechanisms to communicate with the utility companies and SS using a smart meter.

The smart meter receives a price signal from the grid and passes it to the SS, on the other hand, different appliances send on/off requests to a smart meter, which also passes the on to the SS. The SS knows in advance the operation time for each appliance. Taking all these inputs from the grid and appliances through the smart meter, the SS generates energy consumption patterns and schedules all the appliances in the given domain of search space according to the price signals and operation time to decide the optimal time for the smart appliances to minimize energy consumption, minimizing electricity cost, while considering user comfort and reducing the peak to average ratio. Our proposed schemes also consider on-site RES generation and storage systems. During off-peak time intervals when energy cost is minimal, the SS utilizes the grid energy, and when the grid energy cost is at a maximum, the end-user shifts the load from the grid to the RES system to maximize the user comfort level. After performing many simulations, the results demonstrate the minimum cost in terms of electricity bills for the end-user having HEMC compared to those who have no infrastructure and installed architecture for HEMC at their homes.

Energy consumption based on a 24-h equal interval pricing scheme is important for both end-users and utility companies, by distributing the load properly in the H hour horizon to offer the maximum in terms of electricity costs. Given a set of different home-based appliances, a great matter of concern is to carry and distribute the energy power load efficiently in the H hour intervals, such that the installer of the HEMS obtains maximum profit out of the system, i.e., $A = a_1, a_2, a_3, \ldots\ldots\ldots\ldots\ldots, a_{24}$. Each listed appliance requires a different energy level to be operated and their consumption rating is shown in Table 6. Each appliance uses two-way communication with the SS of the HEMC. Smart grids and smart meters continuously exchange the demand and electricity cost frequently, as different utility companies offer different incentives and subsidy-based prices over 24-h time intervals, namely high-peak, low-peak, and off-peak, as listed in Table 7. The end-user tries to operate a maximum number of appliances in off-peak and low-peak hours, to fulfill his requirement, and operate fewer appliances in high-peak hours to minimize electricity bills, respectively. Reviewing the different pricing schemes and energy demand highlights that high-peak energy consumption will charge more to the customer, as compared to low-peak hours in a 24-h interval. Keeping in mind the different pricing schemes, each user optimally consumes energy, thus minimizing his/her electricity bill. We propose our model for the optimization problem given by Equations (1) and (2):

$$H = h_1, h_2, h_3, \ldots\ldots\ldots\ldots\ldots, h_{24} \tag{1}$$

$$A = a_1, a_2, a_3, \ldots\ldots\ldots\ldots\ldots, a_{24} \tag{2}$$

We divide 24 h into equal intervals, $H = h_1, h_2, h_3, \ldots\ldots\ldots\ldots\ldots, h_{24}$, in a fixed horizon of time interval h, the scheduling of every appliance must take into consideration different time bounds such as start time, finish time, and length of the operation time, H_s, H_f, and H_{Iot}, respectively. Each appliance may have a scheduling time interval between $[H_0, H_{max}]$, where each hour has a different price signal. During 24-h time intervals, the energy consumption vector is given in Equation (3):

$$E_T = E_1^{t_1}, E_2^{t_2}, E_3^{t_3}, \ldots\ldots\ldots\ldots\ldots\ldots E_n^{t_n} \tag{3}$$

where $E_1^{t_1}$ is the sum of consumed energy by the first appliance in a fixed time horizon t_1 and so on. The inclusive unbiased and objective function is to decrease the cost of electricity, formulated in Equations (4)–(6):

$$min(C_h) = \sum_{h_1=0}^{h_{24}} C_h \tag{4}$$

Subject to:

$$\sum_{a_1=0}^{a_n} C_h \sum_{h_1=0}^{h_{24}} C_h(E_{h_1} a_1) \leq Egrid \tag{5}$$

where

$$1 \leq h_1 \leq h_{24} \tag{6}$$

where h_1 to h_{24} represents the 24-h horizon from 0 to 24, C_h represents the cost of energy at a particular hour, a_1 represents the set of the appliance, $E_{h_1} a_1$ represents the energy consumed by the appliance a_1 during h_1 time horizon, *Egrid* and means energy from the grid. The end-user pays electricity costs in terms of electricity bills to the utility company for the energy consumption of different home-based appliances in a particular time interval over a 24-h horizon. The cost of appliances is the cost of energy consumption of a particular home-based appliance turned on in a specific time slot h. The cost is estimated mathematically by using Equation (7) and (8):

$$\sum_{a_1=0}^{a_n} C_h \sum_{h1=0}^{h_{24}} (E_{h,load}, *c_h) \tag{7}$$

$$\forall h \in h_1, h_2, h_3, \ldots\ldots\ldots\ldots\ldots, h_{24} \tag{8}$$

where C_h is the electricity cost for the time interval h, $E_{h,load}$ is energy demand by the appliance a in a specific time slot h and is calculated by using Equation (9):

$$\sum_{a_1=0}^{a_n} C_h \sum_{h_1=0}^{h_{24}} (E_{h,load}, *\alpha_{h,a}) \tag{9}$$

where $\alpha_{h,a}$ is a Boolean variable having the value 0 or 1, mathematically defined in Equation (10):

$$\alpha_{h,a} = \begin{cases} 1, & if(applianceisON) \\ 0, & if(ApplianceisOFF) \end{cases} \tag{10}$$

$\alpha_{h,a}$ represents the status of appliance a, the appliance operates and consumes energy in that specific time slot h if $\alpha_{h,a}$ is 1, and off if $\alpha_{h,a}$ is 0. The smart meter receives an on or off signal from the SS, which then further communicates with all the household appliances and sends a control signal, i.e, $\alpha_{h,a}$ to different appliances to change their state. This is mathematically calculated by using Equation (11):

$$E_{h_1,a_1} = \begin{cases} E_a & if(\alpha_{h,a} = 0) \\ 0 & if(\alpha_{h,a} = 1) \end{cases} \tag{11}$$

where h represents the time interval from 0 to 24 and presents appliance 1. In the proposed research work, we have N number of household appliances in our home, so $\alpha_{h,a}$ is the N binary bits pattern. As discussed in the previous sections, the HEMS are also equipped with renewable sources of energy to generate some part of the energy from photovoltaic plates. In our proposed model, we assume that at least 45% of its total energy demand will be generated by the RES and stored. Since RES cannot fulfill all the energy requirements of the end-user, the end-user must be connected to the main grid for the shortage of energy. Thus, the end-user will consume both the grid and on-site RES energy. The hourly energy production of a single photovoltaic module in KW_h is given by Equation (12):

$$E_{RES,h} = \forall h \varepsilon (h_1, h_2, h_3, \ldots\ldots\ldots\ldots\ldots, h_{24}) \tag{12}$$

The RES generated energy added to the HEMC system from a non-site installed RES system is, therefore, using Equation (13):

$$E_{RES,h} = \sum_{a_1=0}^{a_n} \sum_{h_1=0}^{h_{24}} (E_{RES,h}) \tag{13}$$

The peak to the average ratio for GA, BPSO, and GA can be calculated as dividing the maximum energy consumption of all appliances by the average energy consumption in a particular time interval and is given by Equation (14):

$$PAR = \frac{max_{load}}{average_{load}} \tag{14}$$

In the proposed solution, we consider a typical home with N number of appliances, with different power consumption rates and length of operation time. The energy consumption is calculated over a 24-h equal time interval. HEMC controls all household appliances by communicating with the utility which takes energy signals directly from the utility, and different appliances request through the smart meter. The scheduling time, i.e., a whole day is equally divided into slots. HEMC while considering the available energy capacity C_t calculates the time-bound in terms of starting T_s and finishing T_f time intervals, as well as the energy consumption of each appliance in a given time interval. The energy consumption during all time intervals can be calculated by using Equation (15) and (16):

$$E_T = e_1^{t_1}, e_2^{t_2}, e_3^{t_3}, \ldots\ldots\ldots\ldots\ldots\ldots e_n^{t_n} \tag{15}$$

$$T = t_1, t_2, t_3, \ldots\ldots\ldots\ldots\ldots, t_{24} \tag{16}$$

The scheduling time horizon during which appliances can be scheduled is given by Equation (17):

$$T_{sch} = T_{max} - T_{lot} \tag{17}$$

where T_{sch} is the time taken by SS to schedule an appliance, T_{max} is the maximum time available for scheduling, and T_{lot} represents the length of operation time. As WDO and BPSO have binary variables so particles are initialized randomly for binary positions as shown in Equation (18):

$$X_i = [X_i1, X_i2, X_i3, \ldots\ldots\ldots X_in], \forall X_i1, X_i2, X_i3, \ldots\ldots\ldots X_in \in 0,1 \tag{18}$$

Each binary value having a probability of 0.5 is assigned to each particle in each dimension and is given by Equation (19):

$$Xi_d = f(X) = \begin{cases} 1 & if(rand \geqslant 0) \\ 0 & otherwise \end{cases} \tag{19}$$

where $d = 1\ldots\ldots\ldots, N$ represents the position of each particle in the N dimension. To obtain the global best position the position of each particle is updated and is given by Equation (20):

$$X_{(i_d)}^{(k=1)} = f(X) = \begin{cases} 1, & if(rand \geqslant 0) \\ 0, & if(rand < sigmoid(V_{i_d}^{k=1})) \end{cases} \tag{20}$$

where the sigmoid function is calculated by using Equation (21):

$$sigmoid(V_{i_d}^{k=1}) = \frac{1}{1 + exp - (V_{i_d}^{k=1})} \tag{21}$$

After random initialization, each particle in the solution space moves randomly to avoid premature convergence, and the velocity $V_{(i,n)}$ of each particle is updated using the result given by Equation (22):

$$V_{(i,n)} = wv^t_{(i,n)} + (C_1 rand(1) * (p^t_{lb_{(i,n)}} - x^t_{(i,n)})) + (C_2 rand(1) * (p^t_{gb_{(i,n)}} - x^t_{(i,n)})) \quad (22)$$

where C1 and C2 are the weights for the local best and global best of a particle moving with velocity y and position x. $rand(1)$ is a random variable whose value is between $[0,1]$, and w is the inertia factor. The notations used in the proposed scheme are listed in Table 2.

Table 2. Summary of notations and symbols.

Symbol	Description
w	Inertia factor
$rand(1)$	Random variable whose value is between $[0,1]$
C1	Weights for local best of a particle
C2	Weights for global best of a particle
$V_{(i,n)}$	Velocity of each particle
T_{sch}	Time taken by SS to schedule an appliance
T_{max}	Maximum time available for scheduling
T_{lot}	Length of operation time
C_t	Available energy capacity
T_s	Calculates the time-bound in terms of starting and finishing time intervals
T_f	Energy consumption of each appliance in a given time interval
KW_h	Single photovoltaic module
$\alpha_{h,a}$	Household appliances in our home
h_1 to h_{24}	Represents the 24-h horizon from 0 to 24
C_h	Cost of energy at a particular hour
a_1	Set of the appliance
$E_{h_1}a_1$	Energy consumed by the appliance a_1 during h_1 time horizon
$Egrid$	Energy from the grid
E_T	Sum of consumed energy by the first appliance in a fixed time horizon t_1
H_s	Appliance tart time
H_f	Appliance finish time
H_{Iot}	Length of the operation time
$[H_0, H_{max}]$	Appliance scheduling time interval

4. Proposed Algorithm for HEMC

This section presents different algorithms for HEMC. The algorithm for WDO, BPSO, and GA is implemented in this section.

4.1. Genetic Algorithm

We propose and implement three different algorithms WDO, BPSO, and GA to manage the energy consumption and energy consumption patterns for a typical home that has a large number of appliances with different energy requirements and lengths of operation time. The resultant HEMC system not only generates an energy consumption pattern but also calculates the energy consumption and minimizes the electricity bill while considering user comfort and reducing PAR. To address all these problems, the design scheme should be able to tackle all these involutions. In the past, researchers have used different techniques such as LP and DP, but as the complexity of the problem increases, these techniques are not able to handle such a large number of appliances. Algorithms inspired by AI, such as WDO, GA, and BPSO have the potential to solve such types of complex problems. As compared to other algorithms, GA provides the finest solution for the cost optimization problem, for this reason, we use a GA-based scheduling algorithm. The smart meter interactively communicates with utility companies and appliances and sends the input to SS, using GA for scheduling purposes. The HEMC controller cumulatively deals with the appliances in a defined time interval and gives a complete pattern by solving the minimization problem. The SS operates at the beginning of the day, after sending a request from the appliance to the SS controller, the action taken by SS is based on GA techniques used to schedule all the appliance's energy consumption patterns in advance. The chromosome configurations of GA represent the solution, i.e., a schedule for appliances of when and how to operate [47]. In this research work, the ON/OFF status of each appliance is represented by an array of bits. Thus, the length of chromosomes depends on the number of controllable appliances. Here in this work, we use 10 different appliances so:

$$Lengthofchromosomes N = (10) \tag{23}$$

where N represents the total number of appliances. The initial population of chromosomes is randomly initialized, and the initial population is then sent to an objective function, which finds the fitness value for each chromosome.

The GA iterates the population many times, and in every iteration, as a result, a new population is produced by crossover and mutation. As we know that mutation rate and crossover directly affect the convergence of the algorithm, different techniques for crossover such as a uniform crossover, arithmetic crossover, two-point crossover, single-point crossover, and mutation can be used, and here in this work, we use single-point crossover and binary mutation. If we use a larger crossover rate, the algorithm will converge fast, and if using a larger mutation rate, there may be a chance to lose some good solutions, which results in the permute convergence of the algorithm.

It is possible that sometimes in the early population, GA finds an optimal solution but gets missed by crossover and mutation rate. In every population, one finest solution is selected and remembered. The elitism technique is used to record this best solution, which is then forwarded to the next generation. Different techniques exist to merge the population to generate a new population, here we use the tournament-based selection method to make a new parent from the existing population. Different parameters used in GA are shown in Table 3. The Algorithm used in the SS is given in Algorithm 1.

Algorithm 1 GA Algorithm used in the SS

1: Initial generation h = 0
2: Randomly create an initial population representing the appliance patterns
3: Check the termination criteria, i.e., the maximum generation
4: Evaluate the fitness of each individual in the population
5: Select the patterns from the population with the best fitness values; these patterns should represent the chromosomal configuration, which represents the solution
6: Check the on/off status of all the appliances in the chromosomal configuration
7: Repeat steps 1–6 for k = 1 as the population size
8: Select an individual based on fitness and perform mutation
9: If Pm > Rand, then select the next generation
10: Select two individuals based on fitness and perform crossover
11: If Pc > rand, then crossover this pair
12: Create a new population from the off-springs in 9 and 10
13: h = h + 1, go to step 4 and repeat until h = 24

Table 3. Control parameters of GA Algorithm.

Parameters	Values
Population size	100
Maximum generation	300
Crossover rate	0.9
Mutation rate	0.1

4.2. Binary Particle Swarm Optimization Algorithm

Solution for the same problem, energy consumption patterns, minimization of cost, while considering user comfort and PAR is simulated using BPSO [48]. The SS of HEMC based on BPSO operates at the beginning of the day. After sending a request from the appliance to the SS controller, the action taken by the SS is used to schedule all the appliance's energy consumption patterns in advance. Particle Swarm Optimization (PSO) is one of the best replacements for liner and dynamic programming techniques used to solve complex optimization problems, which is inspired by bird flocking and was developed by Kennedy and Eberhart. The working of PSO is based on the food-searching technique of a swarm of birds in a particular search space. In the search space, two important parameters for each bird are noted, the previous position and velocity. Every bird in the group updates his new position and velocity for the previous position and velocity and knows the position and velocity of the nearest bird to the food court. PSO obtains the same behavior and properties from the bird grouping scenario, and places each particle as a bird which is considered a candidate solution in the search space domain.

The total number of particles in the search space is entitled the population or swarm size. Each particle in the solution search space is studied for its velocity, previous and current position, and fitness value, which represent a solution. To find the finest fitness value for the objective function, BPSO is used having a binary value for optimizing the solution. Each particle in the solution search space represents candidate solutions and obtains the optimal solution each particle has to move in the d-dimensional solution search space.

The preliminary position and velocity of each particle are initialized randomly. To form a swarm, N number of particles are combined. After making a swarm, the particles move around the solution space to obtain the optimal solution. At the end of simulations, the overall best solution called the "global best" is taken as the problem solution. The fitness value for each particle is assessed, and if needed, the local and global positions are updated, respectively. After evaluating the fitness values, every particle in the search space flies and dynamically updates its position and velocity by tracking two extremes, i.e., Plbest and Pgbest in each iteration. The control parameters for BPSO are given in Table 4.

Table 4. Control parameters of the BPSO algorithm.

Parameters	Values
Swarm size	10
Maximum velocity	3
No. of iteration	600
C1	2.0
C2	2.0
Wi	1.5
Wf	0.5
Minimum velocity	-3

Algorithm 2 shows the BPSO algorithm used in SS. The steps involved in the BPSO algorithm are:

Algorithm 2 BPSO Algorithm used in SS

Initialize all the parameters, such as swarm size, no. of iteration
Randomly initialize particle (p) for their position and velocity
Evaluate fitness of objective function for particle (p)
Two best positions global best and local best will be obtained
The velocity of each particle is updated, using the inertia weight C1 and C2
The position of each particle is updated
Evaluate the fitness of each particle to obtain global best and local best
Compare the previous best with the current best
Update the global best
Repeat until the termination criteria meet

4.3. Wind Driven Optimization Algorithm

Researchers get inspired by nature to solve complex scientific problems in every field of life. The WDO algorithm is one of the nature-inspired algorithms used to solve optimization problems based on atmospheric motion. WDO is an iterative heuristic global optimization algorithm based on population to cope with multi-dimensional and multi-modal problems, having the aptitude to implement different types of constraints on the search domain, as compared to its counterpart GA and BPSO. In principle, very small and tiny particles of air move in an n-dimensional domain, following the second law of motion also used to describe air particle motion within the earth's atmosphere. One prominent factor of WDO, as compared to its other counterpart heuristic algorithm, is to carry out some additional information for velocity updates, such as gravitational force constant and Coriolis forces to give a global best position of the particle with more freedom and robustness. The control parameters for WDO are given in Table 5. Table 6 is used for appliances, consumption power, and LoT, and Table 7 is for the ToU pricing signal.

Table 5. Control parameters of the WDO algorithm.

Parameters	Values
Swarm size	10
Maximum velocity	5
No. of iteration	500
C1	3.0
C2	3.0
Wi	0.5
Wf	0.5
Minimum velocity	−5

Table 6. Appliances with power rating and LoT.

Device	Power Rating (KWh)	LoT (Hour)
Refrigerator	0.73	21 h
TV	0.50	14 h
Lighting	0.6	21 h
Heater	4.45	3 h
Fan	0.75	20 h
Iron	1.5	3 h
Toaster	0.05	2 h
Dishwasher	3.63	3 h
Washing machine	0.78	2 h
Cloth dryer	4.40	2 h

Table 7. ToU pricing signal.

High-Peak Hour	Low-Peak Hour	Off-Peak Hour
1 am–5 am, 7 pm–10 pm	5 am–03 pm	2 am–6 am

Algorithm 3 shows the WDO algorithm used in SS. The steps involved in the WDO algorithm are:

Algorithm 3 WDO Algorithm used in SS

Initialized different parameters such as swarm size, pop size, no. of iteration, different coefficient
Repeat from h = 0 to h = 24
Randomly generates the population of particles
Randomly assign velocity and position to each particle
Evaluate the fitness of each particle
Obtain the local best and global best value for the particles
Update the velocity and position of the particle, using inertia and gravitational constant
Create a new population
Evaluate the fitness of each particle after updating the velocity and position
Compare the previous best with the current best particle
Update the global best
Continue until the termination condition meets

5. Simulation Results and Discussion

In this section, we are going to discuss the simulation results and graphs for the justification of the proposed HEMC, implemented through Wind Driven Optimization WDO, BPSO, and GA using the ToU pricing scheme. The whole scenario and the proposed model are implemented in the Matlab simulation tool by using the parameters mentioned in Tables 3–7. This research work focuses on calculating energy consumption patterns and electricity costs for different types of appliances with HEMC, without HEMC, and with HEMC using RES, Peak to PAR comparison for different algorithms using schedule load and user comfort while considering appliance waiting time and electricity cost. For our proposed algorithm, we consider different types of appliances having variable length power consumption requirements. HEMC takes the price signal from the utility grid directly through the smart meter and schedules the scheduler according to price signals. Using a two-way communication model, HEMC sends an optimized energy schedule to all the appliances, considering the appliance's consumption patterns and user comfort. The appliances' energy consumption data is taken from a literature review based on reliable data. In our proposed solution, we take 10 different appliances (shiftable, unshiftable, and semi-shiftable) with different energy consumption rates and Lengths of Time (LoT). The attribute: appliance's power rating, number of appliances, and price scheme value are hard coded. For this research paper, an assumption is supposed that household photovoltaic generation must be greater or equal to 35% of its load demand. A time horizon of T = 24 h is considered, which helps the end-user to calculate his/her electricity bill while keeping the constraints of user comfort.

Figure 2 shows the ToU price signal over the 24-h horizon. In the proposed ToU pricing scheme, 24 h of the day are divided into equal intervals. In the ToU pricing model, prices are mostly fixed for a month or season. Based on different incentives and subsidies, the different pricing zone encourages the end-user to schedule and reschedule their daily electricity load to minimize the electricity bill.

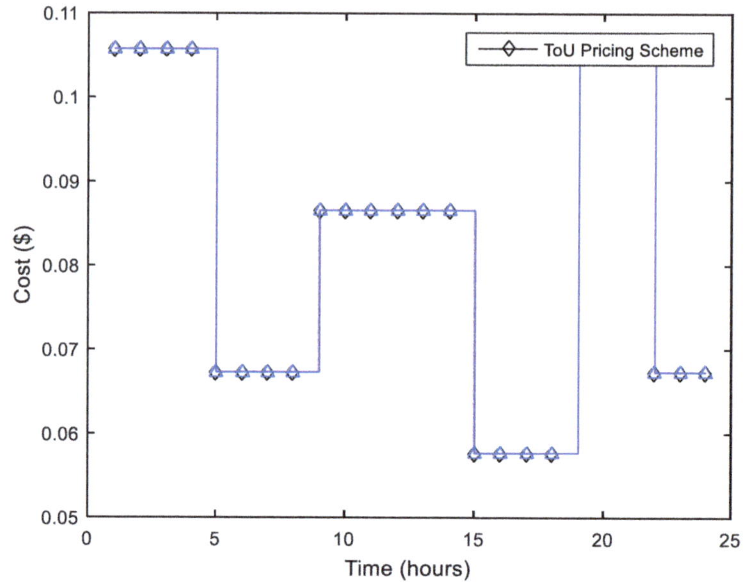

Figure 2. ToU pricing signal.

Figure 2 shows different price ranges in 24-h intervals by giving information to the end-user on how to operate his/her appliances to reduce electricity bills. For example, from (0 to 5) h, (9 to 15) h, and (19 to 22) h. In such high-peak-hour intervals, the end-user tries to operate those appliances which consume minimum energy. Similarly, the customer will turn on his maximum appliances in the interval of (5 to 10) h, (15 to 19) h, and (22 to 24) h, where the electricity prices are minimum. In a more practical environment due to the variable energy demand and supply, electricity price also varies which is the core of Demand Response (DR) programs. Thus, using the TOU pricing scheme, if the end-user wants to obtain maximum benefit by reducing the electricity bill he/she must reschedule their load according to the different pricing peaks.

Figure 3 demonstrates the energy consumption pattern for scheduled and unscheduled loads, from the graph it is clear that the smart scheduler schedules the appliances to operate most of the appliances in the shoulder and low price zones, during the time interval (0 to 5), (10 to 15), and (19 to 23) h, the energy consumption of appliances using WDO and BPSO is low as compared to unscheduled and GA. Moreover, the graph shows some peaks in terms of energy consumption where the price is low, in the time interval (5 to 10) and (15 to 19) h, even at the peak the electricity cost will be minimal depending upon the number of appliances operated at that time based on price signal maximizing the user comfort. On the other hand, an unscheduled load does not consider the price signal and all the appliances operate without any schedule. GA shows some strange behavior at the start of the schedule, even in a high-price zone GA turns on more appliances that consume more energy. In the remaining interval, GA almost shows the best results as compared to both WDO and BPSO, on the other hand, BPSO is more prominent than WDO. From the graph, we can conclude that WDO and BPSO more intelligently operate the appliance, by compromising user comfort and more effectively reducing energy consumption and maximizing PAR in the high-price zone. However, overall, the GA result is a little bit more prominent than WDO and BPSO, as from the graph it is clear that in the interval (5 to 24) h, energy consumption is below 10 kw/h for GA. Figure 4 highlights the cost comparison of electricity using WDO, BPSO, and GA algorithms for the appliances using HEMC and without HEMC, respectively. Figure 3 highlights that WDO and BPSO algorithms more intelligently react in low and mid-peak hour intervals by efficiently scheduling the different home-based appliances, but overall GA gives the best rest as compared to the other two. Since electricity cost is directly related to energy consumption for the same pricing signal, it is clear from the graph that HEMC in low-peak time intervals minimize the energy consumption of different home-based appliances by scheduling them intelligently.

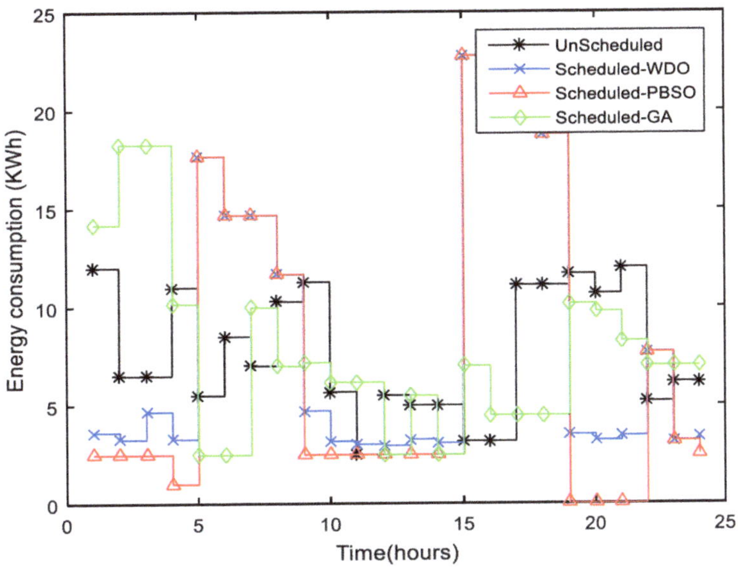

Figure 3. Energy consumption comparison of unscheduled and scheduled using WDO, BPSO, GA.

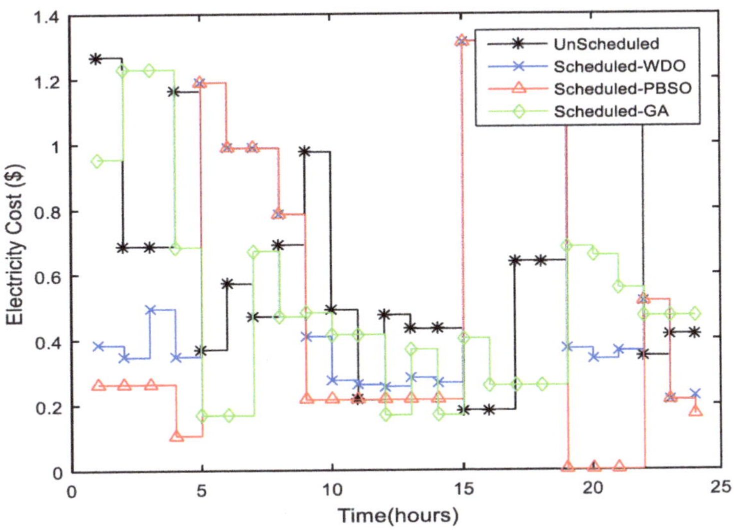

Figure 4. Electricity cost comparison of unscheduled and scheduled using WDO, BPSO, GA.

During hours (0 to 4), the scheduled electricity costs of the HEMS system using WDO and BPSO algorithms are comparatively less than unscheduled and GA because HEMC pays the least attention to the maximum capacity bounds and schedules the different appliances by considering the low pricing time interval. The cost of electricity for the WDO and BPSO algorithms is maximum during the high-price time slot (5 to 10), as compared to unscheduled because HEMC attempts to reschedule a large number of different appliances during these time intervals.

A large number of appliances will be operated in mid-peak hour slots (10 to 15) by GA as compared to WDO and BPSO, resulting in maximum electricity cost. The rest of the cycles for different appliances to be operated are completed during low-price hours (15 to 19). If we compare the cost in discrete quantity: Scheduled with WDO Cost/Day = 12.0748, Scheduled with BPSO Cost/Day = 13.2152, Scheduled with GA Cost/Day = 11.9606,Unscheduled Cost/Day = 15.401. It is clear from the calculations that the electricity bill is higher for unscheduled as compared to WDO, BPSO, and GA, while GA shows better results than WDO and BPSO. If we look at WSO and BPSO, BPSO greatly reduces energy consumption and PAR as compared to WSO. To summarize the discussion, we can conclude that by using the GA algorithm, the electricity cost is effectively reduced, while BPSO and WDO are slightly better than GA in reducing PAR by helping with grid stability and also considering user benefits in terms of electricity bills and comfort levels.

In this scenario, the results of Figures 5 and 6 are compared using WDO, BPSO, and GA optimization techniques for different electrical appliances scheduling by incorporating RES during high-peak hours for smooth and stable functionality of the grid. Cost reduction, peak-to-average reduction, and user comfort are also discussed here. The user's home is equipped with an SS as well as RES and a storage system to store energy. From RES, the user generates and consumes a maximum of 35% of the energy of the daily energy requirement. To reduce energy consumption and minimize electricity bills, appliances considered for this scenario such as washing machines, cloth dryers, iron, and electric vehicle are scheduled in a 24-h time horizon; however, to maximize the comfort level of the end-user, the waiting time for the appliance is also considered, as each appliance has a certain fixed type of interval having a start and finish point. The performance of WDO, BPSO, and GA algorithms is to efficiently consume grid energy as well as the RES is shown in Figure 5.

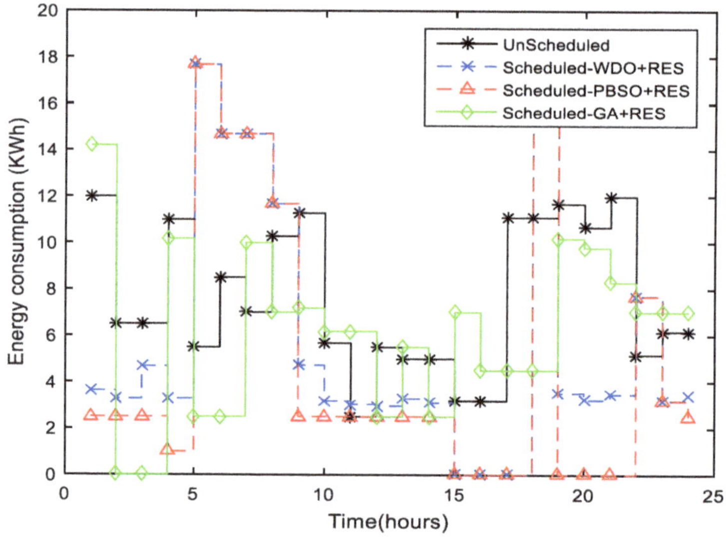

Figure 5. Energy consumption comparison of unscheduled, and scheduled with RES using WDO, BPSO, GA.

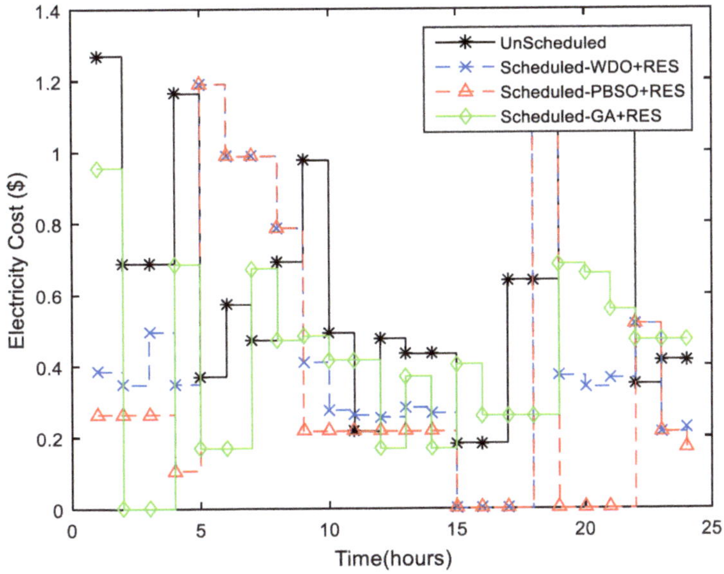

Figure 6. Electricity cost comparison of scheduled with RES, scheduled without RES, and Scheduling using WDO.

To minimize the cost of electricity, the Smart Scheduler (SS) employs the RES stored energy, by shifting the load from the grid to RES energy and maximizing user comfort by a very significant amount. The graph demonstrates the minimum energy consumption of different appliances using the smart scheduler having on-site renewable energy in the time interval (0 to 4), and (9 to 21) h for WDO, GA, and BPSO. From Figure 2, it is clear that the price signal is very low at the time interval (15 to 19) h. Using WDO and BPSO algorithms, the smart scheduler schedules the appliance in such a way that in the same interval, the energy consumption becomes 0 by reducing the electricity bill and maximizing the user comfort, as the appliance waiting time here becomes zero, by shifting the appliance from the grid to RES. On the other hand, GA shows some perks at the start interval, for the rest of the interval, GA shows more stable results than WDO and BPSO, while BPSO is more prominent than WDO. The cost of energy of the SS with Photovoltaic (PV) generation is shown in Figure 6. Figure 5, highlights that the SS optimally and efficiently schedules different appliances in low-price signal intervals by shifting the maximum possible load to the RES storage system during the high-peak hour intervals. Using this strategy, the end-user takes maximum benefit from the RES stored energy during high-peak costs and tries to minimize the high peaks in electricity bills. The cost comparison result using HEMC with RES for WDO, BPSO, and GA, without HEMC is given in Figure 6 which highlights the per-day cost reduction in HEMC using RES is maximum as compared to scheduling without HEMC, even in the interval (14 to 19), the cost is almost 0 by maximizing the user comfort.

Figure 7 demonstrates the peak-to-average ratio of different algorithms, i.e., the Genetic Algorithm, Binary Particle Swarm Optimization, and WDO. In a particular time interval, dividing the maximum energy consumption by the average energy consumption of a PAR in a particular time interval for a single user can be calculated. We start to discuss PAR reduction from the very first graph by concluding that generally, end-users want to minimize the total bill of electricity by somehow compromising comfort, while the utility company tries to make sure the stability of the grid in terms of a balanced energy supply. Figure 7 demonstrates the proposed algorithmic technique efficiency by balancing the energy consumption and minimizing total PAR while paying attention to the capacity constraint of total energy. Figure 7 also shows that WDO and BPSO algorithm remarkably condenses the PAR by 11.48% and 12.48%, respectively, for efficient and optimal home appliance scheduling in low-peak price hours without creating a bottleneck, as compared to the earlier GA algorithm minimizes the PAR by 9.16%, for the same parameter, i.e., many appliances, price signal and power rating for each appliance.

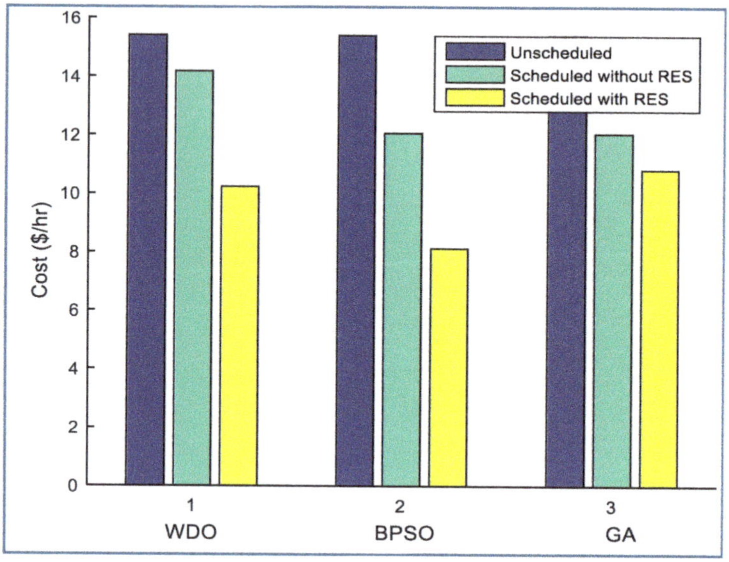

Figure 7. PAR comparison of WDO, BPSO, and GA for a scheduled load.

6. Conclusions and Future Work

This paper proposed and implemented a home energy management system based on GA, WDO, and BPSO algorithms, for different types of appliances, having different power consumption and length of operation time. Our proposed system uses both the grid and on-site RES energy, optimally scheduling all the different appliances by finding energy consumption patterns, and reducing peak demand and electricity costs by maintaining the user comfort level. From the graph, it is clear that HEMC using GA shows better results than WDO and BPSO, in energy consumption and electricity cost, while BPSO is more prominent than WDO and GA by calculating PAR.

The WDO and BPSO reduce the PAR by 11.48% and 12.48%, respectively, in low-peak price hours without creating a bottleneck, as compared to the earlier GA algorithm which minimizes the PAR by 9.16% for the same parameter. The cost comparison result using HEMC with RES for WDO, BPSO, and GA, without HEMC is given in Figure 6 which highlights that the per-day cost reduction in HEMC using RES is maximum as compared to scheduling without HEMC, even in the interval (14 to 19) the cost is almost 0 by maximizing the user comfort. The proposed HEMC system will disturb performance if the number of appliances is increased. In the future, the number of appliances can be increased by maintaining the user comfort level and the electricity cost. In the future, the proposed model will be upgraded to deploy in multiple locations with different desired flexibility of demand response, connected with various sustainable sources. The relevant techno-economic analysis will be conducted in future research.

Author Contributions: This work is carried out in collaboration between all authors. Z.M. and N.A.B. designed a detailed methodology. G.U.R. and M.A. managed the literature searches and detailed analysis. M.Z. wrote the first draft of the manuscript in consultation with G.U.R. The final draft of the paper has been done by A.B. along with M.Z. The funding acquisition is done by B.C. All authors have read and agreed to the published version of the manuscript.

Funding: The authors would like to thanks Jiangsu Key Lab of IoT Application Technology, Wuxi Taihu University, Wuxi, China for supporting this work.

Data Availability Statement: This article has no associated data.

Conflicts of Interest: The authors declare no conflict of interest.

References

1. Ghani, A.; Naqvi, S.H.A.; Ilyas, M.U.; Khan, M.K.; Hassan, A. Energy efficiency in multipath Rayleigh faded wireless sensor networks using collaborative communication. *IEEE Access* **2019**, *7*, 26558–26570. [CrossRef]
2. Castillo, M.S.; Liu, X.; Abd-AlHamid, F.; Connelly, K.; Wu, Y. Intelligent windows for electricity generation: A technologies review. *Build. Simul.* **2022**, *15*, 1747–1773. [CrossRef]
3. İnci, M.; Büyük, M.; Savrun, M.M.; Demir, M.H. Design and analysis of fuel cell vehicle-to-grid (FCV2G) system with high voltage conversion interface for sustainable energy production. *Sustain. Cities Soc.* **2021**, *67*, 102753. [CrossRef]
4. Rehman, A.; Ma, H.; Ozturk, I.; Radulescu, M. Revealing the dynamic effects of fossil fuel energy, nuclear energy, renewable energy, and carbon emissions on Pakistan's economic growth. *Environ. Sci. Pollut. Res.* **2022**, *29*, 48784–48794. [CrossRef]
5. Dey, B.; Márquez, F.P.G.; Panigrahi, P.K.; Bhattacharyya, B. A novel metaheuristic approach to scale the economic impact of grid participation on a microgrid system. *Sustain. Energy Technol. Assess.* **2022**, *53*, 102417. [CrossRef]
6. Roy, P.; He, J.; Liao, Y. Cost minimization of battery-supercapacitor hybrid energy storage for hourly dispatching wind-solar hybrid power system. *IEEE Access* **2020**, *8*, 210099–210115. [CrossRef]
7. Varaiya, P.P.; Wu, F.F.; Bialek, J.W. Smart operation of smart grid: Risk-limiting dispatch. *Proc. IEEE* **2010**, *99*, 40–57. [CrossRef]
8. Ullah, K.; Hafeez, G.; Khan, I.; Jan, S.; Javaid, N. A multi-objective energy optimization in smart grid with high penetration of renewable energy sources. *Appl. Energy* **2021**, *299*, 117104. [CrossRef]
9. Siddiqui, S.A.; Ahmad, M.O.; Ahmed, J. Smart home for efficient energy management. In *Smart Technologies for Energy and Environmental Sustainability*; Springer: Berlin/Heidelberg, Germany, 2022; pp. 97–103.
10. Tostado-Véliz, M.; Arévalo, P.; Kamel, S.; Zawbaa, H.M.; Jurado, F. Home energy management system considering effective demand response strategies and uncertainties. *Energy Rep.* **2022**, *8*, 5256–5271. [CrossRef]
11. Chen, C.R.; Lan, M.J.; Huang, C.C.; Hong, Y.Y.; Low, S.H. Demand response optimization for smart home scheduling using genetic algorithm. In Proceedings of the 2013 IEEE International Conference on Systems, Man, and Cybernetics, Manchester, UK, 13–16 October 2013; pp. 1461–1465.
12. Memiş, S.; Enginoğlu, S.; Erkan, U. Fuzzy Parameterized Fuzzy Soft k-Nearest Neighbor Classifier. *Neurocomputing* **2022**, *500*, 351–378. [CrossRef]
13. Memiş, S.; Enginoğlu, S.; Erkan, U. A classification method in machine learning based on soft decision-making via fuzzy parameterized fuzzy soft matrices. *Soft Comput.* **2022**, *26*, 1165–1180. [CrossRef]
14. Guo, B.; Weeks, M. Dynamic tariffs, demand response, and regulation in retail electricity markets. *Energy Econ.* **2022**, *106*, 105774. [CrossRef]
15. Mohammad, A.; Zuhaib, M.; Ashraf, I. An optimal home energy management system with integration of renewable energy and energy storage with home to grid capability. *Int. J. Energy Res.* **2022**, *46*, 8352–8366. [CrossRef]
16. Deng, Z.; Liu, C.; Zhu, Z. Inter-hours rolling scheduling of behind-the-meter storage operating systems using electricity price forecasting based on deep convolutional neural network. *Int. J. Electr. Power Energy Syst.* **2021**, *125*, 106499. [CrossRef]
17. Jazayeri, P.; Schellenberg, A.; Rosehart, W.; Doudna, J.; Widergren, S.; Lawrence, D.; Mickey, J.; Jones, S. A survey of load control programs for price and system stability. *IEEE Trans. Power Syst.* **2005**, *20*, 1504–1509. [CrossRef]
18. Rasheed, M.B.; R-Moreno, M.D. Minimizing pricing policies based on user load profiles and residential demand responses in smart grids. *Appl. Energy* **2022**, *310*, 118492. [CrossRef]
19. Liang, B.; Yang, J.; Hou, B.; He, Z. A Pricing Method for Distribution System Aggregators Considering Differentiated Load Types and Price Uncertainty. *IEEE Trans. Power Syst.* **2020**, *36*, 1973–1983. [CrossRef]
20. Shewale, A.; Mokhade, A.; Funde, N.; Bokde, N.D. A Survey of Efficient Demand-Side Management Techniques for the Residential Appliance Scheduling Problem in Smart Homes. *Energies* **2022**, *15*, 2863. [CrossRef]
21. Abubakr, H.; Vasquez, J.C.; Mahmoud, K.; Darwish, M.M.; Guerrero, J.M. Comprehensive review on renewable energy sources in Egypt—Current status, grid codes and future vision. *IEEE Access* **2022**, *10*, 4081–4101. [CrossRef]
22. Saoud, L.S.; Al-Marzouqi, H.; Hussein, R. Household Energy Consumption Prediction Using the Stationary Wavelet Transform and Transformers. *IEEE Access* **2022**, *10*, 5171–5183. [CrossRef]
23. Mahapatra, B.; Nayyar, A. Home energy management system (HEMS): Concept, architecture, infrastructure, challenges and energy management schemes. *Energy Syst.* **2022**, *13*, 643–669. [CrossRef]

24. Mahto, N.K.; Jaiswal, S.; Das, D.C. Demand-Side Management Approach Using Heuristic Optimization with Solar Generation and Storage Devices for Future Smart Grid. In *Renewable Energy towards Smart Grid*; Springer: Berlin/Heidelberg, Germany, 2022; pp. 407–420.
25. Duman, A.C.; Erden, H.S.; Gönül, Ö.; Güler, Ö. A home energy management system with an integrated smart thermostat for demand response in smart grids. *Sustain. Cities Soc.* **2021**, *65*, 102639. [CrossRef]
26. Kumar, M.; Sharma, S.C. PSO-based novel resource scheduling technique to improve QoS parameters in cloud computing. *Neural Comput. Appl.* **2020**, *32*, 12103–12126. [CrossRef]
27. Butt, O.M.; Zulqarnain, M.; Butt, T.M. Recent advancement in smart grid technology: Future prospects in the electrical power network. *Ain Shams Eng. J.* **2021**, *12*, 687–695. [CrossRef]
28. Nabuurs, P. Strategic Deployment Document for Eoroupe's Elictricity Networks of the Future. NV KEMA, Draft. 2008. Available online: https://docplayer.net/2818454-European-technology-platform-smartgrids-strategic-deployment-document-for-europe-s-electricity-networks-of-the-future.html (accessed on 1 December 2022).
29. Haq, E.U.; Lyu, C.; Xie, P.; Yan, S.; Ahmad, F.; Jia, Y. Implementation of home energy management system based on reinforcement learning. *Energy Rep.* **2022**, *8*, 560–566. [CrossRef]
30. Menos-Aikateriniadis, C.; Lamprinos, I.; Georgilakis, P.S. Particle Swarm Optimization in Residential Demand-Side Management: A Review on Scheduling and Control Algorithms for Demand Response Provision. *Energies* **2022**, *15*, 2211. [CrossRef]
31. Ahmed, Z.E.; Saeed, R.A.; Mukherjee, A.; Ghorpade, S.N. Energy optimization in low-power wide area networks by using heuristic techniques. In *LPWAN Technologies for IoT and M2M Applications*; Elsevier: Amsterdam, The Netherlands, 2020; pp. 199–223.
32. Zhao, Z.; Lee, W.C.; Shin, Y.; Song, K.B. An optimal power scheduling method for demand response in home energy management system. *IEEE Trans. Smart Grid* **2013**, *4*, 1391–1400. [CrossRef]
33. Pathak, A.K.; Chatterji, D.S.; Narkhede, M.S. Artificial intelligence based optimization algorithm for demand response management of residential load in smart grid. *Int. J. Eng. Innov. Technol. (IJEIT)* **2012**, *2*, 136–141.
34. Adika, C.O.; Wang, L. Autonomous appliance scheduling for household energy management. *IEEE Trans. Smart Grid* **2013**, *5*, 673–682. [CrossRef]
35. Deng, R.; Yang, Z.; Chen, J.; Chow, M.Y. Load scheduling with price uncertainty and temporally-coupled constraints in smart grids. *IEEE Trans. Power Syst.* **2014**, *29*, 2823–2834. [CrossRef]
36. Sarathkumar, D.; Stonier, A.A.; Srinivasan, M.; Senthamil, L.S. Review on Power Restoration Techniques for Smart Power Distribution Systems. In *Renewable Energy towards Smart Grid*; Springer: Berlin/Heidelberg, Germany, 2022; pp. 67–77.
37. Mataifa, H.; Krishnamurthy, S.; Kriger, C. Volt/VAR Optimization: A Survey of Classical and Heuristic Optimization Methods. *IEEE Access* **2022**, *10*, 13379–13399. [CrossRef]
38. Bina, M.T.; Ahmadi, D. Stochastic modeling for the next day domestic demand response applications. *IEEE Trans. Power Syst.* **2015**, *30*, 2880–2893. [CrossRef]
39. Rasheed, M.B.; Javaid, N.; Ahmad, A.; Khan, Z.A.; Qasim, U.; Alrajeh, N. An efficient power scheduling scheme for residential load management in smart homes. *Appl. Sci.* **2015**, *5*, 1134–1163. [CrossRef]
40. Aslam, S.; Javaid, N.; Asif, M.; Iqbal, U.; Iqbal, Z.; Sarwar, M.A. A mixed integer linear programming based optimal home energy management scheme considering grid-connected microgrids. In Proceedings of the 2018 14th International Wireless Communications & Mobile Computing Conference (IWCMC), Limassol, Cyprus, 25–29 June 2018; pp. 993–998.
41. Lokeshgupta, B.; Sivasubramani, S. Cooperative game theory approach for multi-objective home energy management with renewable energy integration. *IET Smart Grid* **2019**, *2*, 34–41. [CrossRef]
42. Salehizadeh, M.R.; Koohbijari, M.A.; Nouri, H.; Taşcıkaraoğlu, A.; Erdinç, O.; Catalao, J.P. Bi-objective optimization model for optimal placement of thyristor-controlled series compensator devices. *Energies* **2019**, *12*, 2601. [CrossRef]
43. Karimianfard, H.; Salehizadeh, M.R.; Siano, P. Economic Profit Enhancement of a Demand Response Aggregator Through Investment of Large-scale Energy Storage Systems. *CSEE J. Power Energy Syst.* **2022**, *8*, 1468–1476.
44. Chamandoust, H. Optimal hybrid participation of customers in a smart micro-grid based on day-ahead electrical market. *Artif. Intell. Rev.* **2022**, *55*, 5891–5915. [CrossRef]
45. Chamandoust, H.; Hashemi, A.; Bahramara, S. Energy management of a smart autonomous electrical grid with a hydrogen storage system. *Int. J. Hydrogen Energy* **2021**, *46*, 17608–17626. [CrossRef]
46. Chen, S.J.; Chiu, W.Y.; Liu, W.J. User Preference-Based Demand Response for Smart Home Energy Management Using Multiobjective Reinforcement Learning. *IEEE Access* **2021**, *9*, 161627–161637. [CrossRef]
47. DY, G.; M, P.; Y, F.; D, S. Optimal power management of residential customers in the smart grid. *IEEE Trans. Evol. Comput.* **2012**, *23*, 1593–1606.
48. Del Valle, Y.; Venayagamoorthy, G.K.; Mohagheghi, S.; Hernandez, J.C.; Harley, R.G. Particle swarm optimization: basic concepts, variants and applications in power systems. *IEEE Trans. Evol. Comput.* **2008**, *12*, 171–195. [CrossRef]

sustainability

MDPI

Article

Smart Diagnosis of Adenocarcinoma Using Convolution Neural Networks and Support Vector Machines

Balasundaram Ananthakrishnan [1,2,*], Ayesha Shaik [2,*], Shubhadip Chakrabarti [2], Vaishnavi Shukla [2], Dewanshi Paul [3] and Muthu Subash Kavitha [4]

1 Centre for Cyber Physical Systems, Vellore Institute of Technology, Chennai 600127, India
2 School of Computer Science and Engineering, Vellore Institute of Technology, Chennai 600127, India
3 School of Electronics Engineering, Vellore Institute of Technology, Chennai 600127, India
4 School of Information and Data Sciences, Nagasaki University, Nagasaki 8528521, Japan
* Correspondence: balasundaram.a@vit.ac.in (B.A.); ayesha.sk@vit.ac.in (A.S.)

Abstract: Adenocarcinoma is a type of cancer that develops in the glands present on the lining of the organs in the human body. It is found that histopathological images, obtained as a result of biopsy, are the most definitive way of diagnosing cancer. The main objective of this work is to use deep learning techniques for the detection and classification of adenocarcinoma using histopathological images of lung and colon tissues with minimal preprocessing. Two approaches have been utilized. The first method entails creating two CNN architectures: CNN with a Softmax classifier (*AdenoCanNet*) and CNN with an SVM classifier (*AdenoCanSVM*). The second approach corresponds to training some of the prominent existing architecture such as VGG16, VGG19, LeNet, and ResNet50. The study aims at understanding the performance of various architectures in diagnosing using histopathological images with cases taken separately and taken together, with a full dataset and a subset of the dataset. The LC25000 dataset used consists of 25,000 histopathological images, having both cancerous and normal images from both the lung and colon regions of the human body. The accuracy metric was taken as the defining parameter for determining and comparing the performance of various architectures undertaken during the study. A comparison between the several models used in the study is presented and discussed.

Keywords: cancer; adenocarcinoma; convolution neural network; CNN; transfer learning; CNN–SVM; medical image processing; deep learning; artificial intelligence; smart cancer diagnosis; *AdenoCanNet*; *AdenoCanSVM*

Citation: Ananthakrishnan, B.; Shaik, A.; Chakrabarti, S.; Shukla, V.; Paul, D.; Kavitha, M.S. Smart Diagnosis of Adenocarcinoma Using Convolution Neural Networks and Support Vector Machines. *Sustainability* **2023**, *15*, 1399. https://doi.org/10.3390/su15021399

Academic Editors: Mourade Azrour, Azidine Guezzaz, Imad Zeroual, Azeem Irshad, Jamal Mabrouki, Said Benkirane and Shehzad Ashraf Chaudhry

Received: 22 November 2022
Revised: 26 December 2022
Accepted: 27 December 2022
Published: 11 January 2023

1. Introduction

The world today witnesses cancer as one of the most dreadful diseases impacting precious human life adversely. While cancer happens to be a generic term to refer to a large variety of diseases, it essentially involves enhanced formation of abnormal cells that propagate through the bloodstream and spread across the body, destroying normal cells, leading to the death of the affected individual. As reported by WHO [1] (World Health Organization), this disease is estimated to have affected around 10 million people in the year 2020. The global cancer trend [2] has been so concerning that an estimated 47% increase in the disease's prevalence worldwide from 2020 to 2040 is predicted.

Adenocarcinoma [3] relates to a common type of cancer found in glandular epithelial cells of human body. Lung, prostate, pancreas, liver, colorectal area, and breast form the primary sites for the adenocarcinoma. Unlike other carcinomas, these cancers do not exhibit any symptoms during their early stages and remain undetected. As data [3] say, adenocarcinoma is responsible for around 40%, 95%, and 96% of non-small-cell lung cancers, pancreatic cancers, and colorectal cancers, respectively. It is also responsible for almost all prostate cancers and most breast cancers.

Histological images continue to be a standard method of cancer diagnosis. The advances made thus far in the area of health informatics have still proven to be unsuccessful in meeting the desired clinical requirements. Most diagnostics to date are still performed manually and rely heavily on the expertise and experience of histopathologists. These diagnostic techniques happen to be very time-consuming and difficult to grade in a reproducible manner.

Computer aided diagnosis using histopathological images has always been a topic of paramount interest in the field of cancer detection. Multiple works have been conducted in this domain using artificial intelligence (AI). Experimentation on AI-based cancer diagnosis using various machine learning and deep learning models has presently evolved as one of the prime areas of interest. Major studies undertaken in this domain include implementation of different AI-based models, improving existing AI-based models, or evaluating existing AI-based models to have an insight into their comparative efficacy. The development of methods to enhance image processing techniques to better extract features is provided as a solution to this problem statement. The development of filters to select the most efficient models among all the multiple machine learning models used as classifiers to improve accuracy is also one of the works conducted in this area. Existing works in this domain have recorded accuracies ranging from 70% to over 90%.

Compared to manual analysis, an AI-based system has the potential to provide rapid and consistent cancer detection and classification results. Therefore, the treatment and analysis of images using advanced machine learning and deep learning techniques need to be introduced to facilitate the increased rate of disease diagnosis in humans.

The primary objective of the present study is to develop an artificial intelligence-based tool that can assist in diagnosing adenocarcinoma from histopathology-based images. The study aims at detecting and classifying adenocarcinoma in the colon and lung regions of the human body using the LC25000 dataset [4] procured from Kaggle.

Literature Study

The use of machine learning (ML), deep learning (DL), and transfer learning (TL) to detect and classify has been the talk of the town for a while and there have been several approaches that performed successfully.

The research work [5] used CNN architectures such as VGG16, VGG19, DenseNet169, and DenseNet201 to extract image characteristics from the LC25000 dataset. The extracted features were put into six widely used ML algorithms including Extreme Gradient Boosting (XGB), Random Forest (RF), Support Vector Machine (SVM), Light Gradient Boosting (LGB), Multi-Layer Perceptron (MLP), and Logistic Regression (LR) to evaluate the performance. Accuracy-based filtering of the findings allowed for the selection of the most effective algorithms. As classifiers, SVM, Logistic Regression, and MLP were chosen because of their superior performance. Using this method, cancers of the lung and colon were found. The authors of [6] used the LC25000 dataset to automate the detection of lung and colon cancer. The pre-processing involved wavelet decomposition and application of 2D Fourier transform on channel-separated images. They used a CNN model with a Softmax classifier for feature extraction and classification tasks, achieving an accuracy of 96.37%. The research in [7] used the LC25000 dataset for classifying histopathology images of lung cancer using CNN. Feature extraction was performed using ResNet50, VGG19, Inception_ResNet_V2, and DenseNet121. The Triplet loss function was used to enhance the performance. CNN having three hidden layers was used to classify the images. In this study, Inception-ResNetv2 performed well, having a test accuracy rate of 99.7%.

The authors in the research work [8] classified adenocarcinoma of the lung region and adenocarcinoma of the colon region using a CNN model. They used the LC25000 dataset for this purpose. The images were first resized to 150 × 150 pixels, then some randomized shear and zoom transformations were applied to the images, followed by the normalization of images. A CNN model was applied separately for the lung dataset and the colon dataset, recording an accuracy of 97% and 96%, respectively. A research work was conducted to perform detection of lung cancer using CNN [9]. The dataset used during the process was LC25000. The images of the dataset were first resized to 180 × 180, and the pixel values were then transformed to a range of (0, 1) to facilitate faster convergence. The CNN model had three hidden layers. The model recorded a training accuracy of 96.11% and validation accuracy of 97.2%. The work [10] aimed at developing a model that classified lung cancer into adenocarcinoma, benign, and squamous cell carcinoma. They used the LC25000 dataset for this purpose. The model consists of a main path responsible for extracting small features and sub-paths which pass the medium- and high-level features to fully connected layers. The model recorded an accuracy of 98.53%.

The work carried out in [11] designed a model for the diagnosis of lung and colon cancer using the LC25000 dataset. A pre-trained AlexNet model, after modifying four of its layers, was used. The model performed well for all classes except the "lung_ssc" class, achieving an accuracy of 89%. To improve the performance, image enhancement techniques such as image contrast enhancement were applied to the underperforming class, improving the accuracy to 98.4%. The research work [12] used the AiCOLON dataset to train a CNN model. transfer learning was applied and results were compared with the built CNN model. The highest accuracy achieved was 96.98% with ResNet. CRC-5000, NCT-CRC-HE-100K, and merged (namely, the CRC-5000, the NCT-CRC-HE-100K) datasets were also used to test the ResNet model, recording an accuracy of 96.77%, 99.76%, and 99.98%, respectively. An accuracy of 98.66%, 99.12%, and 78.39% was achieved by SEGNET for the same datasets. SegNet was concluded to be an efficient model for cancer segmentation.

Both the LC25000 and Kather datasets were used in [13] to develop a super-lightweight plug-and-play module (namely, Pyramidal Deep-Broad Learning (PDBL)), to equip CNN backbones, especially lightweight models, to increase tissue-level classification performance without a re-training burden. Experiments were performed to equip this module on ShuffLeNetV2, EfficientNetb0, and ResNet50, with accuracy of 74.61%, 79.87%, and 85.53% with no re-training. The work [14] used the LC25000 dataset to develop a CNN model to categorize and classify the colon region for adenocarcinoma and benign cells. Lime and DeepLift were used as the optimization techniques to improve the understanding of results predicted via the model. The validation accuracy for diagnosis was found to be higher than 94% for distinguishing adenocarcinoma and benign colonic cells.

Extensive research work was carried out in [15] to utilize the power of machine learning, feature engineering, and image processing for identifying classes of lung and colon cancer. They used the LC25000 dataset in their study. Machine learning models such as XGBoost, SVM, RF, LDA, and MLP were employed. Unsharp masking was used for image preprocessing. The Recursive Feature Elimination (RFE) method was used to eliminate the least important features. XGBoost recorded an accuracy of 99%. The SHAP method was used to show the contributions of each feature in the results predicted by models. The research work [16] used CNN models with max pooling layers and average pooling layers along with MobileNetV2 for analyzing colon cancer using the LC25000 dataset. Using the ImageDataGenerator of the Keras library, images were augmented with flips, shear, zoom, rotation, and width and height range. Images were resized to 224 × 224 pixels and later converted to Numpy arrays for further use. The models were trained with various epochs and reported an accuracy of 97.49%, 95.48%, and 99.67%, respectively, for CNN models with max pooling and average pooling and MobileNetV2.

The research work in [17] developed four different CNN models, varying from two pairs of convolutional layers and max pooling layers to four pairs of convolutional layers and max pooling layers with different number of filters and kernel sizes, using lung images procured from the LC25000 dataset. Three different sizes of input images were taken into consideration. The best result accuracy on the test dataset was 96.6% with input size 768 × 768 pixels with the CNN model having four convolutional layers and max pooling layers. It was observed that, with an increase in input image size and convolutional layers, the accuracy increases. The research carried out in [18] used ResNet18, ResNet30, and ResNet50 architectures for colonic adenocarcinoma diagnosis using histopathological images. Two image datasets were used, LC25000 and CRAG. A total of 10,000 images from the first dataset were used to train and validate the CNN architectures in an 80:20 train and test set ratio. Images from the latter dataset were split into 40% and 60% for training and testing, respectively. This was performed to check how the model behaves using a self-supervised learning step, where networks were trained with the LC25000 dataset. Validation accuracy of 93.91%, 93.04%, and 93.04% was recorded for each of the three CNN architectures, respectively. The authors of [19] proposed the development of homology-based image processing techniques along with conventional texture analysis that gives better results for classification when fed into machine learning models such as Perceptron, Logistic Regression, KNN, SVM with Linear Kernel, SVM with Radial-Basis Function Kernel, Decision Tree, Random Forest, and Gradient Tree Boosting. They primarily used two datasets, private and public. The public dataset was the LC25000. Images were processed by converting them into grayscale, applying binarization, and then finally converting them into betti numbers. The accuracy was 78.33% and 99.43%, respectively, for the private and public dataset.

In the research work [20], a CNN model with three convolutional layers, having 32, 64, and 128 filters, respectively, was designed. The study aimed at the detection of breast cancer using histopathological images from the BreakHis dataset. The images were resized to 350 × 230 pixels and reshaped, maintaining the aspect ratio, achieved using OpenCV library in Python. It achieved an accuracy of 99.86%. The research work [21] used the ACDC@LUNGHP dataset to compare the performance of two prominent CNN architectures, VGG16 and ResNet50. In this case, VGG16 was seen to slightly outperform ResNet50, with an accuracy of 75.41%. The research work [22] used the PatchCamelyon benchmark dataset to compare the performance of 14 existing architectures of CNN. They concluded that DenseNet161 outperformed other architectures, with an AUC score of 0.9924. Three approaches were carried out during the process. The first approach correspondeded to utilizing the weights from the pre-trained network. The second approach related to training only fully connected layers. The last approach corresponded to training the entire model. The work [23] used the BreakHis dataset to detect Breast Cancer using histopathological images using CNN. Image augmentation was applied to improve the performance of the model and it was finally concluded with training accuracy of 96.7% and testing accuracy of 90.4%.

2. Proposed System

2.1. Principles of Diagnosis

This work primarily employs two approaches to achieve the aforementioned objective. The first approach corresponds to designing and developing convolution neural network (CNN) architectures. Two CNN architectures, one with a Softmax classifier (referred to as *AdenoCanNet*) and the other one with an SVM classifier (*AdenoCanSVM*), were developed in the first approach. The second approach corresponds to training some of the prominent architectures (such as VGG16, VGG19) using the concept of transfer learning. The study aims at understanding the performance of various architectures in diagnosing lung and colon adenocarcinoma taken together and individually using histopathological images. We also undertook a study on the performance of the architectures in different subsets of the dataset. The LC25000 dataset used in this work consists of 25,000 histopathological images, having both cancerous and normal images from both the lung and colon regions of the human body, evenly distributed over five classes. The accuracy and loss metrics were taken as the defining parameters for determining and comparing the performance of various architectures undertaken during the study. It was found that, when the entire dataset was taken into consideration, the *AdenoCanNet* model was found to produce the best results, recording a maximum training accuracy and maximum testing accuracy of 99.88% and 99.00%, respectively.

2.2. Dataset Description

The proposed work uses the LC25000 dataset to meet the primary objective of designing an artificial intelligence-based tool for assisting in the diagnosis of adenocarcinoma in the lung and colon regions of the human body. The dataset has five different classes. Histopathological images in two classes are taken from the colon region, while histopathological images in the remaining three classes are taken from the lung region. The dataset has a total of 25,000 histopathological images, with each class having 5000 images in them. Lung adenocarcinoma, lung squamous cell carcinoma, normal lung, colon adenocarcinoma, and normal colon are the five different classes present in the dataset. The LC25000 dataset is derived from an original dataset having 750 lung tissue images and 500 colon tissue images. These images were further augmented and finally presented by the authors of the LC25000 dataset. Figure 1 represents the different classes present in the LC25000 dataset, with one sample image taken from each class.

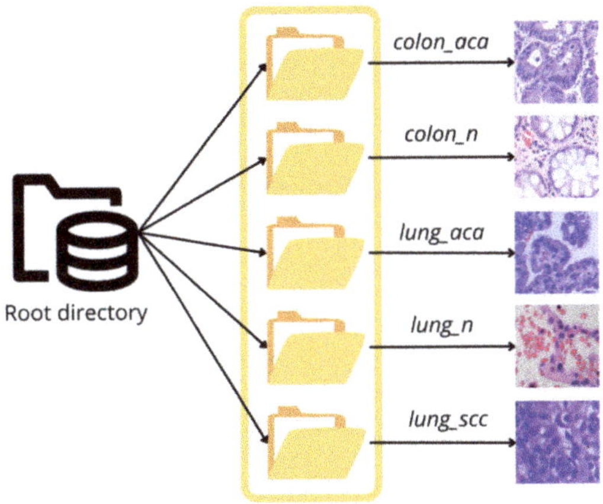

Figure 1. Dataset structure.

2.3. Preprocessing

The initial observations on the histopathological images of the dataset concluded that all the images present in the dataset were colored. It was also noted that each class in the dataset had a total of 5000 images. Careful observations of the image dataset led to the observation of their size. It was found that the images had dimensions of (768, 768, 3). The images required certain preprocessing before we could go ahead with the modeling phase. While preprocessing, the histopathological images were initially denoised to remove any kind of noise that might be present in the image. The denoising was performed by applying the Gaussian Blur. During the study, the kernel size, the SigmaX, and the SigmaY parameters of the Gaussian Blur were set to (3, 3), 90, and 90, respectively. This operation helped in eliminating the noises from every histopathological image present in the dataset. After denoising the images, they were then resized to the dimensions of (64, 64, 3). The processed images were then ready for further processing.

2.4. Architectures Used

Our study essentially employs two different approaches to meet the objective of classification. The first approach corresponds to constructing two convolution neural network (CNN) architectures. Two CNN architectures, one with a Softmax classifier (also referred to as *AdenoCanNet*) and the other one with an SVM classifier (also referred to as *AdenoCanSVM*) were developed. The second approach corresponds to training some of the prominent architectures such as VGG16, VGG19, LeNet, and ResNet50 on the LC25000. Figure 2 is a diagrammatic representation of the workflow adopted in this study.

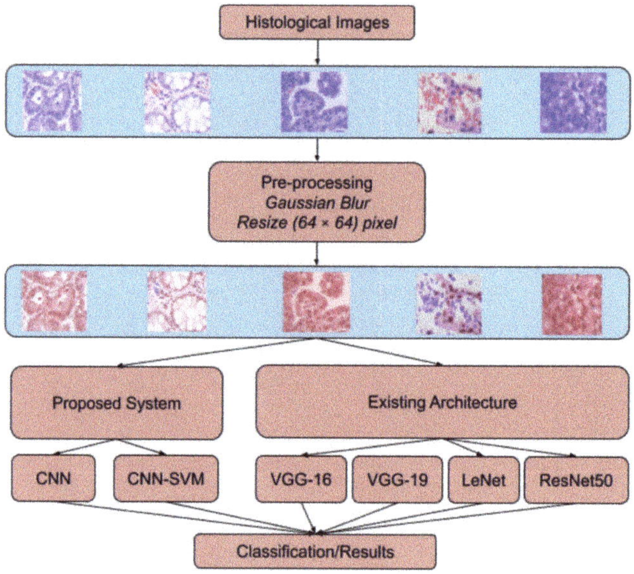

Figure 2. Workflow of the proposed system.

2.4.1. Convolution neural network with Softmax classifier (*AdenoCanNet*)

During the initial phase of the study, a deep CNN architecture was taken into consideration while model building. However, on training and testing the model, the deep neural network was discovered to strongly overfit the dataset. The suggested model is the outcome of several actions that were taken during the process to address the issue of overfitting.

The proposed model has three convolution layers, two max pooling layers and one dropout layer followed by fully connected layers. The proposed model finally used the Softmax classifier for the classification task. Figure 3 represents a graphical form of the proposed *AdenoCanNet* architecture.

The architecture is designed to take an image input of dimensions (64, 64, 3). The first two layers of the proposed CNN architecture correspond to the convolution layers having 32 and 64 channels. The filter size in both these layers is set to (3, 3). A max pooling layer having a stride of (2, 2) is then included in the architecture. The architecture further includes one more convolution layer having 128 channels and with a filter size defined as (3, 3). The output from this layer is then fed into the max pooling layer, having a stride size of (2, 2). A dropout layer is then finally added before flattening it. After a series of convolution, max pooling, and dropout layers, the feature map obtained is flattened to obtain the equivalent feature vector. The feature vector is passed through a series of dense layers before finally throwing the output class of the input image using the Softmax classifier. It is pertinent to note that the architecture uses Sparse Categorical Cross Entropy as the loss function, Adam as the optimizer, and ReLU as its activation function in the entire process. The primary reason of choosing ReLU over sigmoid and hyperbolic tangent activation functions is that it is computationally less time-intensive [24]. Equation (1) corresponds to the mathematical representation of the ReLU activation function. Figure 4 represents the graphical form of representing the ReLU activation function.

$$\text{ReLu}: \ f(x) = \max(0, x) \tag{1}$$

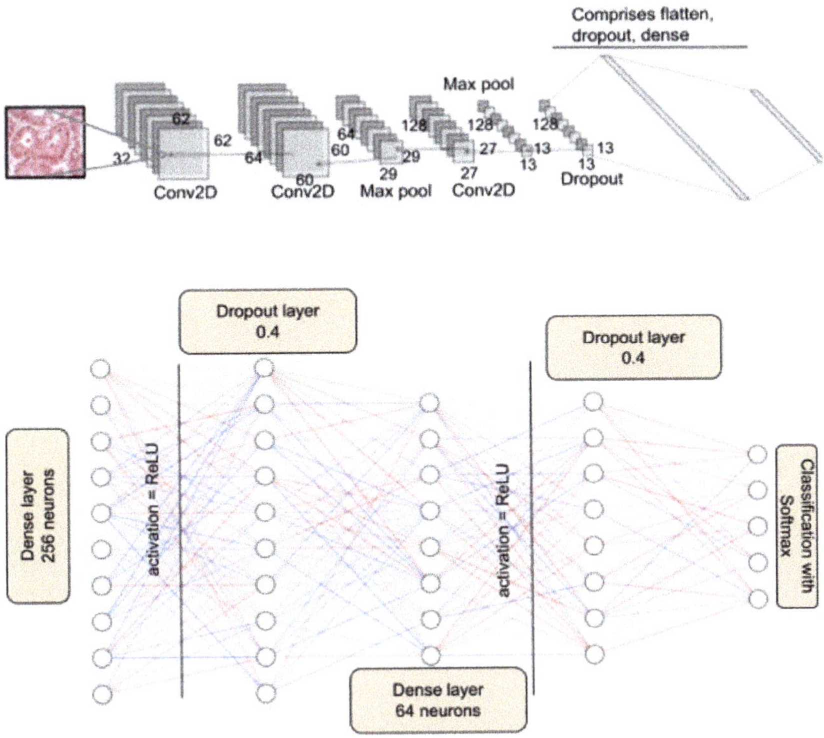

Figure 3. *AdenoCanNet* architecture.

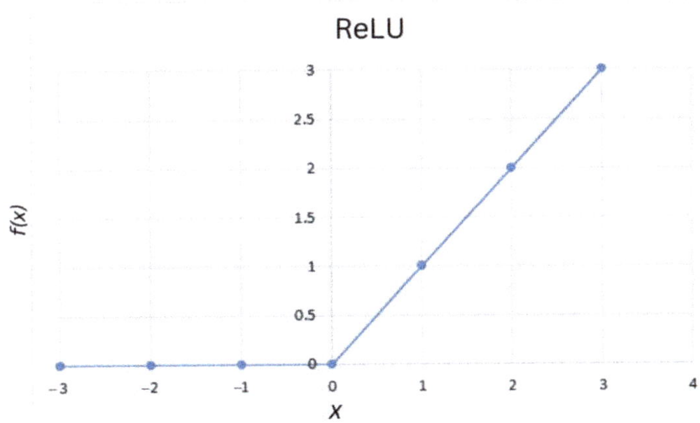

Figure 4. Graph of ReLU activation function.

2.4.2. Convolution neural network with SVM classifier (*AdenoCanSVM*)

During the initial phase of the study, the CNN–SVM architecture design displayed poor performance. Hence, the architecture was altered to make the CNN deeper, enabling better learning of the features of the image. However, developing a deeper neural network led to poor results. Several measures were taken to overcome the problems, including the addition of the data augmentation concept, a result of which was the proposed model.

The proposed model has three convolution layers, two max pooling layers, and one dropout layer, followed by fully connected layers. The proposed model finally used the SVM classifier for the classification task.

The images of the train and test dataset were augmented using the Keras library with the following properties. Table 1 displays the pre-processing done on the train and the test set of the dataset.

Table 1. Pre-processing done on the dataset.

Train Dataset		Test Dataset
➔	Rescale = 1/255	Rescale = 1/255
➔	Zoom Range = 0.2	
➔	Zoom Range = 0.2	
➔	Horizontal Flip = True	

The architecture is designed to take an input image having the dimensions of (64, 64, 3). The first two layers of the proposed CNN architecture correspond to the convolution layers having 32 and 64 channels. The filter size in both the layers is set to (3, 3). A max pooling layer having a stride of (2, 2) is then included, after each layer, in the architecture. Following the max pooling layer, a dropout layer is added. The architecture then includes one more convolution layer having 16 channels and a filter size of (3, 3). The output from this layer is then fed into the max pooling layer, having a stride size of (2, 2), and ends with flattening it. After a series of convolution, max pooling, and dropout layers, the feature map obtained is then flattened to obtain the equivalent feature vector. The feature vector is fed to a series of dense layers before finally throwing the output class of the input image. It is pertinent to note that the architecture uses L2 Regularizers, Squared Hinge as the loss functions, and Adam as the optimizer. It is also noted that ReLU was used as the activation function in the entire process. Figure 5 represents a graphical form of the proposed *AdenoCanSVM* architecture.

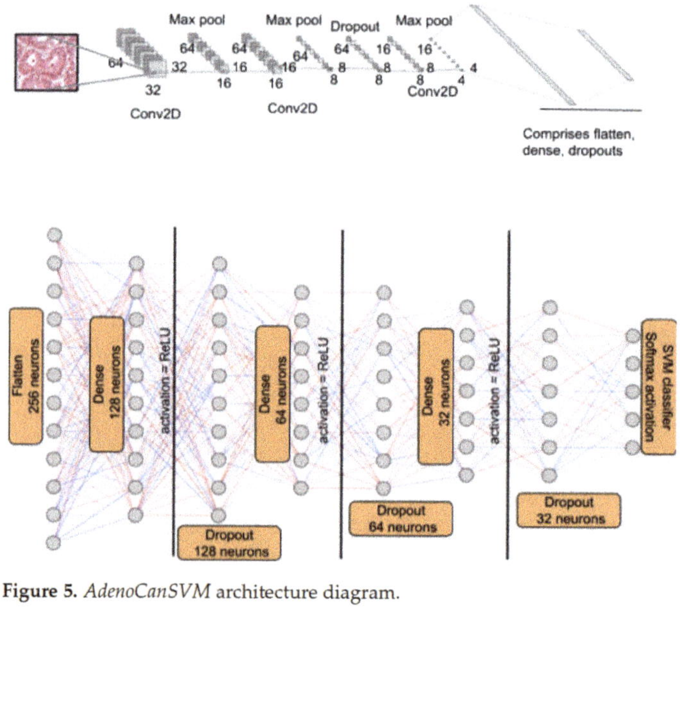

Figure 5. *AdenoCanSVM* architecture diagram.

2.4.3. Pre-existing Architectures

The proposed models were compared with pre-existing architectures for analyzing the performance. For this study, models such as LeNet were trained and concepts of transfer learning were applied to VGG16, VGG19, and ResNet50. Figure 6 summarizes all the existing architectures taken up in this study.

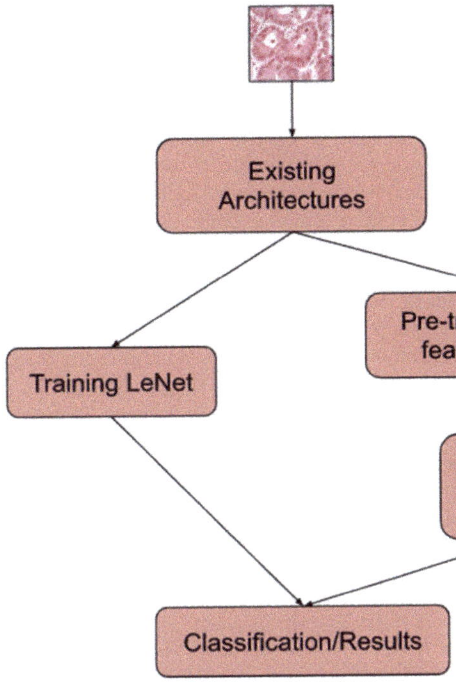

Figure 6. Existing-architecture implementation diagram.

Transfer learning is a method by which models trained on one particular task are re-purposed on a different related task [25]. Transfer learning allows us to use pre-trained models as a starting point instead of building an entirely new model from scratch. These models can further be trained for the required datasets of the problem statement, saving a lot of computational power, hence improving the performance of the second model.

VGG16: VGG16 is a prominent CNN architecture, developed in 2014 [26]. The first convolution layer in the VGG architecture introduced small 3×3 receptive fields instead of having large 11×11 receptive fields found in ALexNet, and a large 7×7 receptive field found in ZFNet. A 3×3 receptive field was used throughout the network in VGG architecture. Additionally, two non-linear activation layers make the decision functions more discriminative. VGG16 has 16 weight layers and is capable of classifying 1000 objects. For this study, VGG16 was used to gather results on the dataset using a Softmax classifier.

VGG19: Similar to VGG16, VGG19 is a 19-layer deep CNN network consisting of 16 convolution layers and 3 fully connected layers [27]. Average pooling layers and dense layers were added to the architecture. The number of neurons in the final dense layer is equal to the number of classes provided for classification. Softmax was used to perform the classification.

ResNet50: ResNet50 model has 48 convolution layers, 1 max pool layer, and 1 average pool layer. In this study, ResNet50 was applied to the dataset and results were obtained using the Softmax classifier.

LeNet: LeNet-5 was introduced in 1989 and has seven layers, including three convolution layers, two subsampling layers, and two fully connected layers. The convolution kernel has a size of 5 × 5. ReLU is the activation function, and Softmax is the classifier.

3. Results and Discussion

The models were built and run using Google Colab. The Colab environment was used to train the models on a large number of data using GPU acceleration.

For the building of the models, several Python libraries were utilized to make the process more efficient and improve the performance. Tensorflow and Keras libraries were essential in building the models, compiling the models, and training the models as per the used dataset. Numpy was used to handle arrays, and OpenCV was used for reading and pre-processing the images. ImageGenerator was used for data augumentation.

In the pre-processing segment, Gaussian Blur with a kernel size of 3 × 3 was used, and SigmaX and SigmaY values were 90 in both cases. The images were initially set to 128 by 128. However, the image size was large for smoother processing in the Colab environment. Hence, the images were resized to 64 by 64.

The model initially designed severely overfit the training set. Hence, the hyperparameters were altered, and the total number of model parameters was reduced. As a result of the above process, we could finally arrive at the proposed model.

In this section, we discuss the results obtained by the CNN models the classification of lung and colon adenocarcinoma from histopathological images. The models were applied to three cases. In the first case, the models trained only on lung classes, the second case is of only colon classes, and the third case has lung and colon classes combined. In each case, 80% of the images are for training the model, and 20% are for testing.

Accuracy metrics were used to compare the results. Accuracy is a metric generally used when all the classes are equally important. Since the dataset used in this study is balanced, and all the classes have the same significance, accuracy metrics seemed reliable. Accuracy is the ratio of correct predictions to the total number of predictions. Equation (2) corresponds to the mathematical formula for the accuracy metric.

$$\text{Accuracy} = \frac{\text{True Positive} + \text{True Negative}}{\text{True Positive} + \text{False Positive} + \text{True Negative} + \text{False Negative}} \tag{2}$$

Table 2 displays the training and testing accuracies obtained for all the discussed models on a subset of the data, each class containing 500 images.

Table 2. Accuracy comparison of models for one subset (20 epochs).

	AdenoCanNet (CNN W/Softmax)	*AdenoCanSVM* (CNN W/SVM)	VGG16	VGG19	LENET	RESNET50
Lung	85.92 89.00	33.75 34.00	95.25 92.67	94.25 89.33	99.75 90.00	99.83 93.67
Colon	51.50 60.00	51.88 43.50	97.25 96.00	98.62 95.00	100 79.50	99.87 94.50
Lung and Colon	75.15 78.00	60.25 59.20	93.30 92.00	92.55 92.20	99.56 92.64	100 95.40

For subsets, each model was trained for 20 epochs. For the lung classes of the dataset, best performance was recorded for the ResNet50 model with the training and testing accuracies noted as 99.83% and 93.67%.

For the colon classes of the subset taken alone, LeNet-5 gave us 100% training accuracy. However, it is pertinent to note that the testing accuracy was only 79.50%, which is a case of overfitting. On the other hand, ResNet50 had a training accuracy of 99.87% and a testing accuracy of 94.50%. It is pertinent to note that VGG16 performed better than the rest of the models on the unseen data, recording the highest validation accuracy of 96.00%.

When histopathological images obtained from both lung and colon regions are analyzed, ResNet50 performed the best, with the highest training accuracy and testing accuracy of 100% and 95.40%.

The pre-existing architectures were found to perform better than the newly proposed system when the number of data used for training the model was drastically reduced.

It is pertinent to note that the concept of data augmentation was not implemented in any model run for 20 epochs.

All the models were then trained on the entire dataset i.e., 25,000 images, evenly distributed across five classes. The same train–test split of 80:20 was used. The models were trained for 50 as well as 100 epochs.

Table 3 displays the results obtained when the models were trained for 50 epochs.

Table 3. Accuracy comparison of models for the entire dataset (50 epochs).

	AdenoCanNet (CNN W/Sotmax)	*AdenoCanSVM* (CNN W/SVM)	VGG16	VGG19	LENET	RESNET50
Lung	99.87 98.80	94.55 95.70	95.57 96.57	93.79 94.67	100 94.97	100 97.67
Colon	99.99 99.90	88.87 68.90	97.45 98.60	96.22 96.95	100 97.40	100 98.90
Lung and Colon	99.88 98.90	92.02 76.59	94.76 96.36	92.27 94.32	100 96.86	100 100

For 50 epochs, in the case of histopathological images of only lungs, both ResNet50 and LeNet were found to record the highest training accuracy of 100%, and the validation accuracy of ResNet50 and LeNet was 97.67% and 94.97%. It is pertinent to note that the training and testing accuracies of the *AdenoCanNet* were 99.87% and 98.80%. The validation accuracy reported by *AdenoCanNet* architecture was the highest among the other models, indicating its better performance on unseen data. Hence, it can be concluded that *AdenoCanNet* performs better in the case of histopathological images in the lung region.

For colon images, as visible from Table 3, ResNet50 and LeNet performed the best on the training dataset with 100% training accuracy. The validation accuracy of ResNet and LeNet was 98.90% and 97.40%. It is pertinent to note that the training and validation accuracy achieved by *AdenoCanNet* were 99.99% and 99.90%. With the training and testing accuracy of *AdenoCanNet* being almost 100%, it can be concluded that *AdenoCanNet* performs better in the case of histopathological images obtained from the colon region.

When histopathological images obtained from both lung and colon regions are taken into consideration, ResNet performed well, the training and testing accuracy being 100%. However, the proposed architecture, *AdenoCanNet*, also recorded a training accuracy of 99.88% and validation accuracy of 98.90%, comparable with the performance of ResNet attained on the complete dataset.

Figures 7 and 8 correspond to the accuracy and loss graphs of *AdenoCanNet* trained for 50 epochs on different sets of the LC25000.

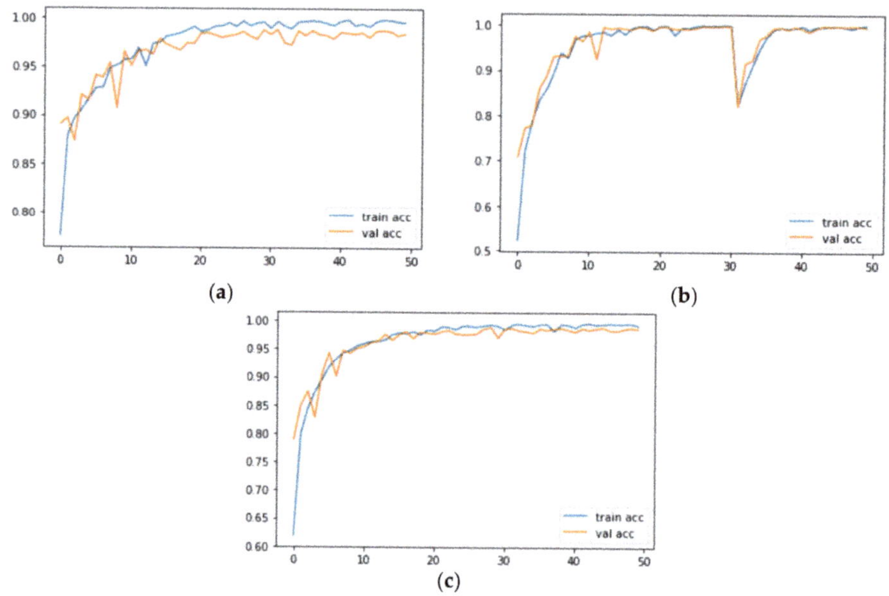

Figure 7. Accuracy graphs of *AdenoCanNet* trained for 50 epochs on: (**a**) lung dataset; (**b**) colon dataset; (**c**) both lung and colon dataset taken together.

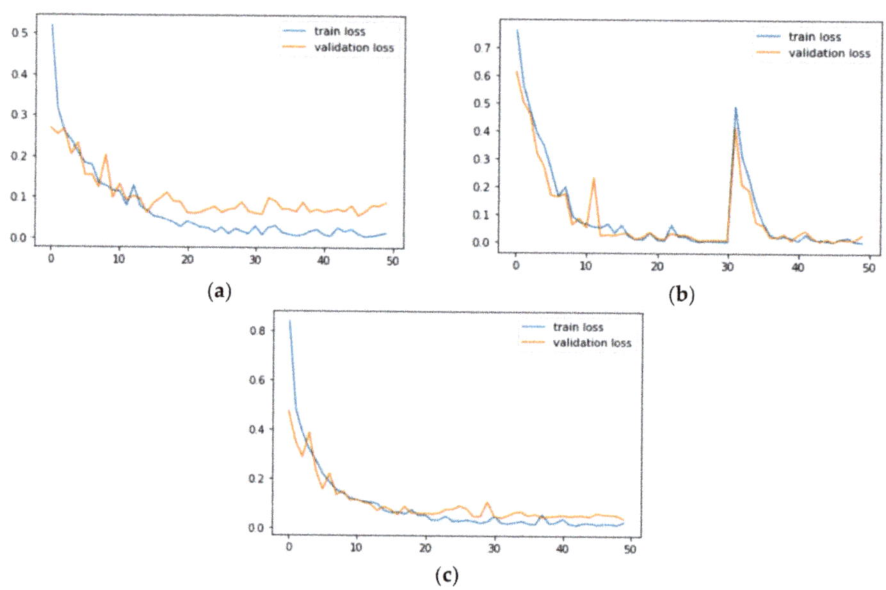

Figure 8. Loss graphs of *AdenoCanNet* trained for 50 epochs on: (**a**) lung dataset; (**b**) colon dataset; (**c**) both lung and colon dataset taken together.

Table 4 discusses the results obtained when the models were trained for 100 epochs.

Table 4. Accuracy comparison of models for the entire dataset (100 epochs).

	AdenoCanNet (CNN W/Softmax)	*AdenoCanSVM* (CNN W/SVM)	VGG16	VGG19	LENET	RESNET50
Lung	99.96 98.77	95.63 96.26	96.76 97.20	95.96 95.90	100 96.00	100 97.60
Colon	100 99.80	95.46 99.40	98.50 99.10	97.46 98.05	100 94.10	100 99.05
Lung and Colon	99.88 99.00	94.52 79.14	96.50 97.10	94.36 95.70	100 96.84	100 100

The results obtained for the models trained on histopathological images in the lung dataset run for 100 epochs show that ResNet50 and LeNet models achieved the highest training accuracy of 100%. The validation accuracy of ResNet and LeNet models correspond to 97.60% and 96.00%. It is worth noting that, among all the pre-existing architectures, VGG16 displayed a better performance on the validation dataset, achieving an accuracy of 97.20%. It is pertinent to note that the accuracy achieved by *AdenoCanNet* achieved a training accuracy of 99.96% and a validation accuracy of 98.77%. Among all the models, the suggested *AdenoCanNet* performs better on the unseen validation dataset. The training accuracy achieved by the CNN model is comparable with the performance of the ResNet and the LeNet models.

For the colon subset taken alone, the suggested *AdenoCanNet* with LeNet and ResNet recorded the highest accuracy of 100%. The validation accuracy obtained using the mentioned *AdenoCanNet* was the highest among all the other models, recording a value of 99.80%. It is worth noting that the proposed model of *AdenoCanSVM* also recorded a good training and validation accuracy of 95.46% and 99.40%. An accuracy of 98.50% and 99.10% reported for training and testing sets for the VGG16 model makes it another good model.

When histopathological images obtained from both lung and colon regions are taken into consideration, ResNet performed well, with both training and testing accuracy being 100%. However, *AdenoCanNet* attained a training accuracy of 99.88% and validation accuracy of 99.00%, which is comparable with the performance of ResNet on the complete dataset. Figures 9 and 10 correspond to the accuracy and loss graphs of *AdenoCanNet* trained for 100 epochs on different sets of the LC25000.

With the entire dataset into consideration, ResNet50 performs better in many cases. However, it is pertinent to note that ResNet50 is a 50-layer deep neural network, and training this model requires intense computation and time. The proposed *AdenoCanNet* consists of three convolution layers, two max pooling layers, and one dropout layer. The accuracies of this model and ResNet turn out to be comparable, and in some cases, *AdenoCanNet* outperforms the ResNet50 model.

When the models were trained on the entire dataset, the testing accuracy of *AdenoCanNet* outperformed the ResNet50 model when only lung and colon classes were considered.

LeNet-5 also displayed substantial results, however the *AdenoCanNet* outperformed the model in terms of testing accuracy for all cases when the entire dataset is considered.

For VGG16 and VGG19, the performance recorded was greatly comparable to the other architectures mentioned; however, *AdenoCanNet* outperformed both the pre-existing models in terms of training and testing accuracy in the case of the entire dataset.

Table 5 compares the results of the existing work on the detection and classification of lung and colon adenocarcinoma. Most works have opted for the LC25000 dataset and opted for CNN models; hence, these works are directly comparable.

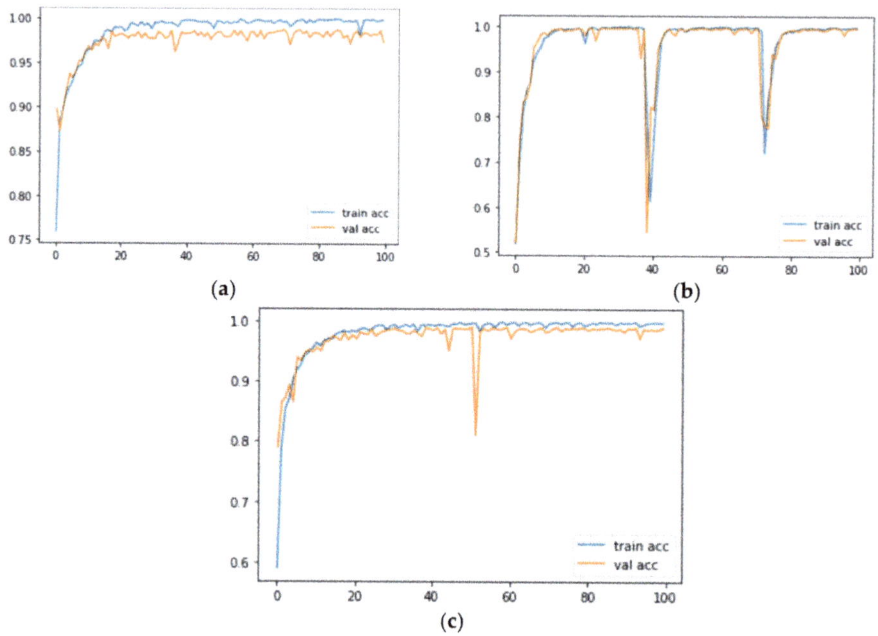

Figure 9. Accuracy graphs of *AdenoCanNet* trained for 100 epochs on: (**a**) lung dataset; (**b**) colon dataset; (**c**) both lung and colon dataset taken together.

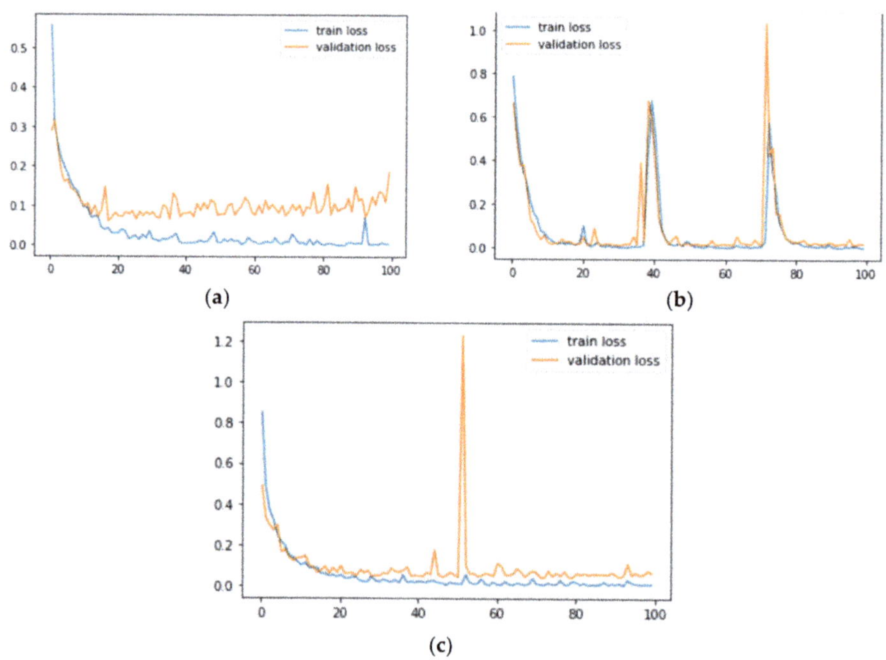

Figure 10. Loss graphs of *AdenoCanNet* trained for 100 epochs on: (**a**) lung dataset; (**b**) colon dataset; (**c**) both lung and colon dataset taken together.

Table 5. Comparison of existing works.

Author	Dataset	Model	Accuracy
Md. Alamin Talukder et al.	LC25000	CNN (Hybrid Model)	96.37%
Neha Baranwal et al.	LC25000	CNN (Inception-ResNetv2)	99.7%
Sanidhya Mangal et al.	LC25000	CNN	97% (Lung) 96% (Colon)
Bijaya Hatuwal et al.	LC25000	CNN	97.2%
Amin Saif et al.	LC25000	CNN	98.53%
Mehmood et al.	LC25000	CNN (AlexNet)	98.4%
Ben Hamida et al.	AiCOLON	CNN (ResNet)	99.8%
Jiatai Lin et al.	LC25000 and Kather	CNN (ResNet)	85.53%
Mainul Hossain et al.	LC25000	CNN	94% (Colon)
Aya Hage Chehade et al.	LC25000	XGBoost	99%
Zarrin Tasnim et al.	LC25000	CNN	99.67% (Colon)
Daria Hlavcheva et al.	LC25000	CNN	96.6% (Lung)
Syed Usama Khalid Bukhari et al.	LC25000 and CRAG	CNN (ResNet)	93.91%
Md. Alamin Talukder et al.	LC25000	CNN	99.30%
Mizuho Nishio et al.	LC25000	Hybrid ML models	99.43%
Sumaiya Dabeer et al.	BreakHis	CNN	99.86%
M. Šarić et al.	ACDC@LUNGHP	CNN (VGG16)	75.41%
AdenoCanNet (Proposed Model)	LC25000 (Lung)	CNN-Softmax	98.77%
AdenoCanNet (Proposed Model)	LC25000 (Colon)	CNN-Softmax	99.8%
Proposed Model	LC25000	CNN (ResNet50)	100%

For lung classes taken alone, Sanidhya Mangal et al. achieved an accuracy of 97%, and for colon classes alone Zarrin Tasnim et al. achieved a 96.6% accuracy rate. For CNN models experimented on the LC25000 dataset, the highest recorded accuracy turned out to be 99.8% achieved by Ben Hamida et al. With the obtained results and in comparison to the works discussed in the earlier section of this study, it can be concluded that, when considering the entire dataset of 25,000 histopathological images, *AdenoCanNet* introduced in this study performs with high accuracy and reliability, outperforming most of the existing models.

Figures 11–13 correspond to performance comparison plots for lung, colon, and lung and colon regions of the LC25000 dataset. As observed from the above comparisons and explanation, the *AdenoCanNet* model exhibited a better performance for the entire dataset for each category considered. Some recent works have used relatively deep neural networks or filtering multiple machine learning models based on accuracy obtained from them. The proposed model, having just three convolution layers, two max pooling layers, and one dropout layer, with minimum image preprocessing, achieved about 99% validation accuracy and 99.88% accuracy for detection and classification on the entire dataset of both lung and colon tissues. A 99.96% accuracy and 98.77% validation accuracy on the entire dataset of lung tissues were achieved, as well as 100% accuracy and 99.80% validation accuracy on the entire dataset of colon tissues with 100 epochs.

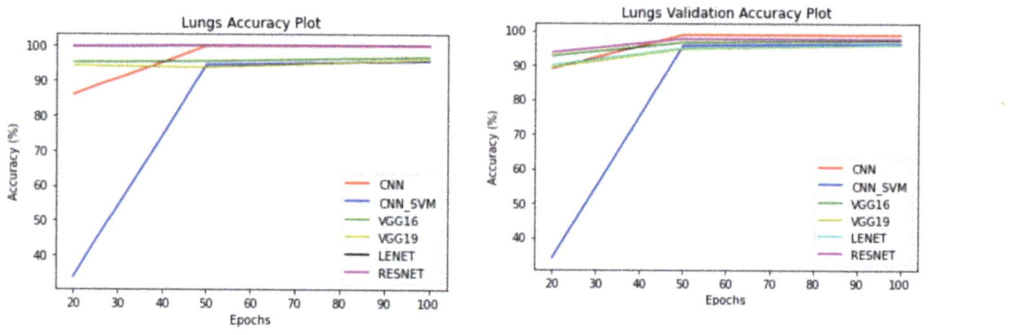

(**a**) (**b**)

Figure 11. Performance-comparison plot for models trained on the lung dataset: (**a**) accuracy comparison; (**b**) validation accuracy comparison.

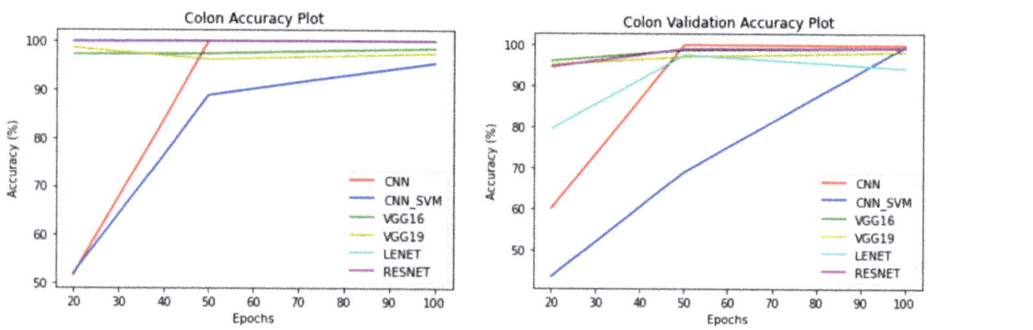

(**a**) (**b**)

Figure 12. Performance-comparison plot for models trained on the colon dataset: (**a**) accuracy comparison; (**b**) validation accuracy comparison.

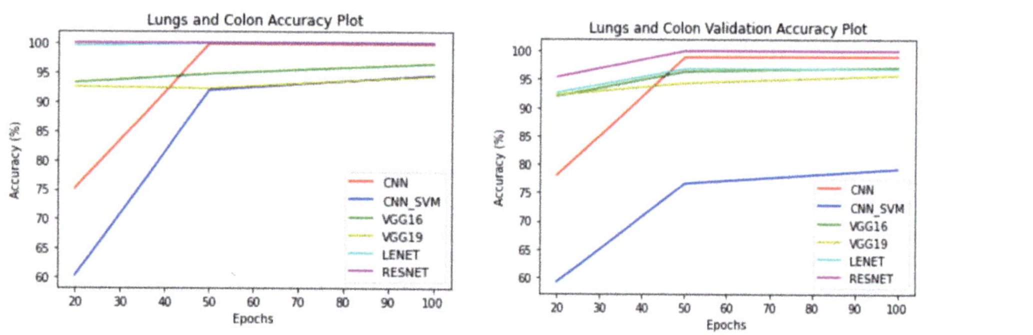

(**a**) (**b**)

Figure 13. Performance-comparison plot for models trained on the lung and colon dataset: (**a**) accuracy comparison; (**b**) validation accuracy comparison.

4. Conclusions

In this study, a deep learning-based supervised learning model was developed and used for the classification of histopathological images obtained from the lung and colon regions of the human body. We introduced *AdenoCanNet* and *AdenoCanSVM* to perform the classification. The experimentation used the LC25000 lung and colon histopathological dataset. Results from pre-existing architectures such as VGG16, VGG19, LeNet-5, and ResNet50 were compared with the results of the devised models. The *AdenoCanNet* proved to outperform the pre-existing models in most of the cases. The experiment performed for only lung classes gave a testing accuracy of 98.77% for *AdenoCanNet*. For the colon dataset, the testing accuracy obtained was 99.80%, and for the combined dataset, the testing accuracy was 99.00%.

With only a limited number of layers in the model, high accuracy was achieved, proving to perform better than most of the existing works.

With outstanding results obtained in diagnosing adenocarcinoma in both the lung and colon regions of the human body, the model may now be extended to test the accuracy in diagnosing adenocarcinoma in other regions of the body too. This work could further be deployed as a website as an easy-to-use platform. Utilization of this work would reduce effort, computation, and time.

There is a need for computer-aided diagnosis systems for adenocarcinoma detection from histopathological images. From this study, we concluded that, for small image datasets, existing architectures of deep learning are better, whereas for larger datasets, the proposed models work better.

Author Contributions: Conceptualization, S.C., B.A. and M.S.K.; methodology, S.C., B.A. and A.S.; proposed architecture design and implementation: S.C.; software, S.C., V.S. and D.P.; validation, B.A., M.S.K., A.S. and S.C.; formal analysis, S.C.; investigation, S.C., V.S. and D.P.; resources, S.C.; data curation, S.C.; writing—original draft preparation, S.C., V.S. and D.P.; writing—review and editing, B.A.; visualization, S.C., V.S. and D.P.; supervision, B.A.; project administration, B.A. and S.C. All authors have read and agreed to the published version of the manuscript.

Funding: This research received no external funding.

Institutional Review Board Statement: Not applicable.

Informed Consent Statement: Not applicable.

Data Availability Statement: The data used to support the findings of this work are available in Kaggle.

Acknowledgments: The authors wish to express their thanks to Vellore Institute of Technology (VIT) management and Centre for Cyber Physical Systems, VIT Chennai for their extensive support during this work.

Conflicts of Interest: The authors declare no conflict of interest.

References

1. WHO. Available online: https://www.who.int/news-room/fact-sheets/detail/cancer (accessed on 14 December 2022).
2. Sung, H.; Ferlay, J.; Siegel, R.L.; Laversanne, M.; Soerjomataram, I.; Jemal, A.; Bray, F. Global Cancer Statistics 2020: GLOBOCAN Estimates of Incidence and Mortality Worldwide for 36 Cancers in 185 Countries. *CA Cancer J. Clin.* **2021**, *71*, 209–249. [CrossRef] [PubMed]
3. Cleveland Clinic. Available online: https://my.clevelandclinic.org/health/diseases/21652-adenocarcinoma-cancers#:~{}:text=Adenocarcinoma%20is%20a%20type%20of,Cancer%20Answer%20Line%20866.223.8100 (accessed on 30 May 2022).
4. Borkowski, A.A.; Bui, M.M.; Thomas, L.B.; Wilson, C.P.; DeLand, L.A.; Mastorides, S.M. Lung and Colon Cancer Histopathological Image Dataset (LC25000). *arXiv* **2019**. [CrossRef]
5. Talukder, M.A.; Islam, M.M.; Uddin, M.A.; Akhter, A.; Hasan, K.F.; Moni, M.A. Machine Learning-Based Lung and Colon Cancer Detection using Deep Feature Extraction and Ensemble Learning. *Expert Syst. Appl.* **2022**, *205*, 117695. [CrossRef]
6. Masud, M.; Sikder, N.; Nahid, A.-A.; Bairagi, A.K.; AlZain, M.A. A Machine Learning Approach to Diagnosing Lung and Colon Cancer using a Deep Learning-Based Classification Framework. *Sensors* **2021**, *21*, 748. [CrossRef] [PubMed]
7. Baranwal, N.; Doravari, P.; Kachhoria, R. Classification of Histopathology Images of Lung Cancer Using Convolutional Neural Network (CNN). *arXiv* **2021**. [CrossRef]
8. Mangal, S.; Chaurasia, A.; Khajanchi, A. Convolution Neural Networks for Diagnosing Colon and Lung Cancer Histopathological Images. *arXiv* **2020**. [CrossRef]
9. Hatuwal, B.K.; Thapa, H.C. Lung Cancer Detection using Convolutional Neural Network on Histopathological Images. *Int. J. Comput. Trends Technol.* **2020**, *68*, 21–24. [CrossRef]

10. Saif, A.; Qasim, Y.R.H.; Al-Sameai, H.A.M.; Ali, O.A.F.; Hassan, A.A.M. Multi Paths Technique on Convolutional Neural Network for Lung Cancer Detection Based on Histopathological Images. *Int. J. Adv. Netw. Appl.* **2020**, *12*, 4549–4554. [CrossRef]
11. Mehmood, S.; Ghazal, T.M.; Khan, M.A.; Zubair, M.; Naseem, M.T.; Faiz, T.; Ahmad, M. Malignancy Detection in Lung and Colon Histopathology Images Using Transfer Learning with Class Selective Image Processing. *IEEE Access* **2020**, *10*, 25657–25668. [CrossRef]
12. Hamida, A.B.; Devanne, M.; Weber, J.; Truntzer, C.; Derangère, V.; Ghiringhelli, F.; Forestier, G.; Wemmert, C. Deep Learning for Colon Cancer Histopathological Images Analysis. In *Computers in Biology and Medicine*; Elsevier: Amsterdam, The Netherlands, 2021. [CrossRef]
13. Lin, J.; Han, G.; Pan, X.; Liu, Z.; Chen, H.; Li, D.; Jia, X.; Shi, Z.; Wang, Z.; Cui, Y.; et al. PDBL: Improving Histopathological Tissue Classification with Plug-and-Play Pyramidal Deep-Broad Learning. *arXiv* **2021**, arXiv:2111.03063. [CrossRef] [PubMed]
14. Hossain, M.; Haque, S.S.; Ahmed, H.; Mahdi, H.A.; Aich, A. Early Stage Detection and Classification of Colon Cancer using Deep Learning and Explainable AI on Histopathological Images. Ph.D. Thesis, Brac University, Dhaka, Bangladesh, 2022. Available online: http://hdl.handle.net/10361/16671 (accessed on 1 June 2022).
15. Hage Chehade, A.; Abdallah, N.; Marion, J.M.; Oueidat, M.; Chauvet, P. Lung and Colon Cancer Classification using Medical Imaging: A Feature Engineering Approach. *Phys. Eng. Sci. Med.* **2021**, in press. [CrossRef] [PubMed]
16. Tasnim, Z.; Chakraborty, S.; Shamrat, F.M.J.M.; Chowdhury, A.N.; Nuha, H.A.; Karim, A.; Zahir, S.B.; Billah, M.M. Deep Learning Predictive Model for Colon Cancer Patient using CNN-based Classification. *Int. J. Adv. Comput. Sci. Appl.* **2021**, *12*, 687–696. [CrossRef]
17. Hlavcheva, D.; Yaloveha, V.; Podorozhniak, A.; Kuchuk, H. Comparison of CNNs for Lung Biopsy Image Classification. In Proceedings of the IEEE Ukraine Conference on Electrical and Computer Engineering (UKRCON), Lviv, Ukraine, 26–28 August 2021. [CrossRef]
18. Bukhari, S.U.K.; Syed, A.; Bokhari, S.K.A.; Hussain, S.S.; Armaghan, S.U.; Shah, S.S.H. The Histological Diagnosis of Colonic Adenocarcinoma by Applying Partial Self Supervised Learning. *MedRxiv* **2020**. [CrossRef]
19. Nishio, M.; Nishio, M.; Jimbo, N.; Nakane, K. Homology-based Image Processing for Automatic Classification of Histopathological Images of Lung Tissue. *Cancers* **2021**, *13*, 1192. [CrossRef] [PubMed]
20. Dabeer, S.; Khan, M.M.; Islam, S. Cancer diagnosis in histopathological image: CNN based approach. *Inform. Med. Unlocked* **2019**, *16*, 100231. [CrossRef]
21. Šarić, M.; Russo, M.; Stella, M.; Sikora, M. CNN-based Method for Lung Cancer Detection in Whole Slide Histopathology Images. In Proceedings of the 4th International Conference on Smart and Sustainable Technologies (SpliTech), Bol and Split, Croatia, 18–21 June 2019. [CrossRef]
22. Aneja, S.; Aneja, N.; Abas, P.E.; Naim, A.G. Transfer learning for cancer diagnosis in histopathological images. *Int. J. Artif. Intell.* **2022**, *11*, 129–136. [CrossRef]
23. Maan, J.; Maan, H. Breast Cancer Detection using Histopathological Images. *Int. J. Comput. Sci. Trends Technol.* **2022**. [CrossRef]
24. DeepAI. Available online: https://deepai.org/machine-learning-glossary-and-terms/relu#:~{}:text=ReLu%20is%20a%20non%2Dlinear,zero%20and%20the%20input%20value (accessed on 1 June 2022).
25. A Gentle Introduction to Transfer Learning for Machine Learning. Available online: https://machinelearningmastery.com/transfer-learning-for-deep-learning/ (accessed on 1 June 2022).
26. Tammina, S. Transfer learning using VGG-16 with Deep Convolutional Neural Network for Classifying Images. *Int. J. Sci. Res. Publ.* **2019**, *9*, 9420. [CrossRef]
27. Machine Learning Blog. Available online: https://blog.techcraft.org/vgg-19-convolutional-neural-network/ (accessed on 2 June 2022).

MDPI
St. Alban-Anlage 66
4052 Basel
Switzerland
www.mdpi.com

Sustainability Editorial Office
E-mail: sustainability@mdpi.com
www.mdpi.com/journal/sustainability

www.ingramcontent.com/pod-product-compliance
Lightning Source LLC
Chambersburg PA
CBHW061620220926
51850CB00005B/555